Modern Digital Signal Processing

An Introduction

Prabhakar S. Naidu

Alpha Science International Ltd.
Pangbourne England

Prabhakar S. Naidu
Electrical Communication Engineering Department
Indian Institute of Science
Bangalore, India

Alpha Science International Ltd
P.O. Box 4067, Pangbourne RG8 8UT, UK

ISBN 1-84265-133-1

Printed in India

Dedication

I am pleased to dedicate this book to the memory of *Prof. Satish Dhawan*, who enabled me to pursue a fruitful career of research and teaching at the Indian Institute of Science, Bangalore.

Prolog

With the advent of digital computers, signal processing has made rapid advances. What was once confined to radar and seismic exploration for oil, signal processing, now known as digital signal processing (DSP), expanded into many diverse areas. Here is a short list:

(1) Speech, audio and video
(2) Channel equalization and echo cancellation
(3) Sonar and underwater detection
(4) Radar and related military applications
(5) Vibration and machine failure studies
(6) Non-destructive testing
(7) Geophysical exploration
(8) Localization (e.g. GPS)
(9) Medical diagnosis
(10) Instrumentation
(11) Financial analysis

The phenomenal growth of DSP has been attributed to the availability of digital signal processor in late seventies and early eighties. This has opened up the possibility of replacing analog filters, correlators, spectrum analyzers, etc., by versatile but inexpensive DSP based equipment. Further, it has now become possible to employ more complex algorithms involving non-linear and iterative steps. All along, the theory of DSP has been a step ahead of the hardware developments. Today we have many new signal processing algorithms, for example, singular value decomposition (SVD) based algorithms, waiting for faster but inexpensive digital signal processor for their exploitation. Digital signal processor technology has finally blossomed.

The digital signal processing has traditionally three main topics, namely,

(i) Discrete signals and systems
(ii) Digital filtering
(iii) Spectrum analysis

The first topic is about signal analysis and modeling of systems. The relation between the input and output of a linear system is of lasting interest in signal processing. The study of discrete signals and systems is also of interest in other areas, such as in communication, control system, circuit theory; also in physical sciences, earth sciences, biological sciences, etc. A discrete signal is often derived by sampling a continuous signal and a digital signal is derived by quantizing the discrete signal; thus sampling rate and quantization step size are of immediate concern. Once the signal is in the digital form it is sufficient to learn how to characterize it without referring to its continuous counterparts. Thus, z-transform and discrete Fourier transform (DFT) are powerful tools for discrete signal analysis in time and frequency domain, respectively. A signal may be a simple waveform like a sinusoidal function or a highly complex time varying stochastic signal. It is imperative to understand the nature of the signal so that the most appropriate tool is selected for its processing. Keeping this in view we have included a discussion on the types of signal and their models. In discrete time linear system the basic operation is that of convolution which may be implemented through delay units (shift registers), adders and multipliers. Chapters one and two (nearly a quarter of the book) are devoted to the above topics.

The second topic of interest deals with digital filters. The digital filters like analog filters are primarily used to enhance a signal in presence of noise. The digital filters are also used for interpolation (signal extraction), extrapolation (signal prediction), equalization, detection, etc. In an introductory book such as this, we shall confine ourselves to the basic types of digital filters meant for signal enhancement. A low pass filter, from which all other filters can be derived, is of main interest. We shall devote considerable effort to learn how to design and implement a low pass filter. Traditionally, lowpass digital filter design has been based on analog lowpass filters, which have been extensively studied before the dawn of digital era. We shall also explore an alternate approach of directly designing the digital filters with the prescribed characteristics. This route is often more difficult and laborious, hence, on account of limited space, it gets a lower priority. The finite impulse response (FIR) and the infinite impulse response (IIR) filters are covered in chapter four and five, respectively.

The third topic is spectrum analysis, which deals with the study of energy/power distribution as a function of frequency (energy for deterministic signal and power for stochastic signal). The spectrum of a signal is essential when we are looking for discrete sinusoids in a signal, as in communication, radar and sonar. Also in the design of optimum filters we need to know the spectra of the signal and the interfering noise. The basic approach for estimation of spectrum is to first compute an autocorrelation function whose Fourier transform then gives us its power spectrum. This classical approach has, in the recent times, undergone many changes. The use of fast Fourier transform, precise modeling of time series, advanced computational tools have lead to better estimation tools. While the classical approach is emphasized, the recent developments are only briefly mentioned. Chapter three is devoted to the study of spectrum analysis.

Applications of signal processing have been growing at a fast rate. In almost every domain of human activity, including economics, signal processing finds a place. The nature of application is so wide, often falling outside the conventional topics listed above. For example, the prediction into future from the past observations is an important requirement in the study of economics. For this, modeling of time series including the fractal models is required. It was therefore felt that even an introductory text on signal processing must contain a brief account of some of the topics of current interest. With this in mind the following two modern topics have been included:

i) Adaptive signal processing
ii) Beamforming

Additionally, a few other modern topics have also been included under the traditional topics. For example, under spectrum analysis we have briefly discussed the higher order spectrum such as bispectrum and the time varying spectrum. The bispectrum is useful in the analysis of non-linear phenomenon and the time varying spectrum is useful in the analysis of non-stationary signals.

Adaptive signal processing deals with estimation of an optimum filter based on the actual observed signal and noise rather than on their unknown statistical structure. Beamforming deals with how to combine the outputs of a group of sensors so as to receive a signal coming from a specific direction and to suppress all other interferences. These two topics are covered in chapter six and seven, respectively. The coverage is at an introductory level. The aim is to expose a reader to the vast resources of the modern digital signal processing.

The final chapter is devoted to digital signal processors, which play a very important role in the application of DSP algorithms to real life problems. The subject is indeed very vast and changing fast. Only an outline of the digital signal processors and the current status have been covered with an emphasis on why a special digital signal processor is required and how the ubiquitous pc (personal computer) is inadequate.

The strength of the present book as compared to some of the modern texts on DSP [1-4] is summarized below:

- Wider selection of topics both from traditional and modern aspects of the DSP.
- All basic concepts are fully derived and illustrated with the help of figures and examples. Please note that each example is separated from the running text by means of a blank line and the end of example is denoted by symbol .
- Precise mathematical language is used in place of verbal description. On the part of the reader the mathematical expertise required is no more than that acquired at the college level.
- At the end of each chapter a selection of problems is provided. The problems range from very simple ones, some of which may be solved even orally, to more complex ones requiring some analysis. (The solution manual will be available to instructors). There are also some computer projects. It is expected that the students have access to a computer with Matlab installed on it. Familiarity with Matlab is a great asset in signal processing.
- The size of the book is less than 425 pages, unlike the DSP books listed below.

A voluminous book is indeed highly intimidating. Ideally a textbook should contain enough material that can be covered in one or utmost in two semesters. I believe the present book can be covered in about two semesters. DSP is generally taught for one semester. Then, it is not possible to cover all eight chapters without a lot of pruning. One other possibility is to cover the core topics, namely, chapters 1-5 and choose one of the remaining three chapters depending upon the interests of the students. My own preference is for the latter. Finally, as for the prerequisite, a course on Signals and Systems would be of great help.

The book is also recommended for self-study by working engineers/scientists. In this case, it is recommended that the Chapters One and Two be first mastered including solving the problems at the end of these chapters. After mastering the first two chapters, one may proceed to study Chapters Three to Five. After a pause, go on with the three remaining chapters. A matured reader may, however, select his/her own scheme of coverage.

May 2003 Prabhakar S. Naidu

Acknowledgement

I wish to thank All India Council for Technical Education (AICTE) for the support received during the preparation of this manuscript. I appreciate the help extended by M/s S. Chandra Sekhar and Raman Chandrasekharan for carefully scrutinizing the manuscript.

Finally, I wish to thank my wife, Madhumati, who has shown exceptional forbearance during the preparation of the manuscript.

Contents

1 Signal Analysis

Signals, continuously time varying waveforms, are encountered in many natural and man-made systems. Examples are seismic waves from an earthquake, radar/sonar signals, communication signals, speech, music, etc. In digital signal processing (DSP) our primary goal is to extract information buried in signal waveforms. Therefore, it is natural to understand the different types of signals and learn how to model them. That is the subject matter of this chapter on Signal Analysis. Although the signals of interest are in continuous time, their processing on a modern digital computer demands that the signals be first sampled and then quantized with minimal loss of information. The sampled signals are processed using digital processing algorithms. This opens up many hitherto unknown possibilities. Digital signal processing is all about these possibilities. In this chapter we shall cover the topics on sampling and quantization; types of signals: deterministic, random, cyclostationary signals; signal transforms: Fourier transform, Laplace transforms and z-transform. The final section is on some commonly used signal models. They are narrowband signals, linear time series models, state space representation and non-linear chaotic signal model. An efficient modeling of a signal is key to the successful extraction of information from the observed data.

§1.1 Continuous and Discrete Signals

A continuous signal is sampled at uniform discrete time instants. A continuous signal is thus converted into a discrete time signal or simply a discrete signal. This is not enough for being able to process on a digital computer. Each continuous valued discrete signal is quantized to integer number of levels producing a digital signal. Here on we begin digital signal processing.

Sampling: A signal defined over a continuous time interval is a continuous signal. An example of continuous signal is a simple sinusoid, $x_c(t) = 4\sin(2\pi 10t + \pi/6)$, having a frequency of 10Hz, amplitude equal to four and phase equal to $\pi/6$. The subscript *c* on *x(t)* denotes that the signal is continuous. Here t is a continuous time variable in units of seconds over any interval, including an infinite interval, $-\infty$ to $+\infty$. A continuous signal has to be digitized for processing on a digital computer. Uniform sampling, that is, samples collected at constant time interval is mathematically stated as

$$x_d(t) = x_c(t) \sum_{t=-\infty}^{\infty} \delta(t - n\Delta t) \tag{1.1}$$

where Δt is the sampling interval which we shall assume for the time being equal to one and $\delta(t)$ is Dirac delta function. $\sum_{n=-\infty}^{\infty} \delta(t - n\Delta t)$ is also known as comb function consisting of a sequence of delta functions. $x_d(t)$ is the discrete version of $x_c(t)$. Note that $x_d(t)$ consists of a sequence of spikes of amplitude $x_c(t)\big|_{t=n\Delta t}$, spaced at interval Δt and null value in between,

$$x_d(t)\big|_{t=n\Delta t} = x_c(n\Delta t)$$
$$= 0 \quad otherwise$$

The sampled signal is given by $x(n) = x_d(t)\big|_{t=n\Delta t}$. The process of sampling a continuous signal is illustrated in fig. 1.1 where we have sampled a continuous sinusoid $x_c(t) = 4\sin(2\pi 10t + \pi/6)$ with a sampling interval equal to 0.04 seconds. The arrow shows the position and magnitude of each sample.

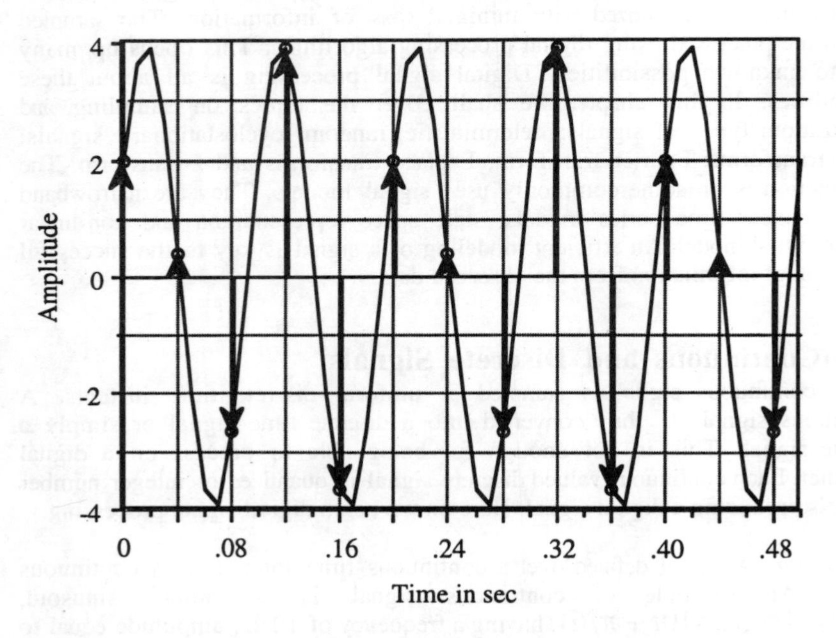

Figure 1.1: A continuous sinusoid of frequency 10 Hz is sampled with sampling interval of 40 ms. Sampled signal is equal to the amplitude of the spikes at sampling instants.

The choice of sampling interval Δt or the sampling rate $(=1/\Delta t)$ is very crucial. A correct sampling rate is one for which it is possible to reconstruct the original continuous signal from the samples by means of a linear interpolation. A theorem, known as Shannon's theorem, tells us how error free sampling can be achieved. If f_b is the highest frequency (in Hertz) component in a signal the sampling rate must be *greater* than (or equal to) $2f_b$ or the sampling interval must be *less* than (or equal to) $1/2f_b$ seconds. Given such samples (known as Nyquist samples) we can reconstruct the original continuous signal by means of interpolation given by

$$\boxed{\hat{x}(t) = \sum_{n=-\infty}^{\infty} x(n)\sin c(t - n\Delta t))} \qquad (1.2)$$

where $\sin c(t) = \sin(2\pi f_b t)/(2\pi f_b t)$ is known as interpolation function. Observe that in the example given above, when the phase is zero, all samples are equal to zero. Evidently no reconstruction is possible. This awkward situation may be avoided by introducing a phase or delay in the signal, or alternatively, by sampling at a slightly higher rate. The error in the reconstruction depends upon the actual sampling rate as against the desired sampling rate, as per Shannon's theorem. We shall return to this topic after we introduce the concepts of spectrum and bandlimited signal. Here we would like point out that the error in the reconstruction can even arise when only a finite number of Nyquist samples are available.

Example 1.1: Let us assume that only a finite number of samples of the sinusoid shown in figure 1.1 are available. In the interpolation formula (equation (1.2)) we can now have only a finite number of terms in the summation. This will result into an error in the interpolation. To estimate this error numerically we consider a random sinusoid of unit magnitude but random phase. The frequency of the sinusoid is 5Hz and it is sampled at a rate of 20Hz. The interpolation operation in (1.2) is expressed as a filter (see chapter 4 on FIR filters) for interpolating at a point midway between two neighbouring samples. The mean square error as a function of the length of the filter as shown in fig. 1.2

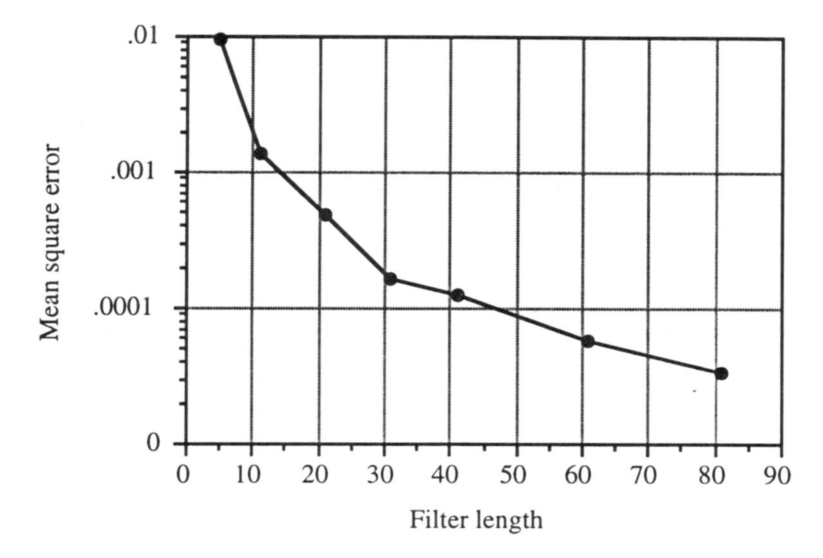

Figure 1.2: Mean square error in the interpolation of a random sinusoid as a function the filter length. The error tends to zero as the length approaches infinity ■.

Interpolating between the given samples can increase the sampling rate. This is known as upsampling. Similarly decimating or discarding a required number of samples periodically can reduce the sampling rate. This is called downsampling. It must be remembered that upsampling and downsampling are different from upscaling and downscaling where there is no interpolation nor decimation. We shall talk about scaling in the next chapter. In computer simulation studies the sampling interval is often set to one, that is, $\Delta t = 1$. The maximum frequency of the simulated signal must then be less than or equal to π, that is, $|\omega| \leq \pi$ (radians/sec), or in the units of Hertz, $|f| \leq 0.5$ Hz. We call this a normalized frequency.

Quantization: Since in computer processing a number is stored using a finite number of bits, all sample values will have to be rounded to the nearest integer multiple of q such that $x(n) \approx q\delta$ where δ is the step size and $|x(n) - q\delta| < 0.5\delta$. This process of converting continuous-valued discrete-time samples into a discrete valued (digital) signal is known as quantization. q is the quantization level and δ is the quantization step size or resolution. The quantization error $(x(n) - q\delta)$ is assumed to be uniformly distributed random variable with zero mean and variance equal to $\delta^2/12$. An electronic device (Analog-to-Digital (A/D) converter) is used to convert a sample into a discrete number. The device is preprogrammed to select any prescribed quantization level, which in practice depends upon signal-to-noise ratio (SNR) and memory capacity of the computer. Ideally, we would like δ to be as small as possible so that the quantization error is negligibly small. But a choice of δ less than the standard deviation of system noise will only make the discrete samples noisier while flooding the computer memory with a large number of bits. An example of quantization of the sinusoidal function shown in fig. 1.1 is shown in table 1.1. The quantization step =1.0 and $(x_{max} - x_{min})$ =8.0. The quantization levels are: <-3.5, -3.5 to <-2.5, -2.5 to <-1.5, -1.5 to <-0.5, -.5 to <0.5, 0.5 to <1.5, 1.5 to <2.5, 2.5 to <3.5 and ≥3.5, in all nine levels.

Table 1.1: Analog and quantized samples. The quantization error (analog sample - quantized sample).

Sample time in sec	Analog sample value	Quantized sample	Quantization error
0	2.000	2.0	0
0.04	0.418	0.0	0.418
0.08	-2.676	-3.0	.324
0.12	3.913	4.0	-0.087
0.16	-3.654	-4.0	.346

For a given resolution the maximum number of quantization levels is given by

$$Q = q|_{max} = \frac{(x_{max} - x_{min})}{\delta} + 1 \tag{1.3}$$

The number of bits required to store Q levels is $\log_2 Q$. The A/D converter must be capable of generating an output word of length of $\log_2 Q$ bits. This is an important characteristic of an A/D converter in addition to the speed of conversion (time taken to complete A/D conversion must be less than Δt, the sampling interval).

The quantization noise can be reduced by increasing the number of levels or the word length of the A/D converter. Sometimes it is possible to reduce the quantization noise by increasing the sampling rate. It is known that the quantization noise is white noise whose spectrum is flat over a range $\pm f_s$ where f_s is the sampling frequency. Let the signal spectrum be in the range $\pm f_b$ and let $f_b << f_s$. The quantization noise will occupy a much wider frequency band than the signal spectrum. Clearly, by means of a lowpass filter we can remove the excess noise power without affecting the signal waveform. We illustrate this possibility in fig. 1.3

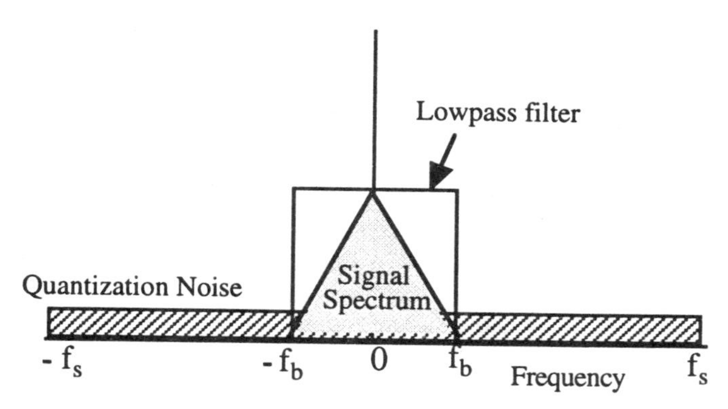

Figure 1.3: The spectrum of the quantization noise occupies a much wider band than the spectrum of the signal. A lowpass filter can remove the noise power lying outside the signal band.

The noise power will be reduced by a factor of f_s/f_b. Note that to achieve the same amount of noise power reduction we will have to increase the number of quantization levels by a factor of $\sqrt{f_s/f_b}$.

There is yet another possibility of reducing the quantization noise. From (1.3) it may be noted that for a fixed Q (quantization levels) the step size d can be made smaller by decreasing the quantity, $(x_{\max} - x_{\min})$, the dynamic range of the signal. This may be achieved by quantizing the prediction error, that is, the current sample minus the one predicted from the past samples (Auto regressive (AR) model to be dealt with later in this chapter). The prediction error will have a very low dynamic range, particularly when the sampling rate is very high. Then, one bit or two bits quantization is quite adequate. Further elaboration on the quantization noise and sampling rate is given in [1].

§1.2 Types of Signals

The signals used in different applications may be classified as deterministic (used in radar and sonar), stationary stochastic (man-made and natural signals), cyclostationary (used in communication).

Deterministic Signal : A sinusoid shown in fig.1.1, but of finite duration, is the simplest signal of great interest in many applications. The duration, amplitude and frequency are the important characteristics of a sinusoidal signal. The frequency may vary with time, often linearly increasing with time. Such a signal is known as a chirp signal after the sound produced by many birds (see fig. 1.4a). In the continuous time domain an example of chirp signal is $x(t) = a_0 \sin(\omega_0 t(1+\alpha t)+\phi)$. The instantaneous frequency (see Exercise 17) at time t=0 is ω_0 and t=τ it is $\omega_0(1+2\alpha\tau)$ where 2α is the coefficient of linear increase. In the discrete domain the above chirp signal takes a form $x(n) = a_0 \sin(\omega_0 n(1+\alpha n)+\phi)$.

The most important property of a chirp signal is that its autocorrelation function is very narrow (see fig. 1.4b). The autocorrelation function of a deterministic signal, that is, correlation with itself, is given by,

$$\boxed{r_{xx}(\tau) = \sum_{n=0}^{N-\tau} x(n)x(n+\tau)} \tag{1.4a}$$

where N stands for the number of discrete samples of a deterministic signal. This property of the chirp signal is very useful in target detection with the help of a beam of radiation as in radar, sonar, seismic exploration, etc. A correlation between two different signals is known as cross-correlation. It is defined as

$$r_{xy}(\tau) = \sum_{n=0}^{N-1} x(n)y(n+\tau) \tag{1.4b}$$

and correlation coefficient is defined as

$$\tilde{r}_{xy}(\tau) = \frac{r_{xy}(\tau)}{\sqrt{r_x(\tau) r_y(\tau)}} \tag{1.4c}$$

The correlation coefficient lies between ± 1; for perfect correlation it is +1, for no correlation it is zero and for reverse perfect correlation it is -1.0.

Stationary Stochastic Signal: A random variable is an outcome of repeated experiments carried out to measure a quantity, say, temperature with an unreliable thermometer which gives a different reading each time an experiment is performed. Let $\theta_1, \theta_2, \theta_3, \ldots, \theta_r$ be the outcomes of r experiments. We denote the quantity being measured as θ, which we treat as a random variable. The observations may be considered as different realizations of a random variable θ taken from a sample space which may be continuous or discrete. In the case of a continuous sample space the random variable may take any value lying in a certain range, for example, +1 to -1; hence, it is a continuous random variable. In the case of a discrete random space, the random variable can take any one of a set of discrete values. We shall confine ourselves to the continuous random

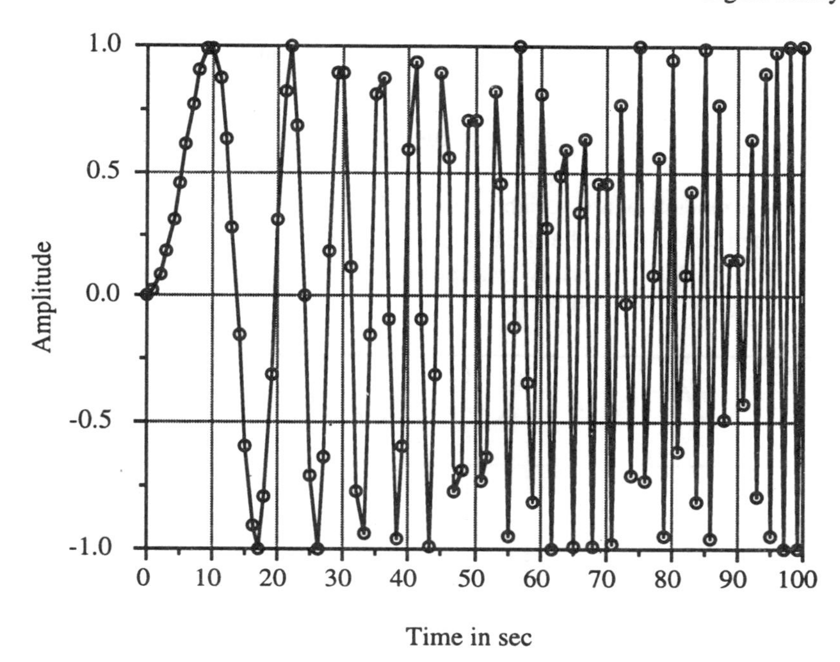

Figure 1.4a: A plot of a chirp signal $x(n) = \sin((0.5\pi/100)n(n+1))$.

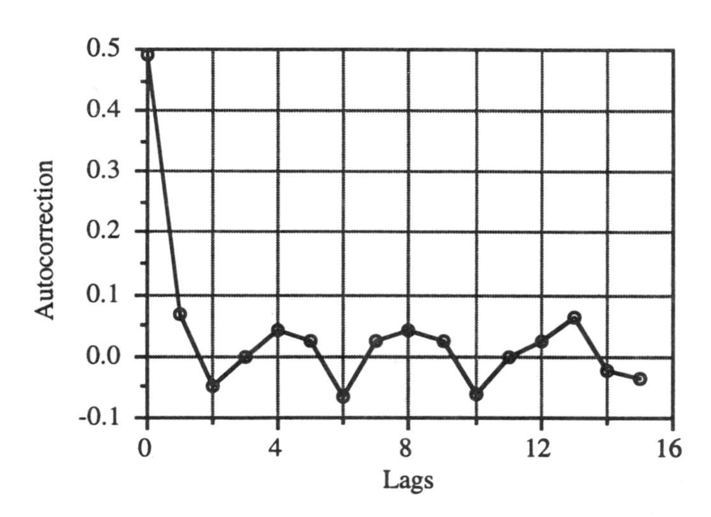

Figure 1.4b: The autocorrelation function of a chirp signal shown in fig. 1.4a.

variable. Formally, the different realizations of a random variable are represented as $x_\lambda, \lambda = 1,2,3,\ldots$ When a random variable is a function of some independent parameter, say time, we call that a random function or simply a stochastic signal. When n is discrete, for example, n=0, 1, 2, 3, ..., we have a random sequence or a stochastic time series. As an elementary example of a continuous-time stochastic signal consider a sinusoidal function with random phase,

$$x_{\varphi}(t) = \sin(\omega t + \varphi)$$

where φ is a random phase variable. For each realization of the random phase we obtain one realization of the stochastic signal. The random variable φ lies in the range $\pm\pi$ and is uniformly distributed. A few realizations of the above stochastic signal are sketched in fig. 1.5.

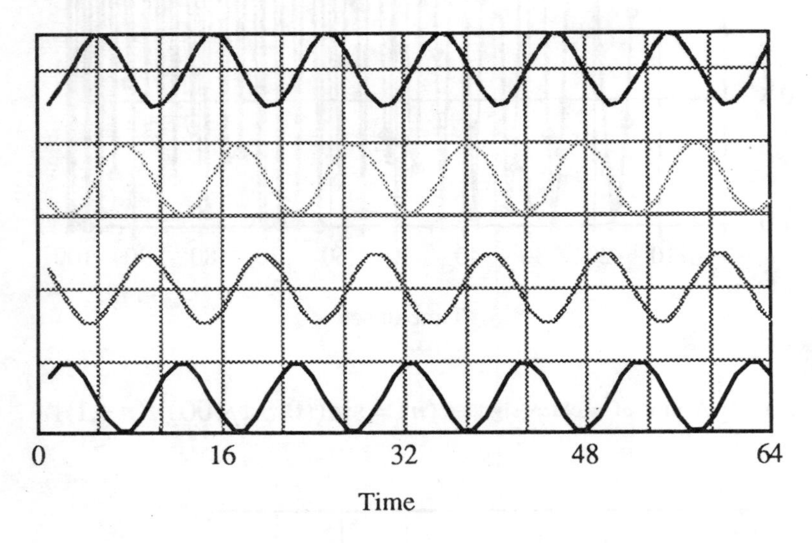

Figure† 1.5: Random sinusoids. The frequency and amplitude are the same but the phase is a uniformly distributed random variable.

A random function may be considered as a vector random variable whose components are obtained by sampling the random function at a large number of time instants, ideally infinite. Let a random function *x(t)* be sampled at time instants $t_1, t_2, \ldots, t_p$ The samples are $x_1, x_2, \ldots, x_p$ which go to make up a vector random variable. A scalar random variable is characterized by probability density function (pdf). Likewise, a vector random may be characterized by a joint probability density function (pdf) $f(x_1, x_2, \ldots, x_p; t_1, t_2, \ldots, t_p)$. Ideally, one must use an infinite number of samples but this leads to extreme analytical complexities. We often consider just two (second order) or three (third order) samples requiring only second or third order pdf. We shall first consider the two-sample case and then go on to the three-sample case. Let x_1 and x_2 be two samples obtained by sampling a stochastic signal at time instants, t_1 and t_2. When the joint probability density function depends upon the difference, $t_2 - t_1$, that is, pdf is of the form $f(x_1, x_2; t_2 - t_1)$, the stochastic signal is said to be stationary with respect to the second order pdf. For such a second order stochastic signal the mean and the variance are independent of time. The covariance function of a zero mean real valued stochastic signal is defined as

---

†Figures 1.5, 1.6, and 1.14-1.21 were taken from [6]

$$c_x(t_2-t_1)=E\{x_1x_2\}$$
$$=\int\int_{-\infty}^{\infty} x_1x_2 f(x_1,x_2;t_2-t_1)dx_1dx_2 \quad (1.5a)$$

and for a pair of real valued stochastic processes a cross-covariance function is defined as

$$c_{xy}(t_2-t_1)=E\{x_1y_2\}$$
$$=\int\int_{-\infty}^{\infty} x_1y_2 f(x_1,y_2;t_2-t_1)dx_1dy_2 \quad (1.5b)$$

where E{.} stands for an expectation operation, defined as $E\{\theta\}=\frac{1}{r}\sum_{i=1}^{r}\theta_i \quad r\to\infty$. The covariance function is a function of the time lag, $\tau=t_2-t_1$. Often, it is difficult to estimate the probability density function from a single sample of stochastic signal. However, the sample covariance function or also known as autocorrelation function, may be defined as a lagged product average of discrete-time signal (see 1.7).

We shall now summarize some of the properties of the discrtete time covariance function:

$$c_x(\tau)=c_x(-\tau)$$
$$c_x(0)\geq|c_x(\tau)| \qquad \tau\geq 0 \quad (1.6a)$$
$$\sum\sum c_x(n_2-n_1)\varphi(n_1)\varphi^*(n_2)\geq 0$$

for any function $\varphi(n)$. The last property mentioned in (1.6a) is known as positive definite property of a covariance function. For $\varphi(n)=e^{j\omega n}$ it takes the following form,

$$\sum\sum c_x(n_2-n_1)e^{-j\omega(n_2-n_1)}$$
$$=\sum_{-\infty}^{\infty} c_x(\tau)e^{-j\omega\tau}=S_x(\omega)\geq 0 \quad (1.6b)$$

where $S_x(\omega)\geq 0$ is known as the spectrum of the stochastic signal. Eq. (1.6b) is also known as the spectral representation of the covariance function. The cross-covariance function has the following properties:

$$c_{xy}(\tau)=c_{yx}(-\tau)$$
$$c_{xy}^2(\tau)\leq c_x(0)c_y(0) \quad (1.6c)$$

A covariance function may also be written in a matrix form as follows:

$$\mathbf{c}_x = \begin{bmatrix} c_x(0)\, c_x(1) \ \dots \ c_x(p) \\ c_x(-1)\, c_x(0) \dots c_x(p-1) \\ \dots \\ c_x(-p) \qquad \dots \ c_x(0) \end{bmatrix}$$

On account of the properties listed in (1.6) the covariance matrix will have a special structure; it is symmetric Toeplitz matrix, that is, all elements on any diagonal are equal and that the matrix is positive semi-definite, that is, all its eigenvalues are real and non-negative (including zero eigenvalue).

A covariance function gives us an idea how rapidly the correlation between two adjacent samples decreases with separation. We define a correlation distance, τ_0, where the covariance function is one tenth of its maximum at zero lag. When the samples are separated by an interval greater than the correlation distance it is safe to assume that they are uncorrelated. As the correlation distance decreases the stochastic signal becomes increasingly uncorrelated. In the limiting case as $\tau_0 \to 0$ we have a completely uncorrelated stochastic signal. The covariance function of the uncorrelated signal is given by

$$C_x(\tau) = \sigma_x^2 \delta_\tau$$

where δ_τ is the Kronecker delta function,

$$\delta_0 = 1$$
$$\delta_\tau = 0 \quad |\tau| > 0$$

Example 1.2: Consider a discrete-time stochastic signal (time series)

$$x(n) = ax(n-1) + \eta(n)$$
$$y(n) = x(n - \tau_0) + \varepsilon(n)$$

where a<1 and $\eta(n)$ and $\varepsilon(n)$ are uncorrelated white noise sequences. The covariance function of *x*(*n*) is given by,

$$c_x(\tau) = \frac{\text{var}_\eta}{1-a^2} a^{|\tau|}$$

and the cross-covariance functions *x*(*n*) and *y*(*n*) are given by $c_{xy}(\tau) = c_x(\tau - \tau_0)$ and $c_{xy}(\tau) = c_x(\tau + \tau_0)$. The covariance and cross-covariance functions for a=0.8 and τ_0=4 are shown in fig. 1.6. Two stochastic sequences are said to be uncorrelated when $c_{xy}(\tau) = 0$ for all τ .

The importance of the covariance function lies in the fact that, when a stochastic sequence is Gaussian, the joint probability density function of any order involves only the covariance matrix.

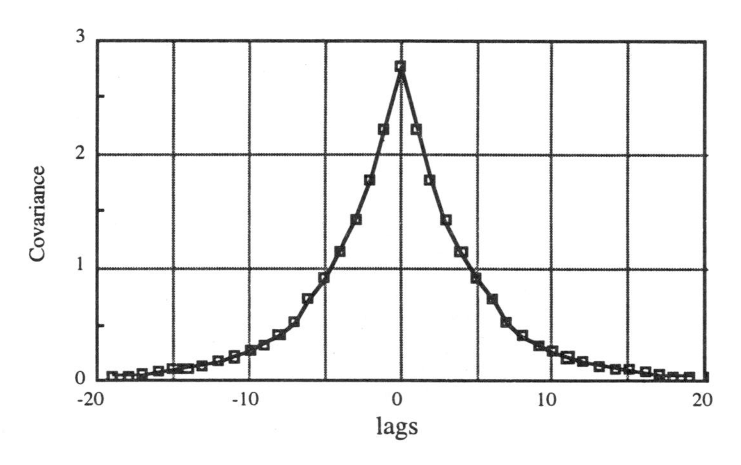

(a) Covariance function

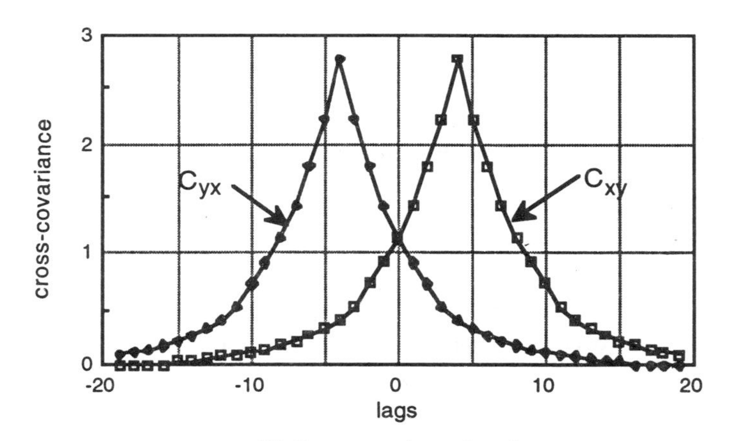

(b) Cross-covariance function

Figure 1.6: Covariance and cross-covariance functions.

In the definition of covariance function (Eq. 1.5) a joint probability density function is used but this is rarely available. In practice, given a single realization of a stochastic signal the covariance function is estimated as below

$$\hat{c}_x(\tau) = \frac{1}{N-\tau} \sum_{n=1}^{N-\tau} x(n)x(n+\tau) \qquad (1.7a)$$

$$N \rightarrow \infty$$

and the cross-covariance function as

$$\hat{c}_{xy}(\tau) = \frac{1}{N-\tau}\sum_{n=1}^{N-\tau} x(n)y(n+\tau) \tag{1.7b}$$
$$N \to \infty$$

The covariance and cross-covariance functions defined in (1.7) turn out to be equal to those defined in (1.5) using expected operation provided the stochastic signals are ergodic (see [5] for more details on ergodicity). From (1.7) it is easy to note the units of the covariance function are energy per unit time, that is, power. The autocorrelation and crosscorrelation function defined for deterministic signals in (1.4) are similar to the covariance and cross-covariance functions except for the normalization factor; hence the units of the autocorrelation and cross-correlation functions (1.4) are simply those of energy.

Cyclostationary signals: We shall now consider a class of non-stationary signals, where the temporal variation in the second order statistical structure, is periodic. They are useful in modeling time series generated by modulating a periodic signal by a stochastic signal. Many modern-day communication signals fall in this group. Consider a stochastic signal with the following first and second order characteristics:

$$E\{x(n)\} = m_x(n) = m_x(n + pN_0), \quad p = 0,1,2,...$$

and

$$E\left\{[x(n+\frac{\tau}{2}) - m_x(n+\frac{\tau}{2})][x(n-\frac{\tau}{2}) - m_x(n-\frac{\tau}{2})]\right\}$$
$$= c_x(n,\tau) = c_x(n+pN_0,\tau), \quad p = 0,1,2,... \tag{1.8}$$

The covariance function is periodic in time. Such a signal has been termed as *cyclostationary* by Gardner [36]. It is also known as a periodically stationary signal. Since the covariance function is periodic in the time variable we can expand it in a Fourier series

$$c_x(n,\tau) = \sum_{p=-\infty}^{\infty} c_x(\frac{2\pi p}{N_0},\tau)e^{j\frac{2\pi p}{N_0}n} \tag{1.9a}$$

with inverse

$$c_x(\frac{2\pi p}{N_0},\tau) = \frac{1}{N}\sum_{n=-\frac{N}{2}}^{\frac{N}{2}} c_x(n,\tau)e^{-j\frac{2\pi p}{N_0}n} \tag{1.9b}$$

where N_0 is the period of repetition of the covariance function in n. Each component in the Fourier series expansion is known as a cyclic covariance function. Substituting in (1.9) for $c_x(n,\tau)$ from (1.8) we obtain for cyclic covariance function,

$$c_x\left(\frac{2\pi p}{N_0},\tau\right)=E\left\{\frac{1}{N}\sum_{-\frac{N}{2}}^{\frac{N}{2}}x\left(n+\frac{\tau}{2}\right)x\left(n-\frac{\tau}{2}\right)e^{-j\frac{2\pi p}{N_0}n}\right\} \tag{1.10}$$

The summation in (1.10) may be evaluated using just a single observed time series. There is no need for the statistical operation, E{.}, which requires a large number of realizations. Analogous to the autocorrelation function, defined in (1.4) using a single time series, we introduce a cyclic autocorrelation function,

$$\hat{c}_x\left(\frac{2\pi p}{N_0},\tau\right)=\frac{1}{N}\sum_{-\frac{N}{2}}^{\frac{N}{2}}x\left(n+\frac{\tau}{2}\right)x\left(n-\frac{\tau}{2}\right)e^{-j\frac{2\pi p}{N_0}n} \tag{1.11}$$
$$N\rightarrow\infty$$

A cyclic covariance function reduces to a simple covariance function for p=0 as can be seen from (1.10) where the exponential term is equal to one and the rest of the expression is the standard covariance function. Again from (1.10) it follows that for a stationary signal the cyclic covariance function is equal to zero, except for p=0, when it is equal to the covariance function of the stationary signal. The magnitude of the cyclic covariance function for $p\neq 0$ may be used as a measure of the degree of cyclostationarity. We give a different explanation to (1.10). For this purpose let us rewrite (1.10) as

$$\begin{aligned}c_x\left(\frac{2\pi p}{N_0},\tau\right)&=E\left\{\frac{1}{N}\sum_{n=0}^{N-1}x\left(n+\frac{\tau}{2}\right)e^{-j\frac{\pi p}{N_0}\left(n+\frac{\tau}{2}\right)}x\left(n-\frac{\tau}{2}\right)e^{-j\frac{\pi p}{N_0}\left(n-\frac{\tau}{2}\right)}\right\}\\&=E\left\{\frac{1}{N}\sum_{n=0}^{N-1}u_p\left(n+\frac{\tau}{2}\right)v_p^{*}\left(n-\frac{\tau}{2}\right)\right\}\end{aligned} \tag{1.12}$$

where

$$\begin{aligned}u_p(n)&=x(n)e^{-j\frac{\pi p}{N_0}n}\\v_p(n)&=x(n)e^{j\frac{\pi p}{N_0}n}\end{aligned} \tag{1.13}$$

We can consider $u_p(n)$ and $v_p(n)$ as frequency shifted versions of $x(n)$, by an amount equal to $-\pi p/N_0$ and $+\pi p/N_0$, respectively. The cyclic covariance function is the cross-covariance between $u_p(n)$ and $v_p(n)$.

Example 1.3: Consider an example of amplitude modulated (AM) sinusoidal carrier signal,

$$x(n)=x_0(n)\cos(\omega_0 n+\theta)$$

where $x_0(n)$ is a zero mean stationary sequence (message) and θ is the phase of the carrier. Note that

$$E\{x(n)\} = E\{x_0(n)\cos(\omega_0 n + \theta)\} = 0.$$

We can represent the signal as a sum of up and down frequency shifted components,

$$x(n) = \frac{1}{2}x_0(n)e^{j(\omega_0 n+\theta)} + \frac{1}{2}x_0(n)e^{-j(\omega_0 n+\theta)}$$

Using the above representation in (1.8) we get the following expression for the covariance function

$$c_x(n,\tau) = \frac{1}{4}E\left\{\begin{array}{l}[x_0(n+\frac{\tau}{2})e^{j(\omega_0(n+\frac{\tau}{2})+\theta)} + x_0(n+\frac{\tau}{2})e^{-j(\omega_0(n+\frac{\tau}{2})+\theta)}] \\ {[x_0(n-\frac{\tau}{2})e^{-j(\omega_0(n-\frac{\tau}{2})+\theta)} + x_0(n-\frac{\tau}{2})e^{j(\omega_0(n-\frac{\tau}{2})+\theta)}]}\end{array}\right\}$$

$$= \frac{1}{4}\left[c_x(\tau)e^{j\omega_0\tau} + c_x(\tau)e^{-j2(\omega_0 n+\theta)} + c_x(\tau)e^{j2(\omega_0 n+\theta)} + c_x(\tau)e^{-j\omega_0\tau}\right]$$

$$= \frac{1}{2}c_x(\tau)\cos(\omega_0\tau) + \frac{1}{2}c_x(\tau)\cos(2(\omega_0 n+\theta))$$

It may be observed that only the second term in the covariance expression is periodic in n with period π/ω_0. Next we compute the cyclic covariance function as defined in (1.11). We obtain, after some straightforward algebraic simplification, the following result:

$$\begin{aligned} c_x(\frac{2\pi p}{N_0},\tau) &= \frac{1}{4}c_x(\tau)e^{\pm j2\theta} && \textit{for } p = \pm 2 \\ &= \frac{1}{2}c_x(\tau)\cos(\omega_0\tau) && \textit{for } p = 0 \\ &= 0 && \textit{otherwise} \end{aligned}$$

where $\omega_0 = 2\pi/N_0$. The most interesting feature of the cyclic covariance function is the presence of terms for $p = \pm 2$, which are not present for a pure stationary signal. It is possible, though beyond the scope of this work, to devise a filtering scheme exploiting such a unique property of the cyclostationary signal for the purpose of extracting a communication signal in the presence of noise or interference which is more likely to be a pure stationary signal. Note the important role of the phase of the carrier. If we treat the phase as unknown or random; a uniformly distributed random variable, we find that the terms corresponding to $p = \pm 2$ simply vanish. The signal behaves more like a pure stationary signal. Therefore, knowledge of the phase or the timing information is mandatory .

§1.3 Fourier and Laplace Transforms

In signal processing, Fourier transform plays a very important role. It enables us to migrate from time domain to frequency domain where some of the signal processing algorithms are more clearly understood and efficiently implemented. Hence, it is essential to be familiar with the mathematical aspects of the Fourier transform and its companion, the Laplace transform. In this section we shall deal with continuous-time functions but later in chapter two we shall return to discrete signals. The coverage is brief but adequate to understand what follows in the remaining chapters. A comprehensive discussion on Fourier transform may be found in [38].

Fourier Transform: Let $x(t)$ be a continuous function with finite energy, that is,

$$\int_{-\infty}^{\infty} |x(t)|^2 \, dt < \infty .$$

The Fourier transform is defined as

$$X(\omega) = \int_{-\infty}^{\infty} x(t) e^{-j\omega t} dt \tag{1.14a}$$

and the inverse Fourier transform is given by

$$x(t) = \frac{1}{2\pi} \int_{-\infty}^{\infty} X(\omega) e^{+j\omega t} d\omega \tag{1.14b}$$

These two operations will be often denoted by FT{.} and FT^{-1}\{.\}, respectively. Note the infinite limits on the integral and the sign in the exponent of exponential function. ω is the angular frequency (radians/sec) which is related to the conventional frequency f in Hertz (cycles/sec) through a relation $\omega = 2\pi f$ (radians/sec). Equation (1.14b) is also known as Fourier integral representation of $x(t)$.

Example 1.4: The Fourier transforms of some well-known functions are derived in this example.

Delta function: $FT\{\delta(t)\} = \int_{-\infty}^{\infty} \delta(t) e^{-j\omega t} dt = 1$

Sinusoid:

$$FT\{\sin(\omega_0 t\} = \int_{-\infty}^{\infty}\left\{\frac{e^{j\omega_0 t} - e^{-j\omega_0 t}}{2j}\right\}e^{-j\omega t}dt$$

$$= \frac{1}{2j}\{\delta(\omega - \omega_0) - \delta(\omega + \omega_0)\}$$

Comb function:

$$FT\left\{\sum_{n=-\infty}^{\infty}\delta(t - n\Delta t)\right\} = \sum_{n=-\infty}^{\infty}\int_{-\infty}^{\infty}\{\delta(t - n\Delta t)\}e^{-j\omega t}dt$$

$$= \sum_{n=-\infty}^{\infty}e^{-j\omega n\Delta t} = \sum_{k=-\infty}^{\infty}\delta(\omega - \frac{2\pi}{\Delta t}k)$$

Box car (rectangular) function: $FT\{r(t)\} = \int_{-\frac{T}{2}}^{\frac{T}{2}} 1\cdot e^{-j\omega t}dt = \dfrac{e^{-j\omega\frac{T}{2}} - e^{j\omega\frac{T}{2}}}{-j\omega}$

$$= \frac{\sin(\omega\frac{T}{2})}{\frac{\omega}{2}}$$

Exponential function: $FT\left\{e^{-\alpha|t|}\right\} = \int_{-\infty}^{\infty} e^{-\alpha|t|}e^{-j\omega t}dt$

$$= \int_{0}^{\infty}\left\{e^{-\alpha|t|}e^{-j\omega t} + e^{-\alpha|t|}e^{j\omega t}\right\}dt$$

$$= \frac{1}{\alpha + j\omega} + \frac{1}{\alpha - j\omega}$$

$X(\omega)$ is generally complex except when *x(t)* is symmetric. We define the energy spectrum and the phase spectrum as follows:

$$\text{Energy spectrum} = |X(\omega)|^2$$

$$\text{Phase spectrum} = \tan^{-1}\frac{\text{Im}\{X(\omega)\}}{\text{Re}\{X(\omega)\}}$$

We shall now state (without proof) some useful properties of the Fourier transform.

Properties	Remarks
1. $X^*(\omega) = X(-\omega)$	when *x(t)* is real
2. $X(0) = \int_{-\infty}^{\infty} x(t)dt$	
3. $\frac{1}{2\pi}\int_{-\infty}^{\infty} \lvert X(\omega)\rvert^2 d\omega = \int_{-\infty}^{\infty} \lvert x(t)\rvert^2 dt$	Parseval's theorem
4. $FT\{x(t-\tau)\} = X(\omega)e^{-j\omega\tau}$	Shift theorem
5. $FT\{x(at)\} = \frac{1}{\lvert a\rvert}X(\frac{\omega}{a})$ where *a* is a real constant and $a \neq 0$	Similarity theorem
6. $FT\{x(t)\cos(\omega_0 t)\} = \frac{1}{2}(X(\omega-\omega_0) + X(\omega+\omega_0))$	Modulation theorem
7. $FT\{x(-t)\} = X^*(\omega)$	Time reversal
8. $FT\left\{\int_{-\infty}^{\infty} h(\tau)x(t-\tau)d\tau\right\} = H(\omega)X(\omega)$	Convolution theorem

Laplace Transform: Closely related to the Fourier transform is the Laplace transform which helps to completely unravel the structure of a signal in terms of its poles and zeros. The Laplace transform is defined as

$$X(s) = \int_{-\infty}^{\infty} x(t)e^{-st}dt \tag{1.15}$$

where $s = \sigma + j\omega$ is a complex frequency. The conditions for the existence of Laplace transform for a given signal are much more stringent than for Fourier transform. The Laplace transform of a signal exists only over a finite interval of the real part of the complex frequency, $\sigma_{min} \leq \sigma \leq \sigma_{max}$ (see fig. 1.7). σ_{min} is controlled by the decay rate of *x(t)* for t >0 and σ_{max} is controlled by the decay rate of *x(t)* for $t < 0$.

$$x(t)e^{\sigma_{min}t} \to 0 \qquad t \to \infty$$

$$x(t)e^{\sigma_{max}t} \to 0 \qquad t \to -\infty$$

The region of convergence (ROC) of Laplace transform in complex frequency plane is a strip containing the imaginary axis as shown in fig.1.7.

Note that the Fourier transform is a special case of the Laplace transform evaluated for $\sigma = 0$ and it exists for most signals with finite energy. The Laplace transform of a signal (or z-transform for discrete signals) is useful in bringing out the poles and zeros of a signal. The poles and zeros play very important role in characterizing a filter response.

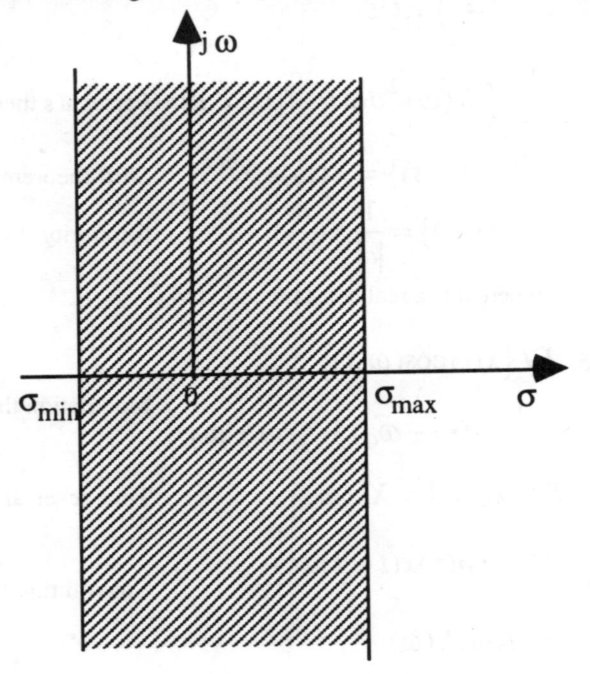

Figure 1.7: The region of convergence of Laplace transform . The imaginary axis is always inside this strip.

Example 1.5: Consider the Laplace transform of an exponential function, $e^{-\alpha_0|t|}$. For $-\alpha_0 < \mathrm{Re}\{s\} < \alpha_0$ the Laplace transform exists.

$$X(s) = \int_{-\infty}^{\infty} e^{-\alpha_0|t|} e^{-st} dt$$

$$= \int_{0}^{\infty} e^{-\alpha_0 t} e^{-st} dt + \int_{0}^{\infty} e^{-\alpha_0 t} e^{st} dt$$

$$= \frac{1}{\alpha_0 + s} + \frac{1}{\alpha_0 - s}$$

The Laplace transform has two poles on the real axis at $s = \pm\alpha_0$. Consider another example of Laplace transform of damped sinusoid, $x(t) = e^{-\alpha_0|t|}\cos(\omega_0 t)$.

$$X(s)=\int_{-\infty}^{\infty} e^{-\alpha_0|t|}\cos(\omega_0 t)e^{-st}dt$$

$$=\int_{0}^{\infty} e^{-\alpha_0 t}\left\{\frac{e^{j\omega_0 t}+e^{-j\omega_0 t}}{2}\right\}(e^{-st}+e^{+st})dt$$

$$=\frac{1}{2}\left[\frac{1}{s+\alpha_0-j\omega_0}+\frac{1}{s+\alpha_0+j\omega_0}-\frac{1}{s-\alpha_0+j\omega_0}-\frac{1}{s-\alpha_0-j\omega_0}\right]$$

Now there are four poles for the damped sinusoid. They are located at $s=\mp\alpha_0 \pm j\omega_0$. The position of poles is shown in fig. 1.8 🍏.

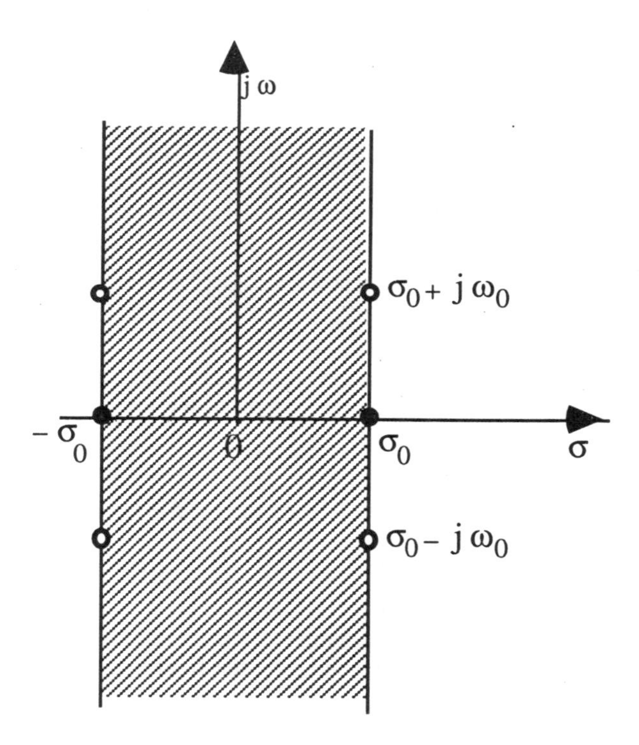

Figure 1.8: Poles of exponential signal (●) and of damped sinusoidal signal (○).

The more familiar Laplace transform is the one-sided transform, that is, the limits on the integral are from 0 to $+\infty$ which is equivalent to assuming that the signal is zero for $t < 0$ or it is a causal signal.

$$X(s)=\int_{0}^{\infty} x(t)e^{-st}dt$$

which exists for $\mathrm{Re}\{s\} > \sigma_{min}$. The Laplace transform of a causal signal has a much wider region of convergence (see fig. 1.9). Consider once again the damped sinusoid but this time it is zero for t < 0.

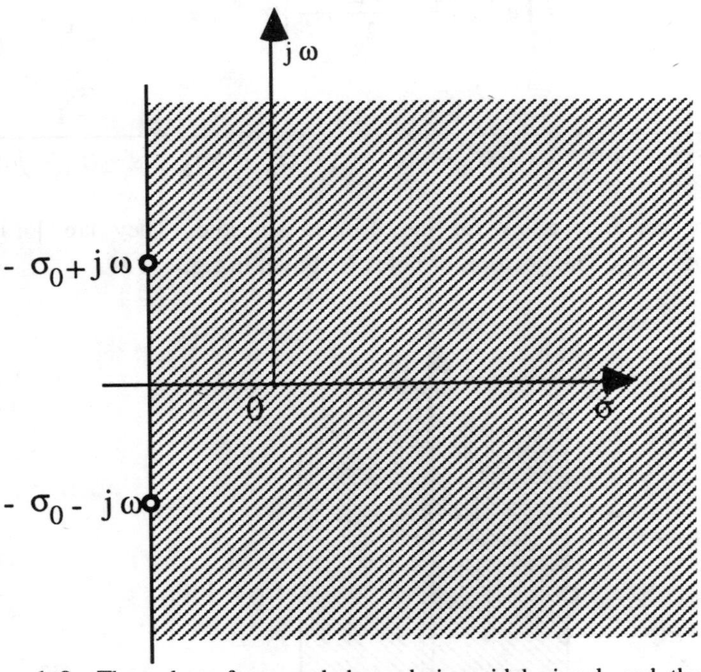

Figure 1.9: The poles of a causal damped sinusoidal signal and the region of convergence is a semi-infinite plane to the right of the line joining the poles.

$$x(t) = \begin{cases} e^{-\alpha t} \cos\omega_0 t & t \geq 0 \\ 0 & t < 0 \end{cases}$$

The Laplace transform of a causal damped sinusoid may be derived following the steps shown in Example 1.5. We shall give the final result,

$$X(s) = \frac{1}{2}\left[\frac{1}{s+\alpha_0 - j\omega_0} + \frac{1}{s+\alpha_0 + j\omega_0}\right]$$

The poles are at $s = -\alpha_0 + j\omega_0$ and $s = -\alpha_0 - j\omega_0$. $\alpha_{min} = -\alpha_0$ and $\alpha_{max} = \infty$. The region of convergence is semi-infinite plane to the right of the line joining the poles The convergence region and pole position are shown in fig. 1.9. The Laplace transform shares many properties of the Fourier transform listed on page 17 Indeed the Fourier transform is a special case of the Laplace transform, that is, the Laplace transform evaluated on the imaginary axis, $s = j\omega$ is equal to the Fourier transform.

$$\left[X(s) = \int_{-\infty}^{\infty} x(t)e^{-st}dt\right]_{s=j\omega} = \int_{-\infty}^{\infty} x(t)e^{-j\omega t}dt = X(\omega)$$

Interestingly the Laplace transform of a signal may be expressed in terms of the Fourier transform of the product, $x(t)e^{-\sigma t}$,

$$X(s) = \int_{-\infty}^{\infty} x(t)e^{-st}dt = \int_{-\infty}^{\infty} x(t)e^{-j\omega t}e^{-\sigma t}dt = \int_{-\infty}^{\infty}\left[x(t)e^{-\sigma t}\right]e^{-j\omega t}dt$$

§1.4 z-Transform

For discrete signals the z-transform appears to be the most convenient transform. In fact, the z-transform is to discrete signals what the Laplace transform is to continuous signal.

Definition: The z-transform is defined as below:

Two-sided: $$X(z) = \sum_{n=-\infty}^{\infty} x(n)z^{-n} \tag{1.16a}$$

One-sided: $$X(z) = \sum_{n=0}^{\infty} x(n)z^{-n} \tag{1.16b}$$

where z is a complex variable, $z=x+jy$. $X(z)$ exists in the region of convergence in the z-plane provided that the signal satisfies the finite energy property, as in the case of Fourier transform (see page 15). The region of convergence (ROC) for two-sided z=transform is bounded by two concentric circles (see fig. 1.10). The z-transform may be related to the Laplace transform through a mapping

$$z = e^{s\Delta t} \tag{1.17}$$

where Δt stands for the sampling interval. Recall that in Laplace transform s stands for complex frequency, $s = \sigma + j\omega$. Hence, equation (1.17) is a mapping between two planes (see fig. 1.11). The $j\omega$ axis maps onto the circumference of the unit circle in the z-plane and the area to the left of $j\omega$ axis and lying between $s = \pm j\pi/\Delta t$ maps into the interior of the unit circle. Examples of mapping are shown in fig. 1.11. The imaginary axis lying between $s = \pm j\pi/\Delta t$ is mapped onto the unit circle in z plane. A point on the imaginary

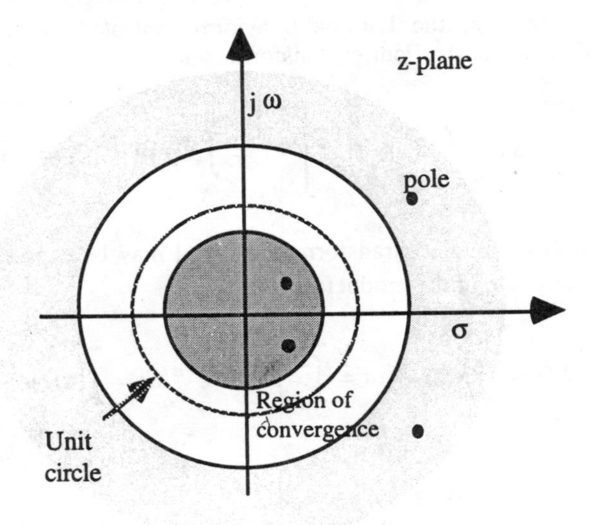

Figure 1.10: Shows the region of convergence (white annular space) of the z-transform of a two-sided discrete signal. Poles lie inside the shaded regions. For reference the unit circle is also shown in the figure.

axis, say, $s = +j\pi/2\Delta t$, is mapped to a point on the unit circle, $z=j$. A vertical line ab lying inside the left strip is mapped onto a circle inside the unit circle shown in fig. 1.11.
As noted earlier all poles of a causal signal lie in the left half plane of s-plane. Upon mapping onto z-plane, all poles of one-sided z-transform will be mapped into the unit circle. There are no poles outside the unit circle. Reverse will be true for an anti causal signal, that is, for a signal which is zero for t>0. The poles of a two-sided signal will lie both inside and outside the unit circle. The region of convergence of the z-transform of a two-sided signal is obtained by mapping the region of convergence in the s-plane (see fig. 1.7) into z-plane as in fig. 1.10).

Example 1.6: The z-transforms of several standard sequences are listed in most books on signal analysis. We note a few examples.
1. Compute the z-transform of an infinite sequence,

$$x(n) = r^n, \quad n = 0,1,2,\dots,\infty$$

It is easily obtained as

$$X(z) = \sum_{n=0}^{\infty} r^n z^{-n} = \frac{1}{1 - rz^{-1}}$$

The sum converges only for $|rz^{-1}| < 1$ or $|z| > r$, hence the region of convergence is outside a circle of radius r. If we now let $r \to 1$, $X(z) \to 1/(1 - z^{-1})$, which is a z-transform of a step function.
2. Compute the z-transform of an infinite sequence,

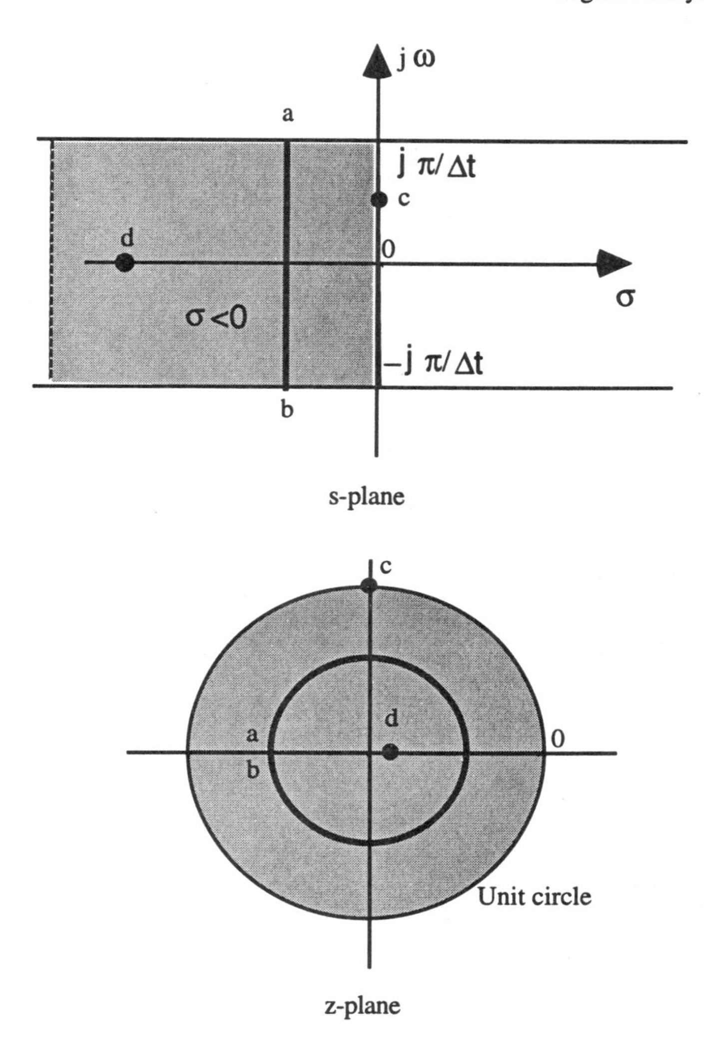

Figure 1.11: The z-transform maps the s-plane onto z-plane. A strip lying within $s = \pm j\,\pi/\Delta t$ in the s-plane is mapped onto z-plane such that the left half goes into the interior of the unit circle and the right half goes to the exterior of the unit circle. Line ab in the s-plane is mapped into a circle within the unit circle. The points c and d are mapped onto the corresponding c and d points in the z-plane.

$$x(n) = q^{-n}, \quad n = 0, -1, -2, \ldots, -\infty$$

$$X(z) = \sum_{-\infty}^{n=-1} q^{n} z^{-n} = \sum_{n=1}^{\infty} q^{-n} z^{n} = \frac{q^{-1} z}{1 - q^{-1} z}$$

The sum converges only for $\left|q^{-1} z\right| < 1$ or $|z| < q$, hence the region of convergence is inside a circle of radius q.

3. Compute the one-sided z-transform of $x(n) = \sin(\omega_0 n)$, $n = 0, 1, 2, \ldots$ Since $x(n)$ is not a finite energy signal, strictly speaking its z-transform does not exist. However, we can obtain a limiting expression as in (1) above.

$$ZT\{r^n \sin(\omega_0 n)\} = \frac{1}{2j}\sum_{n=0}^{\infty}(r^n e^{j\omega_0 n} - r^n e^{-j\omega_0 n})z^{-n}$$

$$|r| < 1$$

$$= \frac{1}{2j}\left[\sum_{n=0}^{\infty}(e^{j\omega_0 n} - e^{-j\omega_0 n})r^n z^{-n}\right]$$

$$= \frac{1}{2j}\left[\frac{1}{1 - e^{j\omega_0} r z^{-1}} - \frac{1}{1 - e^{-j\omega_0} r z^{-1}}\right]$$

$$= \left[\frac{r\sin(\omega_0)z^{-1}}{1 - 2r\cos(\omega_0)z^{-1} + r^2 z^{-2}}\right]$$

Now, if we let $r \to 1$, we get the z-transform of $\sin(\omega_0 n)$, $n = 0, 1, 2, \ldots$ tending to

$$ZT\{\sin(\omega_0 n)\} \to \left[\frac{\sin(\omega_0)z^{-1}}{1 - 2\cos(\omega_0)z^{-1} + z^{-2}}\right]$$

4. Finite sequence, $\{1.0, 0.8, 0.64, 1.2, 0.4, 3.0, 0.1, 1.6\}$

$$X(z) = \begin{Bmatrix} 1.0 + 0.8z^{-1} + 0.64z^{-2} + 1.2z^{-3} + 0.4z^{-4} + 3.0z^{-5} \\ +0.1z^{-6} + 0.6z^{-7-6} + 1.6z^{-7} \end{Bmatrix}$$

5. Recursively defined sequence with initial condition $x(-1), x(-2), \ldots = 0$,

$$x(n) = 0.8x(n-1) + 0.1\,\mathrm{u(n)}$$

where $u(n)$ is a step function, that is, $u(n)=1$ for $n \geq 1$ and $u(n)=0$ for $n<0$.

$$ZT\{x(n)\} = ZT\{0.8x(n-1) + 0.1u(n)\}$$

$$X(z) = 0.8z^{-1}X(z) + \frac{0.1}{1 - z^{-1}}$$

$$X(z) - 0.8z^{-1}X(z) = \frac{0.1}{1 - z^{-1}}$$

$$X(z) = \frac{0.1}{(1 - z^{-1})}\frac{1}{(1 - 0.8z^{-1})}$$

where ZT{.} stands for the z-transform of the quantity within the curly brackets ♠.

Poles and Zeros: A rational z-transform is of great interest in signal modeling and filtering. It is represented by a ratio of two polynomials,

$$H(z)=\frac{\sum_{i=0}^{q} b_i z^{-i}}{\sum_{i=0}^{p} a_i z^{-i}}=\frac{b_0}{a_0} z^{-q+p} \frac{(z-z_1)(z-z_2)\ldots(z-z_q)}{(z-\pi_1)(z-\pi_2)\ldots(z-\pi_p)} \tag{1.18a}$$

The numerator and denominator polynomials are factored into q and p factors, respectively. The roots of the numerator, $z_1, z_2, \ldots, z_q$, are the zeros and the roots of the denominator, $\pi_1, \pi_2, \ldots, \pi_p$, are the poles of *H(z)*. The term z^{-q+p} represents a zero of the order *p-q* (when *p>q*) or a pole of order *q-p* (when *q>p*) at z=0. Note that there are always equal number of poles and zeros. When the coefficients of the numerator and denominator polynomials are real, the zeros and poles must appear as complex conjugate pairs or on the real axis. To show this consider an example *p=q=*2. Eq (1.18a) simplifies to

$$H(z)=\left(\frac{b_0}{a_0}\right)\frac{(z-z_1)(z-z_2)}{(z-\pi_1)(z-\pi_2)}=\left(\frac{b_0}{a_0}\right)\frac{z^2-(z_1+z_2)+z_1 z_2}{z^2-(\pi_1+\pi_2)+\pi_1\pi_2} \tag{1.18b}$$

For the polynomial coefficients to be real it is clear that $z_1 = z_2^*$ and $\pi_1 = \pi_2^*$. Some of the roots of the numerator and denominator polynomials in (1.18a) may repeat or may be shared between the numerator and denominator.

Inverse z-Transform: The z-transform of a sequence can be reversed. The inverse z-transform is defined as below:

$$x(n)=\frac{1}{2\pi j}\oint_{\Gamma}\frac{X(z)z^n}{z}dz \tag{1.19}$$

where Γ is a closed contour lying within the region of convergence (see figure 1.12).

The inverse z-transform is readily obtained when the z-transform can be expanded in partial fractions, that is,

$$X(z)=\sum_{i=1}^{p}\frac{a_i z}{(z-\lambda_i)} \tag{1.20}$$

Each term in (1.20) is a z-transform of $a_i e^{-\lambda_i n}, n = 0,1,2,\ldots\infty$, therefore, following the property of linearity we can evaluate the inverse z-transform of $X(z)$,

$$x(n) = \sum_{i=1}^{p} a_i e^{-\lambda_i n}, n = 0,1,2,\ldots\infty \qquad (1.21)$$

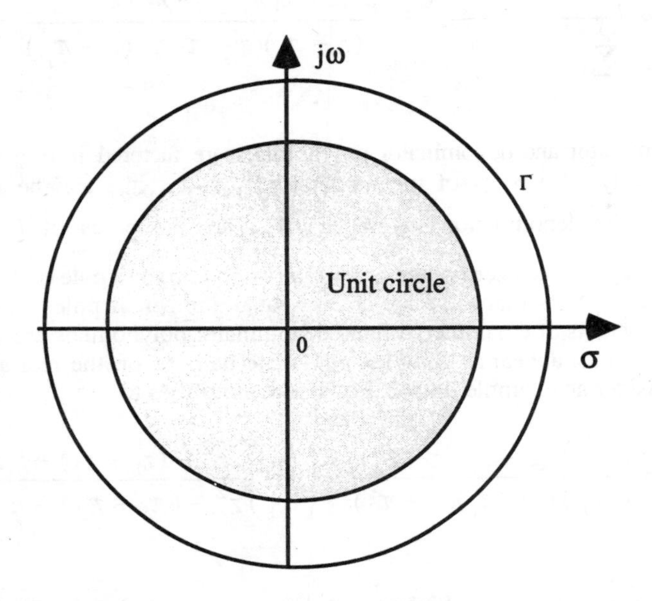

Figure 1.12: Closed contour for integration in the inverse z-transform. The contour must lie within the region of convergence encircling the unit circle. For a causal signal all poles are inside the unit circle.

Properties of z-Transform: Some important properties of the z-transform are shown in table 1.2. The first three properties are straight-forward and hence they are left out as exercises for the reader. The proof of the remaining three properties is given below.

Convolution:

$$y(n) = \sum_{m=0}^{\infty} h(m)x(n-m)$$

It also represents linear filtering operation where *h(m)* is the filter impulse response function, *x(n)* is the input to the filter and *y(n)* is the output of the filter. To show the convolution theorem we evaluate the z-transform on both sides of the convolution sum

Table 1.2: Some properties of the z-transform

Property	Formula	ROC
Linearity:	$ZT\{a_1x_1(n)+a_2x_2(n)\} = a_1X_1(z)+a_2X_2(z)$	Intersection of ROCs of X_1 and X_2
Delay	$ZT\{x(n-l)\}=z^{-l}X(z)$	Same as ROC of X except at z=0 for l>0 and at z=∞ for l<0.
Time reversal	$ZT\{x(-n)\}=X(\frac{1}{z})$	Inverse of ROC of X. Example: Inverse ROC of $\|z\|>r_1$ is $\|z\|<\frac{1}{r_1}$
Convolution theorem	$ZT\{y(n)\}=H(z)X(z)$	Intersection of ROCs of H and X
Modulation	$Y(z)=\frac{1}{2\pi j}\oint_\Gamma \frac{W(\zeta)}{\zeta}X\left(\frac{z}{\zeta}\right)d\zeta$	Let ROC of W be $r_{1l}<\|z\|<r_{1u}$ and ROC of X be $r_{2l}<\|z\|<r_{2u}$. Then, ROC of Y is $r_{1l}r_{2l}<\|z\|<r_{1u}r_{2u}$.
Correlation	$X(z)X(z^{-1})$	Intersection of ROCs of X (z) and X (z^{-1})

$$\sum_{n=0}^{\infty} y(n)z^{-n} = \sum_{n=0}^{\infty}\sum_{m=0}^{\infty} h(m)x(n-m)z^{-n}$$

$$= \sum_{m=0}^{\infty} h(m)z^{-m}\sum_{n=m}^{\infty} x(n-m)z^{-(n-m)}$$

$$\boxed{Y(z)=H(z)X(z)} \tag{1.22}$$

The theorem states that the z-transform of the output of a convolution sum is equal to the product of the z-transforms of the input and the impulse response function of the filter.

Product of two sequences (modulation):

$$ZT\{w(n)x(n)\} = \frac{1}{j2\pi}\oint_\Gamma \frac{W(\zeta)}{\zeta}X\left(\frac{z}{\zeta}\right)d\zeta$$

The proof is as follows:

$$\begin{aligned}\sum_{n=0}^{\infty} y(n)z^{-n} &= \sum_{n=0}^{\infty} w(n)x(n)z^{-n} \\ &= \sum_{n=0}^{\infty} \frac{1}{j2\pi} \oint_{\Gamma} \frac{W(\zeta)\zeta^{n}}{\zeta} d\zeta x(n) z^{-n} \\ &= \frac{1}{2\pi j} \oint_{\Gamma} \frac{W(\zeta)}{\zeta} d\zeta \sum_{n=0}^{\infty} x(n) \left(\frac{z}{\zeta}\right)^{-n} \\ Y(z) &= \frac{1}{2\pi j} \oint_{\Gamma} \frac{W(\zeta)}{\zeta} X\left(\frac{z}{\zeta}\right) d\zeta\end{aligned} \tag{1.23}$$

where Γ is the contour (see fig. 1.12) of integration. Equation (1.23) is an integral relation between the z-transform of the product of two sequences and those of the individual sequences. It is called complex convolution, which may also be expressed in the form of a normal convolution. For this assume that the contour of integration is the unit circle. Further let $z = e^{j\omega}$ and $\zeta = e^{j\theta}$. Substituting in (1.23) we obtain

$$Y(\omega) = \frac{1}{2\pi} \int_{-\pi}^{\pi} W(\theta) X(\omega - \theta) d\theta \tag{1.24}$$

which is the normal convolution between the Fourier transforms, $W(\omega)$ and $X(\omega)$.

Correlation:

$$R_{xx}(z) = X(z)X(z^{-1})$$
$$R_{xy}(z) = X(z)Y(z^{-1})$$

We shall derive the z-transform of the autocorrelation. The autocorrelation of an infinite deterministic sequence is similar to that of finite deterministic sequence (1.4a),

$$r_{xx}(\tau) = \sum_{n=-\infty}^{\infty} x(n)x(n+\tau) \tag{1.25a}$$

Let us compute the z-transform on both sides of (1.25a),

$$\begin{aligned}R_{xx}(z) = \sum_{-\infty}^{\infty} r_{xx}(\tau) z^{-\tau} &= \sum_{\tau=-\infty}^{\infty} \sum_{n=-\infty}^{\infty} x(n) z^{+n} x(n+\tau) z^{-\tau-n} \\ &= \sum_{\tau=-\infty}^{\infty} \sum_{n=-\infty}^{\infty} x(n) z^{+n} x(n+\tau) z^{-\tau-n} \\ &= X(z)X(z^{-1})\end{aligned} \tag{1.25b}$$

The z-transform of the cross-correlation may be derived in the same fashion as above

§1.5 Signal Models

A stochastic signal may be looked upon as an output of a filter driven by random input or as an output of a difference equation driven by random input. Accordingly, it may be modeled either based on its spectrum or the type of the difference equation. The impulse response function of a filter controls the spectrum of the signal, in particular the bandwidth is an important attribute of the signal. The coefficients of the difference equation are useful in prediction or forecasting. The spectral model and the difference equation model are, however, not independent; in fact, they are interchangeable. A spectral model can be expressed as a difference equation model or vice-versa. This complete equivalence enables us to choose any one of the models, depending upon the physics, the purpose of analysis, such as extraction of a signal buried in noise or prediction of time series, computational cost, etc. We shall briefly study in all six signal models; namely, narrowband signal (non-parametric) model and four signal (parametric) models based on difference equation, namely, moving average (MA) model, autoregressive (AR) model and autoregressive-moving average (ARMA) model and state space model. We shall briefly touch upon the chaotic process as a possible signal model. This has already been tried as a model for biomedical and speech signals[10].

Narrowband signal: A signal, when its power is confined to a narrowband of frequencies centered around a central frequency ω_c, is called a narrowband signal. The Fourier transform of a narrowband signal is expressed as

$$X_{nb}(\omega)=0 \qquad |\omega-\omega_c|>\frac{\Delta\omega}{2}$$

where $\Delta\omega$ is the bandwidth and its Fourier representation is given by,

$$x_{nb}(t)=\frac{1}{2\pi}\int_{\omega_c-\frac{\Delta\omega}{2}}^{\omega_c+\frac{\Delta\omega}{2}} X_{nb}(\omega)e^{j\omega t}d\omega+\frac{1}{2\pi}\int_{-\omega_c-\frac{\Delta\omega}{2}}^{-\omega_c+\frac{\Delta\omega}{2}} X_{nb}(\omega)e^{j\omega t}d\omega$$

$$=x_{nb}^{+}(t)+x_{nb}^{-}(t) \tag{1.26}$$

where $x_{nb}^{+}(t)$ is a complex signal, known as complex analytical representation of $x_{nb}(t)$ (continuous-time signal). It is usually computed via Hilbert transform, which is defined as

$$x_{Hilb}(t)=\int_{-\infty}^{\infty}\frac{x(t')}{t-t'}dt'$$

and

$$\boxed{x_{nb}^{+}(t) = x(t) + jx_{Hilb}(t)}$$

Note that, because of the complex conjugate symmetry of the Fourier transform of a real signal, $x_{nb}^{-}(t) = x_{nb}^{+*}(t)$. We can also express $x_{nb}^{+}(t)$ in a different form,

$$\begin{aligned} x_{nb}^{+}(t) &= \frac{1}{2\pi}\int_{\omega_c-\frac{\Delta\omega}{2}}^{\omega_c+\frac{\Delta\omega}{2}} X_{nb}(\omega)e^{j\omega t}d\omega \\ &= \frac{1}{2\pi}e^{j\omega_c t}\int_{-\frac{\Delta\omega}{2}}^{\frac{\Delta\omega}{2}} X_0(\omega')e^{j\omega' t}d\omega' \\ &= x_0(t)e^{j\omega_c t} \end{aligned} \tag{1.27}$$

where

$$x_0(t) = \frac{1}{2\pi}\int_{-\frac{\Delta\omega}{2}}^{\frac{\Delta\omega}{2}} X_0(\omega)e^{j\omega t}d\omega$$

and

$$X_0(\omega) = X_{nb}(\omega_c + \omega).$$

Note that $x_0(t)$ is not necessarily a real function. It may be noted that (1.27) is valid even when $\Delta\omega$ is not small, as in bandpass signal. An interesting property of $x_{nb}^{+}(t)$, for small delays for which $\Delta\omega\delta t << 2\pi$ holds ($x_0(t)$ is a slowly varying function), is that a delay in signal can be expressed as a phase change in $x_{nb}^{+}(t)$.

$$\boxed{x_{nb}^{+}(t-\delta t) = e^{-j\omega_c \delta t}x_{nb}^{+}(t)}$$

This property is central to beamforming in array signal processing. We shall briefly review some basic principles of beamforming in Chapter Seven.
Another way of defining the narrowband property is used in frequency modulation (FM) broadcast. A simple FM signal is given by

$$x(t) = \cos(\omega_c t + \frac{\Delta\omega}{2\omega_m}\sin\omega_m t) \tag{1.28a}$$

where $(\Delta\omega/2)\sin(\omega_m t)$ is the modulating signal and $\Delta\omega/\omega_m$ is the modulation index. When $\Delta\omega/\omega_m << \pi$ (1.28a) may be approximated as

$$x(t) \approx \cos(\omega_c t) - \frac{\Delta\omega}{2\omega_m}\sin\omega_m t\sin(\omega_c t) \tag{1.28b}$$

The Fourier transform of the right hand side of (1.28b) is sketched in fig. 1.13.

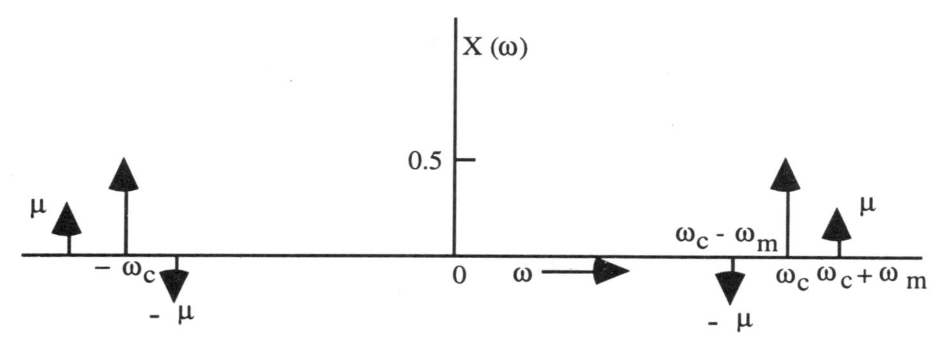

Figure 1.13: A sketch of the Fourier transform of a narrowband frequency modulated signal. where $\mu = \Delta\omega/8\omega_m$.

When the modulation index is large ($\Delta\omega/\omega_m >> \pi$), many peaks swarm around $\pm\omega_c$. Then, the bandwidth of the signal approximately is equal to $\Delta\omega$, making it a wideband signal. Narrowband signals are widely used in communication, sonar, radar, imaging, etc.

Moving Average (MA) Signal: A moving average (MA) signal (time series) is defined as

$$x(n) = \sum_{i=0}^{\infty} b_i \varepsilon(n-i) \tag{1.29}$$

where b_i, i=0,1,..., ∞ are averaging coefficients and $\varepsilon(n)$ is white noise with zero mean and unit variance but not necessarily Gaussian. It is easy to show that

$$E\{x(n)\} = 0$$

$$E\{x^2(n)\} = \sum_{i=0}^{\infty} b_i^2 \sigma_\varepsilon^2 = \sum_{i=0}^{\infty} b_i^2$$

where σ_ε^2 is variance of the white noise presumed to be unity. Further, for stability we require that

$$\sum_{i=0}^{\infty} b_i^2 < \infty.$$

We can look upon $x(n)$ as an output of a filter driven by white noise. The model is illustrated in fig. 1.14. Taking z-transform on both sides of (1.29) we obtain,

$$X(z) = \sum_{i=0}^{\infty} b_i z^{-i} \mathrm{E}(z) = B(z)\mathrm{E}(z)$$

where $B(z)$ is the transfer function. It has zeros in the z-plane and all its poles are at $z=0$, that is, at the origin of z-plane. A moving average (MA) model is also known as all-zero model. The covariance function of a MA signal has some interesting properties.

$$\begin{aligned} c_x(\tau) &= E\{x(n)x(n+\tau)\} \\ &= \sum_i \sum_k b_i b_k E\{\varepsilon(n-i)\varepsilon(n+\tau-k)\} \\ &= \sigma_\varepsilon^2 \sum_i \sum_k b_i b_k \delta(\tau-k+i) \\ &= \sigma_\varepsilon^2 \sum_{i=0}^{\infty} b_i b_{i+\tau} \end{aligned} \tag{1.30a}$$

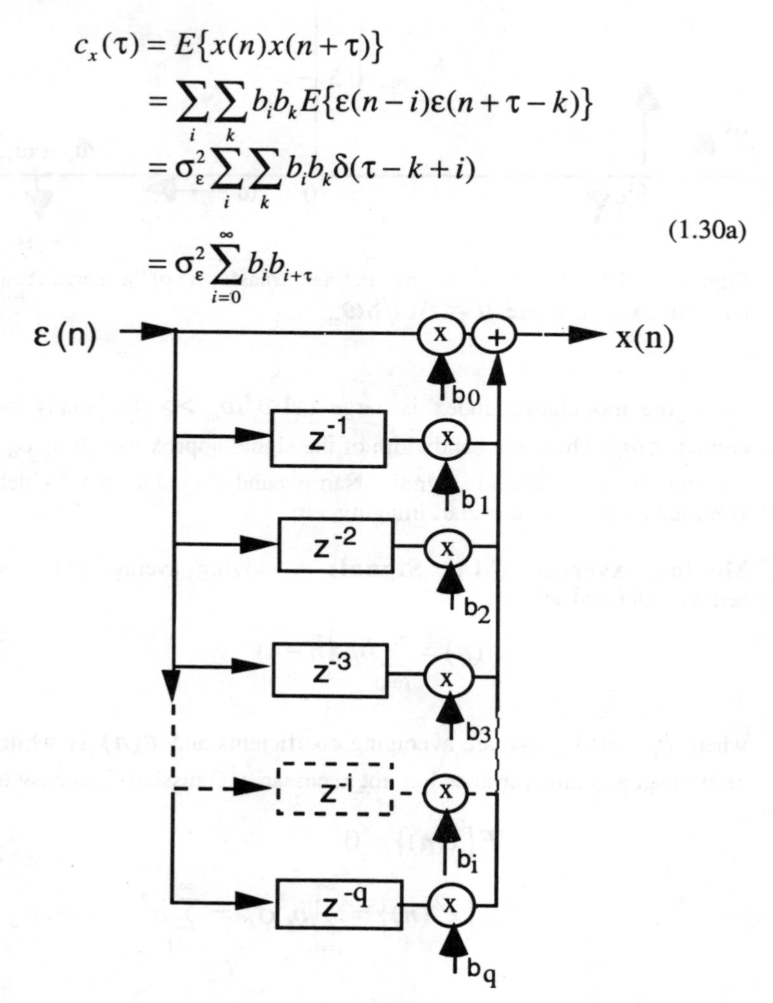

Figure 1.14: q^{th} order moving average model. The input is a white noise and the output is a moving average (MA) signal.

For a finite order MA model the covariance function in (1.30a) becomes

$$\boxed{\begin{aligned} c_x(\tau) &= \sigma_\varepsilon^2 \sum_{i=0}^{q-\tau} b_i b_{i+\tau} & 0 \le \tau \le q \\ &= 0 & \tau > q \end{aligned}} \tag{1.30b}$$

The covariance function vanishes for lags greater than the order of MA signal

Example 1.7: Consider a 2nd order MA signal,

$$x(n) = b_0\varepsilon(n) + b_1\varepsilon(n-1) + b_2\varepsilon(n-2)$$

Let b_0=1.0, $b_1 = -1.5588$ and $b_2 = 0.81$. The transfer function is given by $1.0 - 1.5588z^{-1} + 0.81z^{-2}$ which has zeros at $z = 0.9e^{\pm j\frac{\pi}{6}}$ and a double pole at z=0.0. An example of 2nd order MA signal is shown in fig. 1.15. The covariance function is given by (1.30)

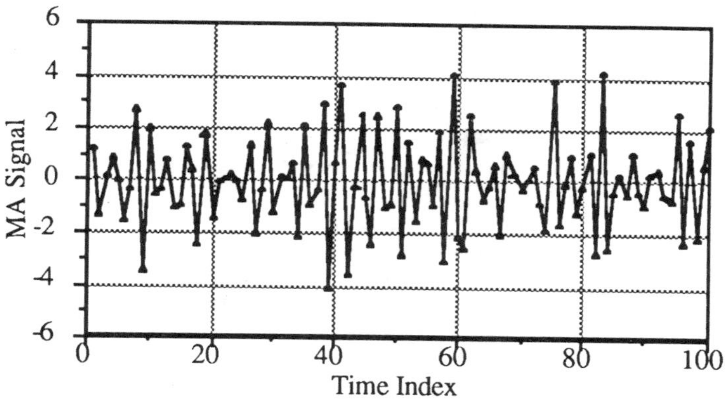

Figure 1.15: A second order MA signal

$$c_x(0) = \sigma_\varepsilon^2(b_0^2 + b_1^2 + b_2^2) \qquad c_x(1) = \sigma_\varepsilon^2(b_0b_1 + b_1b_2)$$

$$c_x(2) = \sigma_\varepsilon^2 b_0 b_2 \qquad c_x(\tau) = 0 \quad \tau \geq 3$$

Autoregressive (AR) Signal: An autoregressive signal (time series) is defined as follows:

$$x(n) + a_1x(n-1) + a_2x(n-2) + \ldots + a_px(n-p) = \varepsilon(n) \tag{1.31}$$

where $\varepsilon(n)$ is zero mean white noise (not necessarily Gaussian) driving a feedback system as shown in fig. 1.16. The coefficients $a_1, a_2, \ldots a_p$ are model parameters and p is the order of the model.

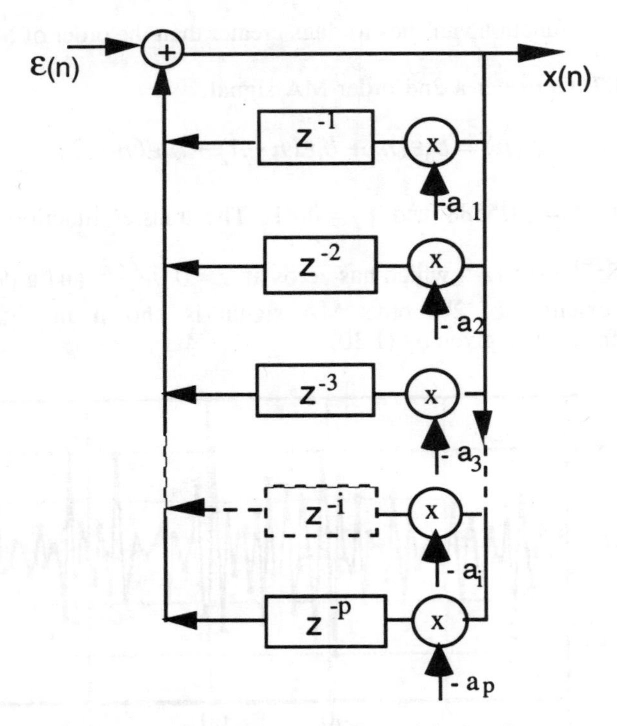

Figure 1.16: A schematic representation of AR model of order p.

Note that since $E\{x(n)\varepsilon(s)\}=0 \quad s>n$, $\varepsilon(n)$ must be a white noise process.

Equation (1.31) may be looked upon as a difference equation driven by $\varepsilon(n)$. Naturally, it has a transient output, which is observed soon after it is excited. The transient output may be obtained by solving the difference equation with $\delta(n)$ as input,

$$x(n)+a_1x(n-1)+a_2x(n-2)+...+a_px(n-p)=\delta(n) \qquad (1.32)$$

whose solution is a sum of complex exponential,

$$x(n)=\sum_{i=1}^{p}\alpha_i e^{\gamma_i n} \quad n\geq 0$$

where $\gamma_i, i=1,2,\ldots,p$ are given by the roots of the polynomial,

$$A(z)=1+a_1z^{-1}+a_2z^{-2}+...+a_pz^{-p}=0 \qquad (1.33)$$

whose i^th^ root is $z_i=\exp(\gamma_i)$. For the transient to decay fast the roots of (1.33) must lie well inside the unit circle. If any one of the roots were to lie close to the unit circle the decay rate will be very slow. As a result the mean of

AR signal will become time dependent, and thus the resulting solution will be non-stationary.

The covariance function of an AR signal can be expressed in terms of its parameters,

$$\begin{aligned} c_x(\tau) &= E\{x(n)x(n+\tau)\} \\ &= E\{x(n)[-a_1x(n+\tau-1)\ldots-a_px(n+\tau-p)+\varepsilon(n+\tau)\} \\ &= -a_1c_x(\tau-1)\ldots-a_pc_x(\tau-p)+E\{x(n)\varepsilon(n+\tau)\} \end{aligned}$$

Since $x(n)$ is naturally uncorrelated with all future inputs but correlated with the current and the past inputs,

$$E\{x(n)\varepsilon(n+\tau)\} = \sigma_\varepsilon^2\delta(\tau) \quad \tau \geq 0$$

Hence

$$c_x(\tau)+a_1c_x(\tau-1)+a_2c_x(\tau-2)+\ldots+a_pc_x(\tau-p) = \sigma_\varepsilon^2\delta(\tau) \quad \tau \geq 0 \tag{1.34}$$

This is the well-known Yule-Walker equation. Observe that the covariance function satisfies the same difference equation as the AR signal itself.

Example 1.8: A second order AR signal is a familiar time series model. The model equation is

$$x(n)+a_1x(n-1)+a_2x(n-2) = \varepsilon(n)$$

Let $a_1 = -1.5588$ and $a_2 = 0.81$. The transfer function of the 2nd order AR signal is given by $A(z) = 1/(1.0-1.5588z^{-1}+0.81z^{-2})$. It has poles at $z = 0.9e^{\pm j\frac{\pi}{6}}$ and a double zero at $z = 0$. The signal is plotted in fig. 1.17. The Yule-Walker equation (1.34) simplifies to

$$\begin{aligned} &c_x(0)+a_1c_x(-1)+a_2c_x(-2) = \sigma_\varepsilon^2 \\ &c_x(1)+a_1c_x(0) \quad +a_2c_x(-1) = 0 \\ &c_x(2)+a_1c_x(1) \quad +a_2c_x(0) = 0 \end{aligned}$$

The covariance function of the AR signal is shown in fig. 1.18 .

Autoregressive-Moving Average (ARMA) Model: The AR model of a time series is an all-pole model while the moving average (MA) model is an all-zero model. There are, however, situations where we require both pole-zero configurations. Such a model of time series is called ARMA model where the current value of the signal is expressed as the weighted sum of the past values and the weighted sum of the past random inputs. An ARMA model of order (p,q) is written as follows:

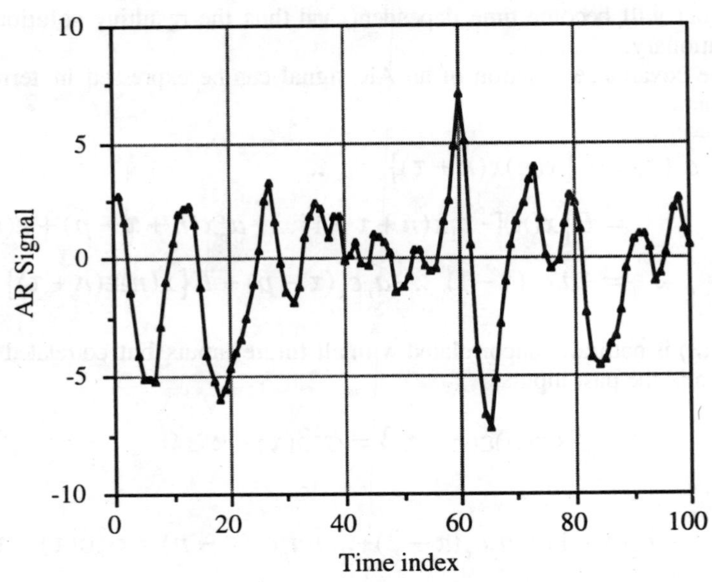

Figure 1.17: A second order AR signal

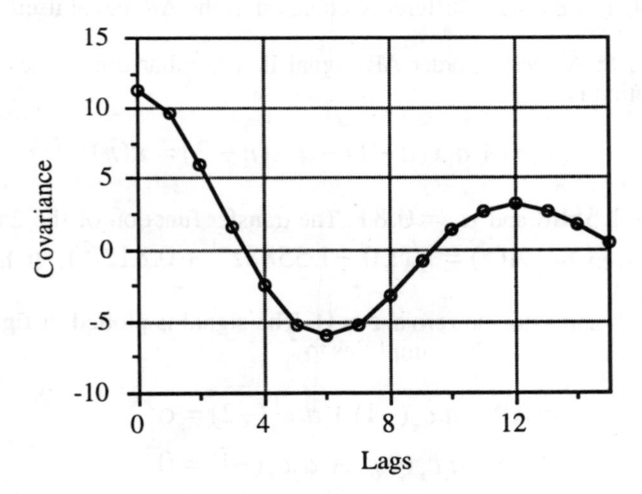

Figure 1.18: Covariance function of a 2nd order AR signal. Notice that the covariance function extends over an infinite lag interval, $-\infty < \tau < \infty$. In contrast the covariance function of moving average (MA) is confined to finite interval.

$$x(n) + \sum_{k=1}^{p} a_k x(n-k) = \sum_{k=1}^{q} b_k \varepsilon(n-k) \tag{1.35}$$

where a_k, k=1,2, ...,p and b_k, k=0,1,2, ..., q are parameters of the ARMA model, and p and q are integers. The right hand side, a moving average of white

noise, is the input to the difference equation in $x(n)$ which is the output. Since it is a combination of MA and AR models, it combines the structure of both. Naturally it is a far more powerful model of a stochastic signal.
Taking the z-transform on both sides of (1.35) we can write

$$X(z)\left[1+\sum_{k=1}^{p} a_k z^{-k}\right] = \mathrm{E}(z)\left[\sum_{k=0}^{q} b_k z^{-k}\right]$$

or

$$X(z) = \frac{\left[\sum_{k=0}^{q} b_k z^{-k}\right]}{\left[1+\sum_{k=1}^{p} a_k z^{-k}\right]}\mathrm{E}(z) \tag{1.36}$$

$$= \frac{B(z)}{A(z)}\mathrm{E}(z)$$

Observe that zeros arise from the numerator (moving average part) and the poles arise from the denominator (autoregressive part) of (1.36). The ARMA model is shown in fig. 1.19, which is obtained basically by combining MA and AR models as illustrated in figs. 1.14 and 1.16, respectively. The ARMA signal lies in between MA and AR signals in so far its smoothness is concerned.

Example 1.9: Consider an ARMA signal with p=q=1. Let us combine the MA and AR signal models considered in Examples 1.7 & 1.8

$$x(n) + a_1 x(n-1) = \varepsilon(n) + b_1 \varepsilon(n-1)$$

where

$$a_1 = -0.9 \text{ and } b_1 = 0.9.$$

The transfer function is given by

$$H(z) = \frac{1+0.9z^{-1}}{1-0.9z^{-1}}.$$

It has a pole at 0.9 and zero at -0.9, both on the real axis. The ARMA signal is shown in fig. 1.20 .

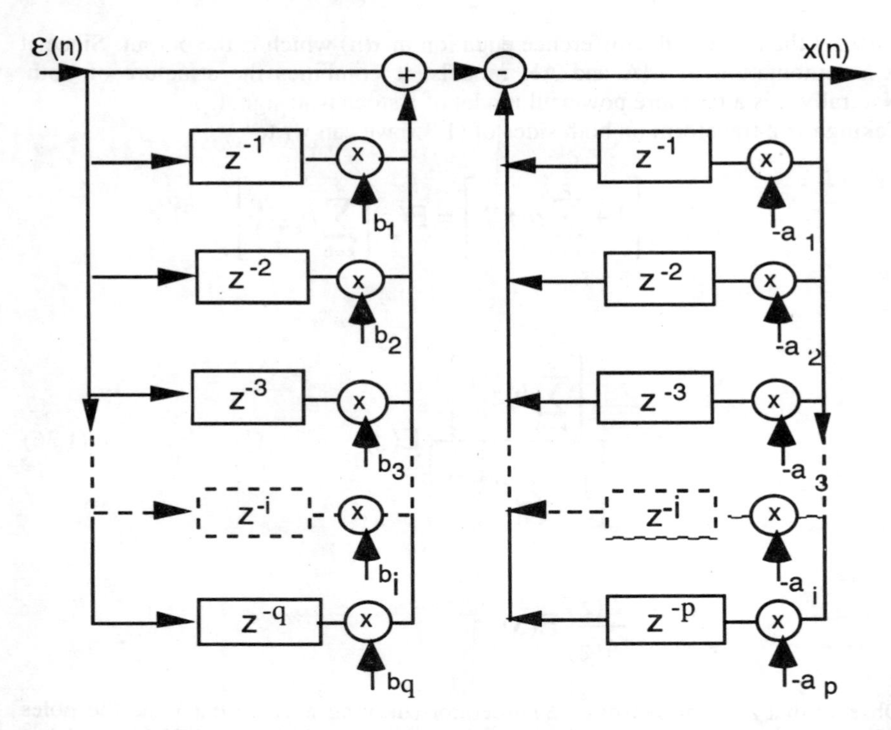

Figure 1.19: A schematic representation of ARMA signal of order (p,q). The left panel is the MA scheme whose output is a weighted average of white noise input. The right panel is the AR scheme, which receives the input from the left panel.

Finally, the covariance function of an ARMA signal satisfies the following Yule-Walker equations,

$$\sum_{k=1}^{p} a_k c_x(\tau - k) = \begin{cases} \sigma_\varepsilon^2 \sum_{k=1}^{q} b_k h(k-\tau) & \tau = 0,1,\ldots,q \\ 0 & \tau > q \end{cases} \tag{1.37a}$$

where $h(n)$ is impulse response function which is given by the solution of the following equation:

$$\sum_{i=1}^{p} a_i h(k-i) = \begin{cases} b_k & 0 \le k \le q \\ 0 & k > q \end{cases} \tag{1.37b}$$

The details of derivation are given in [6].

State Space Representation: In state space representation of a signal, the present and all past inputs are considered as different states of the system generating the signal. We define a state vector at time n as follows:

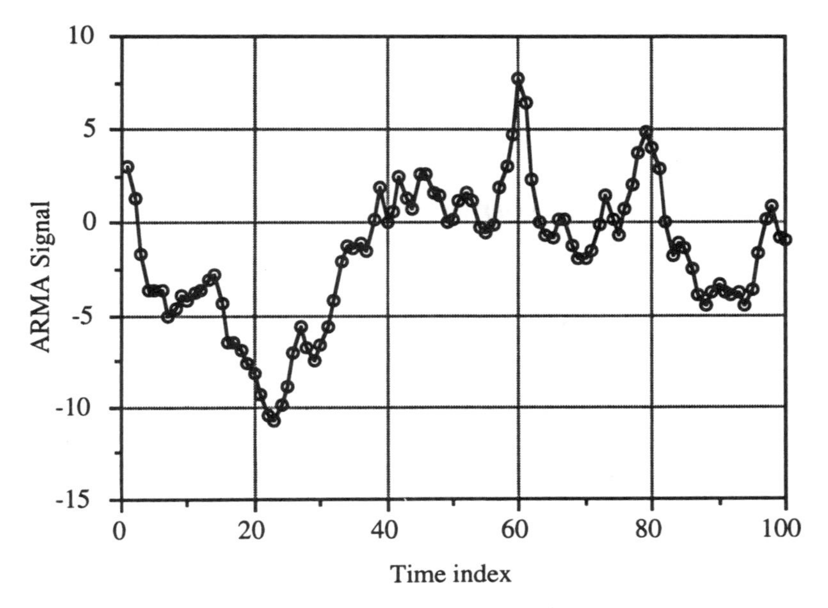

Figure 1.20: An example of ARMA Signal with p=q=1.

$$\mathbf{x}(n) = \{x(n), x(n-1), x(n-2), \ldots, x(n-p+1)\}$$

where $x(n), x(n-1), x(n-2), \ldots, x(n-p+1)$ are different states of the system. Similarly, we can also define another state vector, $\mathbf{x}(n-1)$ at time n − 1. The components of the state vectors are related through the ARMA model, (1.35). Define the following matrices:

$$\phi = \underset{p \times p}{\begin{bmatrix} -a_1, -a_2, -a_3 & \cdots & & -a_p \\ 1 \quad 0 \quad 0 \quad 0 & \cdots & & 0 \\ 0 \quad 1 \quad 0 \quad 0 & \cdots & & 0 \\ \vdots & & \ddots & \vdots \\ 0 \quad 0 \quad 0 & \cdots & 0 \quad 1 & 0 \end{bmatrix}}$$

and

$$\psi = \underset{p \times (q+1)}{\begin{bmatrix} b_0, & b_1 & b_2 & \cdots & b_q \\ 0 & 0 & 0 & \cdots & 0 \\ 0 & 0 & 0 & \cdots & 0 \\ \vdots & & & & \vdots \\ 0 & 0 & 0 & \cdots & 0 \end{bmatrix}}.$$

We are now able to express the ARMA model in terms of the state vectors and matrices ϕ and ψ,

$$\mathbf{x}(n) = \phi\mathbf{x}(n-1) + \psi\varepsilon(\mathrm{n}) \tag{1.38a}$$

$$x(n) = \mathbf{h}^T\mathbf{x}(n) \tag{1.38b}$$

where

$$\varepsilon(\mathrm{n}) = col\{\varepsilon(n), \varepsilon(n-1), \ldots, \varepsilon(n-q)\}$$

is a random vector input and

$$\mathbf{h} = col\{1, 0, \ldots, 0\}.$$

A general solution of the state equation (2.104) is given below:

$$\mathbf{x}(n) = \phi^{\mathrm{n}}\mathbf{x}(0) + \sum_{i=1}^{n} \phi^{\mathrm{n-i}}\psi\varepsilon(\mathrm{i}) \tag{1.39}$$

For large n, $\phi^{\mathrm{n}} \to 0$ (eigenvalues of ϕ are assumed to lie inside the unit circle), hence we can ignore the transient component, $\phi^{\mathrm{n}}\mathbf{x}(0)$. We compute the covariance matrix of $\mathbf{x}(n)$ without the transient component.

$$\begin{aligned}
\mathbf{c}_x = E\{\mathbf{x}(n)\mathbf{x}(n)^H\} &= \sum_{k=1}^{n}\sum_{i=1}^{n} \phi^{\mathrm{n-i}}\psi E\{\varepsilon(\mathrm{i})\varepsilon(\mathrm{k})^H\}\psi^H \phi^{\mathrm{n-k}H} \\
&= \sigma_\varepsilon^2 \sum_{i=1}^{n} \phi^{\mathrm{n-i}}\psi\psi^H \phi^{\mathrm{n-i}H} \\
&= \sigma_\varepsilon^2 \sum_{k=0}^{n-1} \phi^{\mathrm{k}}\psi\psi^H \phi^{\mathrm{k}H}
\end{aligned}$$

For large *n* the covariance matrix becomes independent of n, that is, once the transients have died down. One of the appealing aspects of the state space representation is that it is easy to account for time varying ARMA model.

Chaotic Signal: The discrete signal models described above are linear as they are governed by linear difference equations. A non-linear difference equation, for example, the so called Henon map [7] given by

$$x(n+1) = 1.0 - 1.4x^2(n) + 0.3x(n-1) \tag{1.40}$$

which has no random input, yet it can give rise to a solution with highly random look. The initial conditions have profound effect on the solution. A small change in the initial condition will drastically change the output. However, if the initial conditions are kept exactly same, the solution can be reproduced. Such a signal is known as a deterministic chaotic signal. For example, with initial conditions $x(-1)=x(0)=0.0$, the solution of (1.40) is shown in fig. 1.21.

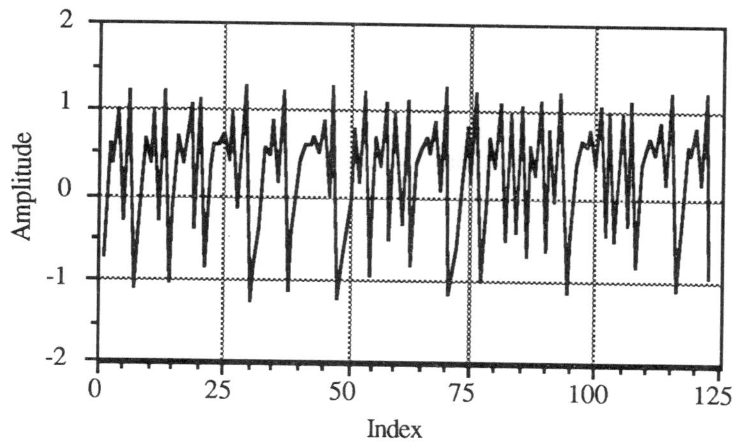

Figure 1.21: Henon map. Initial conditions $x(-1)=x(0)=0.0$

Another well-known example is the pseudo-random noise and related pseudo-random sequences used as signature waveforms in modern wireless communication. Pseudo-random noise is, for example, a solution of the following difference equation,

$$x(n) = 16{,}807x(n-1) \quad \text{mod} \quad 2{,}147{,}483{,}647$$

with the initial condition $x(0)=1$. Note that the mod operation involves finding the remainder of a division operation, for example, 11 mod 3 is 2. The resulting random sequence, $x(n)$, is periodic with a period 2,147,483,646 [5].

A chaotic signal may be characterized in more than one way. The basic idea is to determine the dimension of the state space spanned by the trajectory executed by the tips of data vectors defined as

$$\begin{aligned}
\mathbf{x}_0 &= col\{x(0)\ x(1) \ldots x(P-1)\} \\
\mathbf{x}_1 &= col\{x(1)\ x(2) \ldots x(P)\} \\
\mathbf{x}_2 &= col\{x(2)\ x(3) \ldots x(P+1)\} \\
&\ldots \\
\mathbf{x}_{N-P} &= col\{x(N-P)\ x(N-P+1) \ldots x(N-1)\}
\end{aligned} \tag{1.41}$$

where N stands for number of data points. We represent N-P+1 data vectors in a P (<N) dimensional space as N-P+1 points. These points, which are visited by the signal trajectory, constitute an attractor when the dimension of the space is an integer, or strange attractor when the dimension is a non-integer. The dimension of the space is the number of independent geometric coordinates needed to capture the behaviour of an attractor. Some measure of the closeness of above points is taken as the dimension of the chaotic signal. The most popular measure is the correlation dimension defined as follows:

$$c_P(r) = \lim_{N \to \infty} \frac{1}{(N-P+1)^2} \sum_{i=0}^{N-P} \sum_{j=0}^{N-P} U(r - |\mathbf{x}_i - \mathbf{x}_j|) \tag{1.42}$$

where U(x)=1 for $x \geq 0$ and =0 for $x < 0$ is step function. Observe that $c_P(r) = U(r)$ implies that all points are clustered at r=0. The behaviour of $c_P(r)$ at $r \to 0$ is very important. It is assumed to be $c_P(r) \approx r^D$ for large D and $r \to 0$. The correlation dimension D_c is defined as

$$D_c = \lim_{r \to 0,\ P \to \infty} \frac{\log c_P(r)}{\log r} \tag{1.43}$$

Though the correlation dimension is easy to compute, it requires a large amount of data. The spectrum of a chaotic signal is broadband, like that of white noise which requires a large number of degrees of freedom, but a chaotic signal can be characterized in terms of its dimension by a small number. For example, the Henon map has a correlation dimension of 1.28. The discrete chaotic signal model has been used in a variety of applications; for example, in modeling radar clutter [8], in prediction of time series [9] and in characterization of speech signals [10].

Example 1.10: The effect of a small change in the initial conditions is shown in the plots below (fig. 1.22). A 0.1% change in the initial value produces a signal which increasingly different with time.

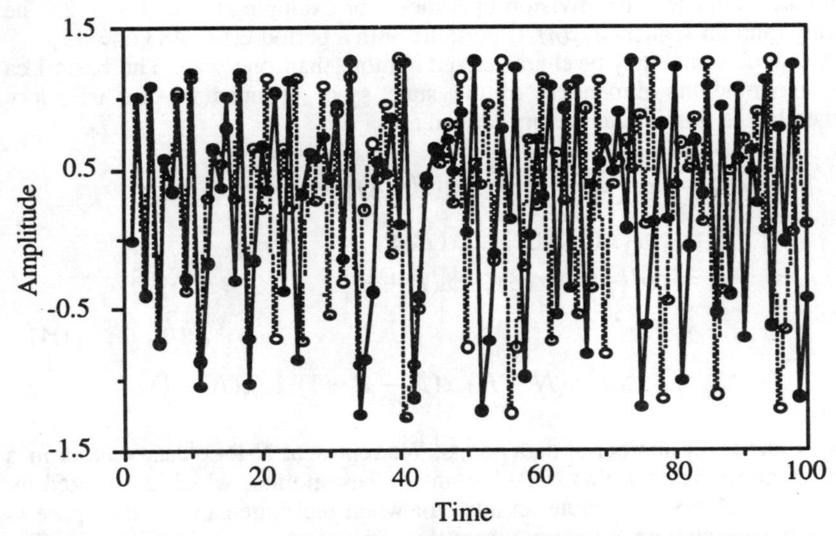

Figure 1.22: Henon map for slightly different initial conditions. Solid curve with filled circles *x(0)*=0 and *x(1)*=1.0. Dashed curve with empty circles *x(0)*=0.0 and *x(1)*=1.001 .

In fact the signals become uncorrelated after some time. This is the basis for our being able to generate independent pseudo-random sequences by merely changing the initial value or the so called seed

§1.6 Exercises

Problems:

1. A complex sinusoid is given by

$$x(n) = e^{-j(\omega_0 n+\varphi)}, n = 0, \pm 1, \pm 2, \ldots$$

where ω_0 ($|\omega_0| \leq \pi$) is the frequency and φ is the phase of the complex sinusoid. Two sinusoids with ω_1 and ω_2 are said to be harmonically related when $\omega_1 = p\omega_0$ and $\omega_2 = q\omega_0$. and p and q are integers. ω_0 is known as fundamental frequency. Consider a signal as a sum of harmonically related sinusoids,

$$x(n) = \sum_{p=1}^{P} e^{-j(p\omega_0 n+\varphi)}$$

Show that $x(n)$ is a periodic function. What is the period?

2. We sampled two sinusoids at 40 samples/sec. The frequencies are 15Hz and 55Hz but the phase is same for both sinusoids. Show that the samples are indistinguishable. We say that frequencies 15 and 55 Hz are aliased at sampling frequency 40Hz. Are there other frequencies which are aliases of 15Hz?.

3. Consider the following analog sinusoidal signal:

$$x(t) = 4.2\sin(64\pi t)$$

Find a sampling rate such that the signal reaches a value of 4.2. Is it possible to find a sampling rate such that the signal reaches any desired value <4.2 ?

4. We have two discrete signals given by

$$x = \{0,0,0,0,1,1,1,1,1,1,1,1,0,0,0,0\}$$
$$y = \{0,0,0,0,-1,-1,-1,-1,1,1,1,1,0,0,0,0\}$$

Compute the autocorrelations function of x and y and cross-correlation function between x and y.

5. Compute the autocorrelation function of the following sequence

$$x = \{1,3,2,5,8,1\}$$

by first computing the z-transform of x and then that of its autocorrelation function. Verify this result by direct computation the autocorrelation function.

6. A sinusoid with random phase is the simplest type of random function (see page 8),

$$x(n) = A\sin(\omega_0 n + \varphi)$$

where φ is a uniformly distributed random phase, that is, its probability density function is given by

$$f(\varphi) = \begin{cases} \dfrac{1}{2\pi} & |\varphi| \le \pi \\ 0 & otherwise \end{cases}$$

Compute mean, $E\{x(n)\}$ and covariance function, $E\{x(n)x(n+\tau)\}$.

7. Use the modulation property of the Fourier transform and obtain the Fourier transform of single cycle of sinusoidal function as shown in fig. 1.23

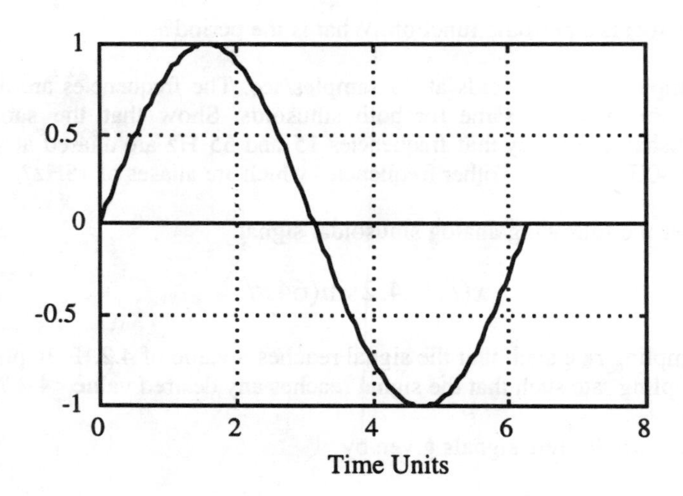

Figure 1.23: A single cycle of sinusoidal function.

8. Compute the Fourier transform of $x(t) = sgn(t)$ where $sgn(t)$ stands for a waveform shown in fig. 1.24. Since this function does not satisfy the condition of finite energy, $\int_{-\infty}^{\infty} |x(t)|^2 dt < \infty$, its Fourier transform does not exist, but,

however a limiting form of expression can be derived. Compute the Fourier transform of $x_a(t) = e^{-a|t|}sgn(t)\;\; a \geq 0$ and then let $a \to 0$. Show that

$$FT\{sgn(t)\} = \underset{a \to 0}{FT\{x_a(t)\}} = \frac{2}{j\omega}$$

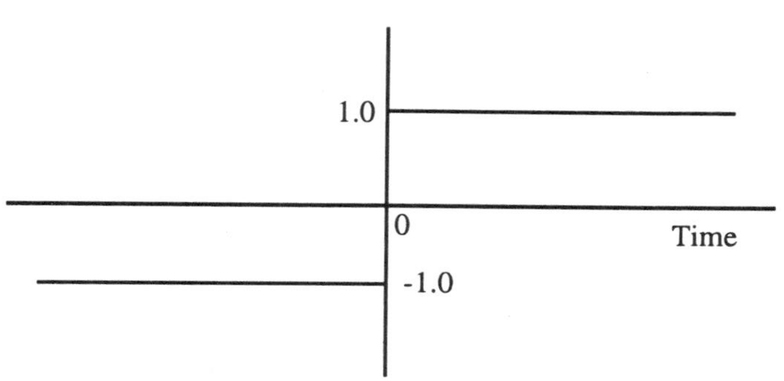

Figure 1.24: A waveform *sgn(t)*, that is, sign of *t*.

9. Given that the poles of a signal are inside a unit circle, show the position of all poles of its autocorrelation function. Let the signal poles move toward the unit circle. In the limiting case as the poles reach the unit circle show that the autocorrelation function tends to a sum of sinusoids.

10. The energy spectrum of a signal *x(t)* is sketched in fig. 1.25. Sketch the energy spectrum of *x(t/2)* and *x(-t/2)*.

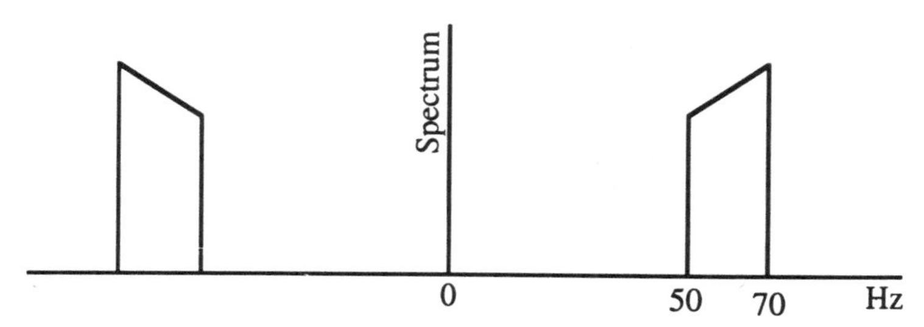

Figure 1.25: Energy spectrum of a signal

11. Define a (one-sided) sequence

$$h(n) = \begin{cases} e^{-(0.5+j0.8)n} + e^{-(0.5-j0.8)n}, & n = 0,1,2,\ldots \\ 0 & n < 0 \end{cases}$$

Let $r_{hh}(\tau)$ be its autocorrelation function. Show the region of convergence of the z-transform of $r_{hh}(\tau)$.

12. Following are the recursive relations between $x(n)$ and $z(n)$ for $n \geq 0$.

$$x(n) = 0.45x(n-1) + 0.2u(n)$$
$$y(n) = y(n-1) + x(n)$$

where u(n) is a step function. Compute the z-transform of x and y. Find the poles and zeros of $Y(z)$, the z-transform of y.

13. Given the z-transforms as below, compute their inverse z-transforms (Expand in partial fractions).

$$X(z) = \frac{5z^2 - 1.6z}{z^2 - 1.2z + 0.32}$$

$$X(z) = \frac{z^2 - 1.2z}{z^2 - 1.8z + 0.8}$$

14. Let $X(z)$ be the z-transform of $x(n)$. Show that

$$ZT\{nx(n)\} = -z\frac{dX(z)}{dz}$$

where $ZT\{.\}$ stands for z-transform of the expression within the braces. What is the region of convergence.

15. Sketch cross-correlation between the functions f(t) and g(t) shown below (fig. 1.26)

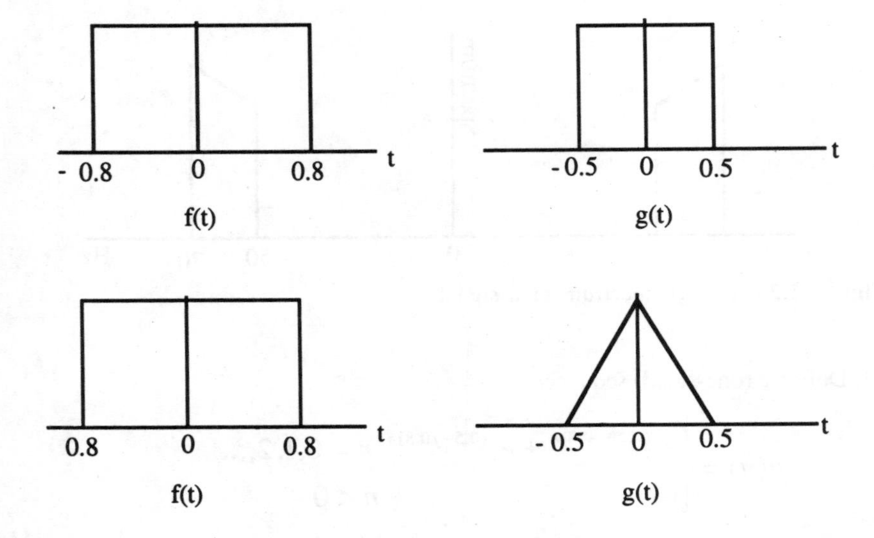

Figure 1.26: Waveforms for cross-correlation

16. The Fourier transform of the derivative of a function is easily derived using the Fourier integral representation of a function,

$$\frac{dx(t)}{dt} = \frac{1}{2\pi}\int_{-\infty}^{\infty} X(\omega)\frac{d}{dt}\{e^{j\omega t}\}d\omega$$

$$= \frac{1}{2\pi}\int_{-\infty}^{\infty} j\omega X(\omega)e^{j\omega t}d\omega$$

$$\therefore FT\left\{\frac{dx(t)}{dt}\right\} = j\omega X(\omega)$$

However, the derivation of the Fourier transform of the integral is a bit involved. We need to note a result

$$\int_{-\infty}^{t} x(t')dt' = \int_{-\infty}^{\infty} x(\tau)u(t-\tau)d\tau$$

where $u(t)$ is a step function. Using this result show that

$$FT\left\{\int_{-\infty}^{t} x(t')dt'\right\} = [\frac{1}{j\omega} + \pi\delta(\omega)]X(\omega).$$

17. Instantaneous frequency is defined as a time derivative of the phase of a sinusoidal waveform, for example, the instantaneous frequency of cos(ωt) is simply ω. The instantaneous frequency of a chirp signal (see p. 5) $\sin(\omega_0 t(t+1))$ is $\omega_0 + 2\omega_0 t$, which is linearly increasing with time. Compute the instantaneous frequency of a frequency modulated signal ,

$$x(t) = \cos(\omega_c t + \frac{\Delta\omega}{2\omega_m}\sin\omega_m t)$$

Show that the instantaneous frequency lies in the range $\omega_c \pm \Delta\omega/2$. What is the role of ω_m?

18. Compute Fourier transform (CFT) and spectrum of the signal shown in the fig. 1.27

19. From (1.26) the analytic signal has zero Fourier transform for negative frequencies. Show that this result also follows from the definition of the analytic signal via Hilbert transform (see page 28). The transfer function of Hilbert transform is given by

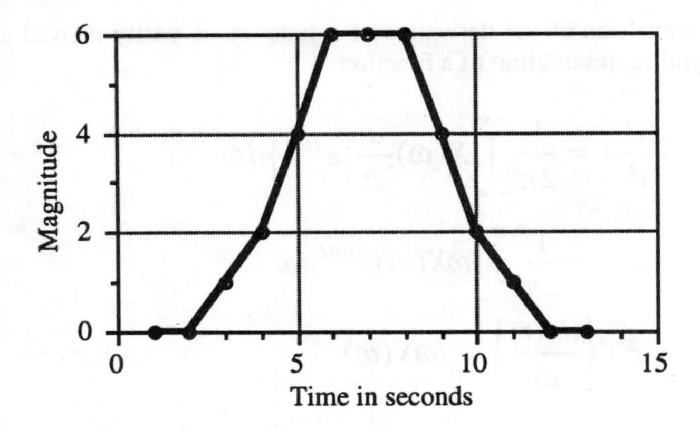

Figure 1.27: A signal whose Fourier transform has to be computed (Hint: Treat the signal as a superposition two overlapping signals).

$$H(\omega) = -jsgn(\omega)$$

where $sgn(\omega)$ stands for sign of (ω).

20. Explain how the quantization noise can be reduced by increasing the sampling rate for a fixed number of quantization levels.

21. Compute the covariance function of the following moving average signal.

$$x(n) = 1.2\varepsilon(n) + 0.8\varepsilon(n-1) + 0.3\varepsilon(n-2)$$

where $\varepsilon(n)$ is white noise. What is the mean and variance of the MA signal?

22. What is normalized frequency? If we sample with 1.0 ms what is the relation between the normalized frequency and the frequency in Hertz.

23. What is an anti causal signal. Let

$$x(n) = \begin{cases} r^n, & n = 0, -1, -2, \ldots \\ 0 & otherwise \end{cases}$$

where $|r| > 1$. Calculate the z-transform of $x(n)$ and show that the poles of $X(z)$ lie outside the unit circle.

24. Derive z-transform of $x(n)$ defined as follows:

$$x(n) = \begin{cases} 1 & n \geq 0 \\ -1 & n < 0 \end{cases}$$

What is the region of convergence? [Hint: In Example 1.6 we have derived the z-transforms of exponentially decaying signals. Consider a limiting case as $r \to 1-\varepsilon$ and $q \to 1+\varepsilon$ where ε is a small positive number tending to zero].

Computer Projects (CP):

1. A discrete sinusoid can naturally be obtained by sampling a continuous sinusoid. But it is also possible to generate the discrete sinusoid by solving a second order AR signal model (1.31) where the right hand side now is a delta function,

$$x(n)+1.7320x(n-1)+x(n-2)=\delta(n)$$

Find the frequency of the discrete sinusoid by computing the poles of the transfer function of the AR model. Check the result by computing a few samples of *x(n)*. Assume the following initial conditions, *x(-1)=x(-2)*=0.0. (See Chapter Five for more related iformation).

2. A chirp signal possesses an interesting property of very narrow autocorrelation function. Another example is a binary phase shift keying (BPSK) coded signal. It consists of a sequence of rectangular waveforms whose amplitude is randomly selected (by throwing a coin) between +1 and -1. An example of such a signal is shown in fig. 1.28. Write a Matlab program to generate a BPSK signal and compute its autocorrelation function. Show that the autocorrelation function is, like that of a chirp signal, is very narrow. BPSK signals find application in wireless communication, global positioning system, radar, sonar, etc.

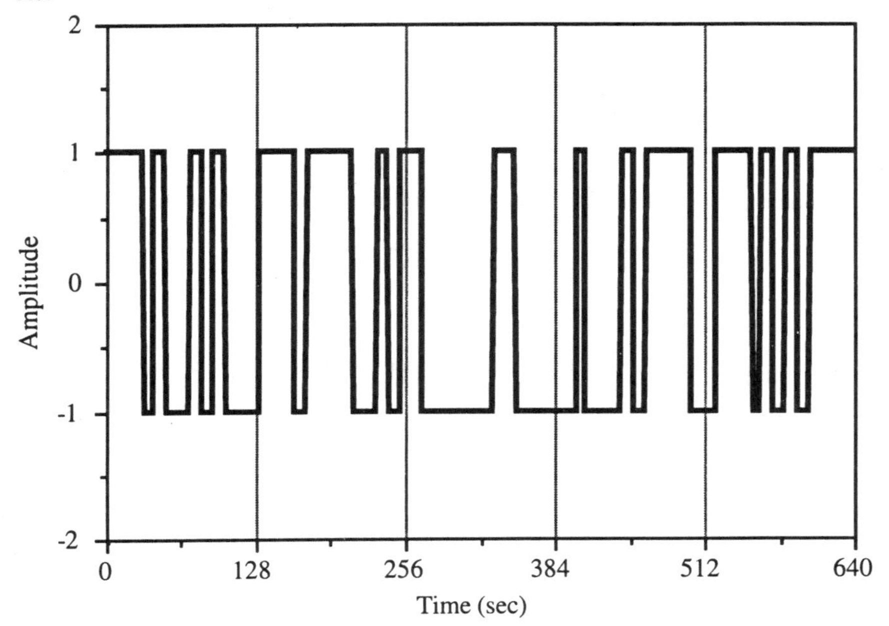

Figure 1.28: An example of BPSK signal.

3. Given a polynomial in z we can compute all its roots. The reverse is also possible. That is, given the roots of a polynomial we can compute the coefficients of the polynomial. Sometimes, in the design of a filter (e.g. IIR filter), it is more convenient to specify the poles and zeros, instead of the rational function. Consider the following poles and zeros of a rational function.

$$\text{Zeros: } 0.95\angle\pm35^0, 0.8\angle\pm45^0$$
$$\text{Poles: } 0.9\angle\pm40^0, 0.91\angle\pm60^0$$

Compute the unknown rational function. Use m-file "poly.m" from Matlab.

4. A random number generator is often required in Monte Carlo experiments (computer simulation). Matlab provides a random number generator with which we like to get familiar. Using the Matlab command "rand('seed', n)" we can set the seed to n, where n is any integer. For example, rand('seed', 11) sets seed to 11. Now use the random number generator to generate, say, 64 random numbers by using the command, rand(1,64). Verify that each time we use rand('seed', 11) followed by rand(1,64) we generate the same set of random numbers. For this reason the computer generated random numbers are known as pseudo-random numbers (PN). They are indeed deterministic but with properties of a stochastic sequence. Generate two PN sequences by setting the seed to 11 and 19, respectively. Let the length be 1024 points. Compute the autocorrelation function of each sequence and the correlation coefficient (see equation (1.4)). Compare the results with those of the ideal white noise.

5. A 20 Hz unit amplitude sinusoidal signal is sampled at a rate of 50 samples per second and quantized to eight levels. Compute the variance of the quantization error. Verify the law that the variance of quantization error is given by $\frac{\delta^2}{12}$ where δ is the quantization step. The same sinusoid is resampled at a rate of 100 samples per second and quantized to eight levels. The quantized samples are subjected to three-point running average. Compute the variance of the quantization error. How does this variance compare with the one computed at slower sampling rate?

6. Compute the cross-correlation between following two sequences:

$$x\text{: } 1.2, 3.1, 3.8, 2.1, 5.7, 4.2, 1.1, 0.5$$
$$y\text{: } 3.1, 6.3, 2.9, 1.1, 6.3, 7.1, ,2.5, 0.3$$

First compute the z-transforms of x and y sequences. Form a product, $X(z)Y(z^{-1})$ where $X(z)$ and $Y(z)$ are the z-transforms of x and y, respectively. Compute the inverse z-transform of the product. Compare the above results with the direct computation of the cross-correlation function.

2 Discrete Fourier Transform and Convolution

In this chapter we focus on discrete Fourier transform (DFT) and finite discrete Fourier transform (FDFT) for sampled signal. Role of sampling in particular on the aliasing error in the discrete Fourier transform is discussed at great length. It is shown how the aliasing error can be avoided for signals whose Fourier transform is confined to a finite frequency band; that is, band limited signal. This naturally leads to the well-known Shannon sampling theorem. In all practical problems the data length is finite, therefore we need finite discrete Fourier transform (FDFT) and fast Fourier transform algorithm used to compute it. While the DFT of an infinite signal is a continuous function, the FDFT of a finite discrete signal is a discrete function. However, the FDFT coefficients are closely related to the DFT of the infinite signal from which the finite signal was derived. Though the signal is finite, the FDFT implicitly converts it into a periodic signal with a period equal to the signal length. This has a profound effect on the discrete filtering, convolution, correlation etc. A clear understanding of this phenomenon is essential in mastering DSP. We also briefly mention a few other transforms which are closely related to FDFT, for example, the discrete cosine transform (DCT) which is of great interest in signal compression, discrete Hartley transform (DHT) which requires fewer multiplications for evaluation of the convolution product.

Finally, the last section is devoted to the study of linear systems and convolution in discrete domain. A linear system is characterized by an impulse response function or by its Fourier transform, a transfer function. Many linear systems use only the past inputs to produce the current output. Such systems are said to be causal and their impulse response function is one-sided, that is, function of the past time. The transfer function of a causal system possesses many interesting properties in terms the position of its poles and zeros. All this has bearing upon the infinite impulse response (IIR) filters which is covered in Chapter Five. Digital filtering, which is the core of the DSP, uses the convolution in discrete-time, which may be implemented in the time domain in a straightforward manner. But, when the signal/filter lengths are long, it turns out that convolution in the frequency domain using the convolution theorem becomes more economical. We talk about fast convolution in this section in some detail.

§2.1 Discrete Fourier Transform

The Fourier transform of a discrete signal is defined as

$$\boxed{X_d(\omega) = \sum_{n=-\infty}^{\infty} x(n) e^{-j\omega n}} \tag{2.1a}$$

and its inverse follows as

$$\boxed{x(n) = \frac{1}{2\pi} \int_{-\pi}^{\pi} X_d(\omega) e^{j\omega n} d\omega} \tag{2.1b}$$

where $x(n)$ is a discrete signal, presumably derived by uniform sampling (with unit time interval) of a continuous signal. We shall call (2.1 a & b) as a discrete Fourier transform (DFT) pair, analogous to the continuous Fourier transform (CFT) pair defined in Chapter One (1.14). The DFT is not an approximate equivalent of CFT as a numerical sum is to an integral. It is a transform on its own merit. However, the DFT shares many properties with CFT. Indeed we can relate the two through an analytical result which we derive next. Also, we shall elaborate on the essential differences between DFT and CFT. Eq.(2.1b) is also known as the Fourier integral representation of a discrete signal.

Relation between DFT and CFT: Let $x_c(t)$ (subscript c denotes continuous) be a signal as a function of continuous time, t. The discrete version of $x_c(t)$ may be represented as in (1.1), reproduced here for convenience

$$x_d(t) = x_c(t) \sum_{n=-N}^{N} \delta(t - n\Delta t), \quad N \to \infty \tag{1.1}$$

where Δt is the sampling interval. $x_d(t)$ consists of N $(N \to \infty)$ spikes with amplitude equal to $x_c(t)$ at $t = n\Delta t, n = 0, \pm 1, \pm 2, \ldots, \pm N$ at other values of t, $x_d(t)$=0. Notice that $x_d(t)$ is a comb function whose teeth are of variable magnitude, controlled by $x_c(t)$ (see fig. 2.1). We shall assume for the time being that the signal duration is infinite. Let us now compute the continuous Fourier transform (CFT) of $x_d(t)$. Multiply both sides of (1.1) by $\exp(-j\omega t)$ and integrate with respect to t.

$$\int_{-\infty}^{\infty} x_d(t) e^{-j\omega t} dt = \int_{-\infty}^{\infty} x_c(t) \sum_{n=-\infty}^{\infty} \delta(t - n\Delta t) e^{-j\omega t} dt$$

$$= \sum_{n=-\infty}^{\infty} \int_{-\infty}^{\infty} x_c(t) \delta(t - n\Delta t) e^{-j\omega t} dt$$

$$= \sum_{n=-\infty}^{\infty} x_c(n\Delta t) e^{-j\omega n\Delta t} = X_d(\omega) \tag{2.2}$$

Replace $x_c(t)$ by its Fourier integral representation (1.15b). The right-hand side of (2.2) becomes

$$X_d(\omega) = \int_{-\infty}^{\infty} x_d(t) e^{-j\omega t} dt$$

$$= \frac{1}{2\pi} \int_{-\infty}^{\infty} X_c(\omega') d\omega' \int_{-\infty}^{\infty} \sum_{n=-\infty}^{\infty} \delta(t - n\Delta t) e^{-j(\omega - \omega')t} dt$$

$$
\begin{aligned}
&= \frac{1}{2\pi}\int_{-\infty}^{\infty} X_c(\omega')d\omega' \sum_{n=-\infty}^{\infty} e^{-j(\omega-\omega')n\Delta t} \\
&= \frac{1}{2\pi}\sum_{k=-\infty}^{\infty}\int_{-\infty}^{\infty} X_c(\omega')d\omega'\delta(\omega-\omega'-\frac{2\pi}{\Delta t}k) \\
&= \sum_{k=-\infty}^{\infty} X_c(\omega-\frac{2\pi}{\Delta t}k)
\end{aligned}
\tag{2.3}
$$

where we have used a result that

$$\sum_{n=-\infty}^{\infty} e^{-j(\omega-\omega')n\Delta t} = \sum_{k=-\infty}^{\infty} \delta(\omega-\omega'-\frac{2\pi}{\Delta t}k).$$

From (2.3) it follows that the Fourier transform of the discrete and continuous signals are related through a relation,

$$
\boxed{
\begin{aligned}
X_d(\omega) &= \sum_{k=-\infty}^{\infty} X_c(\omega-\frac{2\pi}{\Delta t}k) \\
&= X_c(\omega) * \sum_{k=-\infty}^{\infty} \delta(\omega-\frac{2\pi}{\Delta t}k)
\end{aligned}
}
\tag{2.4}
$$

where * stands for convolution.

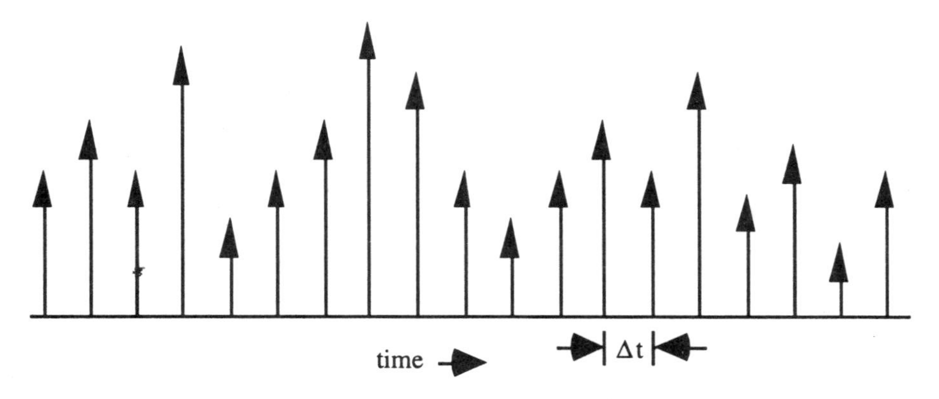

Figure 2.1: $x_d(t)$ as a function of continuous parameter, t. The spike height represents the signal amplitude at the time instant. The sampling interval is Δt.

The Fourier transform of a discrete signal (DFT) has some interesting properties. It is a sum of many copies of the continuous Fourier transform (CFT) which are displaced along the frequency axis from each other by $2\pi/\Delta t$, that is,

$$\ldots, X_c(\omega + 2\frac{2\pi}{\Delta t}), X_c(\omega + \frac{2\pi}{\Delta t}), X_c(\omega), X_c(\omega - \frac{2\pi}{\Delta t}), X_c(\omega - 2\frac{2\pi}{\Delta t}), \ldots$$

Pictorially, the relation between the CFT and DFT is illustrated in fig. 2.2.

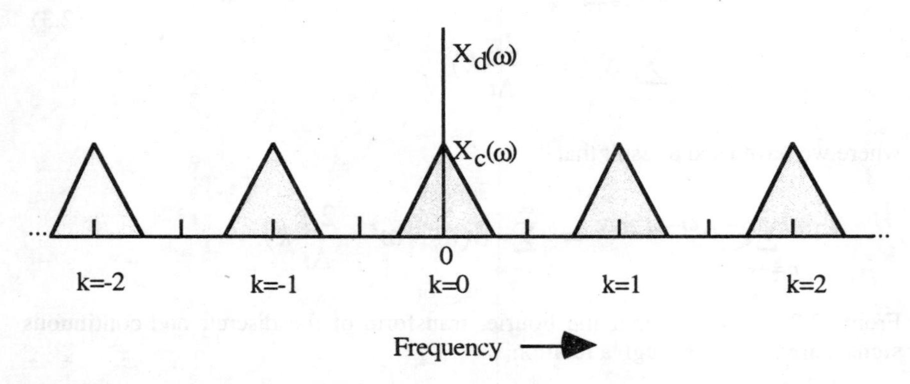

Figure 2.2: Pictorial representation of the relationship between CFT and DFT. The figure corresponding to k=0 is the CFT and all other figures are simply the copies of the CFT.

As a result the discrete Fourier transform becomes periodic with a period $2\pi/\Delta t$. Thus, the discrete Fourier transform is a continuous periodic function. Evidently, the discrete Fourier transform (DFT) will not be equal to the continuous Fourier transform except in a very special case. The error arises on account of the contributions from the adjacent copies to the sum in the frequency range $\pm\pi/\Delta t$, which is our principal band. This error is known as aliasing error. Equation (2.4) is indeed a pair of equations obtained by equating the real and imaginary parts,

$$\boxed{\begin{aligned} \mathrm{Re}\{X_d(\omega)\} &= \sum_{k=-\infty}^{\infty} \mathrm{Re}\left\{X_c(\omega - \frac{2\pi}{\Delta t}k)\right\} \\ \mathrm{Im}\{X_d(\omega)\} &= \sum_{k=-\infty}^{\infty} \mathrm{Im}\left\{X_c(\omega - \frac{2\pi}{\Delta t}k)\right\} \end{aligned}} \tag{2.5}$$

Equivalent expressions involving the magnitude and phase in place of the real and imaginary parts do not exist. However, for a pure stationary stochastic signal where the Fourier transform (or more precisely differential of the generalized Fourier transform) is uncorrelated, the spectrum satisfies a relation quite similar to (2.3).

Example 2.1: Consider a sinusoid $x_C(t)= \sin(2\pi 30t)$ (frequency = 30Hz). Let us sample this signal at rate of 50 samples per second (sampling interval of 20 ms).

$x_d(t) = \sin(2\pi 30t) \sum_{n=-\infty}^{\infty} \delta(t - 0.02n)$. From (2.4) the discrete Fourier transform is given by

$$X_d(\omega) = \frac{1}{2j}\left[\sum_{k=-\infty}^{\infty} \delta(\omega - 2\pi 30 - 2\pi 50k) - \sum_{k=-\infty}^{\infty} \delta(\omega + 2\pi 30 - 2\pi 50k)\right]$$

The peaks are located at $\omega = 2\pi(\pm 30 + 50k)$, that is, at

$$\omega = 2\pi[\ldots, -80, -70, -30, -20, \ 20, 30, 70, 80, \ldots]$$

The peaks along with their correct magnitude are shown in fig. 2.3. Inside the principal band ($\pm 25Hz$) there are two peaks but at wrong location, at $\pm 20Hz$. Additionally, the signs of the peaks are reversed. Observe that if we had sampled at a rate of 60Hz the discrete Fourier transform would have been identically zero (why?).

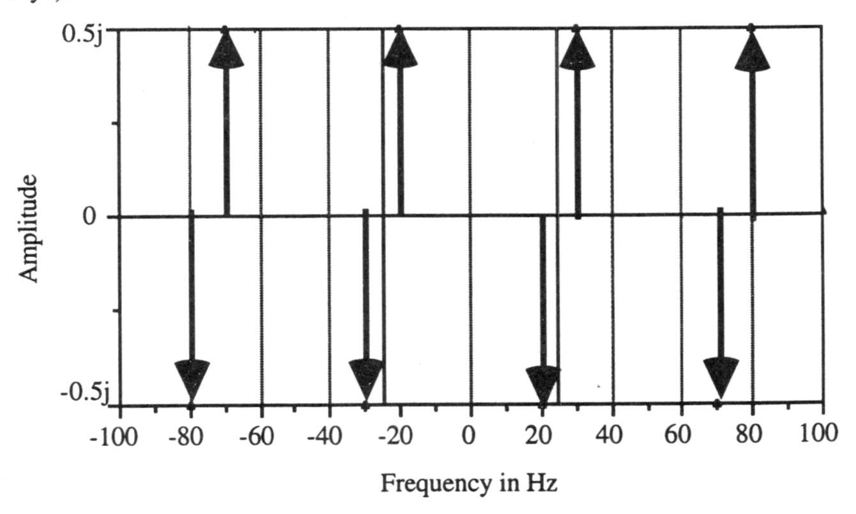

Figure 2.3: Discrete Fourier transform of 30 Hz sinusoidal signal sampled at a rate of 50Hz. Notice the presence of two delta functions at $\omega = \pm 2\pi 20$ lying inside the principal band (shaded region), that is, $-\pi/\Delta t \le \omega \le \pi/\Delta t$.

It may be noted that the DFT of a real signal is a complex function except when the signal is real and symmetric. To show this let us go back to (2.1),

$$X_d(\omega) = \sum_{n=-\infty}^{\infty} x(n) e^{-j\omega n}$$

$$= \sum_{n=-\infty}^{\infty} x(n)[\cos(\omega n) - j\sin(\omega n)]$$

$$= \sum_{n=-\infty}^{\infty} x(n)\cos(\omega n) - j\sum_{n=-\infty}^{\infty} x(n)\sin(\omega n)$$
$$= \sum_{n=-\infty}^{\infty} x(n)\cos(\omega n) \tag{2.6a}$$

For the assumed symmetry of the discrete signal the imaginary component in (2.6a) vanishes leaving the DFT a real function. When the signal is anti symmetric (i.e. $x(n)=-x(-n)$) the DFT is a pure imaginary function,

$$X_d(\omega) = -j \sum_{n=-\infty}^{\infty} x(n)\sin(\omega n) \tag{2.6b}$$

Shift Property: We shall now evaluate the discrete Fourier transform of a delayed signal, $x_c(t-\tau_0)$. We note the shift property of the continuous Fourier transform,

$$\text{FT}\{ x_c(t-\tau_0)\} = X_c(\omega)e^{-j\omega\tau_0}.$$

We use this in (2.4) and obtain the discrete Fourier transform

$$\tilde{X}_d(\omega) = e^{-j\omega\tau_0} \sum_{k=-\infty}^{\infty} X_c(\omega - \frac{2\pi}{\Delta t}k)e^{j\frac{2\pi}{\Delta t}k\tau_0} \tag{2.7}$$

When

$$\tau_0/\Delta t = n \quad (\text{integer}),$$

since $e^{-j\frac{2\pi}{\Delta t}k\tau_0} = 1$, we obtain the shift property for discrete Fourier transform,

$$\boxed{\tilde{X}_d(\omega) = e^{-j\omega\tau_0} X_d(\omega)} \tag{2.8}$$

It is important to remember that (2.8) is applicable only when the delay is an integral multiple of sampling interval. For a fractional delay, $|\tau_0| < \Delta t$, however, we cannot simplify (2.7) further. Compare (2.7) with (2.4), which was derived for zero delay. The phase factor present inside the summation sign is the essential difference. Recall that in Example 2.1 we stated that when the sinusoidal signal was sampled at 60Hz, the sum in (2.4) reduced to zero due to cancellation. If we now delay the sinusoidal signal by an amount less than one sample interval (say, 50/3 milliseconds) each term will be weighted by $e^{j2\pi 60k\tau_0}$,

$$\tilde{X}_d(\omega) = \frac{e^{-j\omega\tau_0}}{2j}\left[\begin{array}{l} \sum_{k=-\infty}^{\infty} e^{j2\pi 60k\tau_0}\delta(\omega - 2\pi 30 - 2\pi 60k) \\ - \sum_{k=-\infty}^{\infty} e^{j2\pi 60k\tau_0}\delta(\omega + 2\pi 30 - 2\pi 60k) \end{array}\right] \tag{2.9}$$

Evaluating the quantity on the right hand side of (2.9) at $\omega = 2\pi 30$ we obtain

$$\begin{aligned} \tilde{X}_d(\omega)\Big|_{\omega=2\pi 30} &= \frac{e^{-j2\pi 30\tau_0}}{2j}\left[-e^{j2\pi 60\tau_0} + 1\right] \\ &= 0 \qquad when\ \tau_0 = 0 \end{aligned}$$

Thus, the cancellation takes place only when $\tau_0 = 0$.

Scaling Property : An analog signal could be time compressed or expanded as desired by simply scaling the time axis (see on page 17 Similarity theorem). A discrete signal can only be expanded by inserting zeros between samples. Let $x(n),\ n = 0, \pm 1, \pm 2, \ldots$ be the discrete signal where we insert p zeros between two adjacent samples. Mathematically, we define an expanded signal $x'(r),\ r = 0, \pm 1, \pm 2, \ldots$ as follows,

$$x'(r) = \begin{cases} x(n) & when\ n = \dfrac{r}{p+1} \\ 0 & otherwise \end{cases} \tag{2.10}$$

Example 2.2: As an example, for let p=2 the expanded signal is given by,

$x'(-6)=x(-2)$	$x'(-3)=x(-1)$	$x'(0)=x(0)$	$x'(3)=x(1)$	$x'(6)=x(2)$
$x'(-7)=0$	$x'(-4)=0$	$x'(-1)=x'(1)=0$	$x'(4)=0$	$x'(7)=0$
$x'(-8)=0$	$x'(-5)=0$	$x'(-2)=x'(2)=0$	$x'(5)=0$	$x'(8)=0$

.

Let us now compute DFT of the expanded signal,

$$\begin{aligned} X'_d(\omega) &= \sum_{r=-\infty}^{\infty} x'(r)e^{-j\omega r} \\ &= \sum_{r=-\infty}^{\infty} x'(r)e^{-j\omega(p+1)n} = \sum_{n=-\infty}^{\infty} x(n)e^{-j\omega(p+1)n} \\ &= X_d((p+1)\omega) \end{aligned} \tag{2.11}$$

$X'_d(\omega)$ is the DFT of the expanded signal and it is equal to the DFT, compressed by a factor of $(p+1)$, of the normal signal. Notice that here we are not considering resampling with interpolation. If the discrete signal was obtained by sampling at the Nyquist rate, it is possible through interpolation to carry out error free resampling and thus obtain a compressed or expanded signal as desired. We have referred to this, in Chapter One, as upsampling (converse is downsampling).

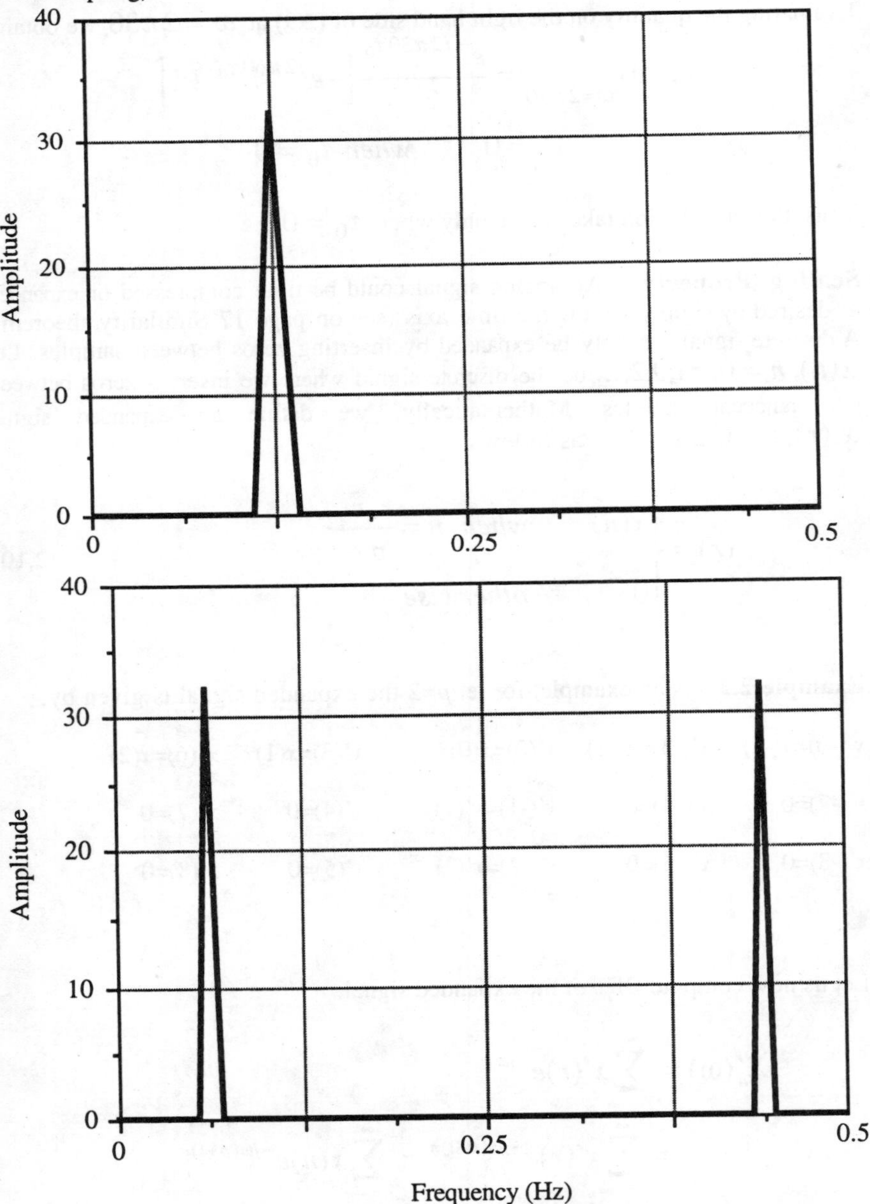

Figure 2.4: A sinusoidal signal of frequency 0.125 Hz is sampled at one sample/sec interval. The amplitude of its Fourier transform is shown in (a). The signals samples were interlaced with zeros, one zero between two signal samples. The amplitude of the

Fourier transform of the expanded signal is shown in (b). How do you explain the presence of another peak at f=0.4375 Hz?.

Example 2.3: A sinusoid

$$x(t) = \sin(0.25 * \pi t + \frac{\pi}{6})$$

is sampled at unit time interval. A signal of 64 samples was Fourier transformed. The magnitude of the Fourier transform is plotted in fig. 2.4a. Next the same signal was interlaced with zeros, one zero between two data samples. The magnitude of the Fourier transform of the expanded signal is plotted in fig. 2.4b .

Properties Common with CFT: The discrete Fourier transform (DFT) shares many properties with the continuous Fourier transform (CFT). We shall merely list below these common properties:

Table 2.1: Properties shared by DFT with CFT

Sr. #	Property	Signal	Fourier transform
1	Linearity	$ax(n)+b(y(n)$	$aX(\omega)+bY(\omega)$
2	Symmetry	$x(n)$ is real	$X(\omega) = X^*(-\omega)$
3	Modulation	$\exp(j\omega_0 n)x(n)$	$X(\omega-\omega_0)$
4	Time reversal	$x(-n)$ (real)	$X^*(\omega)$
5	Convolution	$\sum_{m=0}^{\infty} h(m)x(n-m)$	$H(\omega)X(\omega)$
6	Multiplication	$h(n)x(n)$	$\frac{1}{2\pi}\int_{-\pi}^{\pi} H(\omega')X(\omega-\omega')d\omega'$
7	Parseval's theorem	$\sum_{n=-\infty}^{\infty} \lvert x(n)\rvert^2 = \frac{1}{2\pi}\int_{-\pi}^{\pi} \lvert X(\omega)\rvert^2 d\omega$	

§2.2 Aliasing Error

The discrete Fourier transform (DFT) in the principal band, $\pm\pi/\Delta t$, may be contaminated by the neighbouring copies of the DFT when the bandwidth of the CFT exceeds $2\pi/\Delta t$. The error due to such contamination is known as aliasing error (see fig. 2.5).

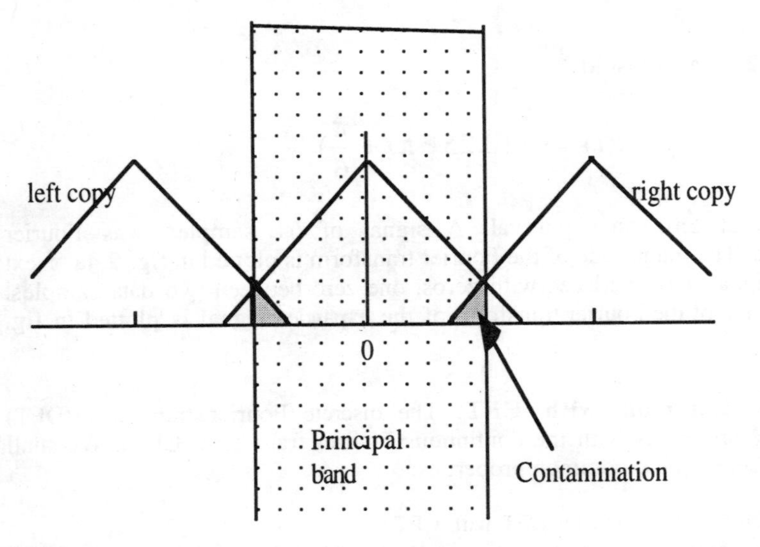

Figure 2.5: The aliasing error is caused by contamination from the neighbouring copies into the principal band. The aliasing error is equal to the area under the curve projected into the principal band.

Folding: Alternate way of visualizing the aliasing error is to fold a plot of the continuous Fourier transform over a series of lines parallel to the y-axis and separated by $\pi/\Delta t$. Then, sum up all segments falling within a frequency range of $\pm\pi/\Delta t$. The process is often referred to as spectral folding (for stochastic signal) and it is illustrated in fig. 2.6. A signal is sampled at 500 Hz, that is, sampling interval is 2 milliseconds. The folding begins at 250Hz. For simplicity we have considered only the real part of the Fourier transform and also we restrict ourselves to a few positive values of k=0, 1, 2, 3. The segments corresponding to these values of k are shown in the figure. The sum of all segments (labeled as "sum") is the discrete Fourier transform and the difference between the "sum" and $X_c(\omega)$ is the aliasing error. The aliasing error is due to the segments folded into the principal band, $\pm\pi/\Delta t$, namely, segment #1, #2 and #3. Evidently, if these segments were to be absent there would have been no aliasing error. For this the signal Fourier transform must be limited to a maximum frequency of $\pm$250Hz. Then, there will be no contribution from any of the successive foldings. This is illustrated in fig. 2.7. Thus, for error free DFT the sampling interval Δt be chosen such that the signal spectrum lies entirely within the frequency range of $\pm\pi/\Delta t$. A signal having such a property is known to be bandlimited. When the spectrum of a signal is $|X(\omega)|^2 = 0$ for $|\omega| > 2\pi f_b > 0$, the signal is said to be bandlimited with a bandwidth of $2f_b$

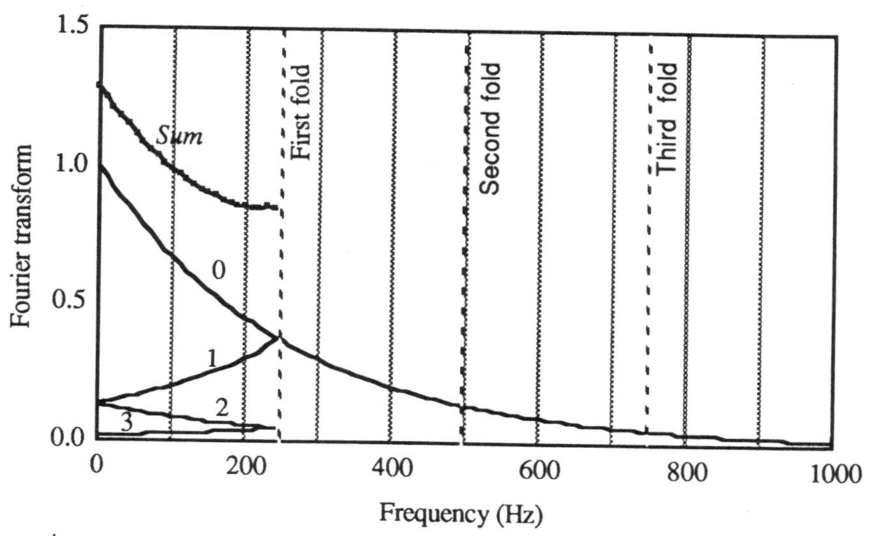

Figure† **2.6**: A discrete Fourier transform is obtained by successively folding the continuous Fourier transform as shown above. The first fold is at half the sampling frequency, $1/2\Delta t$; the second fold is at $1/\Delta t$ and so on. The segments from the successive copies are labeled.

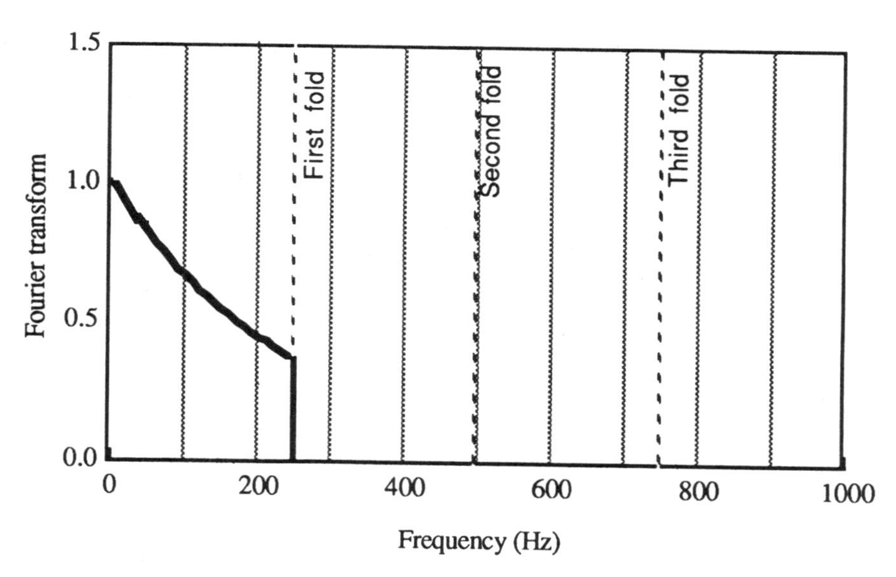

Figure 2.7: When the Fourier transform of a signal is confined to $\pm 250Hz$ and the sampling rate is 500Hz, the successive foldings of the Fourier plane will not cause any aliasing error.

Sampling Theorem: Let the sampling interval Δt be such that

†Figures 2.6, 2.7, and 2.19 were taken from [6]

$$\Delta t \leq \frac{1}{2f_b} \tag{2.12}$$

or the sampling rate be greater than two times the highest frequency present in the signal. In chapter one we had introduced interpolation of sampled signals given by,

$$\hat{x}(t) = \sum_{n=-\infty}^{\infty} x(n) \sin c(t - n\Delta t) \tag{1.2}$$

where

$$\sin c(t) = \frac{\sin(2\pi f_b t)}{2\pi f_b t}.$$

We shall now show that the interpolation error becomes zero when the sampling interval satisfies (2.12). The interpolation error (mean square error, MSE) is given by

$$E\left\{|x(t) - \hat{x}(t)|^2\right\} = E\left\{\left|x(t) - \sum_{n=-\infty}^{\infty} x(n) \frac{\sin(2\pi f_b (t - n\Delta t))}{2\pi f_b (t - n\Delta t)}\right|^2\right\} \tag{2.13a}$$

MSE=

$$c_x(0) + \sum_{-\infty}^{\infty}\sum_{-\infty}^{\infty} c_x(n - n')\Delta t \frac{\sin(2\pi f_b (t - n\Delta t))}{2\pi f_b (t - n\Delta t)} \frac{\sin(2\pi f_b (t - n'\Delta t))}{2\pi f_b (t - n'\Delta t)}$$
$$-2 \sum_{n=-\infty}^{\infty} c_x(t - n\Delta t) \frac{\sin(2\pi f_b (t - n\Delta t))}{2\pi f_b (t - n\Delta t)} \tag{2.13b}$$

By using the Fourier integral representation of $c_x(\tau)$ (see Chapter One, eq.(1.6b)) in (2.13b) we obtain,

$$MSE =$$

$$\frac{1}{2\pi} \int_{-\infty}^{\infty} S(\omega) \left[\begin{array}{l} 1 + \left| \sum_{-\infty}^{\infty} e^{j\omega n\Delta t} \sin c(2\pi f_b (t - n\Delta t)) \right|^2 \\ -2e^{j\omega t} \sum_{n=-\infty}^{\infty} e^{-jn\omega\Delta t} \sin c(2\pi f_b (t - n\Delta t)) \end{array} \right] d\omega \tag{2.13c}$$

To proceed further we need the following Fourier series expansion of $Q(\omega)$,

$$Q(\omega) = \sum_{n=-\infty}^{\infty} e^{-jn\omega\Delta t} \frac{\sin(2\pi f_b (t - n\Delta t))}{2\pi f_b (t - n\Delta t)}$$

where

$$Q(\omega) = \begin{cases} e^{-j\omega t} & for \quad -2\pi f_b \le \omega \le 2\pi f_b \\ 0 & for \quad 2\pi f_b < |\omega| \le \dfrac{\pi}{\Delta t} \end{cases}$$

and it is periodic outside the range $\pm\pi/\Delta t$. Using this result in (2.13c) we obtain

$$MSE = \frac{1}{2\pi}\int_{-\infty}^{\infty} S(\omega)[\Pi(\omega) - 1]d\omega \tag{2.14}$$

where $\Pi(\omega)$ is sketched in fig. 2.8. Clearly $\Pi(\omega)$ becomes one everywhere whenever $f_b = 1/2\Delta t$. Then, it follows from (2.14) that the mean square error in the interpolation becomes zero.

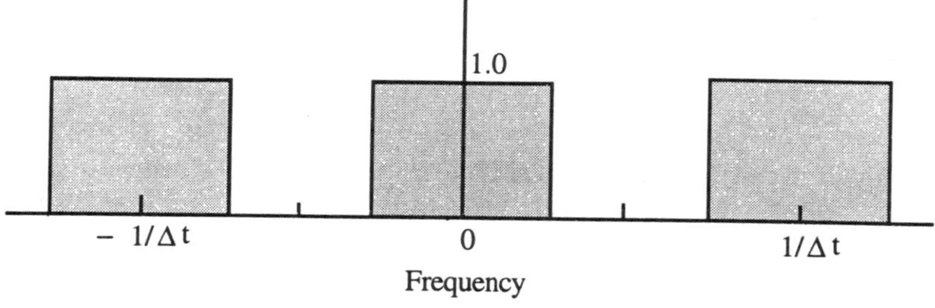

Figure 2.8: $\Pi(\omega)$ function in (2.12). Clearly, $\Pi(\omega)$=1 everywhere whenever $f_b = 1/2\Delta t$.

The *Sampling Theorem for a bandlimited signal states that, when sampled at a rate equal to or greater than two times the highest frequency, there is no reconstruction or interpolation error when (1.2) is used.*

Example 2.4: Let us find out the sampling frequency of signals whose spectrum is shown in fig. 2.9. In (a) the maximum frequency is 120 Hz hence the sampling rate must be at least 240 samples per second or the sampling interval must be must be less than or equal to 0.00416 seconds (What happens if we take rate exactly equal to 240 Hz?). In (b) the highest frequency is 200Hz. The sampling rate must be at least 400 samples per second even though there is null band from 80 to 160 Hz. Now consider a signal of practical interest namely a bandpass signal shown in (c). As per the sampling theorem we must sample at a rate two times the maximum frequency, that is, 400Hz. We shall show in the following section that this need not be done. Sampling may be carried out at a rate equal to two times the bandwidth, that is, at 80 samples per second

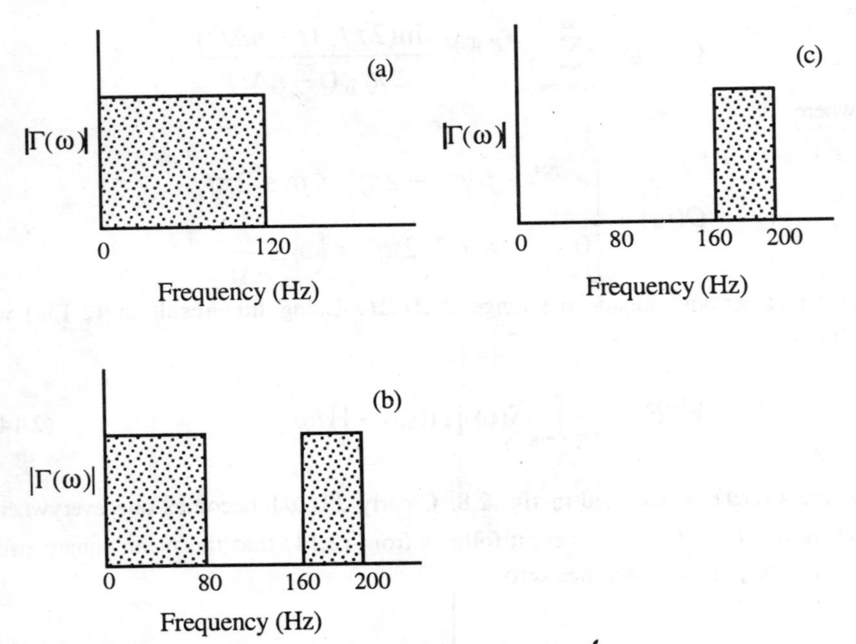

Figure 2.9: Examples of Fourier transform (magnitude) .

Sampling of Bandpass Signal: The Fourier transform of a bandpass signal is finite within two symmetric frequency bands and zero elsewhere. Mathematically, the Fourier transform of a bandpass signal may be expressed as,

$$X_c(\omega) = \Gamma(\omega - \omega_c) + \Gamma^*(-\omega - \omega_c) \tag{2.15}$$

where $\Gamma(\omega)$ is the Fourier transform of a band limited signal (possibly complex), $\Gamma(\omega) = 0$ for $|\omega| > 2\pi f_b$ and ω_c is the center frequency. Figure 2.10 shows an example of the Fourier transform of a bandpass signal. We have shown in chapter one (p.29) that a bandpass signal may be expressed as a complex signal

$$\begin{aligned} x_{bp}^{+}(t) &= x_{bp}(t) + jx_{bp}^{hilb}(t) \\ &= \gamma(t)e^{j\omega_c t} \end{aligned} \tag{2.16}$$

where $\gamma(t) = \dfrac{1}{2\pi}\displaystyle\int_{-2\pi f_b}^{2\pi f_b} \Gamma(\omega)e^{j\omega t}d\omega$, a complex function. From (2.16) we can express a bandpass signal (real) as

$$x_{bp}(t) = \gamma_I(t)\cos(\omega_c t) - \gamma_Q(t)\sin(\omega_c t) \tag{2.17}$$

where $\gamma(t) = \gamma_I(t) + j\gamma_Q(t)$, where $\gamma_I(t)$ is inphase and $\gamma_Q(t)$ is quadrature phase components respectively.
The highest frequency is $f_b + f_c$ and naturally a sampling rate equal to $2(f_b + f_c)$ will suffice. But, in many practical cases since f_c is large, it becomes an expensive proposition to sample at the required high rate. To overcome this problem an easy approach is to demodulate the signal to bring down the center frequency to zero. This must however be done in analog domain. We shall now describe an alternate approach, which does not involve demodulation.
Priest

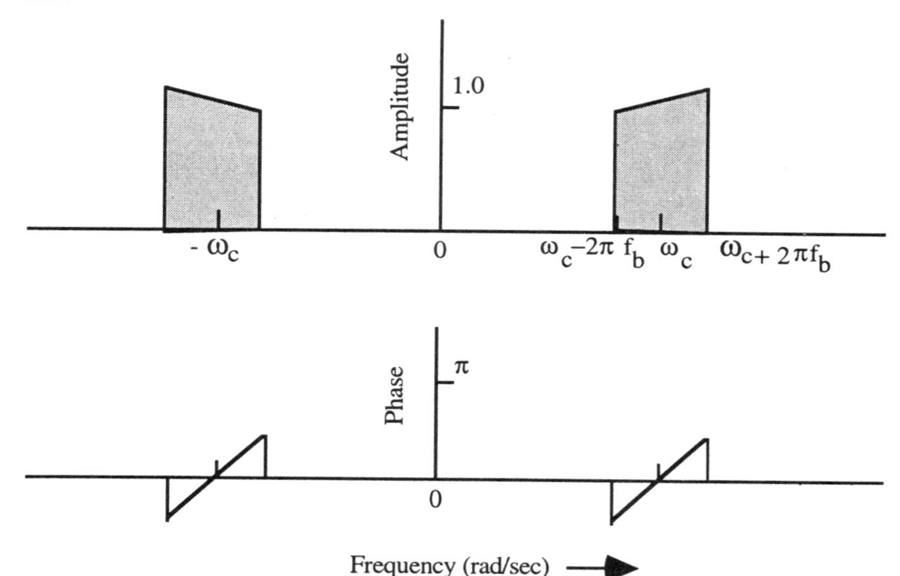

Figure 2.10: The Fourier transform of a bandpass signal. Note the symmetry relations of the Fourier transform of a real signal.

Assume that $f_c + f_b$ is an integer multiple of $2f_b$, that is, let $(f_c + f_b)/2f_b = p$, an integer. Sample $x_{bp}(t)$ at a rate $\boxed{1/\Delta t = 4f_b}$. From (2.17) we get

$$x_{bp}(n\Delta t) = \gamma_I(n\Delta t)\cos(\omega_c n\Delta t) - \gamma_Q(n\Delta t)\sin(\omega_c n\Delta t) \tag{2.18a}$$

and substituting for the sampling interval we obtain

$$\begin{aligned} x_{bp}(n\Delta t) &= \gamma_I(n\Delta t)\cos(\frac{2\pi f_c}{4f_b}n) - \gamma_Q(n\Delta t)\sin(\frac{2\pi f_c}{4f_b}n) \\ &= \gamma_I(n\Delta t)\cos(\frac{\pi n(2p-1)}{2}) - \gamma_Q(n\Delta t)\sin(\frac{\pi n(2p-1)}{2}) \end{aligned} \tag{2.18b}$$

For even n, that is, n=2m (2.18b) reduces to

$$x_{bp}(2m\Delta t) = \gamma_I(2m\Delta t)(-1)^m \tag{2.19a}$$

and for odd n, that is, n=2m-1

$$x_{bp}((2m-1)\Delta t) = \gamma_Q((2m-1)\Delta t)(-1)^{(m+p+1)} \tag{2.19b}$$

Thus, the even numbered samples of $x_{bp}(n\Delta t)$ produce the inphase component and the odd numbered samples produce the quadrature component. We can now reconstruct without error the inphase and quadrature components from the even and odd signal samples,

$$\gamma_I(t) = \sum_{m=-\infty}^{\infty} \gamma_I(m2\Delta t)\frac{\sin(\frac{\pi}{2\Delta t}(t-m2\Delta t))}{(\frac{\pi}{2\Delta t}(t-m2\Delta t))}$$

$$= \sum_{m=-\infty}^{\infty} (-1)^m x_{bp}(2m\Delta t)\frac{\sin(\frac{\pi}{2\Delta t}(t-m2\Delta t))}{(\frac{\pi}{2\Delta t}(t-m2\Delta t))} \tag{2.20a}$$

$$\gamma_Q(t) = \sum_{m=-\infty}^{\infty} \gamma_Q((2m-1)\Delta t)\frac{\sin(\frac{\pi}{2\Delta t}(t-m2\Delta t+\Delta t))}{(\frac{\pi}{2\Delta t}(t-m2\Delta t+\Delta t))}$$

$$= -\sum_{m=-\infty}^{\infty} (-1)^{m+p} x_{bp}((2m-1)\Delta t)\frac{\sin(\frac{\pi}{2\Delta t}(t-m2\Delta t+\Delta t))}{(\frac{\pi}{2\Delta t}(t-m2\Delta t+\Delta t))} \tag{2.20b}$$

Having obtained the in phase and quadrature phase components it is straightforward to compute the bandpass signal from (2.17),

$$x_{bp}(t) = \sum_{m=-\infty}^{\infty} (-1)^m x_{bp}(2m\Delta t)\frac{\sin(\frac{\pi}{2\Delta t}(t-m2\Delta t))}{(\frac{\pi}{2\Delta t}(t-m2\Delta t))}\cos(\omega_c t)$$

$$+ \sum_{m=-\infty}^{\infty} (-1)^{m+p} x_{bp}((2m-1)\Delta t)\frac{\sin(\frac{\pi}{2\Delta t}(t-m2\Delta t+\Delta t))}{(\frac{\pi}{2\Delta t}(t-m2\Delta t+\Delta t))} \tag{2.21a}$$

Note that, since $\Delta t\,\omega_c = \pi(p-1/2)$,

$$(-1)^m \cos(\omega_c t) = \cos(\omega_c (t-2m\Delta t))$$

and $(-1)^{m+p}\sin(\omega_c t) = \cos(\omega_c(t - 2m\Delta t + \Delta t))$. Eq. (2.21a) reduces to

$$x_{bp}(t) = \sum_{m=-\infty}^{\infty} x_{bp}(n\Delta t)\frac{\sin(\frac{\pi}{2\Delta t}(t-n\Delta t))}{\frac{\pi}{2\Delta t}(t-n\Delta t)}\cos(\omega_c(t-n\Delta t)) \qquad (2.21b)$$

Note that in the above reconstruction formula the samples are taken at a rate two times the bandwidth and further we have assumed that the center frequency is $(p-0.5)$ times the bandwidth. Eq.(2.21b) may be reduced to (1.2), a reconstruction formula for lowpass signal. This is left out as an exercise for the reader. Finally the *Sampling Theorem for bandpass signal* states that *when sampled at a rate equal to or greater than two times the bandwidth, there is no reconstruction or interpolation error when (2.21b) is used.*

Example 2.5: A 50 Hertz bandwidth stationary stochastic signal $(f_b = 25Hz)$ riding on a carrier of 1000Hz is sampled at a rate of 100Hz or $\Delta t = 0.01$ seconds. The spectrum of the signal is shown in fig. 2.11. We use (2.21b) to reconstruct the signal from the samples. 1024 samples were used for reconstruction. The actual signal and the reconstructed version are shown in fig. 2.12. The mean square error is very small, 5.9290e-05.

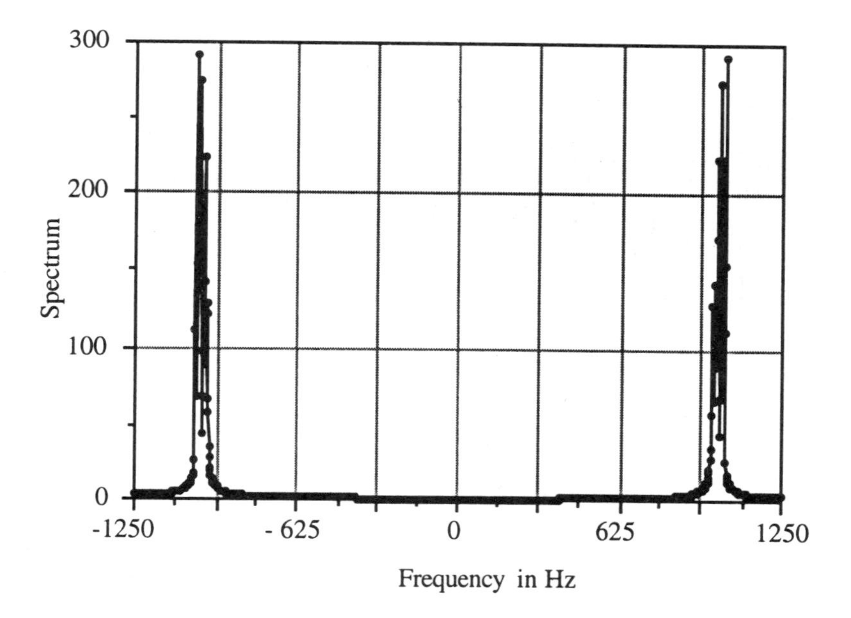

Figure 2.11: Spectrum of a bandpass signal. The center frequency is at 1000 Hz and the bandwidth is 50 Hz.

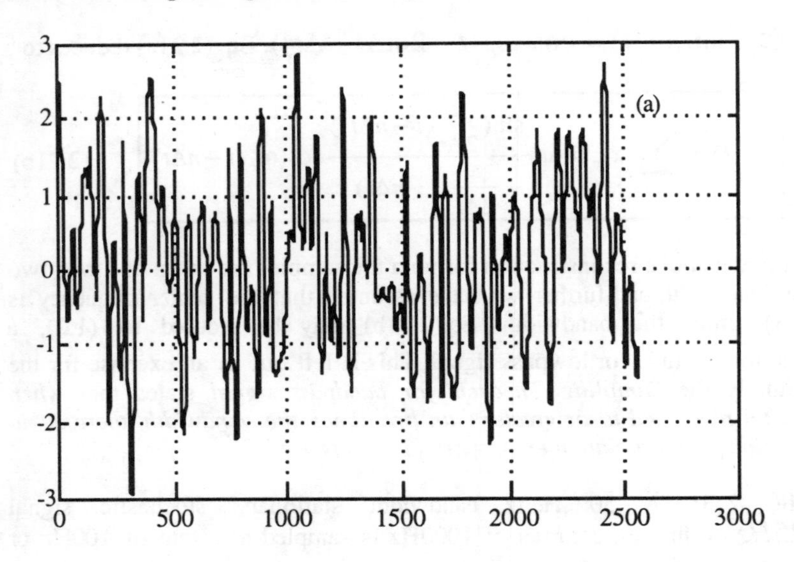

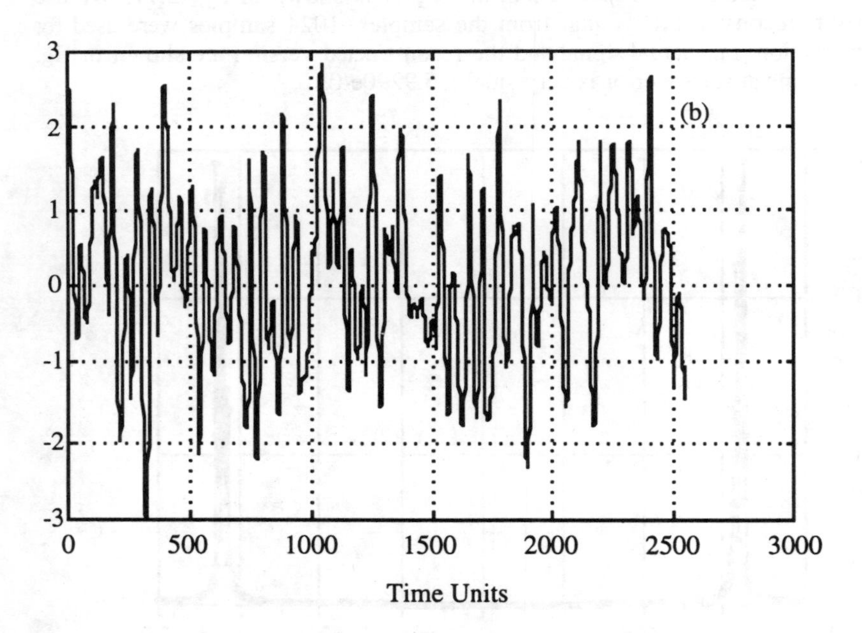

Figure 2.12: Bandpass signal reconstruction. (a) Actual bandpass signal (b) Reconstructed signal from samples obtained by sampling the signal at 100 Hz. The mean square difference is 5.9290e-05 .

Sampling Rate Conversion: In many practical systems the sampling rates may be different. When interfacing two such systems with different sampling rates it is necessary to convert the sampling rate. For example, sampling rate used in audio broadcasting is 32 kHz but for storage in compact disk it is 44.1 kHz and for storage in audiotape it is 48 kHz. A straightforward approach to sampling rate conversion is D/A conversion followed by resampling the analog signal at any desired rate. A major disadvantage of such an approach is the signal distortion introduced by D/A conversion and by quantization due to further A/D conversion of the reconstructed analog signal. In purely digital approach where sampling rate conversion is achieved by means of digital filtering and resampling the above mentioned errors can be avoided. We shall explore this approach. We shall consider three types of rate conversion, upsampling (interpolation) by an integer factor I, downsampling (decimation) by integer factor D and finally resampling at a rate given by a rational number I/D. The bandwidth of the given signal is assumed to be $\pm\pi$ and the samples are one unit apart. In upsampling (interpolation) by a factor I, we must get I-1 samples to fill in the interval between two samples of the given signal. This is signal expansion as defined on page 57. The Fourier transform of the resulting expanded signal is given by (2.11). For a lowpass signal it results into a compressed waveform which repeats I times within the pass band (see fig. 2.13). We need to retain the zeroth order replica and delete all others. This is easily achieved by means of lowpass filtering. All above steps are illustrated in fig. 2.13 (see Example 2.6). The height of the lowpass filter is set equal to I so that the samples of the given signal remain unchanged (see Exercise 17). The Fourier transform of the interpolated sequence (see the scaling property of the Fourier transform) is given by

$$X_I(\omega_I) = I\,X(I\omega_I) \tag{2.22a}$$

Example 2.6: An example will make this clear. Consider a signal whose Fourier transform is a triangular function and let I=3. The samples (one unit apart) are interlaced with two zeros as shown in fig. 2.13a. The Fourier transform is sketched in fig. 2.13b.

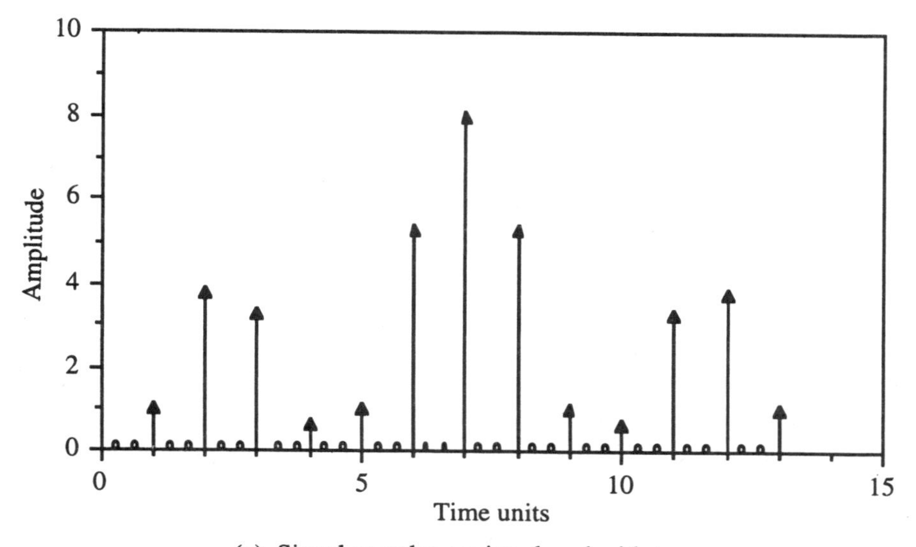

(a) Signal samples are interlaced with two zeros

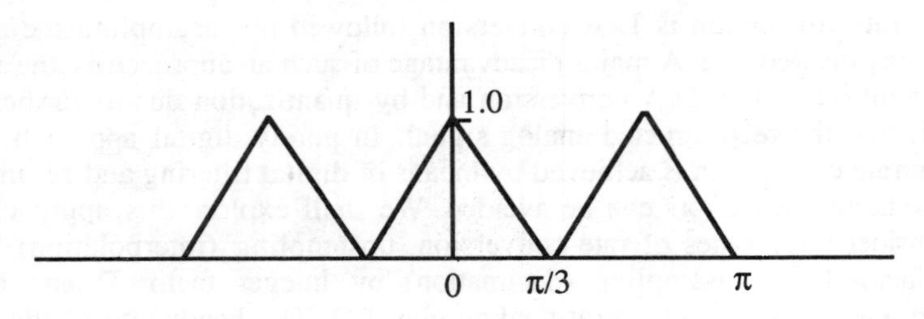

(b) Fourier transform of the signal shown in (a)

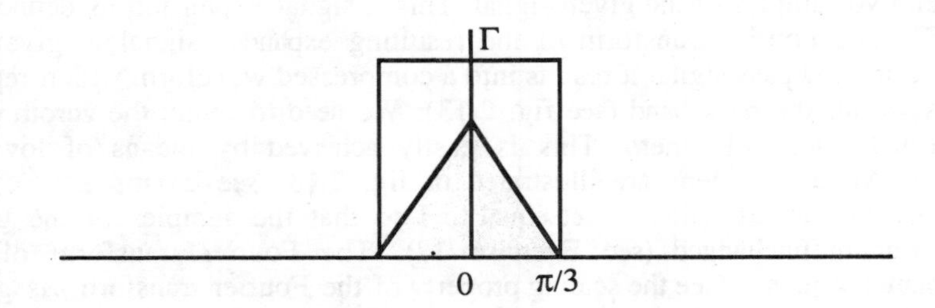

(c) A low pass filter is used to delete unwanted replicas

Figure 2.13: The signal is expanded by inserting I=2 zeros and then filtered with a lowpass filter to retain only the zeroth order replica.

where $\omega_I = \omega/I$ is the new frequency variable pertaining to the interpolated sequence.

In downsampling (decimation) the given signal has to be compressed by deleting D samples. Since compression of a signal results in the expansion of its Fourier transform it is necessary, to avoid aliasing, that the given signal is lowpass filtered before decimation, thus remove the excess energy lying outside the band $\pm\frac{\pi}{D}$. The filtered signal is now resampled by deleting D samples. The procedure is illustrated in Example. 2.7. Computing the Fourier transform of the decimated signal is tricky. We need to represent mathematically the process of decimation. Let

$$\gamma(n) = \sum_{p=-\infty}^{\infty} \delta(n - pD)$$

where D stands for the number of samples to be deleted. The decimated signal may be expressed as $x_D(n) = x_{filtered}(n)\gamma(n)$ where $x_{filtered}(n)$ is lowpass filtered version of the signal. Taking the Fourier transform of $x_D(n)$ we get

$$\begin{aligned} FT\{x_D(n)\} &= FT\{x_{filtered}(n)\gamma(n)\} \\ &= [H(\omega)X(\omega)] * FT\{\gamma(n)\} \end{aligned} \tag{2.22b}$$

But

$$FT\{\gamma(n)\} = \frac{1}{D}\sum_{k=-\infty}^{\infty} \delta(\omega - \frac{2\pi k}{D})$$

which we use in (2.22b) and evaluate the convolution operation. We obtain

$$\begin{aligned} X_D(\omega) &= [H(\omega)X(\omega)] * \left\{\frac{1}{D}\sum_{n=-\infty}^{\infty} \delta(\omega - \frac{2\pi k}{D})\right\} \\ &= \frac{1}{D}\sum_{n=-\infty}^{\infty} H(\omega - \frac{2\pi k}{D})X(\omega - \frac{2\pi k}{D}) \end{aligned} \tag{2.22c}$$

Here the frequency variable, ω, refers to the sampling rate prior to decimation. The frequency variable after decimation, ω_D, is related to ω as $\omega_D = D\omega$. Further, since the lowpass filter has unit response in the band $\pm\pi/D$ only zeroth order term in (2.22c) will be retained

$$X_D(\omega) = \frac{1}{D}X(\frac{\omega_D}{D}) \tag{2.22d}$$

Finally, resampling at a fractional rate is possible when the fraction is a rational fraction, that is, equal to I/D where I and D are positive integers. The approach is to first upsample (interpolate) at a rate I followed by downsample (decimate) at the rate D. We need to cascade an interpolator and a decimator as shown in fig. 2.14. The two lowpass filters may be combined into a single lowpass filter whose pass band is $0 \le \omega \le \min\{\pi/D, \pi/I\}$. The Fourier transform of the output may be obtained by combining (2.22a) and (2.22d). The result is

$$\begin{aligned} X_{I/D}(\omega_D) &= \frac{I}{D}X(\frac{I}{D}\omega_D) \qquad 0 \le |\omega_D| \le \min(\pi, \frac{D}{I}\pi) \\ &= 0 \qquad \text{otherwise} \end{aligned} \tag{2.22e}$$

§2.3 **Finite Discrete Fourier Transform** (FDFT)

The finite discrete Fourier transform (FDFT) plays a very important role in signal processing. A note of clarification that we have used DFT for infinite sequence and FDFT for finite sequence. This distinction is not made in many texts. The DFT and FDFT are sufficiently different (DFT is continuous function but FDFT is a discrete function) warranting the above distinction. Along with fast Fourier transform (FFT), described in the next section, FDFT is the core of many signal processing tools, such as, convolution, correlation, spectrum analysis, beamforming etc.

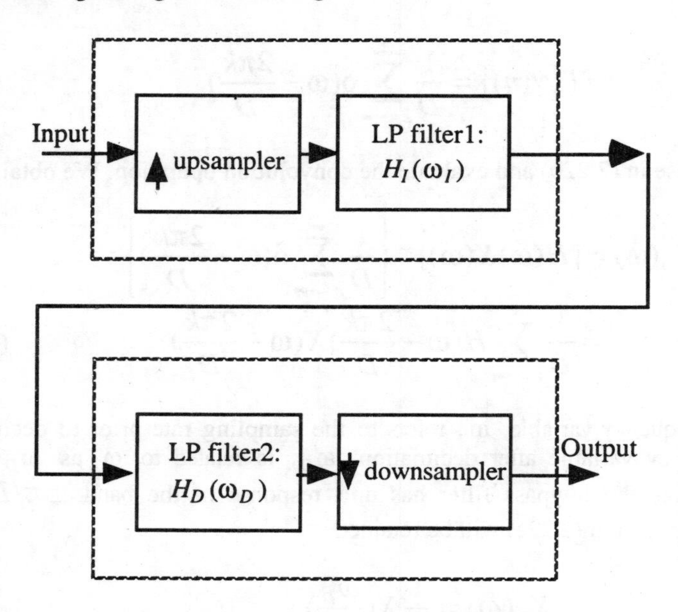

Figure 2.14: Fractional sampling rate converter In fig.2.16 LP stands for lowpass filter. The passband of $H_I(\omega_I)$ is $\pm\pi/I$ and that of $H_D(\omega_D)$ is $\pm\pi/D$. Further $H_I(\omega_I) = I,\ |\omega| \le \pi/I$.

Definition: We define the Fourier transform of a finite discrete signal,

$$X(k) = \sum_{n=0}^{N-1} x(n) e^{-j\frac{2\pi k}{N}n}, \quad k = 0,1,2,\ldots,N-1 \tag{2.23a}$$

The inverse of (2.23a) exists and is given by

$$x(n) = \frac{1}{N}\sum_{k=0}^{N-1} X(k) e^{j\frac{2\pi k}{N}n}, \quad n = 0,1,2,\ldots,N-1 \tag{2.23b}$$

Example 2.7: Consider a signal whose Fourier transform is a triangular function with a base $\pm\pi$. Let D=3. The signal is first lowpass filtered with a pass band of $\pm\pi/3$ (see fig. 2.15). The filter output is compressed by deleting two samples but retaining the third. The resulting compressed signal will have a bandwidth of $\pm\pi$, which is same as that of the given signal

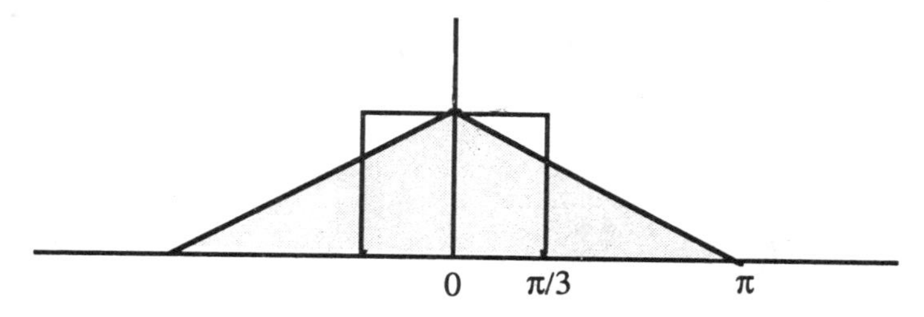

(a) Lowpass filtering of the given signal. Pass band is from - π/3 to π/3

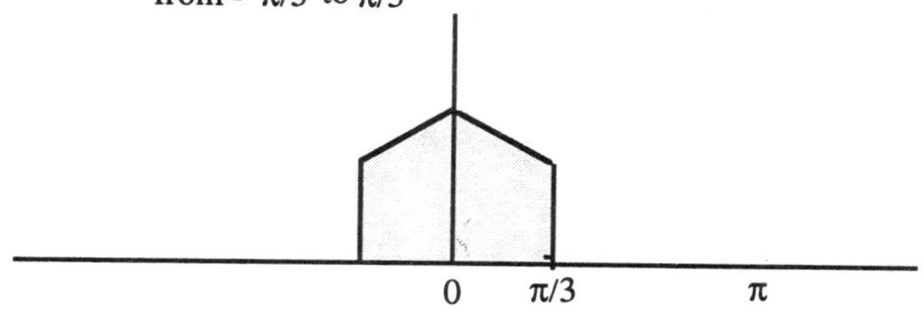

(b) Fourier transform of the filtered signal

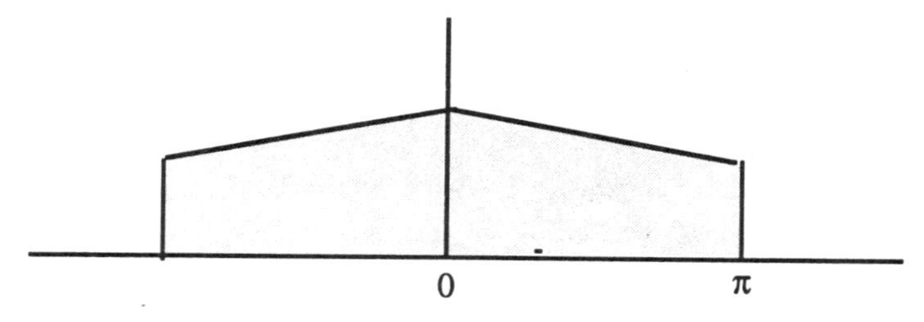

(c) Fourier transform of the compressed signals

Figure 2.15: In downsampling (decimation) the given signal is first lowpass filtered followed by signal decimation

$$X(0) = \sum_{n=0}^{N-1} x(n) = N \times \textit{mean of the signal} \qquad (2.24a)$$

Example 2.8: As mentioned earlier the sampling frequency used for broadcast is 32 kHz and that for the storage on audiotape is 48 kHz. Hence, it is necessary to convert the sampling rate from 32 kHz to 48kHz. We illustrate this conversion through a scaled down example. Consider a stochastic lowpass signal with bandwidth ± 50Hz sampled at 0.8k samples per second. We like to convert this into a digital signal sampled at 1.2k samples per second. We shall first upsample the signal three times (p=2) and then down sample by a factor of 2 (q=1). The result is shown in fig. 2.16

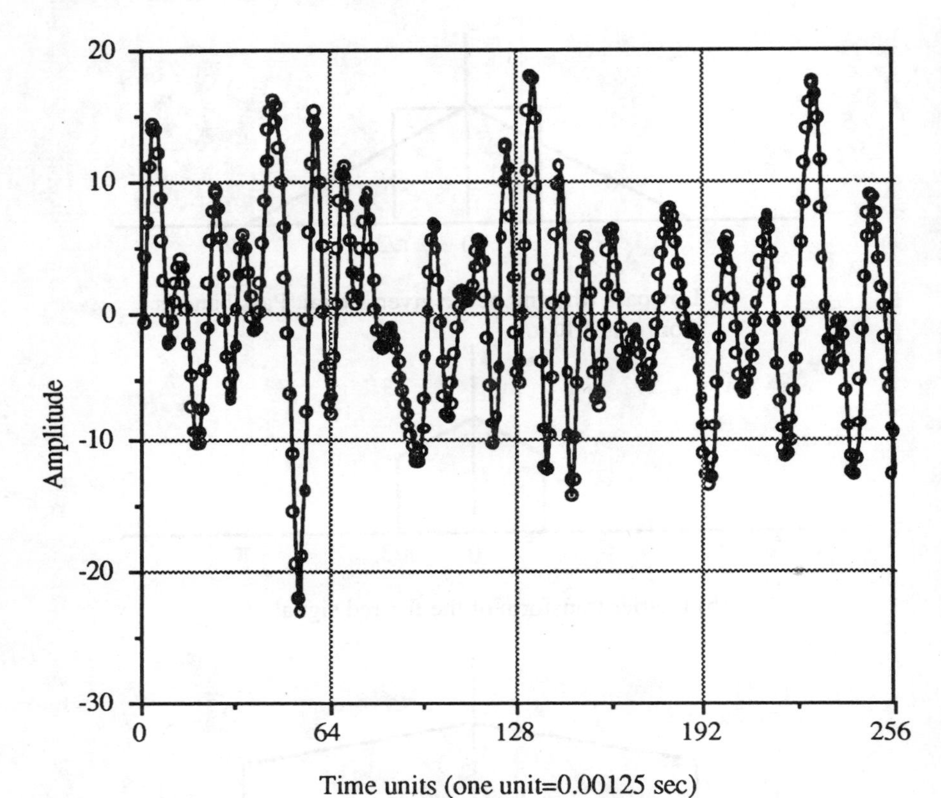

Figure 2.16: A stochastic lowpass signal (continuous curve) is sampled at 0.8KHz. The sampling rate is next converted into 1.2 kHz. Samples empty circles ●.

and at $k = N/2$ (N even)

$$X(\frac{N}{2}) = \sum_{n=0}^{N-1} (-1)^n x(n) \tag{2.24b}$$

Note that for real signal the FDFT coefficients $X(k)$ are complex except $X(0)$ and $X(N/2)$ which are real numbers. The FDFT of N point signal yields N Fourier coefficients. The first coefficient corresponds to zero frequency. The next $\frac{N}{2}$ coefficients correspond to +ve frequencies and the last $N/2-1$ coefficients correspond to -ve frequencies. To prove the last statement let us evaluate X(N-k)

$$\begin{aligned} X(N-k) &= \sum_{n=0}^{N-1} x(n) e^{-j\frac{2\pi(N-k)}{N}n} \\ &= \sum_{n=0}^{N-1} x(n) e^{j\frac{2\pi k}{N}n} = X(-k), \quad k = 1,2,\ldots,\frac{N}{2}-1 \end{aligned} \tag{2.24c}$$

Further, for real signal the FDFT coefficients corresponding to -ve frequencies are equal to complex conjugate of those corresponding to +ve frequencies (see table 2.2).

Table 2.2: The coefficients corresponding to negative frequencies are shown in shaded box. The coefficients in the column two clearly satisfy the properties shown in equation (2.24).

Real signal	FDFT coefficients	Complex Signal	FDFT coefficients
2.190e-1	*6.4571*	0.2190 + 0.6868i	*6.4571 + 9.1217i*
4.700e-2	**-0.415 -1.5159i**	0.0470 + 0.5890i	**0.152 - 1.2685i**
6.789e-1	-1.2322 - 0.2232i	0.6789 + 0.9304i	-0.0047 + 1.0226i
6.793e-1	0.2308 + 0.1286i	0.6793 + 0.8462i	-0.3733 - 0.9489i
9.347e-1	-0.5989 + 1.7315i	0.9347 + 0.5269i	-0.4288 + 1.0190i
3.835e-1	0.5671 - 0.8744i	0.3835 + 0.0920i	0.2234 + 0.3969i
5.194e-1	0.1455 + 1.0214i	0.5194 + 0.6539i	0.3389 + 1.4027i
8.310e-1	0.3547 + 1.1891i	0.8310 + 0.4160i	-0.7357 + 0.6902i
3.460e-2	*-0.4756*	0.0346 + 0.7012i	*-0.4756 + 0.1525i*
5.350e-2	0.3547 - 1.1891i	0.0535 + 0.9103i	1.4451 - 1.6880i
5.297e-1	-0.1455 - 1.0214i	0.5297 + 0.7622i	0.0479 - .6401i
6.711e-1	0.5671 + 0.8744i	0.6711 + 0.2625i	0.9108 + 2.1458i
7.700e-3	-0.5989 - 1.7315i	0.0077 + 0.0475i	-0.7690 - 2.4439i
3.834e-1	0.2308 - 0.1286i	0.3834 + 0.7361i	0.8349 - 1.2061i
6.680e-2	-1.2322 + 0.2232i	0.0668 + 0.3282i	-2.4597 + 1.4690i
4.175e-1	**-0.415 + 1.5159i**	0.4175 + 0.6326i	**-0.982 +1.7633i**

The arrangement of the FDFT coefficients is shown in fig. 2.17. This is often known as DFT symmetry which is not the same as convention plot where the dc terms is placed at the center with positive frequencies to the right and the negative frequencies to the left. To and fro mapping between DFT symmetry and the conventional arrangement can be achieved with 'fftshift' command in MATLAB.

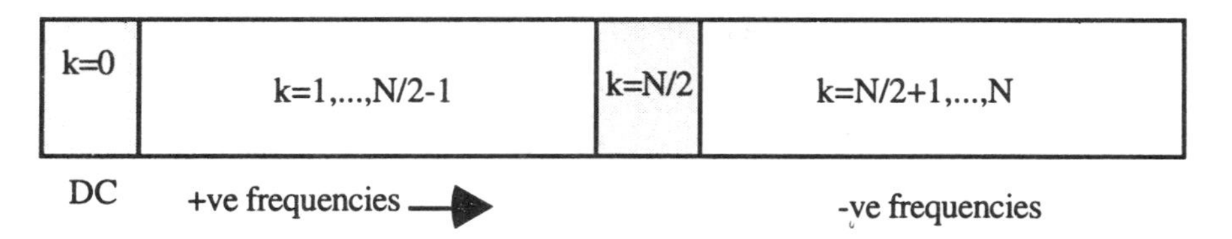

Figure 2.17: The arrangement of FDFT coefficients is shown above. Observe the position and the order of negative frequencies. This arrangement is often known as DFT symmetry.

It is important to note that the FDFT of a finite sequence implicitly assumes that the sequence is periodic, that is, for all p (integer),

$$x(pN+n) = x(n) \tag{2.25a}$$

To show this important property, which has a significant effect on finite discrete convolution, let us evaluate $x(n)$ from (2.25a) at $pN+n$,

$$x(pN+n)=\frac{1}{N}\sum_{k=0}^{N-1}X(k)e^{j\frac{2\pi k}{N}(pN+n)}$$

$$=\frac{1}{N}\sum_{k=0}^{N-1}X(k)e^{j2\pi kp}e^{j\frac{2\pi k}{N}n}$$

$$=\frac{1}{N}\sum_{k=0}^{N-1}X(k)e^{j\frac{2\pi k}{N}n}=x(n) \tag{2.25b}$$

Likewise, the FDFT coefficients are also periodic, that is, $X(pN+k)=X(k)$. The FDFT of a (circularly) shifted sequence is equal to the FDFT of the sequence before shifting multiplied by a phase factor. This is known as the shift property of FDFT.

$$FDFT\{x(n-\tau)\}=\sum_{n=0}^{N-1}x(n-\tau)e^{-j\frac{2\pi k}{N}n}$$

$$=\sum_{n'=-\tau}^{N-\tau-1}x(n')e^{-j\frac{2\pi k}{N}(n'+\tau)}$$

$$=e^{-j\frac{2\pi k\tau}{N}}\sum_{n'=-\tau}^{N-\tau-1}x(n')e^{-j\frac{2\pi k}{N}n'} \tag{2.26a}$$

$$=X(k)e^{-j\frac{2\pi k\tau}{N}}$$

Because of the implied periodicity of $x(n)$ and the periodicity of $e^{-j\frac{2\pi k}{N}n'}$, the summation in (2.26a) may be replaced by $X(k)$. Another way of looking at the symmetry induced by the periodicity of $x(n)$ and $e^{-j\frac{2\pi k}{N}n'}$ is to represent the finite sequence on a circumference of a circle. A shift by τ units is equivalent to rotating the circle anti-clockwise by $(2\pi/N)\tau$ radians.

Let us now evaluate the sum $\sum_{n'=-\tau}^{N-\tau-1}x(n')e^{-j\frac{2\pi k}{N}n'}$ in (2.26a)

$$\sum_{n'=-3}^{12}x(n')e^{-j\frac{2\pi k}{16}n'}=x(-3)e^{j\frac{2\pi k}{16}3}+x(-2)e^{j\frac{2\pi k}{16}2}+x(-1)e^{j\frac{2\pi k}{16}1}+x(0)+ \tag{2.26b}$$

$$x(1)e^{-j\frac{2\pi k}{16}1}+x(2)e^{-j\frac{2\pi k}{16}2}+\ \ldots\ +x(12)e^{-j\frac{2\pi k}{16}12}$$

From fig. 2.18(b) we note that x(-1)= x(15), x(-2)=x(14) and x(-3)=x(13). Further, we note that

Example 2.9: We represent a 16 point sequence on a circle in fig.2.18(a) and the same sequence circularly shifted by three points in fig. 2.18(b).

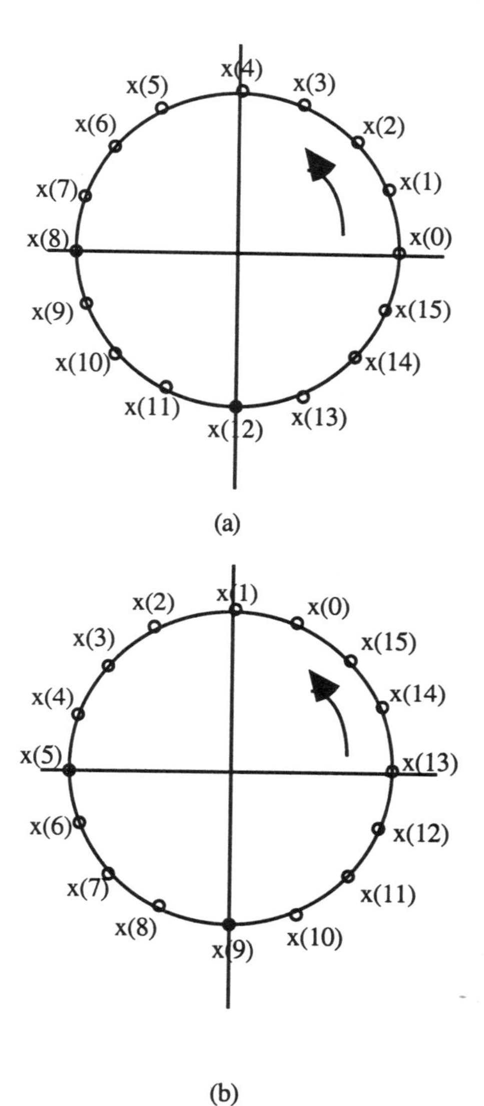

Figure 2.18: A 16 point periodic sequence is represented on a circle (anti-clockwise). A cyclic shift of the sequence by three points is shown in (b) .

$$e^{j\frac{2\pi k}{16}3} = e^{j\frac{2\pi k}{16}(16-13)} = e^{-j\frac{2\pi k}{16}13}$$

and similarly

$$e^{j\frac{2\pi k}{16}2} = e^{-j\frac{2\pi k}{16}14} \quad \text{and} \quad e^{j\frac{2\pi k}{16}1} = e^{-j\frac{2\pi k}{16}15}$$

Using these relations in (2.26b) we get

$$\sum_{n'=-3}^{12} x(n')e^{-j\frac{2\pi k}{16}n'} = x(13)e^{-j\frac{2\pi k}{16}13} + x(14)e^{-j\frac{2\pi k}{16}14} + x(15)e^{-j\frac{2\pi k}{16}15} + x(0) +$$

$$x(1)e^{-j\frac{2\pi k}{16}1} + x(2)e^{-j\frac{2\pi k}{16}2} + \ \ldots \ + x(12)e^{-j\frac{2\pi k}{16}12}$$

$$= \sum_{n'=0}^{15} x(n')e^{-j\frac{2\pi k}{16}n'} = X(k)$$

Time reversal of a finite sequence is obtained by reading the sequence mapped on a circle (fig. 2.18(a)) in a clockwise direction. Thus, $x(-n) = x(N-n)$, n=1,..., N. The FDFT of a time reversed sequence is

$$FDFT\{x(-n)\} = FDFT\{x(N-n)\}$$

$$= \sum_{n=1}^{N} x(N-n)e^{-j\frac{2\pi}{N}nk} = \sum_{m=0}^{N-1} x(m)e^{-j\frac{2\pi}{N}(N-m)k}$$

$$= \sum_{m=0}^{N-1} x(m)e^{j\frac{2\pi}{N}mk} = X(-k) = X(N-k)$$

Some of the properties of the FDFT coefficients are listed in table 2.3:

Table 2.3: A summary of the properties of the FDFT coefficients

Sl. #.	Property	Equation
1	Linearity	$FDFT\{a_1x_1(n) + a_2x_2(n)\}$ $= a_1X_1(k) + a_2X_2(k)$
2	Periodicity	$X(pN+k) = X(k)$
3	Mapping	$X(-k) = X(N-k)$
4	For real signal	$X(-k) = X^*(k)$
5	Shift (circular)	$FDFT\{x(n-\tau\} = X(k)e^{-j\frac{2\pi}{N}\tau}$
6	Time reversal	$FDFT\{x(-n)\} = X(-k)$
7	Parseval's Theorem	$\sum_{n=0}^{N-1} x(n)y^*(n) = \frac{1}{N}\sum_{k=0}^{N-1} X(k)Y^*(k)$

Relation between FDFT and DFT: We shall now establish a relationship between the FDFT coefficients of finite discrete signal and discrete Fourier transform (DFT) of an infinite discrete signal. In eq. (2.23) let us assume that the finite discrete signal is selected by windowing an infinite discrete signal. We can then replace $x(n)$ by its Fourier integral representation (1.57). We obtain

$$X(k) = \sum_{n=0}^{N-1} x(n) e^{-j\frac{2\pi k}{N} n}$$

$$= \frac{1}{2\pi} \int_{-\pi}^{\pi} X(\omega) d\omega \sum_{n=0}^{N-1} e^{j(\omega - \frac{2\pi k}{N} n) n}$$

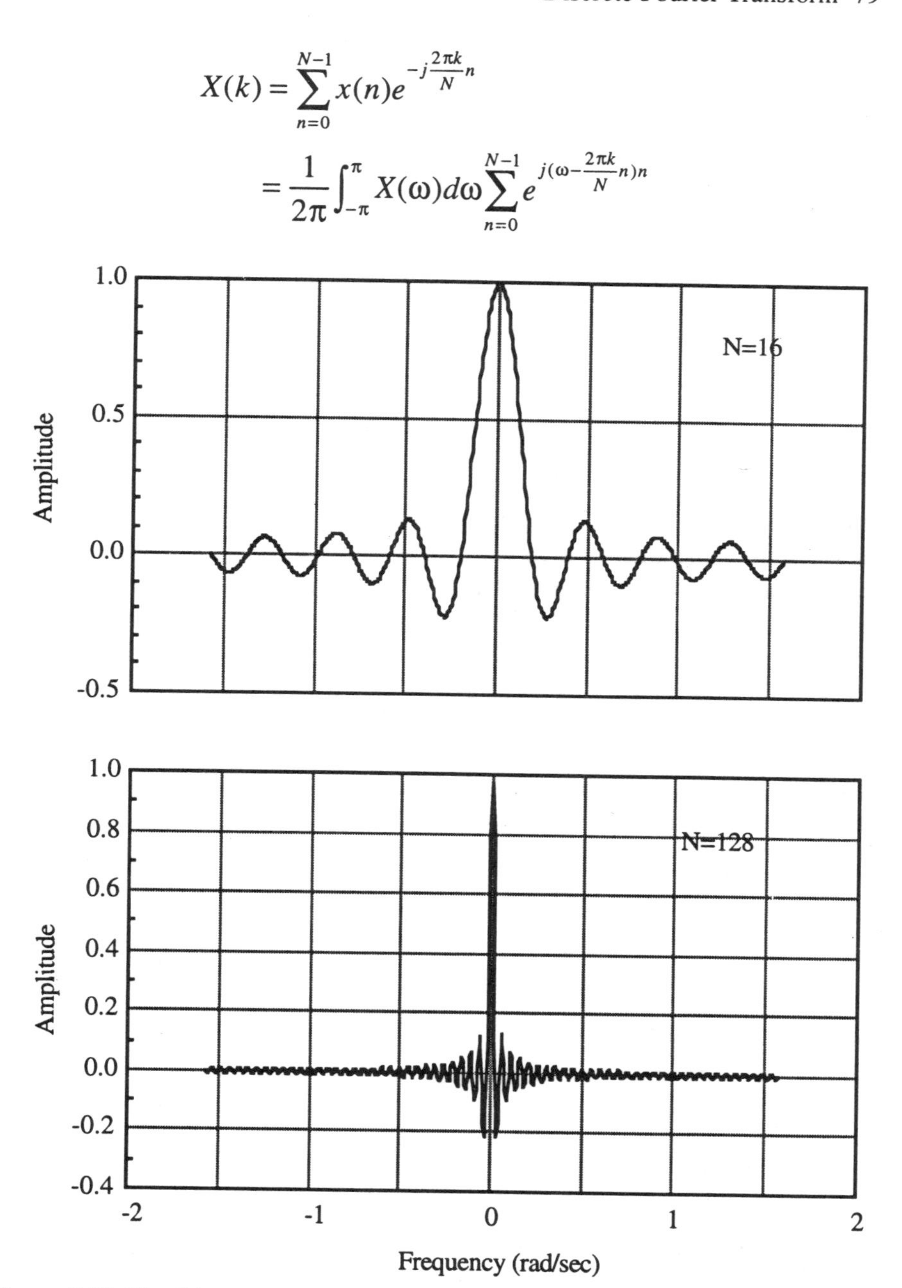

Figure 2.19: The Dirichlet function divided by N is plotted as a function of ω with k=0 for two different values of N. The function is periodic with a period of 2π. Notice that the main lobe becomes narrow for large N.

$$= \frac{1}{2\pi} \int_{-\pi}^{\pi} X(\omega) \frac{\sin((\omega - \frac{2\pi k}{N}) \frac{N}{2})}{\sin((\omega - \frac{2\pi k}{N}) \frac{1}{2})} e^{j(\omega - \frac{2\pi k}{N}) \frac{(N-1)}{2}} d\omega \qquad (2.27a)$$

where

$$\frac{\sin((\omega - \frac{2\pi k}{N})\frac{N}{2})}{\sin((\omega - \frac{2\pi k}{N})\frac{1}{2})}$$

is Dirichlet function (see fig. 2.19). It has a peak at $\omega = 2\pi k/N$ in the frequency range $-\pi \le \omega \ge \pi$. This peak becomes sharper tending to a delta function as $N \to \infty$.

In this limiting case eq. (2.27a) yields

$$\frac{1}{N}X(k) \to \frac{1}{2\pi}X(\omega)\big|_{\omega=\frac{2\pi k}{N}} \quad N \to \infty \tag{2.27b}$$

The significance of this result is that for large enough data length, we can, in all practical computations involving real-life data, replace $X(\omega)$ by the corresponding FDFT coefficient. The FDFT coefficient at $k = N/2$ corresponds to the DFT frequency $\omega = \pi$ and the FDFT coefficient at $k = N/2 + 1$ corresponds to the DFT frequency $\omega = -\pi + 2\pi/N$ (use eq.(2.24c) to show this). Thus, the FDFT coefficients k=0, 1,..., N-1 cover a frequency range $\pi\ to - \pi + 2\pi/N$ in steps of $2\pi/N$.

Relation Between FDFT and z-Transform: In Chapter One z-transform of a discrete signal was introduced in (1.16b) which we reproduce here for convenience,

$$X(z) = \sum_{n=0}^{\infty} x(n) z^{-n} \tag{1.16b}$$

where $z = e^{s\Delta t}$ and $s = \sigma + j\omega$. Let $\sigma = 0$ and evaluate the z-transform for finite data on the unit circle $|z| = 1$. Eq. (1.16b) now reduces to

$$X(z)\big|_{z=e^{j\omega}} = X(e^{j\omega}) = \sum_{n=0}^{N-1} x(n) e^{-j\omega n} \tag{2.28a}$$

Further, we shall evaluate (2.28a) on the unit circle on equispaced N points, $\omega = (2\pi/N)k$. We obtain,

$$\begin{aligned} X(z)\big|_{z=e^{\frac{2\pi k}{N}}} &= \sum_{n=0}^{N-1} x(n) e^{-j\frac{2\pi k}{N}n} \\ &= X(k) \end{aligned} \tag{2.28b}$$

which indeed is FDFT (2.23b). Thus, the z-transform evaluated on the unit circle is equal to FDFT.

Zero padding: A finite signal is often required to be padded with zeros so as to avoid circular convolution effects (discussed later in this chapter) in digital filtering or in autocorrelation estimation (Chapter Three) or simply to obtain more FDFT coefficients so that we get a smooth Fourier transform. We like to show that zero padding helps to interpolate between two FDFT coefficients and thus create more coefficients. Let $x_e(n) = x(n)$ for $0 \le n \le N-1$ and $x_e(n) = 0$ for $N \le n \le 2N-1$. We shall now compute the FDFT of x_e,

$$X_e(k) = \sum_{n=0}^{2N-1} x_e(n) e^{-j\frac{2\pi kn}{2N}}$$

$$= \sum_{n=0}^{N-1} x(n) e^{-j\frac{2\pi kn}{2N}}$$

(2.29a)

Express

$$k = 2k_1 + k_0$$

where $k_1 = 0,1,\ldots N-1$ and $k_0 = 0,1$. On substituting in (2.29a) we obtain

$$X_e(k) = \sum_{n=0}^{N-1} x(n) e^{-j\frac{2\pi k_0 n}{2N}} e^{-j\frac{2\pi k_1 n}{N}}$$

$$= FDFT\left\{ x(n) e^{-j\frac{2\pi k_0 n}{2N}} \right\}$$

$$= X(k_1) * \left[\frac{\sin(\pi(\frac{k_0}{2} + k_1)}{\sin(\frac{\pi}{N}(\frac{k_0}{2} + k_1))} e^{-j2\pi(\frac{k_0}{2}+k_1)\frac{N-1}{2}} \right] \quad (2.29b)$$

where $\sin(\pi(\frac{k_0}{2} + k_1) \Big/ \sin(\frac{\pi}{N}(\frac{k_0}{2} + k_1))$ is a Dirichlet function shown in fig. 2.19. For $k_0 = 0$, that is, all *even* FDFT coefficients in $X_e(k)$ are equal to $X(k_1)$. For $k_0 = 1$, that is, all *odd* FDFT coefficients in $X_e(k)$ are given by

$$X_e(2k_1 + 1) = X(k_1) * \left[\frac{-je^{j\frac{\pi}{N}(\frac{1}{2}+k_1)}}{\sin(\frac{\pi}{N}(\frac{1}{2} + k_1))} \right] \quad (2.29c)$$

Statistical Properties: Let $x(n)$, n=0, 1, 2, ..., $N-1$ be a finite segment of a stochastic signal. We would like to evaluate the statistical properties of the Fourier coefficients.

Mean:

$$E\{X(k)\}=\sum_{n=0}^{N-1}E\{x(n)\}e^{-j\frac{2\pi k}{N}n}=\sum_{n=0}^{N-1}m_x e^{-j\frac{2\pi k}{N}n}$$

$$=m_x N \quad for \quad k=0$$

$$=0 \quad for \quad k\neq 0 \tag{2.30a}$$

Except the first coefficient, all other FDFT coefficients are of zero mean. To remove this asymmetry, it is a common practice to subtract the mean from the stochastic signal before computing its discrete Fourier transform.
Covariance:

$$E\{X(k)X^{*}(l)\}=\sum_{n=0}^{N-1}\sum_{n=0}^{N-1}E\{x(n)x(n')\}e^{-j\frac{2\pi k}{N}n}e^{j\frac{2\pi l}{N}n'}$$

$$=\sum_{n=0}^{N-1}\sum_{n=0}^{N-1}c_x(n-n')e^{-j\frac{2\pi k}{N}(n-n')}e^{-j\frac{2\pi(k-l)}{N}n'}$$

$$=\sum_{\tau=-(N-1)}^{N-1}c_x(\tau)e^{-j\frac{2\pi k\tau}{N}}\frac{\sin(\frac{\pi(k-l)}{N}(N-|\tau|))}{\sin(\frac{\pi(k-l)}{N})}e^{-j\frac{\pi(k-l)}{N}(N-|\tau|-1)} \tag{2.30b}$$

For $N >>$ correlation interval (an interval beyond which the covariance function is negligible),

$$\frac{\sin(\frac{\pi(k-l)}{N}(N-|\tau|))}{\sin(\frac{\pi(k-l)}{N})}\approx N \qquad \text{when} \quad k=1$$

$$\approx 0 \qquad \text{when} \quad k\neq 1$$

Hence,

$$E\{X(k)X^{*}(l)\}=N\sum_{\tau=-(N-1)}^{N-1}c_x(\tau)e^{-j\frac{2\pi k\tau}{N}}\delta_{k,l} \tag{2.30c}$$

Example 2.10: Consider a stochastic discrete signal whose covariance function is a triangular function as shown in fig. 2.20a. It has a correlation distance equal to two. The weighting function is plotted in fig. 2.20b.

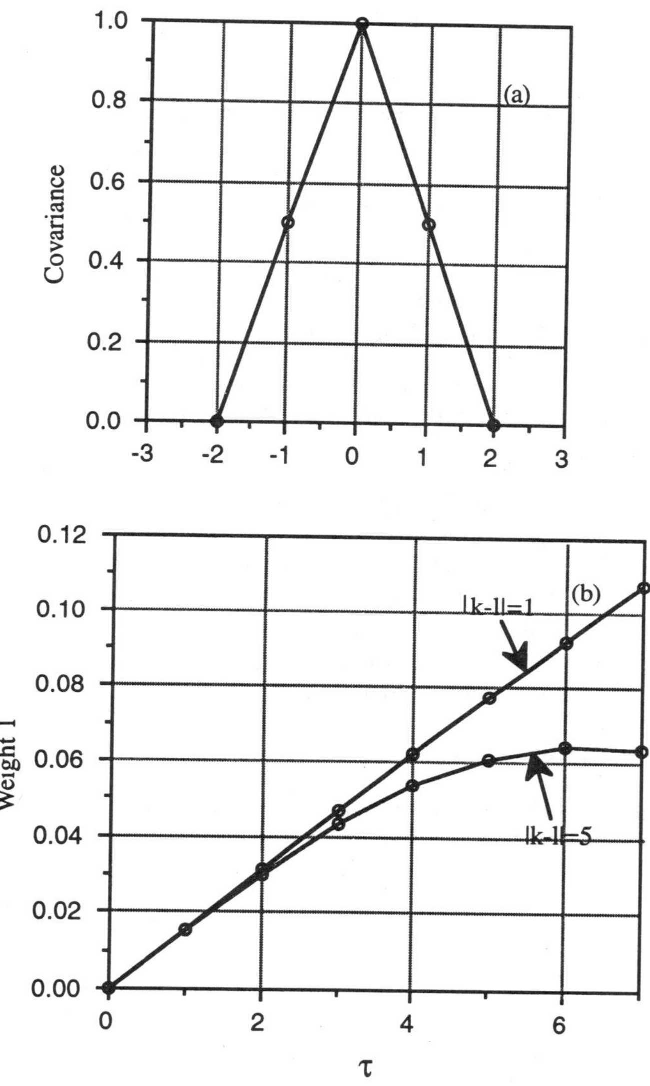

Figure 2.20: (a) Covariance function of a stochastic signal and (b) weighting function as a function of τ for $|k-l|=1$ and $|k-l|=5$. N=64 .

To understand the role of the approximation used in deriving (2.30c) let us take a closer look at $\sin(\frac{\pi(k-l)}{N}(N-|\tau|)) \Big/ (N\sin(\frac{\pi(k-l)}{N}))$, the weighting function. For small $|\tau|$ it is $O(|\tau|/N)$; thus, magnitude of the covariance function will be strongly attenuated by the weighting function when N>> correlation distance. From (2.30c) two facts emerge, namely, X(k) and X(l) are uncorrelated for large data lengths (much larger than the correlation distance) and

$$\boxed{\frac{1}{N}E\left\{|X(k)|^2\right\} = \sum_{\tau=-(N-1)}^{(N-1)} c_x(\tau)e^{-j\frac{2\pi k}{N}\tau}} \tag{2.31}$$

The right-hand side of (2.31) is by definition (see Chapter Three, eq.(3.5)) the spectrum of a stochastic signal. Thus, (2.31) provides a basis for estimation of the spectrum directly from the FDFT coefficients.

§2.4 Fast Fourier Transform

The central concept underlying the fast Fourier transform, as observed by Runge and Konig [11], is that one could use the periodicity of the sine-cosine functions to compute a N-point Fourier transform from two *N*/2-point transforms with only slightly more than N/2 operations. This is known as doubling algorithm. To compute N/2-point transforms we divide it further into two smaller sequences each of length *N*/4 and compute their Fourier transforms. By applying the above-mentioned doubling algorithm we combine them to obtain the Fourier transforms of N/2 points. This process of divide and conquer may be applied repeatedly until an elementary sequence of length one is obtained. When $N = 2^\gamma$, where γ is an integer, there will be γ stages. Applying recursively the doubling algorithm γ times we obtain the N-point Fourier transform. A brief historical background of fast Fourier transform is given in [12]. Current intense interest in FFT algorithm started after its discovery in 1965 by Cooley and Tukey [13] who also gave an algorithm for its implementation on a modern digital computer.

Doubling Algorithm: Consider a discrete signal with N samples. We pick the even and odd samples and form two half sequences,

$$y(n) = x(2n) \qquad n = 0,1,\ldots,\frac{N}{2}-1$$

$$z(n) = x(2n+1)$$

The FDFT of $x(n)$ may be expressed as follows,

$$X(k) = \sum_{n=0}^{N-1} x(n)e^{-j\frac{2\pi}{N}kn}$$

$$= \sum_{n=0}^{\frac{N}{2}-1} y(n)e^{-j\frac{4\pi}{N}kn} + \sum_{n=0}^{\frac{N}{2}-1} z(n)e^{-j\frac{(2n+1)\pi}{N}k}$$

$$= \sum_{n=0}^{\frac{N}{2}-1} y(n) e^{-j\frac{2\pi}{N/2}kn} + e^{-j\frac{2\pi}{N}k} \sum_{n=0}^{\frac{N}{2}-1} z(n) e^{-j\frac{2\pi}{N/2}kn}$$

(2.32a)

$$= Y(k) + e^{-j\frac{2\pi}{N}k} Z(k) \quad for\ 0 \le k \le \frac{N}{2} -$$

and for $k > N/2 - 1$, since FDFT is periodic (property #2 in table 2.3),

$$Y(k + \frac{N}{2}) = Y(k)$$

and

$$Z(k + \frac{N}{2}) = Z(k)$$

we obtain by substituting $k + N/2$ for k in (2.32a),

$$X(k + \frac{N}{2}) = Y(k) - e^{-j\frac{2\pi}{N}k} Z(k) \qquad (2.32b)$$

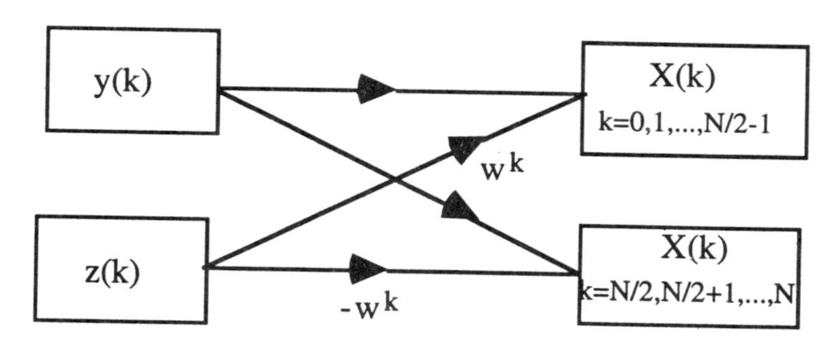

Figure 2.21: Doubling algorithm is illustrated in above signal flow diagram.

The doubling algorithm is given by eq. (2.32). We illustrate the doubling algorithm by means of a signal flow diagram in Equations (2.32a & b) are compactly represented by

$$\boxed{X(k) = Y(k) + W^k Z(k) \quad for\ 0 \le k \le N-1} \qquad (2.33)$$

where $w = e^{-j\frac{2\pi}{N}}$, known as a twiddle factor

In the signal flow graph each square block symbolically represents Fourier transformation operation. The input to each block is a discrete signal and the output of the block is its Fourier transform. The outputs of two blocks are then combined to yield the Fourier transform of the composite sequence. The arithmetic operation given by (2.33) is implemented in the signal flow graph (see fig. 2.22). An arrow starting from a node carries along with it the quantity at the node. It is multiplied by a constant appearing near the arrow mark and deposited at the next node. For example, the horizontal arrow starting from node three (fourth from the top, fig. 2.22) carries Y(3) and it is multiplied by one (not shown). The product is deposited at node three. Two quantities deposited at the same node are added together. Thus, the quantity in the sixth node of second array (the left most array is the zeroth array) is equal to $Y(2)+w^6Z(2)$. The basic computation step is the so-called Butterfly signal flow graph, which is repeated many times with different exponents on the twiddle factor. The Butterfly signal flow graph is shown in fig. 2.23.

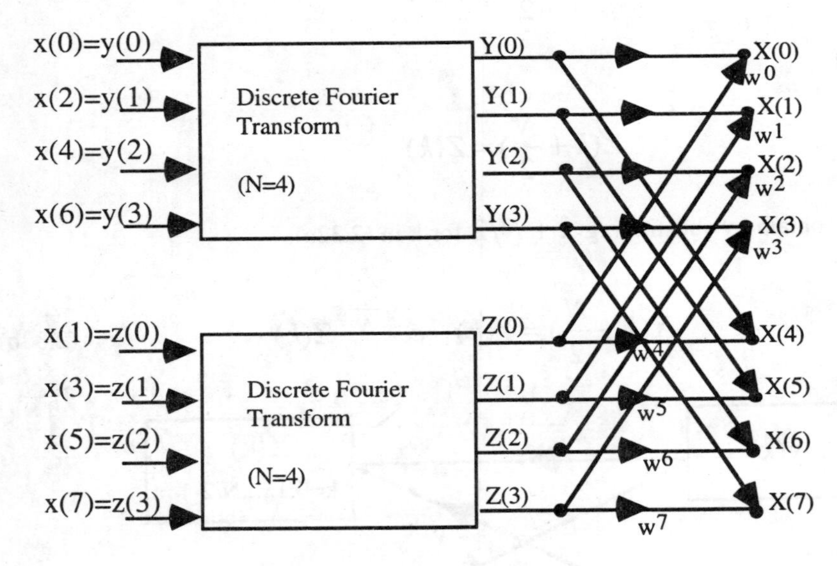

Figure 2.22: First application of the doubling algorithm. DFT's of two half sequences are combined to give the DFT of the composite sequence.

We can easily determine the number of complex arithmetic operations (one complex operation includes one complex multiplication and one complex addition) involved in executing a doubling algorithm. Fourier transformation of each half sequence requires $N^2/4$ operations. Evaluation of (2.33) requires N complex operations. Thus, the total number of operations adds up to $N(1+N/2)$. For large N this amounts to a saving of nearly 50% just by single application of the doubling algorithm.

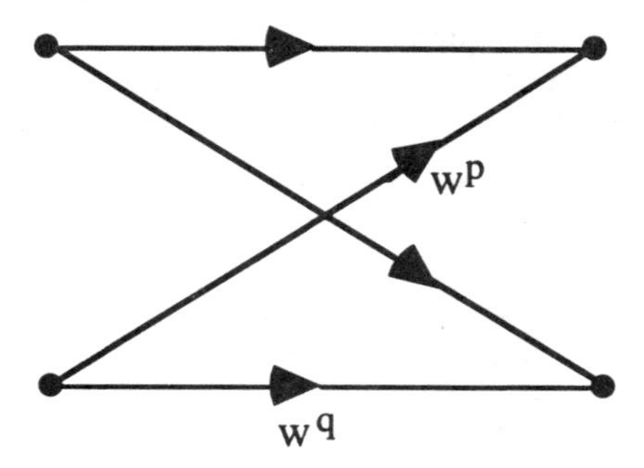

Figure 2.23: Basic computation step is illustrated in above signal flow graph, often known as butterfly. W is a twiddle factor and *p* and q are integers.

Next let us further break each half sequence into two-quarter sequences each of length two numbers. Given the Fourier transform of quarter sequences we obtain the Fourier transform of half sequence by applying the doubling algorithm once again. This is illustrated in a signal flow graph shown in fig. 2.24

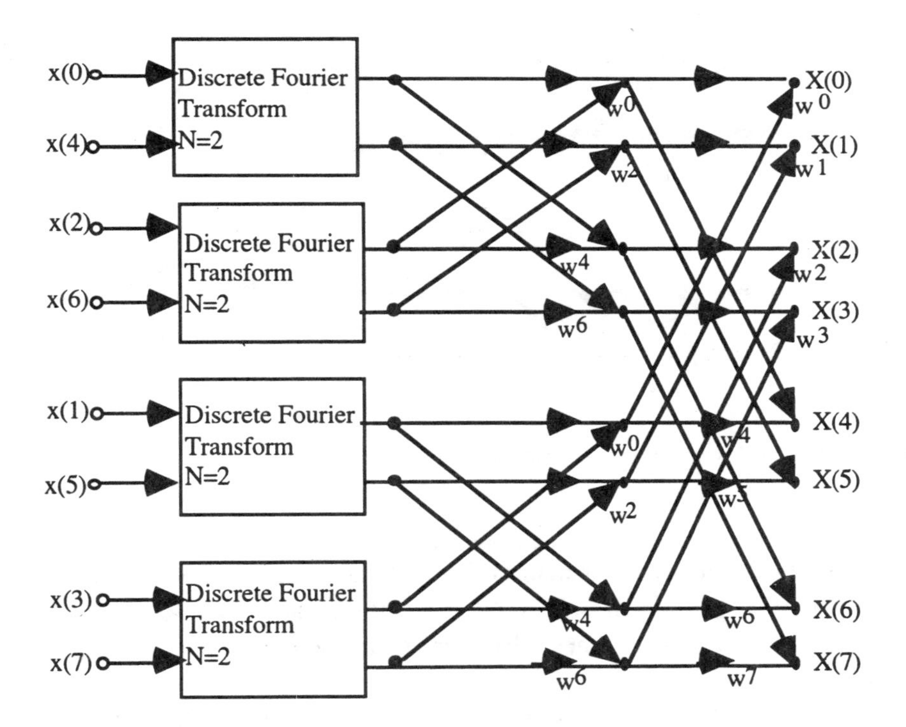

Figure 2.24: Second application of the doubling algorithm. The DFT of four quarter sequences are combined to give the DFT of one composite sequence.

The number of operations required to execute the second application of the doubling algorithm is $4N^2/16+N$. In the final stage a Fourier transform of length two is computed. When a quarter sequence is further broken we get single number sequence whose Fourier transform is itself. Applying the doubling algorithm once again we compute the Fourier transforms of all quarter sequences (see fig. 2.25).

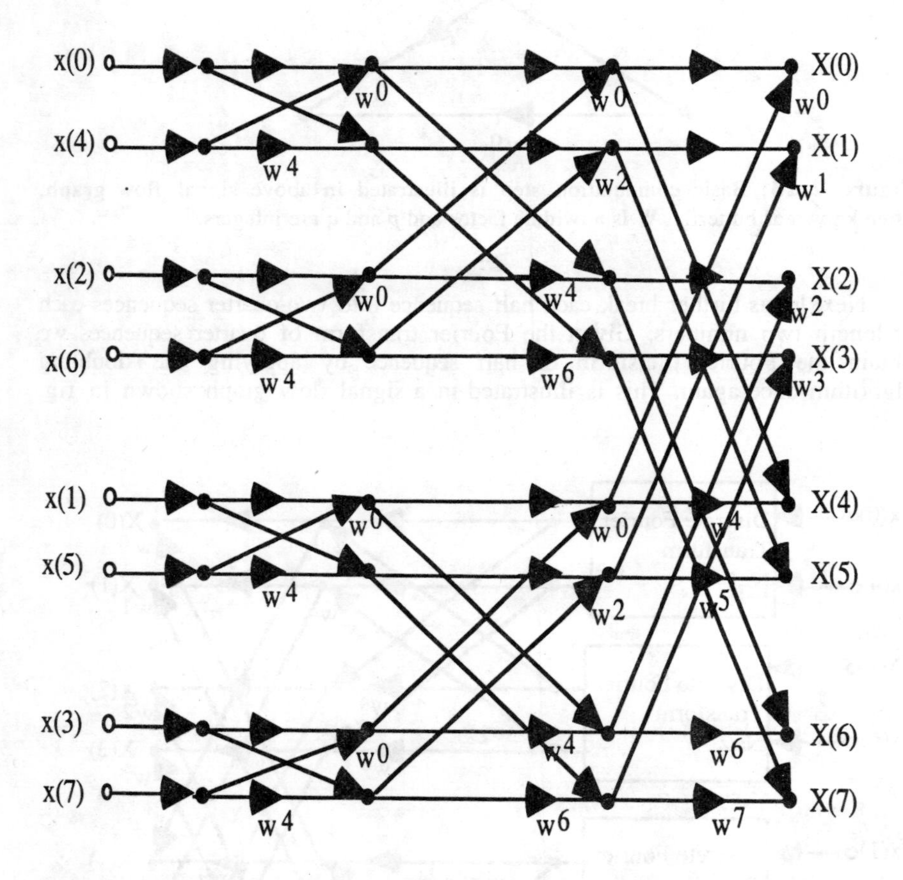

Figure 2.25: Third and final application of the doubling algorithm

The number of complex operations required for the execution of this step is N. Thus, the total number of operations from all three applications is equal to $N+N+N$. In general when $N=2^{\gamma}$ where γ is an integer we need to apply the doubling algorithm γ times. The number of operations will be equal to γN.

It is interesting to note that in course of breaking of the original sequence into eight smaller sequences the order of the sequence is destroyed. The input sequence appears to be completely shuffled. This is known as decimation in time. Fortunately, the shuffling may be described as reversing the binary bits (by interchanging zeros and ones) in the binary representation of the original index. For example, take *x(6)* the position index 6 in binary form is 110. After the bit reversal we get 011 or the new position index of X(6) is 3 which is what we find in fig.2.25. The output, however, is in the normal order. It is also possible to arrange the scheme of computation such that the input data is in the normal order but the output is in the shuffled form (bit reversed order). The doubling algorithm would then take a different form. Such a scheme is known as decimation in frequency. It is also possible to devise a scheme of computation wherein both input and output are in natural order (no need of bit reversing). But, then we do not have the facility of in-place computation. An extra storage space will be required to store the intermediate results. Some of these issues are discussed in [14].

In the algorithm described above the data sequence was broken into two halves each time until we reached a state with single data point. The discrete Fourier transform was constructed recursively using the doubling algorithm. There are obviously other possibilities. For example, the data sequence can be broken into three parts and their Fourier transforms can be combined by means of an algorithm similar to the doubling algorithm. Thus, we have a number of fast Fourier transform algorithms using different radixes, namely, 2, 3, 4 and 8. Although the higher radix algorithms are more efficient, the most popular one is the radix two algorithm as there are many more data lengths of powers of two in a given range. For example, in the range 1 to 1024 we have 10 data lengths (2, 4, 8, 16, 32, 64, 128, 256, 512, 1024) which are powers of two, but only six data lengths (3, 9, 27, 81, 243, 729) which are powers of three.

Data Length Not a Power of Two: In some practical problems the signal length may not be a power of two. However, it may be possible to break the sequence into two or more smaller sequences, each of length equal to a power of two. Take an example of data length of 48 which is not a power of two but it can be broken into three smaller sequences each of length 16, a power of two. Now we can compute the discrete Fourier transform of each sequence and combine all of them into a FDFT of the composite sequence. A sequence $x(n)$, n=0, 1, ..., N-1 is broken into P smaller sequences each of length ΔN, a power of two and $N=P*\Delta N$.

$$\begin{aligned}
X(k) &= \sum_{n=0}^{N-1} x(n) e^{-j\frac{2\pi n}{N}k} \\
&= \sum_{p=0}^{P-1} \sum_{n'=0}^{\Delta N-1} x(p*\Delta N+n') e^{-j\frac{2\pi(p*\Delta N+n')}{N}k} \\
&= \sum_{p=0}^{P-1} e^{-j\frac{2\pi p*\Delta N}{N}k} \sum_{n'=0}^{\Delta N-1} x(p*\Delta N+n') e^{-j\frac{2\pi n'}{N}k} \\
&= \sum_{p=0}^{P-1} e^{-j\frac{2\pi p(k'+q\Delta N)}{N}} X_p(k'),
\end{aligned} \tag{2.34}$$

where

$$X_p(k') = \sum_{n'=0}^{\Delta N-1} x(p*\Delta N+n') e^{-j\frac{2\pi n'}{N}k'}$$
$$k = k' + q\Delta N$$

$$k' = 0,1,\ldots,\Delta N - l \text{ and } q = 0,1,\ldots,P-1$$

There may be some situations where above approach may not be of help. For example, let the data length be, say, 75. It is short by five points from 80, for which above approach (ΔN=16 and P=5) may be used. An easy solution is to append five zeros to the given data sequence. The resulting DFT coefficients will, however, differ from those obtained by taking the actual data sequence.

§2.5 Related Transforms

There are some transforms that are closely related to the DFT yet possessing certain advantages; for example, the transformation results into a real function. We shall briefly mention two such popular transforms.

Discrete Cosine Transform: The discrete cosine transform (DCT) is defined as

$$X_c(k) = 2\sum_{n=0}^{N-1} x(n)\cos(\pi\frac{2n+1}{2N}k), \qquad 0 \le k \le N-1 \tag{2.35a}$$

The DCT coefficients $X_c(k)$, unlike the FDFT coefficients, are real. The inverse DCT is defined as

$$x(n) = \frac{1}{N}\sum_{k=0}^{N-1} \alpha(k)X_c(k)\cos(\pi\frac{2n+1}{2N}k), \qquad 0 \le n \le N-1 \tag{2.35b}$$

where

$$\alpha(k) = \frac{1}{2} \quad for\ k = 0$$
$$= 1 \quad for \quad 1 \le k \le N-1$$

We shall now relate DCT to FDFT. For this let us rewrite (2.35a) as follows:

$$\begin{aligned} X_c(k) &= \sum_{n=0}^{N-1} x(n)\left[e^{j\pi\frac{2n+1}{2N}k} + e^{-j\pi\frac{2n+1}{2N}k}\right] \\ &= e^{j\frac{2\pi\frac{k}{2}}{2N}}\sum_{n=0}^{N-1} x(n)e^{j2\pi\frac{n}{2N}k} + e^{-j\frac{2\pi\frac{k}{2}}{2N}}\sum_{n=0}^{N-1} x(n)e^{-j2\pi\frac{n}{2N}k} \\ &= e^{j\frac{2\pi\frac{k}{2}}{2N}}Y(-k) + e^{-j\frac{2\pi\frac{k}{2}}{2N}}Y(k) \end{aligned} \tag{2.36}$$

where Y(k) is FDFT of

$$y(n) = x(n), \quad 0 \le n \le N-1$$
$$= 0 \qquad N \le n \le 2N-1$$

From the properties of the FDFT coefficients

$$Y(-k) = FDFT\{y(2N-1-n), 0 \le n \le 2N-1\}$$

Using this property in (2.36) we obtain a relation between DCT coefficients and FDFT of the (real) signal.

$$X_c(k) = e^{-j\frac{2\pi\frac{k}{2}}{2N}} FDFT\{y(n)\} + e^{j\frac{2\pi\frac{k}{2}}{2N}} FDFT\{y(2N-1-n)\} \qquad (2.37)$$

Discrete Hartley Transform: Another transform that yields real coefficients is the discrete Hartley transform (DHT) as defined below:

$$X_{Hart}(k) = \sum_{n=0}^{N-1} x(n)\left(\cos(\frac{2\pi n}{N}k) + \sin(\frac{2\pi n}{N}k)\right), \quad 0 \le k \le N-1 \qquad (2.38a)$$

and the inverse DHT is given by

$$x(n) = \frac{1}{N}\sum_{k=0}^{N-1} X_{Hart}(k)\left(\cos(\frac{2\pi n}{N}k) + \sin(\frac{2\pi n}{N}k)\right), \quad 0 \le n \le N-1 \qquad (2.38b)$$

Like DCT the DHT is also real. It has a close resemblance to FDFT. Naturally, we like to explore this possibility and try to relate DHT to FDFT. First, let us express

$$\cos(\frac{2\pi n}{N}k) = \frac{1}{2}(e^{j\frac{2\pi n}{N}k} + e^{-j\frac{2\pi n}{N}k})$$

and

$$\sin(\frac{2\pi n}{N}k) = \frac{1}{2j}(e^{j\frac{2\pi n}{N}k} - e^{-j\frac{2\pi n}{N}k}).$$

On simplifying (2.38a) we get

$$\begin{aligned} X_{Hart}(k) &= \sum_{n=0}^{N-1} x(n)\left(\cos(\frac{2\pi n}{N}k) + \sin(\frac{2\pi n}{N}k)\right), \quad 0 \le k \le N-1 \\ &= \frac{1}{2}[X^*(k) + X(k)] + \frac{1}{2j}[X^*(k) - X(k)] \\ &= \frac{1}{2}(1+j)X(k) + \frac{1}{2}(1-j)X^*(k) \\ &= \frac{X(k)+X^*(k)}{2} + j\frac{X(k)-X^*(k)}{2} \end{aligned} \qquad (2.39a)$$

Note that $(X(k)+X^*(k))/2 = \mathrm{Re}\{X(k)\}$ and $j(X(k)-X^*(k))/2 = \mathrm{Im}\{X(k)\}$. Hence, (2.39a) can be expressed as

$$X_{Hart}(k) = \mathrm{Re}\{X(k)\} - \mathrm{Im}\{X(k)\} \tag{2.39b}$$

Given the Hartley transform it is possible to compute the real and imaginary parts of $X(k)$ from (2.39b) and remembering that $\mathrm{Re}\{X(k)\}$ is even and $\mathrm{Im}\{X(k)\}$ is odd. In fact $X_{Hart}(k)$ can be split into even and odd parts,

$$X_{Hart}(k) = X_{Hart}^{e}(k) + X_{Hart}^{o}(k) \tag{2.39c}$$

Comparing (2.39c) with (2.39b) we obtain

$$X_{Hart}^{e}(k) = \mathrm{Re}\{X(k)\} = \sum_{n=0}^{N-1} x(n)\cos(\frac{2\pi nk}{N})$$

$$X_{Hart}^{o}(k) = -\mathrm{Im}\{X(k)\} = -\sum_{n=0}^{N-1} x(n)\sin(\frac{2\pi nk}{N}) \tag{2.39d}$$

The convolution theorem for Hartley transform takes a slightly different form. Let us evaluate the Hartley transform of the convolution product.

$$\begin{aligned}
Y_{Hart}(k) &= \sum_{\tau=0}^{N-1} y(\tau)\left(\cos(\frac{2\pi\tau}{N}k) + \sin(\frac{2\pi\tau}{N}k) \right) \\
&= \sum_{n=0}^{N-1} h(n) \sum_{\tau=0}^{N-1} x(\tau-n)\left(\cos(\frac{2\pi\tau}{N}k) + \sin(\frac{2\pi\tau}{N}k) \right) \\
&= \sum_{n=0}^{N-1} h(n) \sum_{\tau'=-n}^{N-1-n} x(\tau')\left(\cos(\frac{2\pi(\tau'+n)}{N}k) + \sin(\frac{2\pi(\tau'+n)}{N}k) \right) \\
&= \sum_{n=0}^{N-1} h(n)\left[X_{Hart}(k)\cos\frac{2\pi n}{N}k + X_{Hart}(-k)\sin\frac{2\pi n}{N}k \right] \\
&= X_{Hart}(k)H_{Hart}^{e}(k) + X_{Hart}(-k)H_{Hart}^{o}(k)
\end{aligned} \tag{2.40}$$

It may be observed that the Hartely transform of the convolution product requires just two real multiplication, where as the FDFT of the convolution product (2.47) requires one complex multiplication which is equal to four real multiplications. Hence, the DHT has computational advantage over FDFT.

§2.6 Linear Systems and Convolution

The input and output of a linear system such as an electronic device are related through a convolution integral where the system itself is characterized by an impulse response function. A digital filter is a linear system and likewise it is characterized by an impulse response function. The output of a digital filter is given by convolution sum of the impulse response function and the input signal. Because of this close relationship between the digital filters and the linear systems, a brief account of linear system theory is helpful in the understanding of digital filters.

Linear System Theory: The output of a system, when excited by an impulse, is known as its impulse response function. Pictorially, this is shown in fig. 2.26. A system is excited by an impulse at time t=0. The system output, which was zero until the impulse was applied, rapidly goes high, thereafter it decays to zero after sufficient time. The impulse response function, h(t), is one-sided function, that is, h(t)=0 for t<0 and it must go to zero as $t \longrightarrow \infty$. The impulse response function is said to be causal.

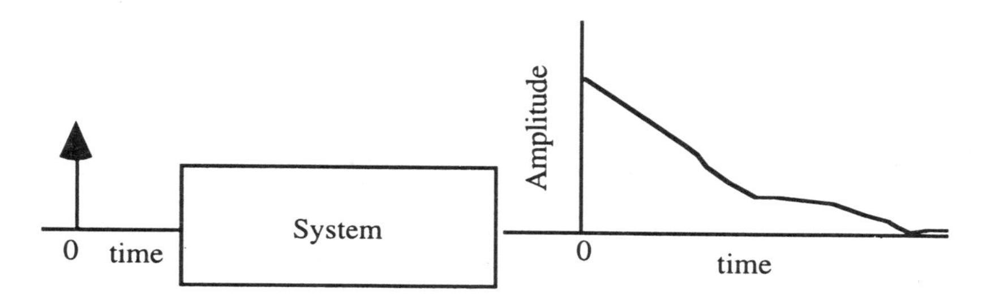

Figure 2.26: An impulse input to a system produces an output known as impulse response function.

Instead of an impulse input let us input an arbitrary function, $x(t)$. The system output at a given time instant is a sum total of the system response due to the past input, which may be treated as a train of closely spaced impulses. In the limiting case the sum may be replaced by an integral,

$$y(t) = \int_{0}^{\infty} h(\tau)x(t-\tau)d\tau \tag{2.41}$$

A system is said to be *linear* when the output of the system to an input, which consists of a sum of two or more functions, is equal to the sum of the individual outputs. Let $z_0(t), z_1(t), \ldots, z_{P-1}(t)$ be the outputs of a system to inputs $x_0(t), x_1(t), \ldots, x_{P-1}(t)$, respectively; then the output of the system to an input $a_0x_0(t) + a_1x_1(t) + \ldots + a_{P-1}x_{P-1}(t)$ is $a_0z_0(t) + a_1z_1(t) + \ldots + a_{P-1}z_{P-1}(t)$. A system is said to be *time invariant* when for a delay in the input there is a corresponding delay in the output. Let $x(t-\tau_0)$ be the input to a time invariant system the output is $z(t-\tau_0)$. The most commonly referred systems are both linear and time invariant (LTI). A consequence of linearity and time invariance is the convolution integral (2.41) connecting the input to output of a linear system. We shall in this text confine ourselves to LTI systems.

Next, we shall study the transfer function of a system. For this let us evaluate the Laplace transform on both sides of (2.41).

$$\int_{-\infty}^{\infty} y(t)e^{-st}dt = \int_{0}^{\infty}\int_{-\infty}^{\infty} h(\tau)x(t-\tau)d\tau e^{-st}dt$$

$$= \int_{0}^{\infty} h(\tau)e^{-s\tau}d\tau \int_{-\infty}^{\infty} x(t-\tau)e^{-s(t-\tau)}dt$$

$$= \int_{0}^{\infty} h(\tau)e^{-s\tau}d\tau \int_{-\infty}^{\infty} x(t-\tau)e^{-s(t-\tau)}dt$$

$$Y(s) = H(s)X(s) \tag{2.42}$$

$H(s)$ is the Laplace transform of the impulse response function. It is known as a transfer function. Since $h(t)=0$, $t<0$ its Laplace transform will have all its poles in the left half of the s-plane (see Chapter One). The Fourier transform of the impulse response function, that is, $H(s)|_{s=j\omega}$ possesses an interesting property. Let

$$H(\omega) = P(\omega) + jQ(\omega).$$

It may be shown that, on account of the causal nature of the impulse response function, the real and imaginary parts are related to each other through Hilbert transform [15].

$$P(\omega) = -\frac{1}{\pi}\int_{-\infty}^{\infty} \frac{Q(\omega')}{\omega'-\omega}d\omega'$$

$$Q(\omega) = \frac{1}{\pi}\int_{-\infty}^{\infty} \frac{P(\omega')}{\omega'-\omega}d\omega' \tag{2.43a}$$

Let us express the transfer function in terms of the amplitude and phase,

$$H(\omega) = A(\omega)e^{j\phi(\omega)}$$

where

$$A(\omega) = \sqrt{P^2(\omega) + Q^2(\omega)}$$

and

$$\phi(\omega) = \tan^{-1}\frac{Q(\omega)}{P(\omega)}.$$

Further, take a logarithm of the transfer function,

$$\ln H(\omega) = \ln A(\omega) + j\phi(\omega).$$

The real and imaginary parts of $\ln H(\omega)$ will be related to each other as $P(\omega)$ and $Q(\omega)$ are [15] when all its poles lie in the left half of the s-plane.

$$\ln A(\omega) = -\frac{1}{\pi}\int_{-\infty}^{\infty} \frac{\phi(\omega')}{\omega' - \omega} d\omega'$$

$$\phi(\omega) = \frac{1}{\pi}\int_{-\infty}^{\infty} \frac{\ln A(\omega')}{\omega' - \omega} d\omega' \tag{2.43b}$$

Since a zero of $H(\omega)$ becomes a pole of $\ln H(\omega)$ we require that not only all poles but also all zeros of $H(\omega)$ must also lie in the left half of s-plane. Such a system is said to be minimum phase system. The amplitude and phase response functions of a causal transfer function are NOT independent; given any one the other is completely determined. The necessary and sufficient condition for the amplitude characteristics $A(\omega)$ to be physically realizable, that is, the resulting impulse response function is causal, is that the following integral must exists (known as Paley-Wiener condition [15, p. 51]):

$$\int_{-\infty}^{\infty} \frac{|\ln A(\omega)|}{1+\omega^2} d\omega < \infty$$

The Paley-Wiener condition will not be satisfied whenever $A(\omega)$ vanishes over a range of frequencies.

Discrete-Time LTI Systems: Analogous to continuous time LTI system we have discrete-time LTI system, either as a sampled version of a continuous time LTI system or intrinsically a digital system, for example a digital filter. We shall view a discrete-time system as of latter type and thus avoid the implications of sampling a continuous time impulse response function. A discrete-time system impulse response function is denoted by h(n), n=0,1,... and the input output relation is given by a discrete convolution sum,

$$y(m) = \sum_{n=0}^{\infty} h(n)x(m-n) \tag{2.44a}$$

The sampling interval, Δt, in the event the system is analog, is assumed to be unit in a suitable time units depending upon the system bandwidth. A discrete-time LTI system shares many of the properties with the corresponding analog LTI system. The essential differences arise in the frequency domain, on account of its discrete nature. We shall emphasize this aspect. Consider the z-transform of $y(n)$,

$$\begin{aligned} Y(z) &= \sum_{m=-\infty}^{\infty} y(m) z^{-m} \\ &= \sum_{m=-\infty}^{\infty} \sum_{n=0}^{\infty} h(n) x(m-n) z^{-m} \\ &= \sum_{m=-\infty}^{\infty} \sum_{n=0}^{\infty} h(n) z^{-n} x(m-n) z^{-(m-n)} \\ &= H(z) X(z) \end{aligned} \tag{2.44b}$$

where

$$H(z) = \sum_{n=0}^{\infty} h(n) z^{-n}$$

is the z-transform of the impulse response function. It is known as transfer function of the system. Since the system is assumed to be causal, that is, $h(n)=0$ for $n<0$ and stable, the poles of $H(z)$ must lie inside the unit circle. The region of convergence lies in the entire z-plane outside the unit circle. Further, the system is said to be minimum phase when the zeros of $H(z)$ also lie inside the unit circle. Additionally, for a real system (where both input and output are real) the poles and zeros of the transfer function must be located in complex conjugate position within the unit circle as shown in fig. 2.27. For a real system $H(z^{-1})$ is a transpose of $H(z)$. The poles and zeros of the transpose are at reciprocal location, that is, if $r_i e^{j\theta_i}$ is a pole (or a zero) of $H(z)$, then $r_i^{-1} e^{-j\theta_i}$ is a pole (or a zero) of $H(z^{-1})$. If $H(z)$ is a minimum phase system, $H(z^{-1})$ becomes a maximum phase system. Further,

$$H(z)H(z^{-1})\Big|_{z=e^{j\omega}} = |H(\omega)|^2.$$

Interestingly, $H(z)H(z^{-1})$ remains unchanged even after we replace some of its poles (or zeros) by their reciprocal.

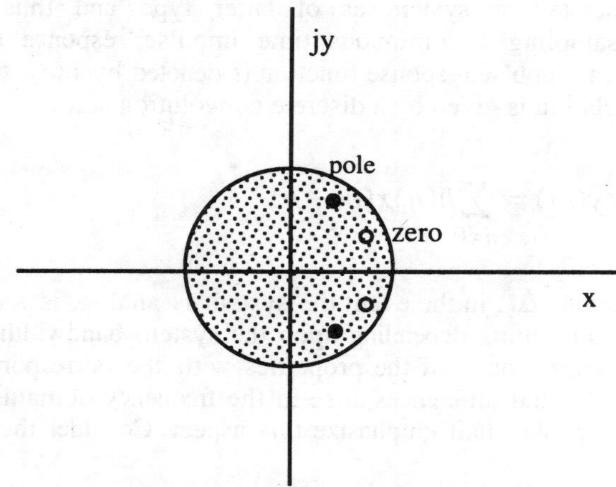

Figure 2.27: The location of poles and zeros of a real causal minimum phase LTI system.

Example 2.11: Let the poles be at

$$z=0.8\, e^{\pm j\frac{\pi}{4}}$$

and the zeros be at

$$z=0.88\, e^{\pm j\frac{\pi}{6}}$$

The transfer function of such a system is given by

$$H(z)=\frac{(z-0.66e^{+j\frac{\pi}{6}})(z-0.66e^{-j\frac{\pi}{6}})}{(z-0.6e^{+j\frac{\pi}{4}})(z-0.6e^{-j\frac{\pi}{4}})}$$

Note that $H(z)$ is a minimum phase system as all of its poles and zeros are inside the unit circle. The impulse response function be obtained by expressing above transfer function as a polynomial in inverse powers of z. The transfer function and impulse response function (first ten terms only) are listed below:

$$H(z)\approx\begin{Bmatrix}1.0-.294z^{-1}-.1744z^{-2}-.0419z^{-3}+.0272z^{-4}+.0382z^{-5}\\+.0226z^{-6}+.0054z^{7}-.0035z^{-8}-.0192z^{-9}\end{Bmatrix}$$

$$h(n)=\begin{Bmatrix}1.0,-.2946,-.1744,-.0419,.0272,.0382,.0226\\.0054,-.0035,-.0192,...\end{Bmatrix}\quad n\geq 0$$

$$=0 \qquad\qquad \text{otherwise}$$

Let us next compute $H(z^{-1})$, the transpose of $H(z)$,

$$H(z^{-1})=\frac{(z^{-1}-0.66e^{+j\frac{\pi}{6}})(z^{-1}-0.66e^{-j\frac{\pi}{6}})}{(z^{-1}-0.6e^{+j\frac{\pi}{4}})(z^{-1}-0.6e^{-j\frac{\pi}{4}})}$$

$$H(z^{-1})=1.0-.2946z^{1}-.1744z^{2}-.0419z^{3}+.0272z^{4}+.038z^{5}$$
$$+.0226z^{6}+.0054z^{7}-.0035z^{8}-.0192z^{9}$$

The impulse response of $H(z^{-1})$ is given by

$$h(n) = \begin{Bmatrix} 1.0, -.2946, -.1744, -.0419, .0272, .0382, \\ .0226, .0054, -.0035, -0192, ... \end{Bmatrix} \quad \text{for } n \leq 0$$

$$= 0 \qquad\qquad \text{for } n > 0$$

$H(z^{-1})$ is now a maximum phase linear system as all its poles and zeros are outside the unit circle. Its impulse response function is non zero for negative time. ♠.

Group Delay: The phase response $\theta(\omega)$ of a transfer function is important in the understanding of the signal shape distortion as the signal passes through the system. Each frequency component of the signal is delayed by an amount equal to $-d\theta(\omega)/d\omega$, known as group delay which from the point of realizability must be positive. Let

$$\theta(\omega) = -\omega\tau$$

then the group delay will be equal to τ for all frequencies. This would mean that the output will be delayed by τ. However, the signal shape remains unaltered. When there is a non-linear phase variation, the group delay will be frequency dependent causing different delays to different frequency components. This will result in severe signal shape distortion.

Linear Convolution: The input and output of a LTI system are related through a convolution sum (2.44a). In practical situation where we have finite data the infinite limit on the sum in (2.44a) has to be replaced by a finite number

$$y(n) = \sum_{l=0}^{L-1} h(l)x(n-l) \tag{2.45}$$

where L stands for the finite length of the impulse response function. There are two approaches for the evaluation of finite convolution (2.45). In the first approach we express (2.45) as a matrix product and then use the power of the Matlab for computation of matrix expression. The expression (2.45) can be written as a series of expressions, one for each output point,

$$y(0) = h(0)x(0)$$

...

$$y(L-2) = h(0)x(L-2) + h(1)x(L-3) + ... + h(L-2)x(0)$$

$$y(L-1) = x(L-1)h(0) + x(L-2)h(1) + ... + x(0)h(L-1)$$

$$y(L) \quad = x(L)h(0) + x(L-1)h(1) + ... + x(1)h(L-1)$$

...

$$y(N-1) = x(N-1)h(0) + x(N-2)h(1) + ... + x(N-L+1)h(L-1)$$

$$y(N) = h(1)x(N-1) + h(2)x(N-2) + \ldots + h(L-1)x(N-L+1)$$

$$\ldots$$

$$y(N+L-2) = h(L-1)x(N-1)$$

or in a compact matrix form,

$$\begin{bmatrix} y(0) \\ \ldots \\ y(L-2) \\ ----- \\ y(L-1) \\ y(L) \\ y(L+1) \\ \cdot \\ \cdot \\ \cdot \\ y(N-1) \\ ----- \\ y(N) \\ \ldots \\ y(N+L-2) \end{bmatrix} = \begin{bmatrix} x(0) & 0 & 0 & \ldots & 0 \\ & & & \ldots & \\ x(L-2) & x(L-3) & \ldots & x(0) & 0 \\ ----- & ----- & ----- & ----- & ----- \\ x(L-1) & x(L-2) & \ldots & & x(0) \\ x(L) & x(L-1) & \ldots & & x(1) \\ & & \ldots & & \\ & & \ldots & & \\ & & \ldots & & \\ x(N-1) & x(N-2) & \ldots & x(N-L) & \\ ----- & ----- & ----- & ----- & ----- \\ 0 & x(N-1) & \ldots & x(N-L+1) & \\ & & \ldots & & \\ 0 & 0 & 0 & x(N-1) & \end{bmatrix} \begin{bmatrix} h(0) \\ h(1) \\ h(2) \\ \cdot \\ \cdot \\ \cdot \\ h(L-1) \end{bmatrix}$$

$$(N+L-1)\times 1 \qquad\qquad (N+L-1)\times L \qquad\qquad L\times 1 \tag{2.46a}$$

$$\mathbf{y} = \mathbf{xh} \tag{2.46b}$$

We have assumed that the signal is zero outside the duration in which it is defined. The input signal is arranged as a rectangular matrix $\mathbf{x}$ and that the impulse response function as a column vector $\mathbf{h}$. The system output is simply the product of $\mathbf{x}$ and $\mathbf{h}$. There are N+L-1 output points. Note that, out of N+L - 1 output points, the beginning L-1 and ending L-1 outputs are incomplete as the entire impulse response function could not be used in the computation of these points. The reason being that the signal was assumed to be zero outside the duration. Only the central N-L+1 output points, as marked by the dashed lines in (2.46a), are complete and hence useful. The incomplete 2(L-1) output points represent the so-called edge effect.

Polynomial Product: For finite duration sequences, a linear convolution may be implemented through multiplication of the z-transform of the impulse response and the z-transform of the discrete signal, both of which are expressed as polynomials in inverse power of z. A polynomial product is obtained by straight-forward multiplication or through the use of Matlab m-file, conv. The inverse z-transform of the resulting polynomial is simply the coefficients of the product polynomial.

Example 2.12: Consider a signal $x(n)$=1.0, n=0,1,...7 and zero outside the duration. Let $h(n)$=1.0, n=0,1, 2. Eq. (2.46a) takes the form

$$\begin{bmatrix} 1 \\ 2 \\ 3 \\ 3 \\ 3 \\ 3 \\ 3 \\ 3 \\ 2 \\ 1 \end{bmatrix} = \begin{bmatrix} 1 & 0 & 0 \\ 1 & 1 & 0 \\ 1 & 1 & 1 \\ 1 & 1 & 1 \\ 1 & 1 & 1 \\ 1 & 1 & 1 \\ 1 & 1 & 1 \\ 1 & 1 & 1 \\ 1 & 1 & 0 \\ 1 & 0 & 0 \end{bmatrix} \begin{bmatrix} 1 \\ 1 \\ 1 \end{bmatrix}$$

The result of linear convolution is shown in fig. 2. 28. Observe the edge effect caused by finite data. Only the central part is the correct convolution output.

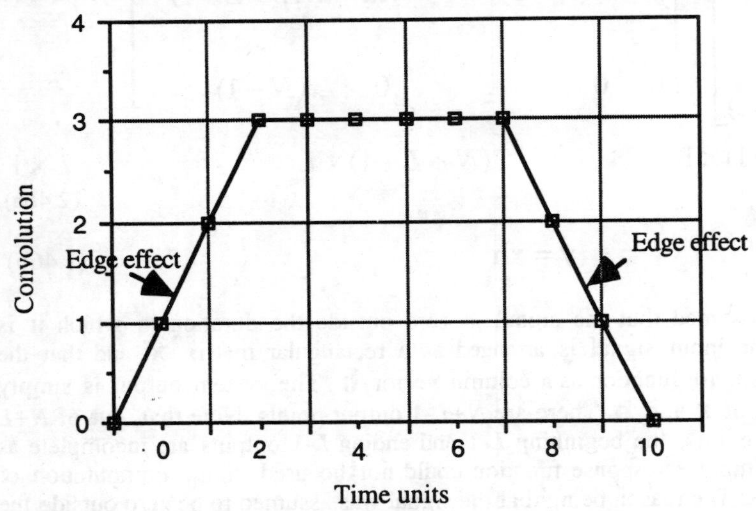

Figure 2.28: Convolution output as a function of time index ♠.

Let

$$H(z) = h(0) + h(1)z^{-1} + h(2)z^{-2} + ...$$
$$X(z) = x(0) + x(1)z^{-1} + x(2)z^{-2} + ...$$

The product of z-transforms,

$$
\begin{aligned}
Y(z) &= X(z)H(z) \\
&= h(0)x(0) + (h(0)x(1) + x(0)h(1))z^{-1} \\
&\quad + (h(0)x(2) + x(0)h(2) + h(1)x(1))z^{-2} + \ldots
\end{aligned}
$$

and the inverse z-transform of the product polynomial is

$$
y = \begin{bmatrix} h(0)x(0), & (h(0)x(1) + x(0)h(1)), & \\ (h(0)x(2) + x(0)h(2) + h(1)x(1)), & & \ldots \end{bmatrix}
$$

Circular Convolution: In the second approach the convolution is carried out via FDFT. For this we compute the FDFT on both sides of (2.45). We obtain

$$
\begin{aligned}
Y(k) &= \sum_{n=0}^{N-1} h(n) e^{-j\frac{2\pi}{N}nk} \sum_{l=-n}^{N-1-n} x(l) e^{-j\frac{2\pi}{N}lk} \\
&= H(k)X(k)
\end{aligned}
\tag{2.47}
$$

Notice that the length of the impulse response function has been increased to N, that is, equals to the length of the input signal. This is easily achieved by appending required number of zeros to the impulse response function. We get the convolved output by computing the inverse FDFT of Y(k). The inverse FDFT of Y(k) is

$$
\begin{aligned}
y(n) &= \sum_{k=0}^{N-1} Y(k) e^{j\frac{2\pi nk}{N}} = \sum_{k=0}^{N-1} H(k)X(k) e^{j\frac{2\pi nk}{N}} \\
&= \sum_{k=0}^{N-1} \left[\sum_{m=0}^{N-1} h(m) e^{-j\frac{2\pi mk}{N}} \sum_{l=0}^{N-1} x(l) e^{-j\frac{2\pi lk}{N}} \right] e^{j\frac{2\pi nk}{N}} \\
&= \left[\sum_{m=0}^{N-1} h(m) \sum_{l=0}^{N-1} x(l) \right] \sum_{k=0}^{N-1} e^{j(n-m-l)\frac{2\pi k}{N}}
\end{aligned}
\tag{2.48a}
$$

Note that

$$
\begin{aligned}
\sum_{k=0}^{N-1} e^{j\frac{2\pi}{N}(n-m-l)k} &= N \qquad (n-m-l) = pN, \\
&= 0 \qquad \textit{otherwise}
\end{aligned}
\tag{2.48b}
$$

where p is an integer. From (2.48b) it follows that

$$
l = n - m - pN = (n-m)_{\text{mod } N}.
$$

Eq (2.48a) reduces to

$$y(n) = \sum_{m=0}^{N-1} h(m)x(n-m)_{\text{mod } N} \qquad (2.48c)$$

The modulo N operation for N=8 on (n-m) for $0 \le m \le 15$ and $0 \le n \le 3$ is shown in table 2.4. Observe that the modulo operation, $(n-m)_{\text{mod } N}$, results into mapping of a number sequence onto a circle with periodicity equal to the base. A shift of the sequence by m steps results into m circular shifts. Both results are illustrated in fig. 2.29.

Table 2.4: The modulo operation on (*n-m*) for $0 \le m \le 15$ and $0 \le n \le 3$ with base *N*=8.

m →	0	1	2	3	4	5	6	7	8	9	10	11	12	13	14	15
$(0-m)_{\text{mod } N}$	0	7	6	5	4	3	2	1	0	7	6	5	4	3	2	1
$(1-m)_{\text{mod } N}$	1	0	7	6	5	4	3	2	1	0	7	6	5	4	3	2
$(2-m)_{\text{mod } N}$	2	1	0	7	6	5	4	3	2	1	0	7	6	5	4	3
$(3-m)_{\text{mod } N}$	3	2	1	0	7	6	5	4	3	2	1	0	7	6	5	4

We will now evaluate circular convolution, which involves modulo operation on the index of the signal. Consider the first element of the convolved output (n=0),

$$\begin{aligned} y(0) &= h(0)x(0) + h(1)x(-1) + h(2)x(-2) + \ldots + h(L-1)x(-N+1) \\ &= h(0)x(0) + h(1)x(N-1) + h(2)x(N-2) + \ldots + h(L-1)x(1) \end{aligned} \qquad (2.49)$$

Because of the modulo operation we have *x*(-1)=*x*(N-1), *x*(-2)=*x*(*N*-2) and so on. The circular convolution is best visualized by representing the impulse response sequence and the signal on two circles, impulse response on the first circle in anti-clockwise order and the signal on the second circle in clockwise order. The arrangement is shown in fig. 2.30. The first circle is held fixed and the second circle is turned anti-clockwise in discrete steps equal to the number n. The corresponding numbers are multiplied and then summed to give us the circular convolution. To eliminate the error terms (e.g. $h(1)x(N-1) + h(2)x(N-2) + \ldots$ in eq.(2.49)) from the circular convolution, the signal length is extended by appending *L*-1 zeros to both *x* and also to *h*, making each of them of length *N*+ *L*-1 (see fig. 2.32). The error terms are now set to zero and the edge effect as described in (2.46) will be correctly reproduced.

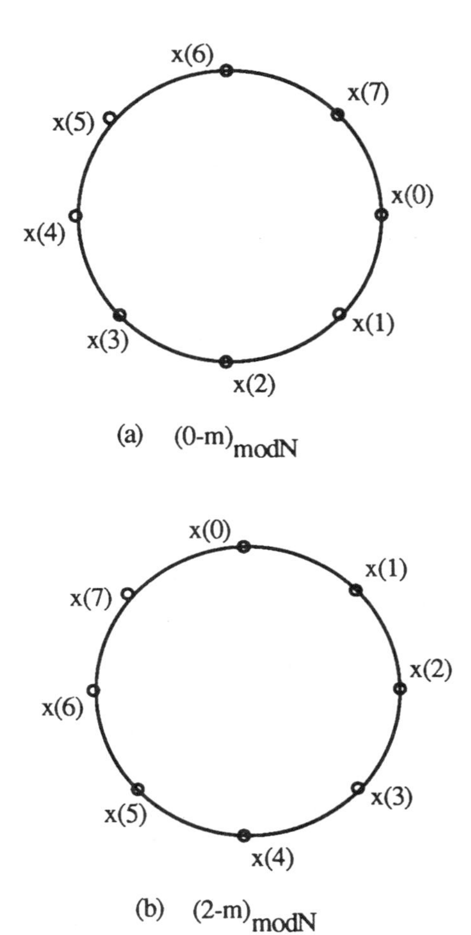

Figure 2.29: A modulo operation results into mapping of a sequence onto a circle.(a) without any shift (2nd row in table 2.4) and (b) with two unit linear shift (4th row).

Example 2.13: As an example, we illustrate circular convolution between an eight point impulse response sequence and a sixteen point discrete signal. Let n=3. Turn the second circle in fig.2.30(b) by three steps. The result is shown in fig. 2.31b. The sum of the products of the corresponding numbers on two circles is

$$y(3) = \begin{Bmatrix} h(0)x(3) + h(1)x(2) + h(2)x(1) + h(3)x(0) \\ h(4)x(15) + h(5)x(14) + h(6)x(13) + h(7)x(12) \end{Bmatrix}$$

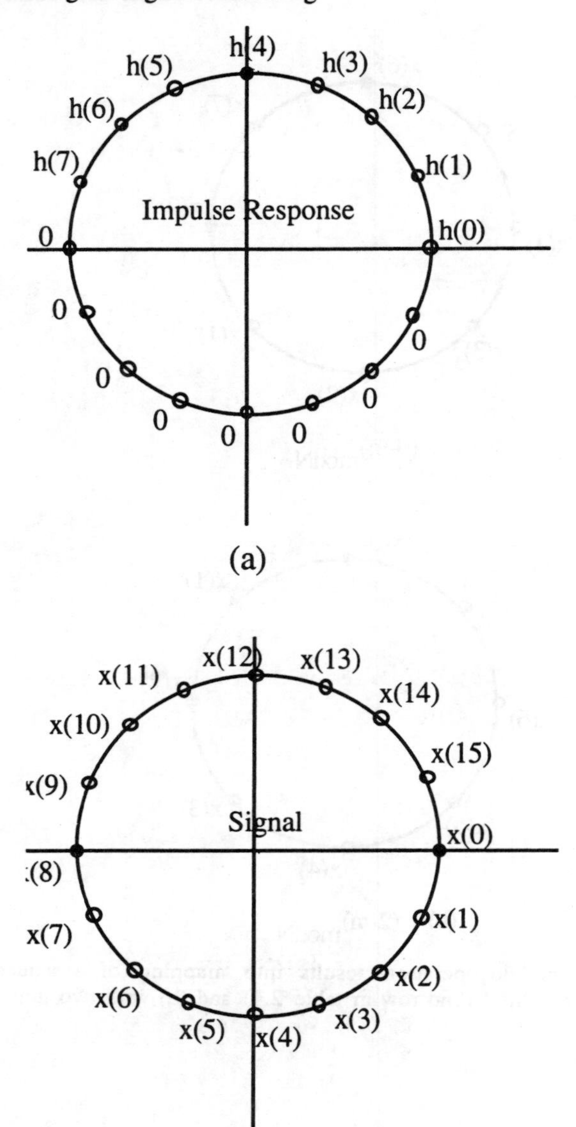

Figure 2.30: (a) Representation of impulse response sequence (8 points) in the anti-clockwise order and (b) representation of signal sequence (16 points) in the clockwise order.

The last four terms in the sum are error terms caused by the periodicity imposed by FDFT. Similarly, the convolution values at n=0, 1, and 2 are easily verified by redrawing the second circle in fig. 2.31 with different rotations

$$y(0) = \begin{Bmatrix} h(0)x(0) + h(1)x(15) + h(2)x(14) + h(3)x(13) + h(4)x(12) \\ +h(5)x(11) + h(6)x(10) + h(7)x(9) \end{Bmatrix}$$

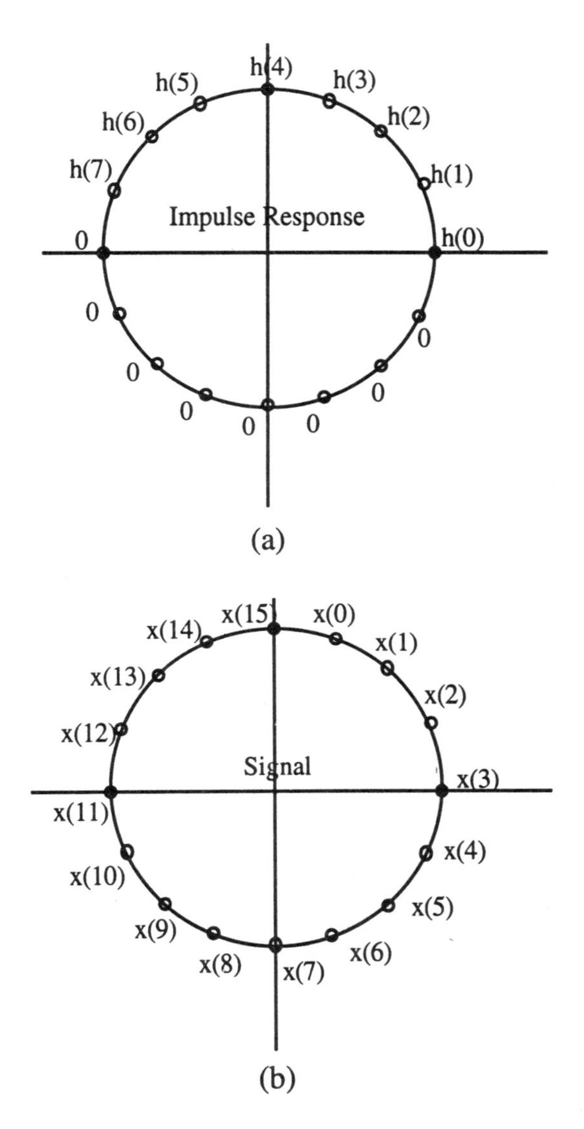

Figure 2.31: Representation of the signal for computation of convolution at n=3.

$$y(1) = \begin{Bmatrix} h(0)x(1) + h(1)x(0) + h(2)x(15) + h(3)x(14) + h(4)x(13) \\ h(5)x(12) + h(6)x(11) + h(7)x(10) \end{Bmatrix}$$

$$y(2)=\begin{Bmatrix} h(0)x(2)+h(1)x(1)+h(2)x(3)+h(3)x(15)+h(4)x(14) \\ h(5)x(13)+h(6)x(12)+h(7)x(11) \end{Bmatrix}$$

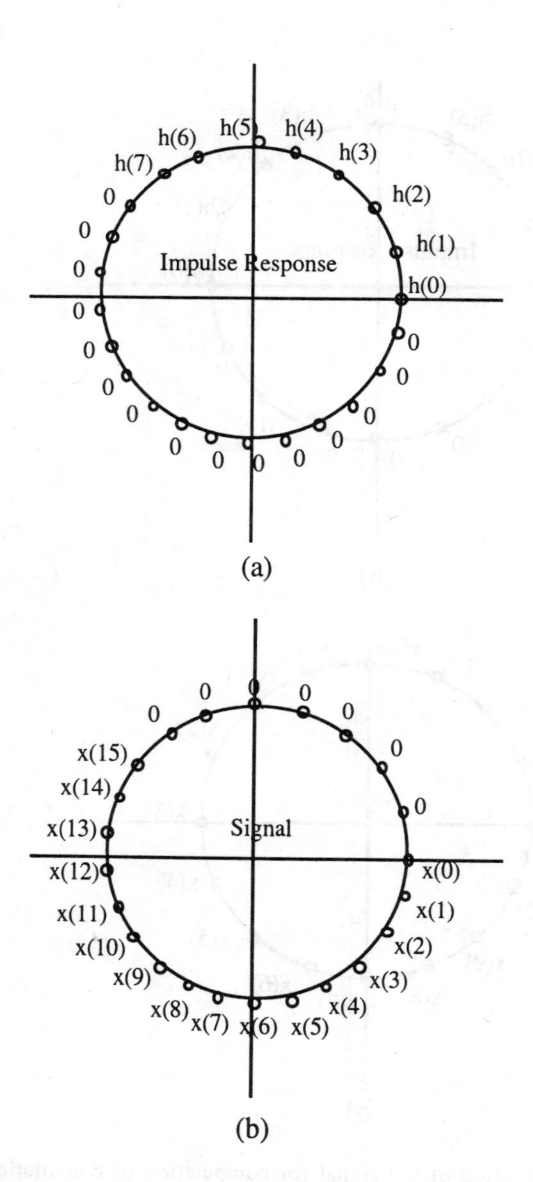

Figure 2.32: The signal length is extended by appending L-1 zeros to both x and also to h. It is easy to verify that the error terms arising out of periodic extension of the signal in FDFT computation disappear. The circular convolution is now equal to linear convolution.

The length of the output will now be equal to $N+L$-1. Out of these $N+L$-1 output points L-1 output points at the beginning and equal number at the end represent the so-called edge effect. There are only N-L+1 useful output points.

Fast Convolution: When the signal length is very long requiring a large amount of computer memory it is convenient to break the signal into many small segments and carry out the desired convolution on each segment separately and finally combine them in a suitable manner. We shall describe two well known overlap-and-add and overlap-and-save methods of fast convolution [1]. In overlap-and-add method a long signal is broken into q contiguous segments each of length **N**. Let the length of the impulse response function be L. Each segment is padded with L-1 zeros making its length equal to N+L-1. The impulse response function is also appended with N-1 zeros to make its length also equal to N+L-1. All segments are then convolved with the impulse response function using the FDFT approach, which we have described earlier. All processed segments are lined up but with a delay equal to N and then summed to yield a single sequence, which represents the convolution between the entire signal and the impulse response function. A pictorial layout of the overlap-and-add method is shown in fig. 2.33.

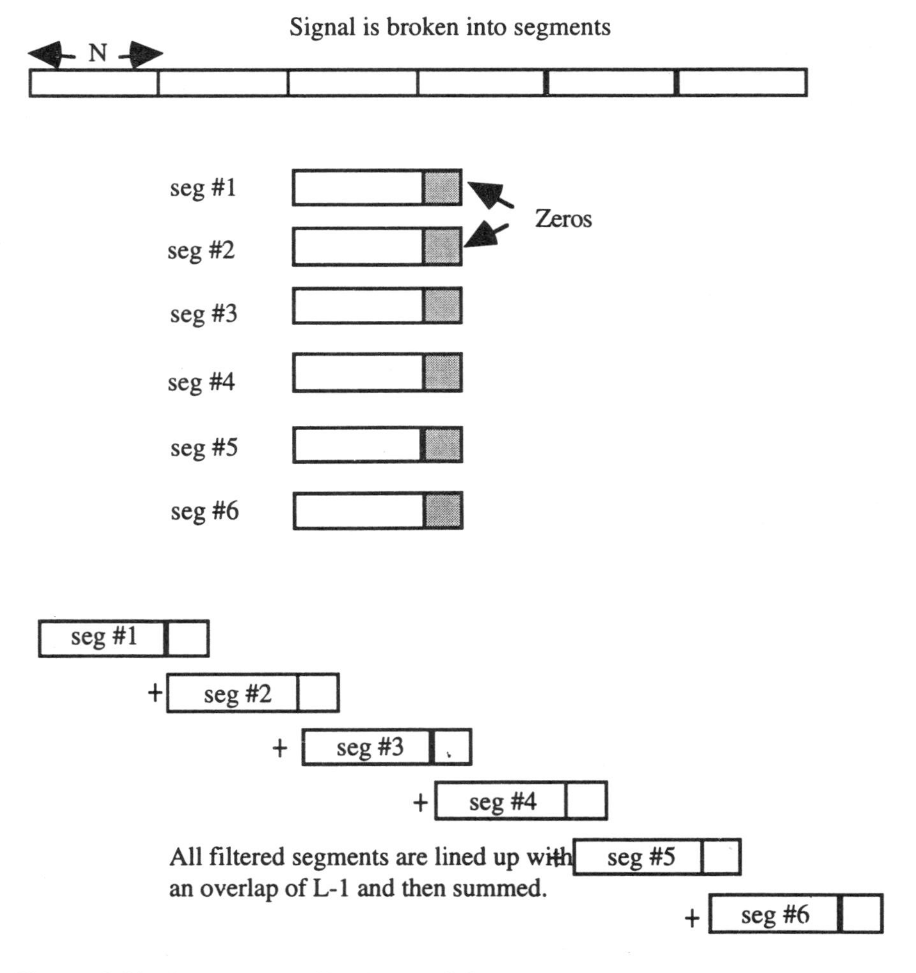

Figure 2.33: Overlap-and-add scheme of fast convolution.

In overlap-and-save method a long signal is broken into q overlapping segments, each of length N+L-1 with an overlap of L-1 ($N = N_0/q$). Length of non-overlapping part contains N points. The impulse response function as before is padded with *N*-1 zeros making its length equal to N+L-1. A circular convolution between each segment and the impulse response function is computed. The beginning L-1 points of the circular convolution are erroneous hence they are dropped. The last *N* output points from each segment are saved and placed contiguously with those from the neighbouring segments. The pictorial layout of the overlap-and-save method is illustrated in fig. 2.34. The essential difference between the two methods lies in what is contained in the overlapping part, zeros in overlap-and-add but a part of the neighbouring segment in the overlap-and-save. Both schemes are about equal in so for as computational load is concerned. Both need equal number of real multiplications.

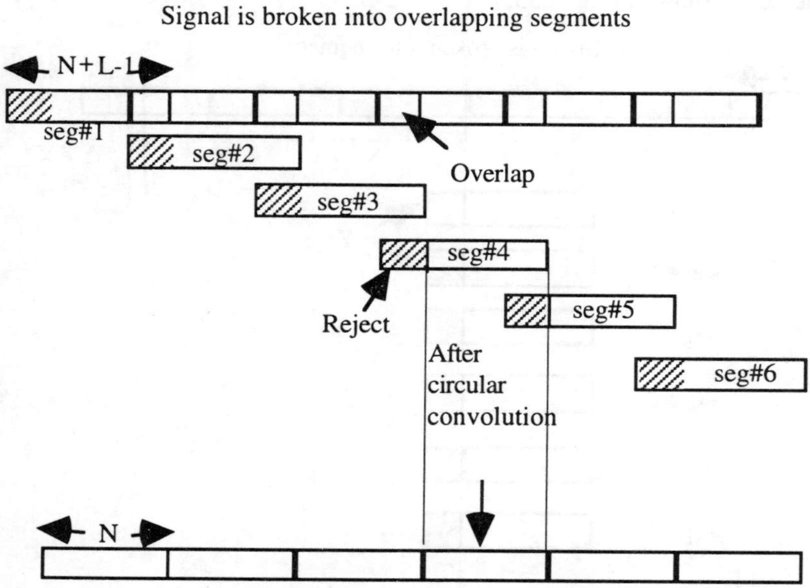

Figure 2.34: Overlap-and-save scheme of fast convolution.

We shall briefly enumerate computational load of four different types of convolutions. The objective is to provide an approximate comparative computational load involving only multiplications. We assume that both the signal and impulse response function are real. To compute the FDFT we shall place two real signals as real and imaginary parts of a complex signal. After Fourier transformation signals are separated as suggested in Exercise #7. With no extra multiplications it is possible to compute the FDFT of two real signals at the same time. Let N_0 be the length of the signal and let L be that of the impulse response. Forward FDFT requires $4(N_0+L-1)\log_2(N_0+L-1)$ real multiplications. To multiply the FDFTs of the signal and the impulse response we need $4(N_0+L-1)$ multiplications. For inverse FDFT we need the same number multiplications as for forward transformation. The total number of multiplications required in circular convolution is as shown in table 2.5.

sNext we enumerate multiplications required in overlap-and-add method. Assume that the signal has been divided into q segments where q is even and $\geq$ 2. For forward FDFT we need $2q(N_0/q+L-1)\log_2(N_0/q+L-1)$ real multiplications where we assume that two real segments are paired into a single complex sequence. FDFT of the impulse response sequence requires $4(N_0/q+L-1)\log_2(N_0/q+L-1)$ real multiplications. For multiplication of the Fourier transform of segments and the Fourier transform of the impulse response we need $4q(N_0/q+L-1)$ real multiplications. Finally, for inverse FDFT we will need the same number of multiplications as for forward FDFT. The total number of multiplications is as shown in table 2.5. For relative comparison we form a ratio of the number of multiplications,

$$\text{Ratio}=\frac{2\left[1+(2+\frac{2}{q})\log_2(\frac{N_0}{q}+L-1)\right]q(\frac{N_0}{q}+L-1)}{4\left[1+2\log_2(N_0+L-1)\right](N_0+L-1)} \tag{2.50}$$

Table 2.5: Number of real multiplications in different types of convolution. N_0 stands for the total length of the discrete signal and L for the length of the impulse response function. We have assumed that q is even and greater than 2.

Type of convolution	Number of multiplications (real)
Linear convolution (Eq. 2.45)	N_0L
Circular convolution (Eq. 2.48c)	$4[1+2\log_2(N_0+L-1)]$ (N_0+L-1)
Overlap-and-add/ Overlap-and-save	$2[1+(2+\frac{2}{q})\log_2(\frac{N_0}{q}+L-1)]$ $q(\frac{N_0}{q}+L-1)$

Example 2.14: Consider an example where N_0=4096 and L=64. The ratio (eq. 2.50) of the number of multiplications required in fast convolution (i.e. overlap-and-add) to the number of multiplication required in circular convolution is plotted in fig. 2.35. Notice the rapid decrease of the ratio as the number of segments increases from one to six and there after it approaches a limiting value depending on N_0 and L

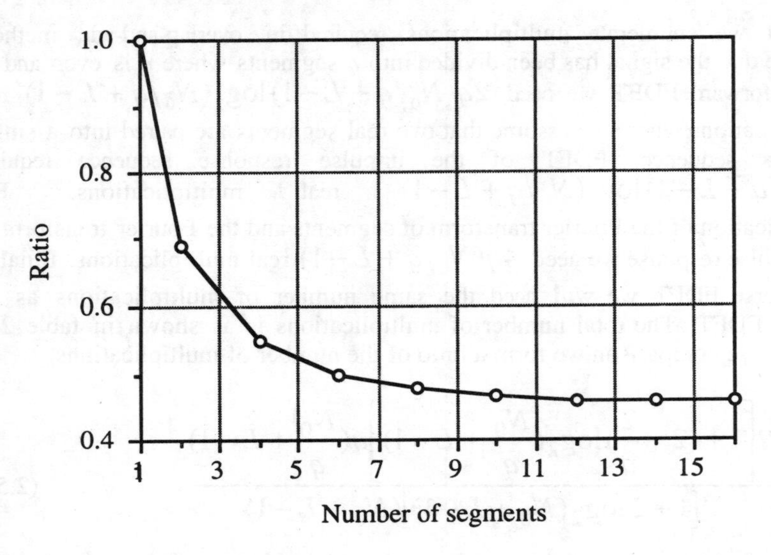

Figure 2.35: Ratio of number of real multiplication in Overlap-and-add method to those in circular convolution (N_0=4096 and L=64) .

§2.7 Exercises

Problems:

1. Let $X(\omega)$ be the DFT of a sequence,

$$x(n) = \left\{ \begin{matrix} ..00...0001.0, 3.5, 6.7, 4.0, 3.3, 2.8, 000...00.. \\ \uparrow \\ n=0 \end{matrix} \right\}$$

Evaluate the following quantities without evaluating $X(\omega)$:

(a) $X(0)$

(b) $\dfrac{1}{2\pi}\int_{-\pi}^{\pi} |X(\omega)|^2 \, d\omega$

(c) $X(\pi)$

(d) $\dfrac{1}{2\pi}\int_{-\pi}^{\pi} X(\omega) d\omega$

2. Let $x(n)$ be a discrete sequence whose DFT is illustrated in fig. 2.36

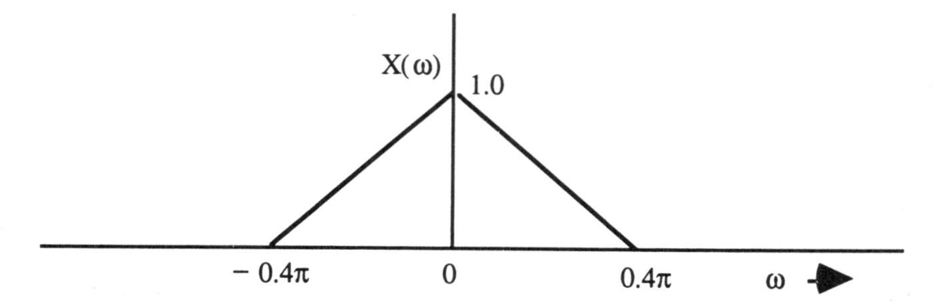

Figure 2.36: Fourier transform of a discrete signal.

Define another sequence $y(n)$=x(n)w(n) where w(n) is a window function. Sketch the Fourier transform of $y(n)$ for the following choice of window functions:

(a) $\cos(0.4\pi n)$

(b) $\sin(0.4\pi n)$

(c) $\sum_{k=-\infty}^{\infty} \delta(n-2k)$

3.$x(t)$ is a continuous time signal whose CFT is shown in fig. 2.37. The signal is sampled with two different sampling intervals, $\Delta t = 1.0$ and $\Delta t = 2.0$. Sketch the discrete Fourier transform (DFT) of the sampled sequences over a frequency range $\pm 3\pi$.

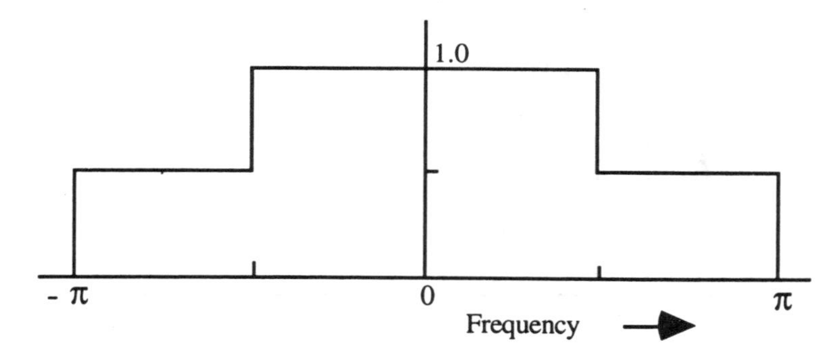

Figure 2.37: Continuous Fourier transform (CFT) of a signal referred to in problem 3.

4. Let $x_1(t)$ and $x_2(t)$ be two bandlimited signals, that is, their Fourier transforms are $X_1(\omega) = 0 \quad |\omega| > 0.3\pi$ and $X_2(\omega) = 0 \quad |\omega| > 0.4\pi$. What should be the sampling interval so that there is no aliasing error in the following functions:

$$y(t) = x_1(t)x_2(t)$$

$$z(t) = \int_{-\infty}^{\infty} x_1(\tau)x_2(t-\tau)d\tau$$

5. A bandlimited signal which is sampled at a Nyquist rate ($>1/2f_b$) may be reconstructed using equation (1.2). There are other simple but suboptimal alternatives, namely, (a) signal is held constant between two adjacent samples and (b) signal is linearly varying between the samples [16]. This is illustrated in fig. 2.38. Show that above suboptimum interpolations may be achieved by passing the discrete signal through a filter with an impulse response function given by a rectangular function and a triangular function (see fig. 2.39), respectively,

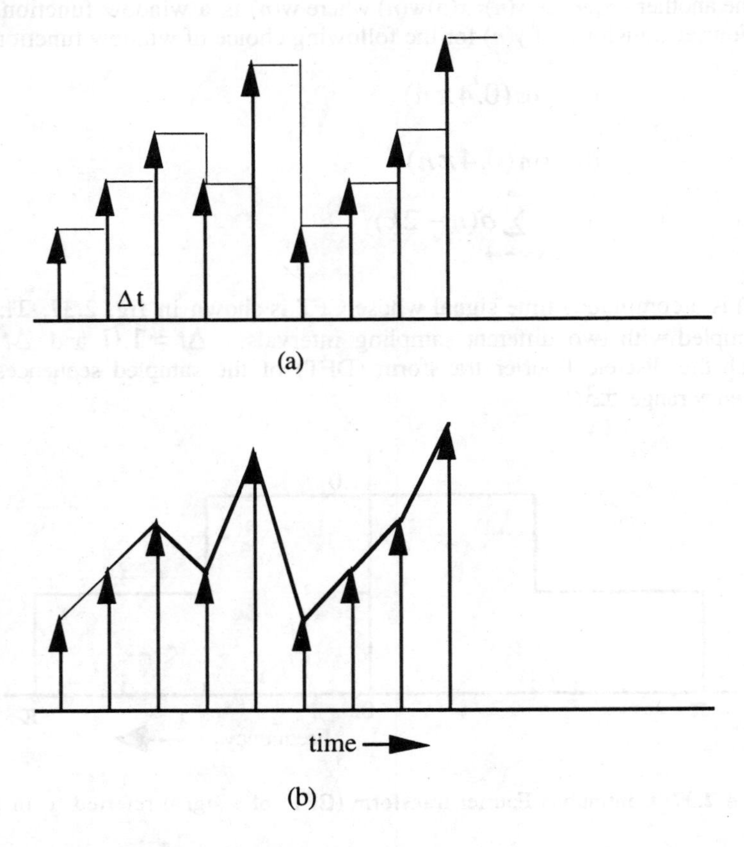

Figure 2.38: (a) signal is held constant between two adjacent samples and (b) signal is linearly varying between the samples.

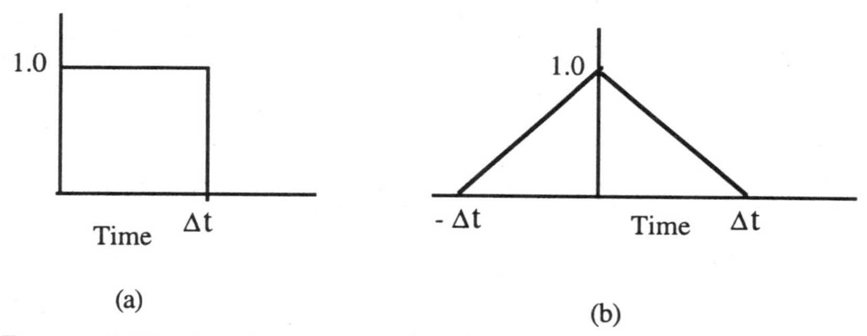

Figure 2.39: Impulse response functions of constant and linear response interpolators.

6. Compare the frequency response functions of the sub-optimal interpolation filters given in problem 5 with that of the exact interpolation filter given in equation (1.2). What is the essential difference?

7. Let $x(n) = x_1(n) + jx_2(n)$, $n = 0,1,\ldots,N-1$ where $x_1(n)$ and $x_2(n)$ are real discrete signals. Given the FDFT of $x(n)$ find the FDFT of $x_1(n)$ and $x_2(n)$. (Hint: $x_1(n) = (x(n) + x^*(n))/2$ and $x_2(n) = (x(n) - x^*(n))/2j$).

8. Given the FDFT of any one of the discrete signals shown in fig. 2.40 below, how do you get the FDFT of the remaining signals?

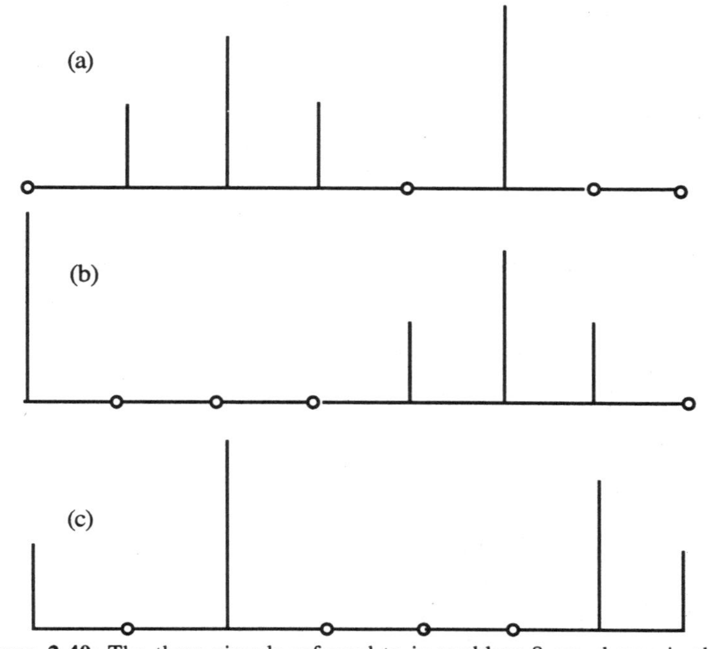

Figure 2.40: The three signals refereed to in problem 8 are shown in the fig.9.

9. Let

$$H(\omega) = -j\, sgn(\omega)$$
$$= 0 \qquad \omega = 0$$

where $sgn(\omega)$ stands for sign of ω. Hilbert transform of a real discrete signal $x(n)$ is defined as

$$x_H(n) = \frac{1}{2\pi}\int_{-\pi}^{\pi} H(\omega)X(\omega)e^{j\omega n}d\omega$$

Show that the Fourier transform of a complex signal defined as $x_c(n) = x(n) + jx_H(n)$ is zero on negative frequency axis.

10. Show that the Hilbert transform of a signal (defined in Exercise 9) is orthogonal to that signal, that is,

$$\sum_{n=-\infty}^{\infty} x(n)x_H(n) = 0$$

11. Circular autocorrelation function, like circular convolution, is defined as

$$y(n) = \sum_{m=0}^{N-1} x(m)x(n+m)_{\mathrm{mod}\,N}$$

Let **x**={1, 4, 3,-2]. Compute the circular autocorrelation function of **x**. Represent $x(m)$ and $x(n+m)_{\mathrm{mod}\,N}$ on a circle. Then, form a sum of the products of the corresponding numbers. Show that the circular autocorrelation function is periodic.

12. Derive a tripling algorithm along the lines of the doubling algorithm derived on page 85. In particular show the following result: Let $x(n)$, n=0,1, ..., N-1 where N is a integer multiple of three (e.g. N=3M). Let

$$x_0(m) = x(3m), \qquad m = 0,1,\ldots,M-1$$
$$x_1(m) = x(3m+1), \quad m = 0,1,\ldots,M-1$$
$$x_2(m) = x(3m+2), \quad m = 0,1,\ldots,M-1$$

Show that

$$X(k) = X_0(k) + e^{-j\frac{2\pi}{N}k} X_1(k) + e^{-j\frac{2\pi}{N}2k} X_2(k) \quad 0 \le k \le N-1$$

where $X_0(k) = FDFT\{x_0(m)\}$, $X_1(k) = FDFT\{x_1(m)\}$ and

$X_2(k) = FDFT\{x_2(m)\}$.

13. Consider an eight point sequence, $x = \{1, 2.3, 5.3, 3, 6.9, 1.0, 8.3, 0.0\}$ and let X(k) be its FDFT. Let $Z_R(k) = \text{Re}\{X(k)\}$, $Z_I(k) = \text{Im}\{X(k)\}$ and $A(k) = |X(k)|^2$. Compute the inverse FDFT of $Z_R(k), Z_I(k), A(k)$ without ever computing X(k).

14. $x(n)$ is a sampled band limited signal. The sampling interval Δt satisfies the Shannon's sampling theorem. $\hat{x}_1(t)$ and $\hat{x}_2(t)$ are interpolated signals obtained with different interpolating functions. Compute the mean square difference between $\hat{x}_1(t)$ and $\hat{x}_2(t)$.

$$\hat{x}_1(t) = \sum_{n=-\infty}^{\infty} x(n) \sin c(t - n\Delta t))$$

$$\hat{x}_2(t) = \sum_{n=-\infty}^{\infty} x(n) \Pi(t - n\Delta t))$$

where

$$\Pi(\frac{|t|}{\Delta t/2}) = 1 \quad |t| \le \Delta t/2$$

$$= 0 \quad otherwise$$

15. Let $X(k)$ be FDFT of the sequence shown in row 1. Without actually computing the FDFT obtain the FDFT of the sequences shown in rows 2 and 3, which are all derived from row 1 linear transformation.

1	2.0	1.8	1.2	1.6	1.0	1.2	2.2	0.9
2	2.9	4.0	2.4	2.6	2.6	2.4	4.0	2.9
3	1.1	-.40	0.0	0.6	-0.6	0.0	0.4	-1.1

16. Let $H_1(z)$ and $H_2(z)$ transfer functions of two LTI systems. The transfer functions are given by

$$H_1(z) = 1 + 2.0784z^{-1} + 1.44z^{-2}$$

$$H_2(z) = 1 + 1.5588z^{-1} + 0.81z^{-2}$$

Compute the impulse response functions of different combinations of above two LTIs.

(a) $H_1(z) + H_2(z)$ (b) $H_1(z) * H_2(z)$

(c) $\dfrac{H_1(z)}{H_2(z)}$ and (d) $\dfrac{H_2(z)}{H_1(z)}$

17. A lowpass filter is used for upsampling (interpolator). Show that the magnitude of the lowpass filter must be equal to I (upsampling factor) in order that the samples of the original signal remain unchanged.

18. Show that the Hartley transform becomes symmetric when the discrete signal has a symmetry,

$$X(n) = x(N-n), \quad n = 1,2,\ldots N-1$$

On account of this property the Hartley transform of a convolution product will require just one real multiplication (instead of two as in (2.40)).

19. Show how to reduce the reconstruction formula (2.21b) for bandpass signal to that for a lowpass signal, eq.(1.2). (Assume that p=1).

20. Let

$$z(t) = \int_{-\infty}^{\infty} x(\tau)h(t-\tau)d\tau.$$

$z(t)$ is now sampled with a sampling interval Δt. The discrete signal thus obtained is Fourier transformed. Show that

$$DFT\{z(n\Delta t)\} = \sum_{m=-\infty}^{\infty} X(\omega + \frac{2m\pi}{\Delta t})H(\omega + \frac{2m\pi}{\Delta t})$$

(Hint: Use folding theorem)

21. In Example 2.3 let us insert two zeros between adjacent data samples. The DFT of the expanded signal is then computed. How many spectral peaks will you find? Find their position.

22. Define a new sequence, $x'(n) = x(n)e^{\sigma n}, \quad n = 0,1,2,\ldots,N-1$ where σ is a constant and $x(n)$ is any signal. Show the following result:

$$FDFT\{x'(n)\} = X(z)\big|_{z=e^{(\sigma + j\frac{2\pi k}{N})}}$$

where $FDFT\{.\}$ stands for the FDFT of the signal within braces.

23.Let

$$g(n) = \frac{1}{2}x(2n) + \frac{1}{2}x(2n+1)$$

$$h(n) = \frac{3}{4}x(2n) - \frac{1}{8}x(2n+1), \qquad n = 0,1,\ldots,\frac{N}{2}-1$$

where $x(n)$, n=0,1,...,N-1 is any signal. Given the FDFT of g(n) and $h(n)$ compute the FDFT of $x(n)$.

24. A continuous band limited signal, whose spectrum is shown in fig. 2.41 below, is required to be sampled. What should be the sampling rate?

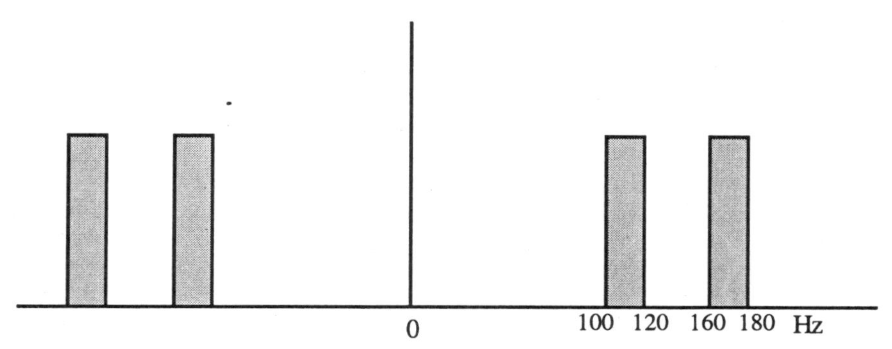

Figure 2.41: The spectrum of a band limited signal to be sampled. The starting and ending frequency in Hz of each band is shown.

25.Consider the following sequences:

$$x_1(n) = \{0,1,2,3\}, \quad x_2(n) = \{0,1,0,0\}, \quad s(n) = \{1,0,0,0\}$$

and their FDFTs. Do the following:

(b) Determine a sequence y(n) so that $Y(k)=X_1(k)X_2(k)$

(c) Is there a sequence $x_3(n)$ such that $S(k)=X_1(k)X_3(k)$?

Computer Projects:

1. Consider a pulse shown in fig. 2.42. Sample this pulse at a rate of 10 samples per second. Compute the discrete Fourier transform of the discretized signal. Do not use the FFT m_file from the Matlab. Verify some of the properties of DFT, in particular, verify the properties 2, 3, 4, and 7 given in table 2.1.

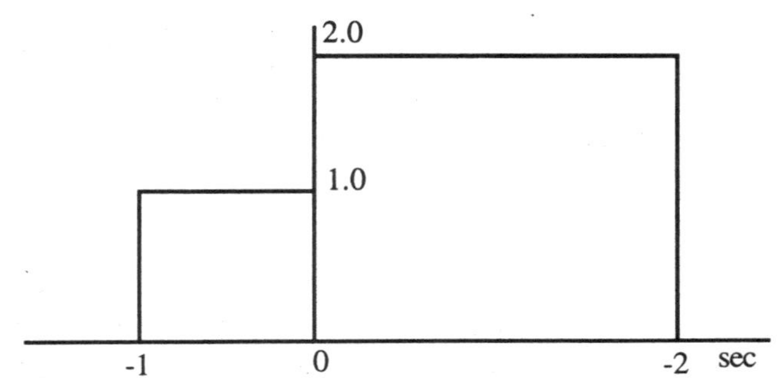

Figure 2.42: A continuous time pulse for sampling and DFT computation

2. A discrete convolution (2.45) has been expressed as a matrix product in (2.46). The result was verified through a constant signal and constant impulse response function. In this project we would like to verify the same result through a more complex signal and impulse response function. Let the impulse response be given by,

$$h=\{1, 0.8,0.5, 0.85,0.25\}$$

and let the signal be a Gaussian pseudo-random sequence generated by m-file, randn. Generate 64 Gaussian random numbers. Arrange them in a matrix form as shown in (2.46) and compute the discrete convolution. Next, using the same signal carry out the discrete convolution given by (2.45). Compare the results.

3. Generate a sequence of 64 pseudo-random numbers. Compute its FDFT. Let X(k), k=0,...,63 be the FDFT coefficients. Multiply the FDFT coefficients with a phase function, $e^{-j\frac{2\pi k}{64}\tau_0}$ where τ_0 is an integer constant (Caution: Keep in mind the FDFT symmetry). Now compute the inverse FDFT of the product sequence. Compare the result with the input. Do this for τ_0 a positive constant and a negative constant, and also a fraction. In each case how does the output compare with the input?

4. The autocorrelation function of a deterministic signal is given by (1.4a). Taking FDFT on both sides of (1.4a) it is easy to show that

$$FDFT\{r_{xx}(\tau)\} = FDFT\{\sum_{n=0}^{N-\tau} x(n)x(n+\tau)\}$$

$$= X(k)X^*(k)$$

Hence, the autocorrelation may be computed as inverse FDFT of $X(k)X^*(k)$. It turns out that the result is a circular autocorrelation in the same sense as a circular convolution discussed in this chapter. Verify this result for a simple eight point sequence, for example,

$$x=\{1, 2.3, 1.8, 4.3, 1.2, 0.5, 0.8, 0.1\}$$

Compute the autocorrelation using FDFT technique as well as using eq.(1.4a). It is possible to overcome the problem by simply padding equal number of zeros to sequence *x*. Verify this statement.

5. Let *x* be a 32 point sequence, (1,2,...,32) and *h*=[1,2,3,4] an FIR filter. As in overlap-and-add method of fast convolution (see §2.6) we divide *x* into two sub-sequences, *x1* and *x2* each of length 16. Each sub-sequence is padded with four zeros (which is the length of the filter). We use FDFT to compute the convolution between a sub-sequence and the filter. The convolved subsequences are then added up with overlap as shown in fig. 2.33. Observe what happens in the region of overlap. We note that two edge effects when added up produce a correct result.

6. Repeat the above project with overlap-and-save method in mind. Observe that the edge effect is now encountered on the left hand side of each segment. In the overlap-and-save method this edge is rejected and the rest of the output is used for building up the total convolution output.

3 Power Spectrum Analysis

The distribution of power, or power spectrum density (PSD), of a stochastic signal is an important attribute of practical interest in many problems. In addition to power spectrum, or simply spectrum, we have cross-spectrum, coherence, time varying spectrum, bispectrum for non-Gaussian signals. In this chapter we shall study the spectrum, cross-spectrum and coherence for a stationary stochastic signals and the time varying spectrum for non-stationary stochastic signals. Though Gaussian distribution is commonly assumed for a stochastic signal, there are situations where such a model is inappropriate, particularly, where some non-linear physical interactions are involved. For such non-Gaussian signals bispectrum and other higher order spectra are the right tools. Another interesting feature of the spectrum is its variation with time. Human speech and sounds produced by many animals and birds possess interesting time-frequency dependence. Spectrogram, an example of time varying spectrum, is a basic tool in speech applications. Spectrum has many uses in widely unrelated areas. For example, spectrum and other related quantities have been used in astrophysics, geophysics, telecommunication, biomedical electronics, etc. Just to cite one example of its application is for classification of the roughness of seafloor. A curious reader may wish to look up the reference [37].

The chapter is prepared to emphasize the basic principles of spectrum estimation of stationary stochastic signals with some introductory material on more advanced topics such as bispectrum and time varying spectrum. More information on the advanced topics is available in recent texts on spectrum analysis [6, 17]. Estimation of spectrum from a finite length stochastic signal is fraught with many uncertainties. It often happens that an observed peak may be simply an artifact caused by insufficient degrees of freedom. A careful analysis of the variability of the estimated spectrum is mandatory in order to make the estimated spectrum practically meaningful. However, in an introductory book such as this we can only give an outline of spectrum variability.

§3.1 Spectrum, Cross-spectrum and Coherence

We shall first introduce the definitions of (power) spectrum, cross-spectrum and coherence of a pair of stationary stochastic discrete signals. For deterministic signals the corresponding quantity is energy spectrum which is simply the magnitude square of the DFT of a discrete signal.

Spectral Representation of Covariance Function: Wiener established the existence of a bounded non-decreasing function $F(\omega)$ called the distribution function such that

$$c_x(\tau) = \frac{1}{2\pi} \int_{-\pi}^{\pi} \exp(j\omega\tau) dF(\omega) \tag{3.1}$$

where $F(\omega)$ is the power lying between $-\pi$ and ω and $dF(\omega)$ is the differential of $F(\omega)$, that is, the power lying in a small frequency interval $\Delta\omega$ around ω. $c_x(\tau)$ is covariance function already introduced along with cross-covariance in Chapter One. $F(\omega)$ can be expressed as a sum of two components, the discrete and continuous parts,

$$F(\omega) = F_d(\omega) + F_c(\omega) \tag{3.2}$$

where $F_d(\omega)$ is the discrete part, given by

$$F_d(\omega) = \sum_i p_i U(\omega - \omega_i)$$

where $U(\omega)$ is a unit step function and the continuous part

$$F_c(\omega) = \frac{1}{2\pi}\int_{-\pi}^{\omega} S_x(\omega')d\omega' \tag{3.3}$$

$S_x(\omega)$ is the power spectral density (PSD) function of a stochastic process with continuous spectrum distribution function. Assuming that $F_d(\omega)=0$ the spectral representation in (3.1) reduces to

$$c_x(\tau) = \frac{1}{2\pi}\int_{-\pi}^{\pi} S_x(\omega)\exp(j\omega\tau)d\omega \tag{3.4}$$

Note that since the unit of covariance function is power (see Chapter One), the unit of $S_x(\omega)$ will be power per Hertz. It must satisfy, being power, a condition that

$$S_x(\omega) \geq 0 \text{ (Khintchin's theorem).}$$

The converse of (3.4) is

$$S(\omega) = \sum_{n=-\infty}^{\infty} c_x(\tau)\exp(-j\omega\tau) \tag{3.5}$$

When the discrete part is not zero, that is, $F_d(\omega) \neq 0$, the covariance function can be expressed as follows:

$$c_x(\tau) = \sum_i p_i \cos(\omega_i \tau) + \frac{1}{2\pi}\int_{-\pi}^{\pi} S_x(\omega)\exp(j\omega\tau)d\omega \tag{3.6}$$

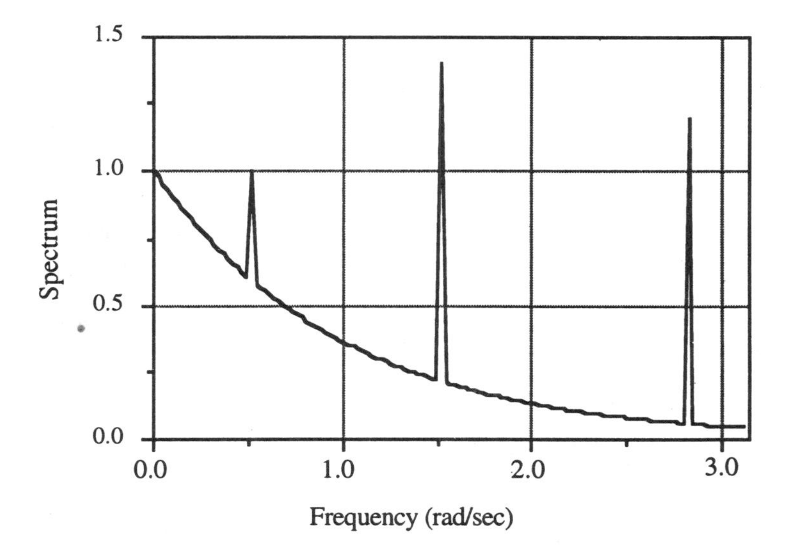

Figure 3.1: An example of mixed spectrum, discrete and continuous components. The discrete component is attributed to the presence of random sinusoidal signals (see eq.3.7).

Corresponding to the discrete and continuous parts of a spectral density (fig. 3.1), a stochastic signal can be expressed as a sum of random sinusoids and a stochastic process with continuous spectrum,

$$x(n) = \sum_i \alpha_i \sin(\omega_i n + \phi_i) + x_c(n) \tag{3.7}$$

where $p_i = \alpha_i^2$ and ϕ_i is random phase. The sinusoidal component is fully predictable from the past samples; hence, it is also known as the deterministic component.

Spectra of Continuous and Discrete Signals: We have in (2.4) noted a relationship between the discrete and continuous Fourier transforms. For stochastic signal we have a similar relation between the spectra of the discrete and continuous stochastic signals [6],

$$S_d(\omega) = \sum_{i=-\infty}^{\infty} S_c(\omega - \frac{2\pi}{\Delta t} i) \tag{3.8}$$

where $S_c(\omega)$ is spectrum of continuous signal and $S_d(\omega)$ is that of discrete signal.

Cross-Spectrum: The cross-spectrum is defined as the Fourier transform of the cross-covariance function,

$$S_{xy}(\omega) = \sum_\tau c_{xy}(\tau) \exp(-j\omega\tau) \tag{3.9}$$

Since $c_{xy}(\tau)$ is not symmetric, $S_{xy}(\omega)$ will be a complex function. From the symmetry properties of the cross-covariance function (1.6c) it follows that $S_{xy}(\omega) = S'^{*}_{yx}(\omega)$. Let us express it in terms of its real and imaginary parts,

$$S_{xy}(\omega) = P(\omega) + jQ(\omega)$$

where $P(\omega)$ is known as co-spectrum and $Q(\omega)$ is known as quadrature spectrum. Alternatively, we can also express the cross-spectrum in terms of the amplitude spectrum and the phase spectrum

$$S_{xy}(\omega) = A(\omega)\exp(j\phi(\omega))$$

where $A(\omega) = \sqrt{P^2(\omega) + Q^2(\omega)}$ and $\phi(\omega) = \tan^{-1}(Q(\omega)/P(\omega))$

The spectrum and cross-spectrum of a pair of signals can be expressed in a matrix form

$$\mathbf{S} = \begin{bmatrix} S_x(\omega) & S_{yx}(\omega) \\ S_{xy}(\omega) & S_y(\omega) \end{bmatrix} \tag{3.10}$$

The spectral matrix, $\mathbf{S}(\omega)$, is always Hermitian positive semi-definite. Hence, its determinant must be greater than or equal to zero,

$$\det[\mathbf{S}] = S_x(\omega)S_y(\omega) - \left|S_{xy}(\omega)\right|^2$$

$$= S_x(\omega)S_y(\omega)[1 - \frac{\left|S_{xy}(\omega)\right|^2}{S_x(\omega)S_y(\omega)}] \geq 0$$

Hence

$$0 \leq \frac{\left|S_{xy}(\omega)\right|^2}{S_x(\omega)S_y(\omega)} \leq 1 \tag{3.11}$$

Coherence: Coherence is a measure of the relationship that exists between the two signals, *x(n)* and *y(n)* and is defined as

$$coh_{xy}(\omega) = \frac{S_{xy}(\omega)}{\sqrt{S_x(\omega)S_y(\omega)}} \tag{3.12}$$

where $S_x(\omega) \neq 0$. and $S_y(\omega) \neq 0$ Equation (3.11) expressed in terms of $coh_{xy}(\omega)$ becomes

$$0 \leq \left|coh_{xy}(\omega)\right|^2 \leq 1 \tag{3.13}$$

It may be shown that the fraction of energy in the first time series that can be predicted from the knowledge of the second signal is equal to the magnitude square of the coherence [6].

Example 3.1: Let us take a pair of stationary stochastic signals being input and output of a linear time invariant (LTI) system (fig. 3.2) whose impulse response function is given by $h(\tau)$. The input and output of a LTI are related through the convolution integral (2.44a)

$$y(n) = \sum_{\tau=0}^{\infty} h(\tau)x(n-\tau) + \eta(n)$$

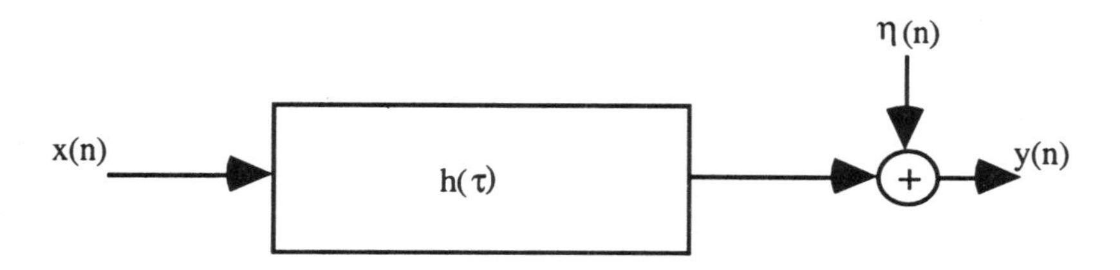

Figure 3.2: A pair of stationary stochastic signals being input and output of a linear time invariant (LTI) system with measurement noise.

We shall compute the cross-covariance between the input and output signals,

$$\begin{aligned} c_{xy}(\tau) &= E\{x(n)y(n+\tau)\} + E\{x(n)\eta(n+\tau)\} \\ &= \sum_{\tau'=0}^{\infty} h(\tau')E\{x(n)x(n+\tau-\tau')\} + E\{x(n)\eta(n+\tau)\} \\ &= \sum_{\tau'=0}^{\infty} h(\tau')c_x(\tau-\tau') \end{aligned}$$

where we have assumed that the measurement noise is uncorrelated with the input signal. Similarly, the covariance of the output may be shown to be

$$\begin{aligned} c_y(\tau) &= E\{y(n)y(n+\tau)\} \\ &= \sum_{\tau'=0}^{\infty} h(\tau')E\{x(n-\tau')y(n+\tau)\} + E\{\eta(n)y(n+\tau)\} \\ &= \sum_{\tau'=0}^{\infty} h(\tau')c_{xy}(\tau+\tau') + \sigma_\eta^2\delta_\tau \end{aligned}$$

Applying the convolution theorem of discrete convolution (Table 2.1) we obtain

$$S_{xy}(\omega) = H(\omega)S_x(\omega)$$
$$S_y(\omega) = H^*(\omega)S_{xy}(\omega) + \sigma_\eta^2$$
$$= |H(\omega)|^2 S_x(\omega) + \sigma_\eta^2$$

Note that the spectrum of the output is related to magnitude square of the filter transfer function. The phase of the transfer function plays no role.
The coherence function as defined in (3.13) reduces to

$$Coh_{xy}(\omega) = \frac{S_{xy}(\omega)}{\sqrt{S_x(\omega)S_y(\omega)}}$$

$$= \frac{H(\omega)S_x(\omega)}{\sqrt{S_x(\omega)|H(\omega)|^2 S_x(\omega) + \sigma_\eta^2 S_x(\omega)}}$$

$$= \frac{H(\omega)}{|H(\omega)|}\frac{1}{\sqrt{1+\frac{1}{SNR}}}$$

where $SNR = |H(\omega)|^2 S_x(\omega)/\sigma_\eta^2$, the signal-to-noise ratio (SNR) at the output of LTI system ●.

Thus, in absence of measurement noise the magnitude of the coherence between the input and the output of a LTI system is one. It is independent of the transfer function. However, the phase of the coherence function is equal to the phase of the transfer function. For example, let the LTI be a simple delay device, that is, the output is $y(t) = x(t - \tau_0)$; the coherence between the input and output will be given by $\exp(-j\omega\tau_0)$. In the presence of measurement noise, however, the coherence magnitude depends on the SNR at the output. Indeed, given the coherence we can estimate the SNR at the output.

Spectrum of Deterministic Signal: The Fourier transform of the autocorrelation function of a deterministic signal (1.4a) yields the energy spectrum of the signal. We shall assume that the signals are defined over an infinite interval.

$$R_x(\omega) = \sum_{\tau=-\infty}^{\infty} r_{xx}(\tau)e^{-j\omega\tau} = \sum_{\tau=-\infty}^{\infty}\sum_{n=-\infty}^{\infty} x(n)x(n+\tau)e^{-j\omega\tau}$$

$$= \sum_{\tau=-\infty}^{\infty}\sum_{n=-\infty}^{\infty} x(n)x(n+\tau)e^{-j\omega(\tau+n)}e^{j\omega n}$$

$$= \sum_{\tau=-\infty}^{\infty} x(n+\tau)e^{-j\omega(\tau+n)} \sum_{n=-\infty}^{\infty} x(n)e^{j\omega n} \quad (3.14a)$$

$$= X(\omega)X^{*}(\omega) = |X(\omega)|$$

Similarly the Fourier transform of the cross-correlation gives us the cross-spectrum,

$$R_{xy}(\omega) = \sum_{\tau=-\infty}^{\infty} r_{xy}(\tau)e^{-j\omega\tau} = X(\omega)Y^{\bullet}(\omega) \tag{3.14b}$$

$R_x(\omega)$ is a real and positive quantity having unit of energy per Hertz. $R_{xy}(\omega)$ is however a complex quantity whose magnitude is energy per hertz and the phase is the phase difference between the two signals at a specific frequency. Further, it is easily verified that the coherence between two deterministic signal is always one.

§3.2 Estimation of Spectrum and Cross-Spectrum

The classical approach to estimating spectrum and cross-spectrum (of a pair of time series) has been via Fourier transformation of autocorrelation and cross-correlation functions, respectively. This method is often called the indirect method. After the discovery of fast Fourier transform, however, the direct method of first computing the magnitude square of the Fourier transform of the time series followed by averaging over different samples of the signal or over a band of frequencies has gained considerable popularity. There are different variations of the direct method. The spectral estimate is essentially a random variable since it is derived from a random process. Therefore, the statistical properties of the spectral estimate, such as its mean, variance and probability density, are of great interest but outside the scope of this introductory text. We shall, however, give the final results taken from [6]. We shall restrict our coverage to two most commonly used estimation techniques, namely, Blackman & Tukey and Welch. These are known as low resolution but linear methods. There is a vast literature on high resolution spectrum estimation techniques which are outside the scope of this introductory text. Interested reader may like to refer to [6] for high resolution spectrum estimation.

Blackman-Tukey (BT): The Fourier transform of the estimated covariance function straight away gives us an estimate of the spectrum, and similarly, the Fourier transform of the estimated cross-covariance function gives us an estimate of the cross-spectrum. Let $\hat{c}_x(\tau)$, $\tau = 0, \pm 1, \pm 2, \ldots, \pm(N-1)$ be the estimate of the covariance function. The estimated spectrum and cross-spectrum are given by

$$\begin{aligned} \hat{S}_x(\omega, N) &= \sum_{\tau=-N+1}^{N-1} \hat{c}_x(\tau)\exp(-j\omega\tau) \\ \hat{S}_{xy}(\omega, N) &= \sum_{\tau=-N+1}^{N-1} \hat{c}_{xy}(\tau)\exp(-j\omega\tau) \end{aligned} \tag{3.15}$$

The dependence on the maximum lag, for which the covariance function is available, is explicitly shown. We can relate $\hat{S}_x(\omega, N)$ to $\hat{S}_x(\omega)$, that is, when the covariance function is available over infinite lags. For this, define a window function $w_1(\tau)$ smoothly decreasing to zero for lags greater than L where

$L \le N$. The window length is equal to M=2L+1. The window used with the estimated covariance function is sometimes known as the quadratic window. We shall now rewrite (3.15) in the form of convolution product.

$$\tilde{S}_x(\omega) = \sum_{\tau=-\infty}^{\infty} w_1(\tau)\hat{c}_x(\tau)\exp(-j\omega\tau)$$

$$= \sum_{\tau=-\infty}^{\infty} \frac{1}{2\pi}\int_{-\pi}^{\pi} W_1(\omega')e^{-j\omega'\tau}d\omega'\hat{c}_x(\tau)\exp(-j\omega\tau)$$

$$= \frac{1}{2\pi}\int_{-\pi}^{\pi} W_1(\omega')d\omega'\sum_{\tau-\infty}^{\infty}\hat{c}_x(\tau)\exp(-j(\omega-\omega')\tau) \tag{2.16a}$$

$$= \frac{1}{2\pi}\int_{-\pi}^{\pi} W_1(\omega')\hat{S}_x(\omega-\omega')d\omega' = W_1(\omega)*\hat{S}_x(\omega)$$

where $\tilde{S}_x(\omega)$ is a windowed spectrum estimate and $W_1(\omega)$ is the Fourier transform of the window. $w_1(\tau)$ is known as quadratic spectral window. Similarly, for cross-spectrum we obtain

$$\tilde{S}_{xy}(\omega) = \sum_{\tau=-\infty}^{\infty} w_1(\tau)\hat{c}_{xy}(\tau)\exp(-j\omega\tau) \tag{3.16b}$$

$$= W_1(\omega)*\hat{S}_{xy}(\omega)$$

The window function has another important role, namely, it underweights those covariance estimates at large lags where the estimation error is likely to be large. The window function is a symmetric function with $w_1(0) = 1.0$ so that $\tilde{S}_{xy}(\omega) \approx \hat{S}_{xy}(\omega)$ in the region where the spectrum is smooth. To show this we shall assume that the spectrum is practically constant over the range of rapid variations in $W_1(\omega)$. Then, we may approximate (3.16) as

$$\tilde{S}_x(\omega) = W_1(\omega)*\hat{S}_x(\omega)$$

$$\approx \hat{S}_x(\omega)\frac{1}{2\pi}\int_{-\pi}^{\pi} W_1(\omega)d\omega = \hat{S}_x(\omega)$$

where we have set $\frac{1}{2\pi}\int_{-\pi}^{\pi} W_1(\omega)d\omega = 1$; therefore, $w_1(0) = 1.0$.

A simple window is the uniform window, also known as a default window,

$$w_1(\tau) = 1 \qquad for\ |\tau| \le L$$
$$= 0 \qquad otherwise$$

whose Fourier transform is $W_1(\omega) = \dfrac{\sin(\omega L)}{\sin\dfrac{\omega}{2}}$. The uniform window (L=64) and its Fourier transform are shown in fig. 3.3,

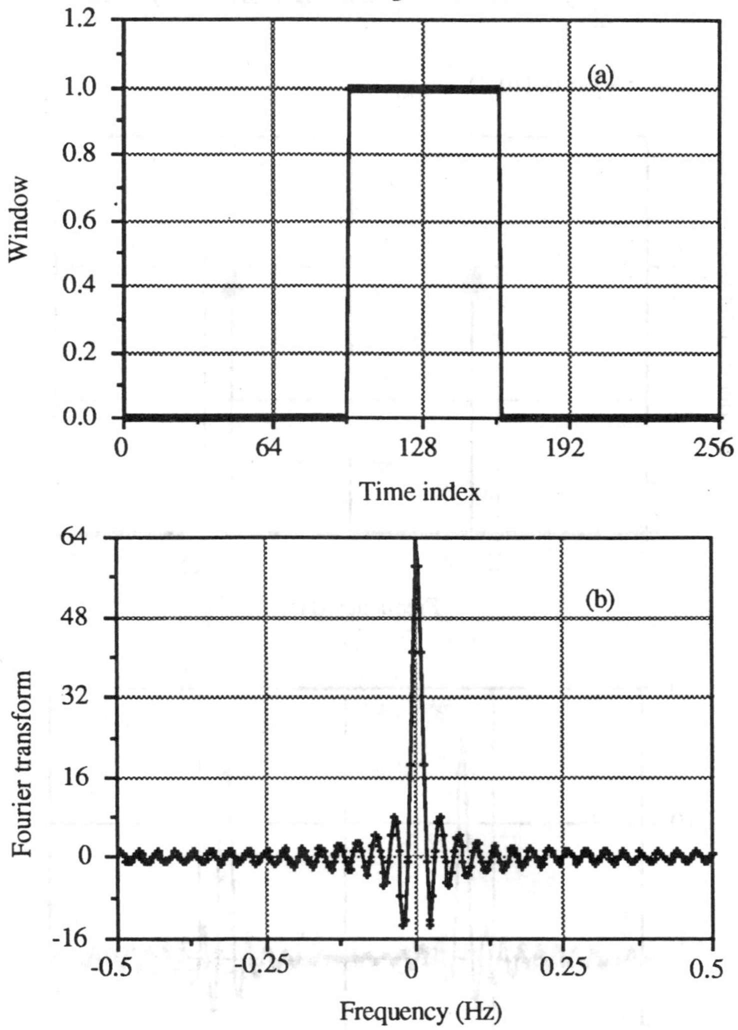

Figure† **3.3**: Uniform or default window and its Fourier transform. L=64. The 3 dB width of the main lobe is approximately equal to $3.15\pi/L$ radians.

Example 3.2: Consider a random sinusoid $x(n) = A_0 \sin(\omega_0 n + \phi)$ where ϕ is random phase uniformly distributed in the range $\pm\pi$. Further, its exact covariance function is given by,

---

†Figures 3.3, 3.4, 3.7, 3.8, 3.10, 3.13, 3.15 were taken from [6]

$$
\begin{aligned}
c_x(\tau) &= E\{x(n)x(n+\tau)\} \\
&= A_0^2 E\{\sin(\omega n+\phi)\sin(\omega(n+\tau)+\phi)\} \\
&= \frac{A_0^2}{2}[\cos(\omega\tau) - E\{\cos(\omega n+\omega(n+\tau)+2\phi\} \\
&= \frac{A_0^2}{2}\cos(\omega\tau)
\end{aligned}
$$

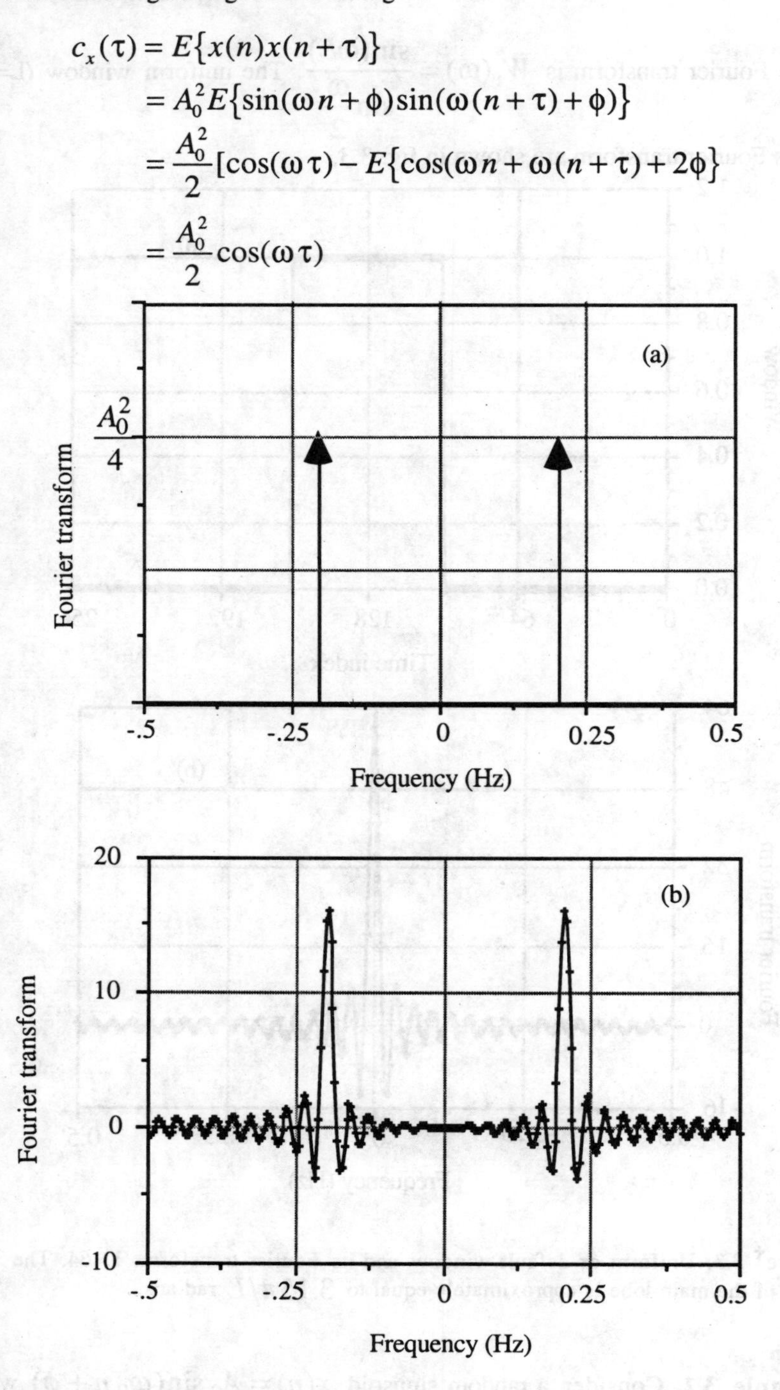

Figure 3.4: The effect of a quadratic window on the estimated spectrum. (a) An ideal spectrum of a random sinusoid (frequency 0.2 Hz). (b) Estimated spectrum with a quadratic window (L=64) applied to the exact covariance function. Note the windowed spectrum appears to go negative.

A uniform window of length 64 (-31 to 32 lags) is used to truncate the covariance function. The truncated covariance function (padded with zeros to increase its length to 256) is then Fourier transformed. The BT spectrum thus computed is shown in fig. 3.4. The BT spectrum can become negative on account of the Fourier transform of the window function, $W_1(\omega)$, not being non-negative everywhere. An example of this phenomenon is illustrated in fig. 3.4(b). In this case the window is entirely responsible for the negative spectrum which is not an acceptable estimate as it violates the basic principle that the power must be a positive quantity ●.

Other popular windows Hanning and Hamming windows also suffer from this problem [39]. The window function must then be slightly altered so that its Fourier transform becomes positive everywhere. A recursive procedure exists to generate such a closest positive window sequence [6].

Mean and Variance of BT Spectral Estimate: The BT spectrum estimate given by (3.16a) may be expressed in a different form as

$$
\begin{aligned}
\tilde{S}_x(\omega) &= \sum_{\tau=-N+1}^{N-1} w_1(\tau)(1-\frac{|\tau|}{N})\hat{c}_x(\tau)\exp(-j\omega\tau) \\
&= \sum_{\tau=-N+1}^{N-1} w_1(\tau)(1-\frac{|\tau|}{N})\frac{1}{N-|\tau|}\sum_{n=0}^{N-|\tau|-1} x(n)x(n+\tau)\exp(-j\omega\tau) \\
&= \frac{1}{N}\sum_{\tau=-N+1}^{N-1} w_1(\tau)\sum_{n=0}^{N-|\tau|-1} x(n)x(n+\tau)\exp(-j\omega\tau)
\end{aligned}
\tag{3.17}
$$

where we have introduced an additional window, $(1-|\tau|/N)$, a triangular window. Observe that the covariance function is estimated to a maximum possible lag of $N-1$ where N is data length. Hence the limits on the summation in (3.17) are from -N+1 to N-1.

$$
\tilde{S}_x(\omega) = \frac{1}{N}\sum_{k=0}^{N-1} W_1(\frac{2\pi}{N}k)\left|X(\omega-\frac{2\pi}{N}k)\right|^2 \tag{3.18}
$$

We shall now use (3.18) for evaluating the mean and variance of the spectral estimate.

$$
E\{\tilde{S}_x(\omega)\} = \frac{1}{N}\sum_{k=0}^{N-1} W_1(\frac{2\pi k}{N})E\left\{\left|X(\omega-\frac{2\pi k}{N})\right|^2\right\} \tag{3.19}
$$

From the properties of the DFT coefficients of a finite stationary stochastic process (2.31) we note that

$$
\frac{1}{N}E\left\{\left|X(\omega-\frac{2\pi k}{N})\right|^2\right\} = S_x(\omega-\frac{2\pi k}{N})
$$

and hence (3.19) reduces to

$$E\left\{\tilde{S}_x(\omega)\right\} = \sum_{k=0}^{N-1} W_1(\frac{2\pi k}{N}) S_x(\omega - \frac{2\pi k}{N}) \tag{3.20}$$

The BT spectrum will be unbiased, that is,

$$E\left\{\tilde{S}_x(\omega)\right\} = S_x(\omega)$$

only when $S_x(\omega)$ is a smoothly varying function and

$$\sum_{k=0}^{N-1} W_1(\frac{2\pi k}{N}) = N w_1(0) = 1$$

that is, $w_1(0) = 1/N$.

The variance of the BT spectral estimate, taken from [6], is

$$Var\left\{\tilde{S}_x(\omega)\right\} = \sum_{k=0}^{N-1} W_1^2(\frac{2\pi k}{N}) S_x^2(\omega - \frac{2\pi k}{N}) \tag{3.21a}$$

Assuming that the spectrum is a smooth function such that it is almost a constant over the mainlobe of the window spectrum we can express the variance as follows:

$$\frac{Var\left\{\hat{S}_x(\omega)\right\}}{S_x^2(\omega)} = N \sum_{n=0}^{N-1} w_1^2(n) \tag{3.21b}$$

Let $w_1(\tau) = 1/N$ for $\tau = 0,1,2,\ldots,L_{eff} - 1$ and zero otherwise where L_{eff} is the effective width of the window; then (3.21b) reduces to

$$\frac{Var\left\{\hat{S}_x(\omega)\right\}}{S_x^2(\omega)} \approx \frac{L_{eff}}{N}$$

from which it follows that we must have $L_{eff} << N$ so that the variance of the estimate is small compared to the spectrum. In general, however, when the condition of smoothness is not satisfied, the spectrum of the window plays a significant role in determining the bias and variance. In particular, in the neighborhood of a large spectral peak the bias and variance are likely to be affected by the sidelobes of the window spectrum. Inverse of the ratio, L_{eff}/N, is often known as degrees of freedom.

Welch Method: When the length of the time series is large it is computationally efficient to segment the signal and compute the square of the finite discrete Fourier transform (FDFT), which is also known as periodogram and then average them over all segments. The method of segmentation, first suggested by Welch [38], exploits the basic statistical properties of DFT coefficients (see eq. 2.31),

$$
\begin{aligned}
S_x(k) &= \frac{1}{L} E\{X(k)X^*(k)\} \\
S_{xy}(k) &= \frac{1}{L} E\{X(k)Y^*(k)\} \\
L &\rightarrow \infty
\end{aligned}
\tag{3.22}
$$

where L stands for the length of each segment. There are basically two difficulties in the implementation of the procedure indicated in (3.22). First, given just a single time series how does one carry out the expected operation and second, since all practical data records are finite how does one carry out the limiting operation, i.e., $L \rightarrow \infty$? By segmenting a signal, P uncorrelated (independent when the process is Gaussian) segments, the expected operation is carried out by taking an average over all uncorrelated segments. Alternatively, since the DFT coefficients are uncorrelated, the expected operation may also be replaced by an arithmetic average over a narrowband of frequencies where the spectrum variation is small.

Segmentation: Let us segment a signal $x(n)$, n=0, 1, 2, ... $N-1$ into P equal length partially overlapping segments, (fig. 3.5).

$$
\begin{aligned}
&x_i(t) = x(i(1-\alpha)L + n), \\
&i = 0,1,2,\ldots,P-1, \\
&n = 0,1,2,\ldots,L-1
\end{aligned}
\tag{3.23}
$$

where α is the fraction of overlap and L is the length of each segment. Note that the number of segments, $P = N/(1-\alpha)L$, increases as the overlap increases; consequently, more computational effort is required for the same data length. Each segment is multiplied by a window function, $w_0(\tau)$, and DFT of the product is computed. Note that here the window is applied in the data domain. Compare this with that in BT spectrum method where the window is applied in the covariance domain. The window applied in the data domain is known as a linear window and the one applied in the covariance domain as a quadratic window. The direct estimation of spectrum involves evaluation of the following:

$$
\begin{aligned}
\hat{S}_x(k) &= \frac{1}{LP} \sum_{i=0}^{P-1} X_i(k) X_i^*(k) \\
\hat{S}_{xy}(k) &= \frac{1}{LP} \sum_{i=0}^{P-1} X_i(k) Y_i^*(k)
\end{aligned}
\tag{3.24}
$$

It is assumed that the segments are uncorrelated.

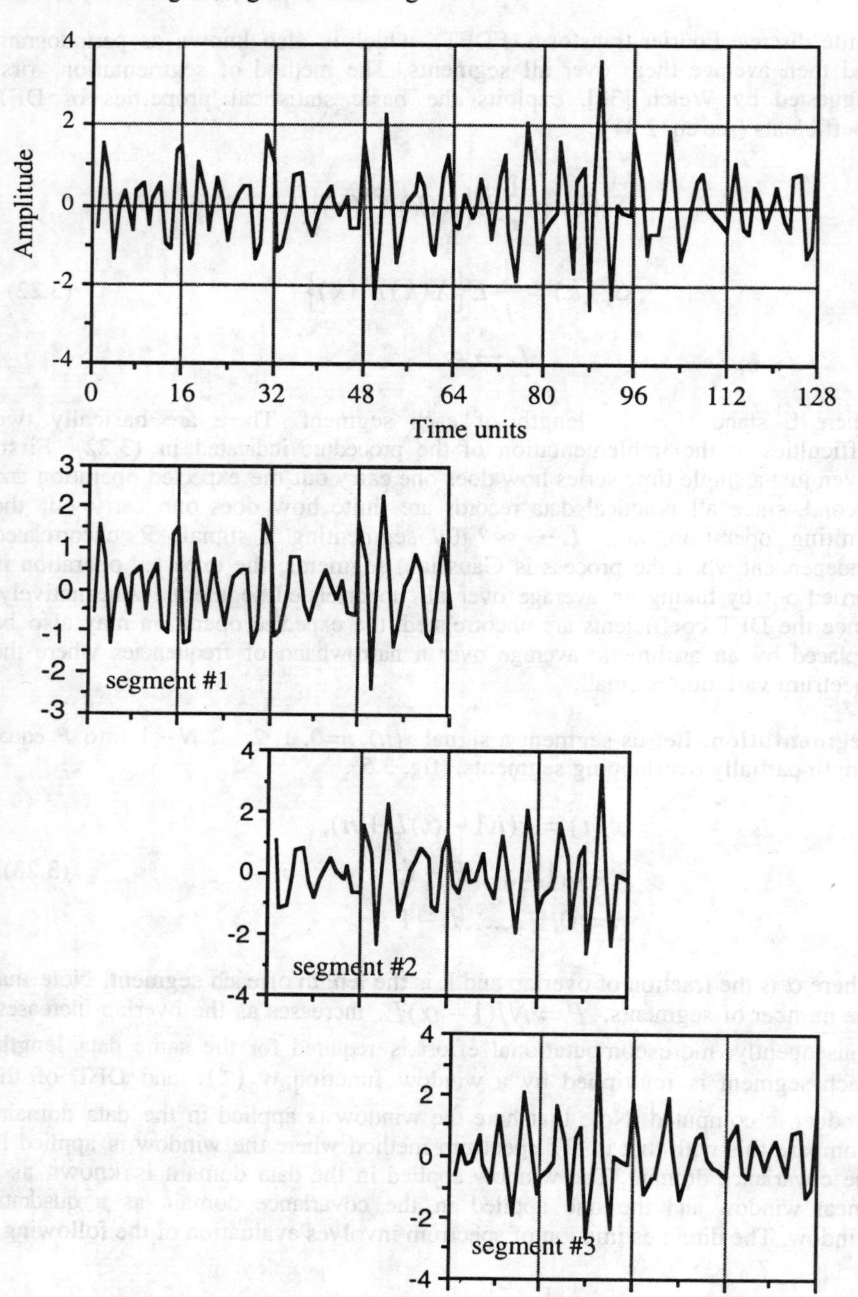

Figure 3.5: Segmentation of a long time series (bandlimited signal) with 50% overlap.

Example 3.3: In estimating the spectrum of a stochastic signal it is essential that the estimate must be averaged over a number of segments. When the averaging is done over a few segments it is possible that the estimate may show some peaks which really do not exist, they are mere processing artifacts. In this example we consider a bandlimited stochastic signal where the signal band is from 0.3π to 0.7π and from -0.3π to -0.7π. A sample of the stochastic signal is shown in fig. 3.5. The signal is divided into three overlapping segments as shown in fig.3.5. The periodograms of all three segments are shown in fig. 3 6(a-c). The average of three periodograms is shown in fig. 3.6(d) where the actual spectrum is also shown by a thick line. Evidently, the peaks appearing in the averaged spectrum are artifacts, which not present in the actual spectrum. Therefore, some caution must be exercised while interpreting the spectrum of a stochastic signal If we could divide the signal into more number of segments and then averaged the periodograms we would have obtained a spectrum resembling the true spectrum. This, unfortunately, cannot be carried on indefinitely as the segments will not remain uncorrelqated. Moreover, the resolution will be greatly reduced ♠.

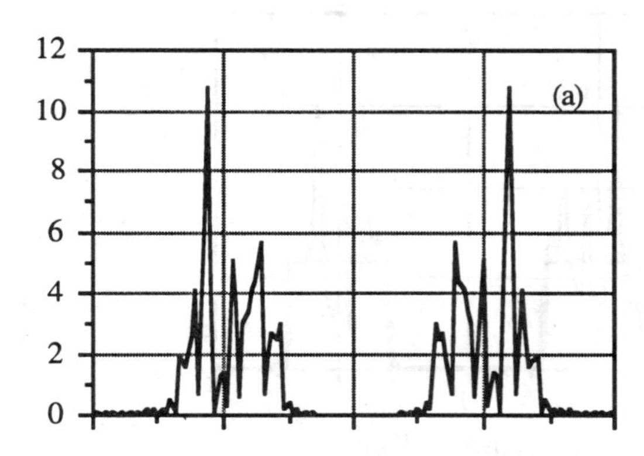

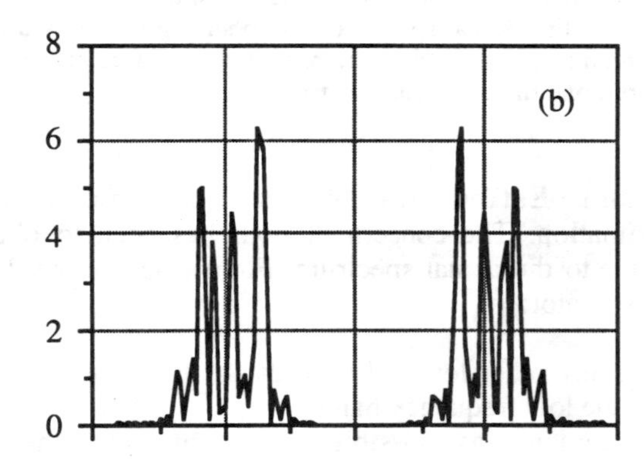

(Continued on next page)

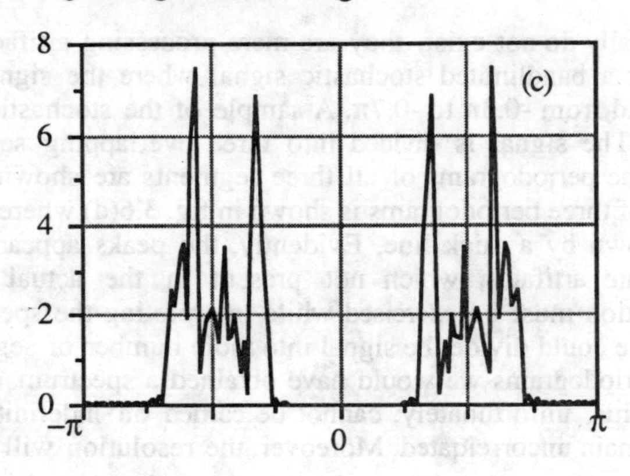

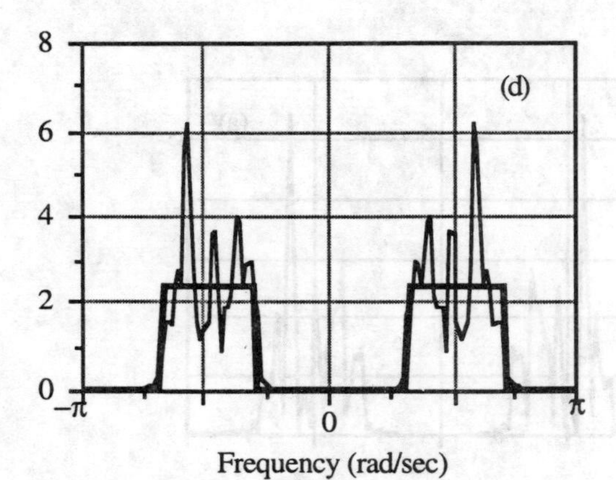

Figure 3.6: Periodograms of three overlapping segments are shown in (a, b, c) and the average is shown in (d). The theoretical spectrum is also shown in (d) by a thick line (figures c & d are shown on next page). Note the presence of a large number of spectral peaks when there are none in the actual spectrum.

Properties of Spectrum Estimators: We have described above two methods of spectrum estimation. The concern in both has been to obtain an estimate as close as possible to the actual spectrum. Following are the desirable properties of a spectrum estimator:

- Remove the mean and any other low order trends in the time series. This will minimize error in the low frequency band.
- Use as many data segments as possible consistent with the desired resolution. In BT estimator the maximum lag should be of the order of 10% of the data length.
- Use a window (linear or quadratic) with small sidelobes. This will reduce the bias error and improve the resolution
- Both spectrum estimators are unbiased only in the limiting case of $N \to \infty$ Further, a smooth spectrum has low bias.

Circular Autocorrelation: On performing an inverse Fourier transform of the discrete spectrum obtained from (3.24) we shall get a discrete covariance function but with a difference. It is a periodic function with period equal to L. The reason for this periodicity lies in the fact that the DFT operation implicitly assumes that the signal is periodic with a period of L. Indeed we can express the circular covariance function as an aliased form of the regular covariance estimate (3.1) windowed by $(1-|\tau|/L)$,

$$\hat{c}_{cir}(\tau) = \sum_{i=-\infty}^{\infty} \hat{c}_x(\tau + iL) \tag{3.25}$$

where $\hat{c}_x(\tau)$ is an estimate of the covariance function of windowed segments and $\hat{c}_{cir}(\tau)$ is inverse Fourier transform of Welch spectrum (3.24). The nature of aliasing is similar to the aliasing arising in the discrete spectrum (see (2.4)).

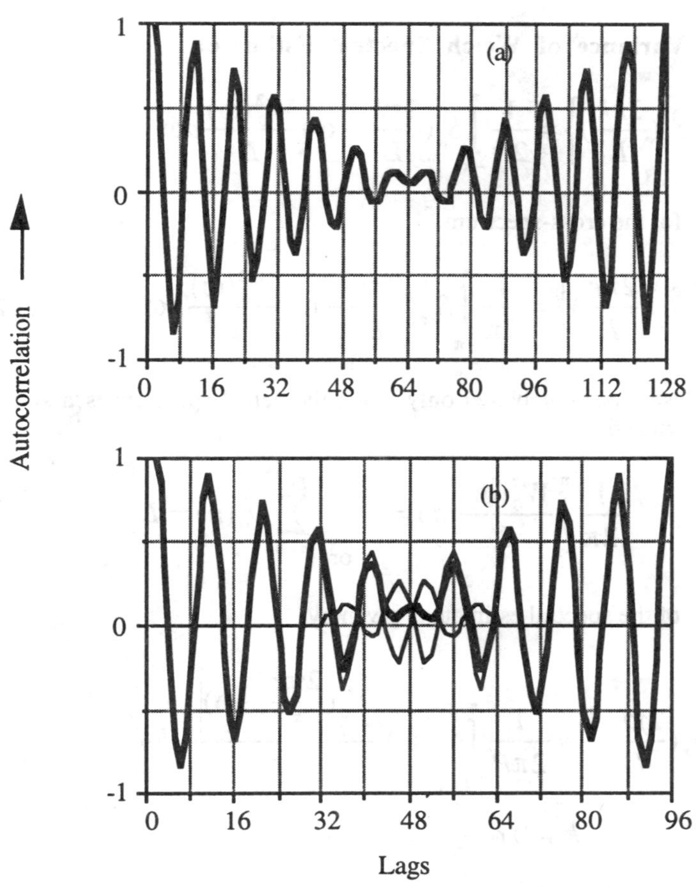

Figure 3.7: Aliasing arises in the estimation of covariance function. (a) 64 zeros were padded. There is no aliasing. (b) 32 zeros were padded. Aliasing error, on account of partial overlapping, is seen in the central part.

However, since the signal is limited to 0 to $(L-1)$, $\hat{c}_x(\tau)$ is confined to a lag interval of $\pm L$. The aliasing can be avoided only if each segment of the signal is padded with as many zeros as its length. By doing so we would indeed be increasing the period of the circular covariance function to $2L$, then there will be no aliasing. In the example given below we demonstrate the role of zero padding on the autocorrelation function obtained via inverse DFT of the spectrum.

Example 3.4 To demonstrate the aliasing error in the estimation of covariance (autocorrelation) function consider a simple random sinusoid, $x(n) = \cos(\omega n + \phi)$,, $n = 0,1,\ldots,N-1$ where $\omega = 0.2\pi$, ϕ is the random phase and L=64. The data segment is padded with zeros, 64 and 32 zeros, and then its DFT is computed. The inverse DFT of the Fourier coefficient magnitude square is shown below (fig. 3.7). There is no aliasing when the number of zeros padded is equal to the segment length. The role of the implicit triangular window is also seen ●.

Mean and Variance of Welch Spectral Estimate:

$$E\left\{\hat{S}_x(\frac{2\pi k}{L})\right\} = \frac{1}{2\pi}\int_{-\pi}^{\pi} S_x(\frac{2\pi k}{L} - \omega)\frac{W_0^2(\omega)}{L}d\omega \tag{3.26a}$$

and similarly for the cross-spectrum

$$E\left\{\hat{S}_{xy}(\frac{2\pi k}{L})\right\} = \frac{1}{2\pi}\int_{-\pi}^{\pi} S_{xy}(\frac{2\pi k}{L} - \omega)\frac{W_0^2(\omega)}{L}d\omega \tag{3.26b}$$

The spectral estimate is unbiased only when the actual spectrum is a smoothly varying function and

$$\frac{1}{2\pi}\int_{-\pi}^{\pi}\frac{W_0^2(\omega)}{L}d\omega = 1 \quad \text{or} \quad \sum_{n=0}^{L-1} w_0^2(n) = L$$

The variance of the spectral estimate is given by,

$$Var\left\{\hat{S}_x(\frac{2\pi k}{L})\right\} \approx \frac{1}{2\pi P}\int_{-\pi}^{\pi} S_x^2(\omega)\frac{\left|W_0(\frac{2\pi k}{L} - \omega)\right|^4}{L^2}d\omega \tag{3.27a}$$

$$k \neq 0 \; or \; \frac{L}{2}$$

Assume that $S_x(\omega)$ is smoothly varying function, approximately constant over the bandwidth of $W_0(\omega)$; then (3.27a) may be reduced to

$$\frac{Var\left\{\hat{S}_x(\frac{2\pi k}{L})\right\}}{S_x^2(\frac{2\pi k}{L})} \approx \frac{1}{P}\left[\frac{1}{2\pi}\int_{-\pi}^{\pi}\frac{|W_0(\omega)|^4}{L^2}d\omega\right] \tag{3.27b}$$

Table 3.1: Mean and variance the BT spectrum and Welch spectrum

BT Method

Formula	$\hat{S}_x(\omega) = \sum_{\tau=-\infty}^{\infty} w_1(\tau)\hat{c}_x(\tau)\exp(-j\omega\tau)$	**Comments:** A quadratic window is used.
Mean, (3.20)	$E\left\{\tilde{S}_x(\omega)\right\} = \sum_{k=0}^{N-1} W_1(\frac{2\pi k}{N})S_x(\omega - \frac{2\pi k}{N}$	Bias is zero only for white noise.
Variance, (3.21a)	$\frac{Var\left\{\hat{S}_x(\omega)\right\}}{S_x^2(\omega)} \approx \frac{L_{eff}}{N}$	For low variance, the effective width of the window must be a small fraction of the data length.

Welch Method

Formula	$\hat{S}_x(k) = \frac{1}{PL}\sum_{i=0}^{P-1} X_i(k)X_i^*(k)$	**Comments:** A linear window is used
Mean (3.26a)	$E\left\{\hat{S}_x(\frac{2\pi k}{L})\right\} =$ $\frac{1}{2\pi}\int_{-\pi}^{\pi} S_x(\frac{2\pi k}{L} - \omega)\frac{W_0^2(\omega)}{L}d\omega$	Bias is zero only for white noise.
Variance, (3.27a)	$Var\left\{\hat{S}_x(\frac{2\pi k}{L})\right\} \approx$ $\frac{1}{2\pi P}\int_{-\pi}^{\pi} S_x^2(\omega)\frac{\left\vert W_0(\frac{2\pi k}{L} - \omega)\right\vert^4}{L^2}d\omega$ $k \neq 0 \; or \; \frac{L}{2}$	The variance is inversely proportional to the number of segments over which the spectrum is averaged.

At $k=0$ or $L/2$ the variance of the spectral estimate will be twice that given in (3.27a). The variability in Welch spectral estimate is inversely proportional to the number of uncorrelated segments that we are able to create from the available signal length. This is often referred to as degrees of freedom (actually 2P for uncorrelated segments).

We have looked at two different methods of spectrum estimation and their bias and variability. A proper choice of the window is required in order that the bias is minimal and the variance is as low as possible. There are other properties of a window from the point of resolution and leakage of power. These will be discussed in the next section. A short summary of important formulas is given in table 3.1. Finally, we compare in table 3.2 the important differences between the BT and Welch spectra.

Table 3.2: A comparison of the BT and Welch spectrum estimation methods

Blackman-Tukey (BT)	**Welch**
Fourier transform of the estimated covariance function or the autocorrelaion	Averaged magnitude square of the Fourier transform of all data segments,
The autocorrletion function is windowed (quadratic window)	Each data segment is windowed (linear window)
The estimated spectrum may become negative.	It is always positive
The mean of the estimated spectrum is a convolution of the Fourier transform of a window and the actual spectrum	The mean of the estimated spectrum is a convolution of the spectrum of a window and the actual spectrum

§3.3 Spectral Windows

From (3.20 & 3.26a) we note that the expected value of a spectral estimate is equal to the convolution of the true spectrum of the signal and the spectrum of the linear window or the Fourier transform of the quadratic window. In the last section we have described the role of a window on bias and variance of the spectral estimate. Here we shall look into the role of a window on the leakage of power and on the resolution of spectrum.

Spectral Window: A spectral window is characterized by two attributes, namely, a mainlobe of finite width and a series of sidelobes. While the former is responsible for blurring sharp spectral features, if any, the latter is responsible for leakage of power, particularly from a strong peak into a weak peak. Let us first consider a simple uniform window (or default window) whose spectrum is illustrated in fig. 3.8. The 3dB (half) width of the main lobe is $1.75\pi/L$. The maximum sidelobe level is -13.5 dB. The first null occurs at $\pm 2\pi/L$ and the successive nulls are spaced $2\pi/L$ apart.

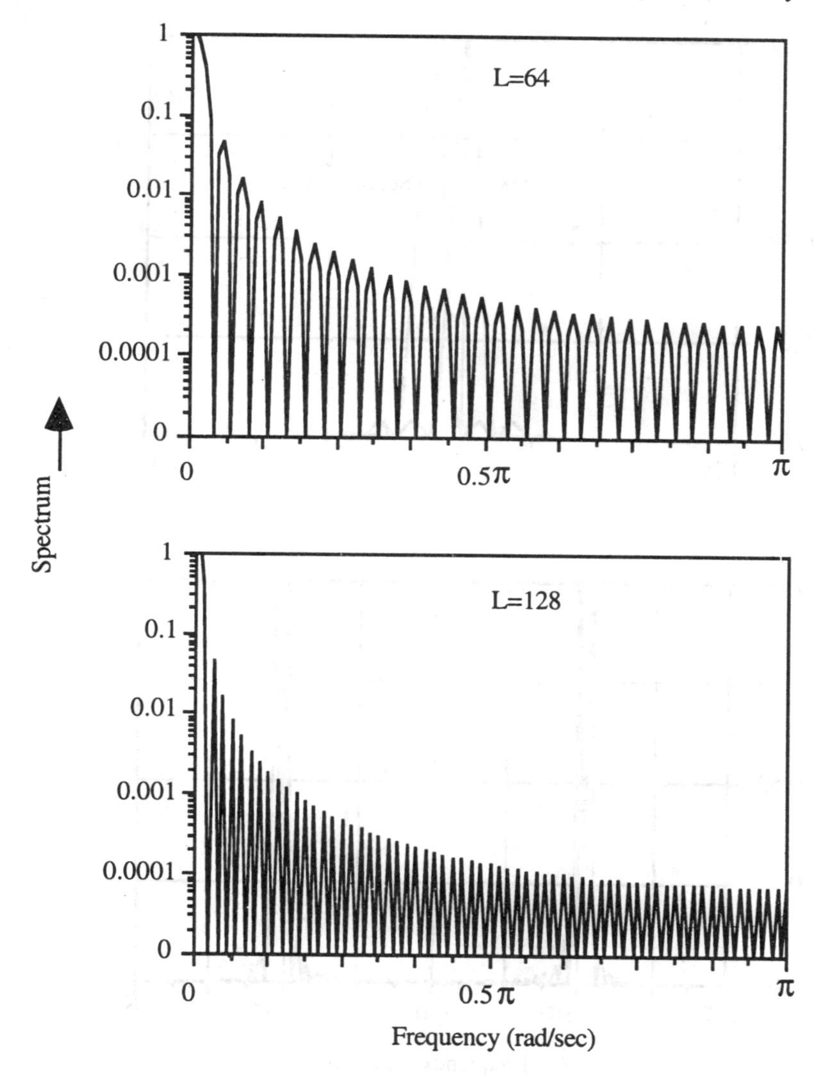

Figure 3.8: Spectrum of uniform window. Each window was padded with zeros to increase its length to 256. The sampling interval is assumed to be one second.

Rayleigh Resolution:
How close can two sinusoids get, yet they are capable of being resolved from their spectrum? Surely, the width of the main lobe must play the decisive role. We must first quantitatively define the resolution. According to the Rayleigh resolution criterion, two sinusoids are said to be resolved when the spectral peak of the first sinusoid falls on the first null of the second sinusoid (see fig. 3.9a). The spectrum of two random sinusoids with frequencies 0.25 and 0.2656 Hz, just satisfying the Rayleigh criterion, is shown in fig. 3.9b.

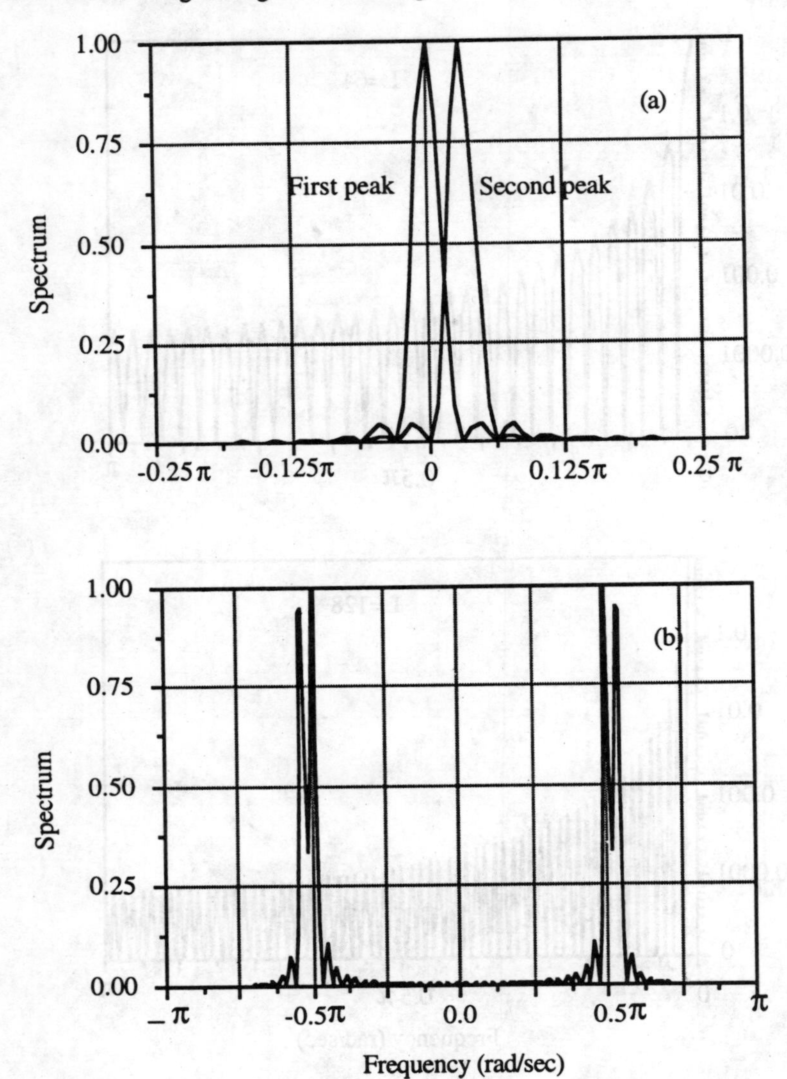

Figure 3.9: Rayleigh resolution criterion. (a) two sinusoids are said to be resolved when the spectral peak of the first falls on the first null of the second peak (b) The spectrum of two random sinusoids with frequencies 0.25 and 0.2656 Hz, just satisfying the Rayleigh criterion. Data length=64. The first zero is at $2\pi/64$.

Power Leakage: Leakage of power takes place through sidelobes of a window. Consider a signal, which is a sum of two random sinusoids with normalized frequencies 0.5076π and 0.5546π. The power in the first sinusoid is ten times that in the second sinusoid. The spectra of the sinusoids are individually shown in fig. 3.10 a & b respectively where the arrow represents the ideal spectrum.

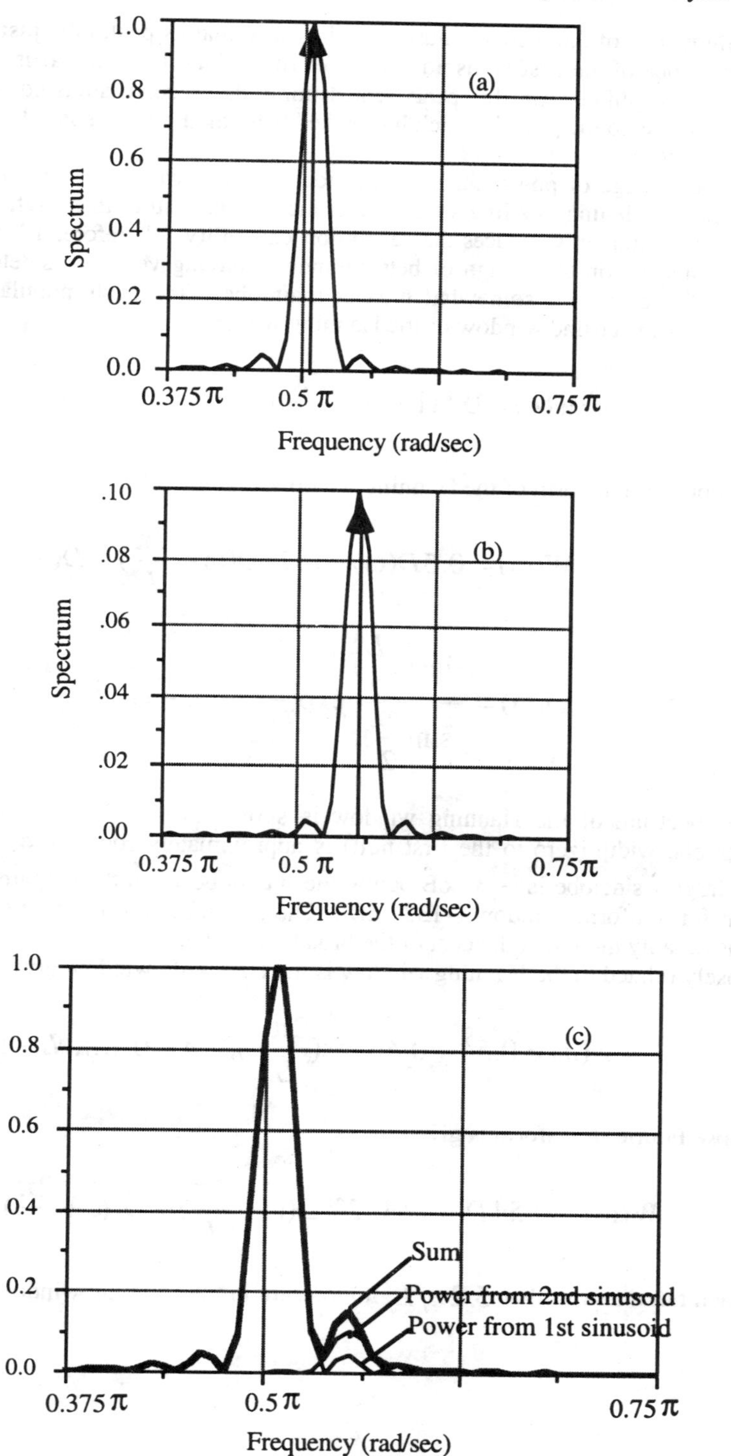

Figure 3.10: Phenomenon of power leakage from sidelobe (See text for explanation)

The frequency of the second sinusoid is chosen so that its peak falls just over the first sidelobe of the first sinusoid. The spectrum of the total signal is shown in (c). The magnitude of the peak representing the second sinusoid has been increased due to the power, which has leaked from the first sinusoid. The increase is nearly 46%.

The leakage of power through the sidelobes is perhaps by far more serious than the broadening of a line spectrum as the former results in a high bias error while the latter only reduces the resolution capability. Therefore, a lot of effort has been spent on the design of better windows having very low sidelobe level, but at the cost of a somewhat broader mainlobe. One such popular spectral window is the cosine window or the Hanning window,

$$w(n) = 0.5(1-\cos(\frac{2\pi}{L}n)), \quad n = 0,1,\ldots,L-1$$

The Fourier transform of the Hanning window is given by

$$W(\omega) = 0.5D(\omega) + 0.25D(\omega - \frac{2\pi}{L}) + D(\omega + \frac{2\pi}{L})$$

where

$$D(\omega) = \frac{\sin(\omega\frac{L}{2})}{\sin(\frac{\omega}{2})}\exp(-j\frac{\omega}{2}(L-1))$$

The spectrum of the Hanning window is shown in fig. 3.11 for L=64. The mainlobe width (zero to the first null) is approximately equal to $4.0\pi/L$ and the largest sidelobe is -32 dB below the mainlobe height (compare this with that for uniform window, -13.5 dB). The sidelobe level has been reduced considerably though at the cost of the broad mainlobe.
Closely related to the Hanning window is the Hamming window given by

$$w(n) = 0.54 - 0.46\cos(\frac{2\pi}{L}n), \quad n = 0,1,\ldots,L-1$$

whose Fourier transform is given by

$$W(\omega) = 0.54D(\omega) + 0.23[D(\omega - \frac{2\pi}{L}) + D(\omega + \frac{2\pi}{L})]$$

which has a new null at $5.2\pi/L$, not present in the Hanning window. The

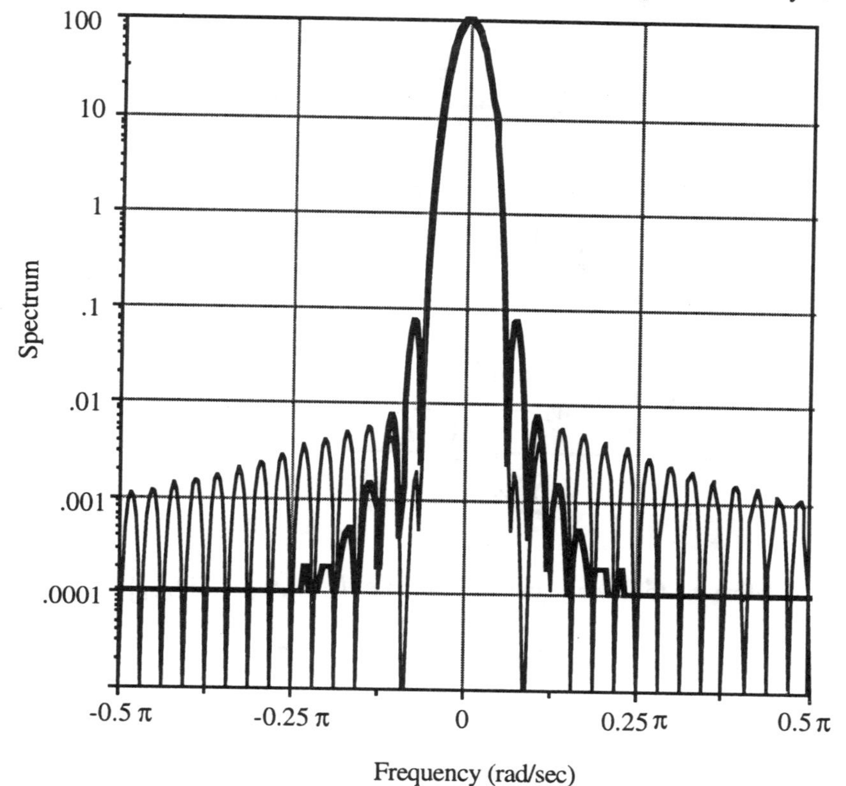

Figure 3.11: The spectra of Hanning (thick line) and Hamming (thin line) windows. Notice the difference in the sidelobe structure of the two windows in the neighbourhood of the first sidelobe. On account of extra zero at 0.08125π the first sidelobe has been pulled down.

spectrum of the Hamming window is shown in fig. 3.11. An important difference between the Hanning and Hamming windows is in their sidelobe structure. Notice that on account of the null at $2.6(2\pi/L)$ the first sidelobe of the Hamming window spectrum is considerably smaller than that of the Hanning window.

Optimum Windows: A window is optimized in the sense that the area under the mainlobe of its spectrum is maximum in relation to the total area [39] Let $w_0, w_1, \ldots, w_{L-1}$ be the window coefficients. The area under the mainlobe is given by $\frac{1}{2\pi}\int_{-\frac{2\pi}{L}}^{\frac{2\pi}{L}} |W(\omega)|^2 d\omega$ and the total area is given by $\frac{1}{2\pi}\int_{-\pi}^{\pi} |W(\omega)|^2 d\omega$ where $W(\omega)$ is the discrete Fourier transform of the window sequence. Substituting for $W(\omega)$ in terms of the window coefficients we obtain the area in the mainlobe and the total area as follows:

$$\frac{1}{2\pi}\int_{-\frac{2\pi}{L}}^{\frac{2\pi}{L}}|W(\omega)|^2 d\omega = \sum_{k=0}^{L-1}\sum_{l=0}^{L-1} w_k w_l \frac{\sin(\frac{2\pi}{L}(k-l))}{\pi(k-l)}$$

and $\frac{1}{2\pi}\int_{-\pi}^{\pi}|W(\omega)|^2 d\omega = \sum_{k=0}^{L-1} w_k^2$. The ratio of the area under the mainlobe to the total area is given by

$$\beta = \frac{\sum_{k=0}^{L-1}\sum_{l=0}^{L-1} w_k w_l \frac{\sin(\frac{2\pi}{L}(k-l))}{\pi(k-l)}}{\sum_{k=0}^{L-1} w_k^2} \tag{3.28a}$$

Next, we define the following quantities: $\mathbf{w} = col\{w_0, w_1, \ldots, w_{L-1}\}$

$$\mathbf{M} = \begin{bmatrix} \frac{2}{M} & \frac{\sin(\frac{2\pi}{L})}{\pi} & \cdots & \frac{\sin(\frac{2\pi}{L}(L-1))}{\pi(L-1)} \\ \frac{\sin(\frac{2\pi}{L})}{\pi} & \frac{2}{L} & \cdots & \frac{\sin(\frac{2\pi}{L}(L-2))}{\pi(L-2)} \\ & & & \\ \frac{\sin(\frac{2\pi}{L}(L-1))}{\pi(L-1)} & \frac{\sin(\frac{2\pi}{L}(L-2))}{\pi(L-2)} & \cdots & \frac{2}{L} \end{bmatrix}$$

In terms of $\mathbf{w}$ and $\mathbf{M}$ defined above we can rewrite the power ratio defined in (3.80a) as follows:

$$\beta = \frac{\mathbf{w}^T\mathbf{M}\mathbf{w}}{\mathbf{w}^T\mathbf{w}} \tag{3.28b}$$

We would like to maximize the power ratio under the constraint that $\sum_{k=0}^{L-1} w_k^2 = 1$. The solution to the above constrained maximization problem is that the window sequence must be equal to the eigenvector corresponding to the largest eigenvalue of matrix **M** and β is the corresponding largest eigenvalue. The optimum window has the property of concentrating energy in a narrowband, $\pm\frac{2\pi}{L}$, just like the well-known prolate spheroidal functions, introduced by Slepian, Pollack, and Landau [18]. The energy of the prolate spheroidal functions is concentrated in a specific band, $\pm\Delta\omega$. In fact, if we let $\Delta\omega = \frac{2\pi}{L}$, the optimum window turns out to be the discrete version of the prolate spheroidal function. A few examples of optimum window are listed in table 3.3. The frequency response of an optimum filter of length 64 is shown in fig. 3.12. Finally, in table 3.4 we give a comparison of all four types of windows. It may be observed that the sidelobe magnitude can be reduced only at the cost of increased mainlobe width which will then reduce the resolution of the spectrum. The optimum window however provides some relief in this regard.

Table 3.3: Optimum window coefficients taken from [39]. Note that only the right half of the window is shown above

L=16, β=0.98155523

1.0	0.9647692	0.8969838
0.8017354	0.6860384	0.5581594
0.4268421	0.3005089	

L=32, β=0.9811725

1.0	0.9911078	0.9734944
0.9474967	0.9136096	0.8724740
0.8248617	0.7716553	0.7138285
0.6524216	0.5885176	0.5232161
0.4576095	0.3927591	0.3296615
0.2694880		

L=64, β=0.9811725

1.0	0.9977716	0.9933259
0.9866837	0.9778773	0.9669485
0.9539502	0.9389438	0.9220010
0.9032019	0.8826353	0.8603978
0.8365931	0.8113318	0.7847307
0.7569119	0.7280011	0.6981291
0.6674292	0.6390368	0.6040891
0.5717239	0.5390787	0.5062901
0.4734933	0.4408209	0.4084023
0.3763631	0.3448242	0.3139016
0.2837055	0.2543395	

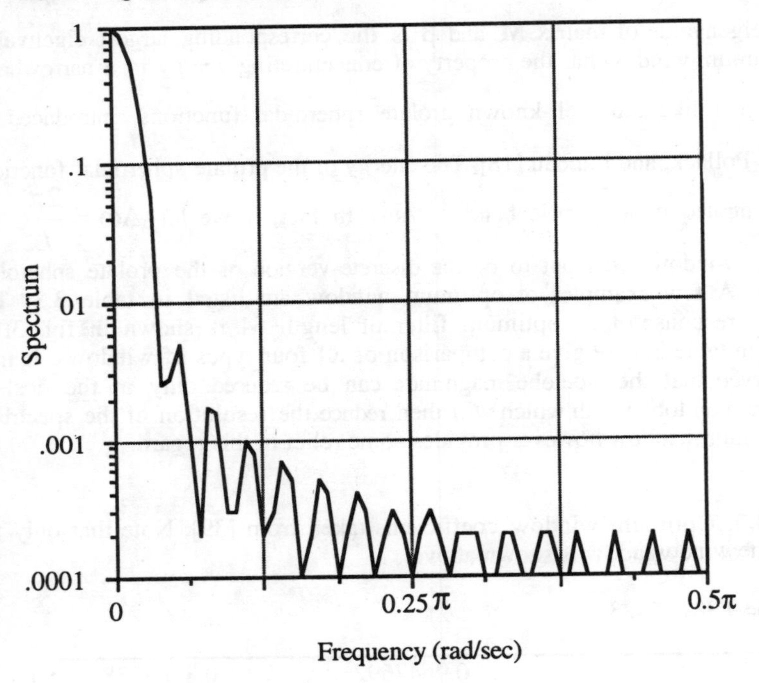

Figure 3.12: Spectrum of the optimum window. L=64. To express the ordinate in dB we need to compute 10*log of the numbers on the ordinate.

Table 3.4: A relative comparison of four different types of commonly used spectral (linear) windows L stands for the length of the window

Type of window	3dB (full) width	Zero to first null	First sidelobe
Uniform	$\frac{1.75\pi}{L}$	$\frac{2\pi}{L}$	-13 5dB
Hanning	$\frac{2.75\pi}{L}$	$\frac{4\pi}{L}$	-32 dB
Hamming	$\frac{2.75\pi}{L}$	$\frac{4\pi}{L}$	-44 dB
Optimum	$\frac{2.134\pi}{L}$	$\frac{2.75\pi}{L}$	-23 dB

§3.4 **Higher Order Spectrum** (Bispectrum)

The covariance function or the spectrum completely characterizes a Gaussian stationary stochastic signal. However, for a non-Gaussian signal, higher order spectral density function would be required for complete characterization of the process. Since the third moment is an important measure of non-Gaussian character, we shall concentrate on the third order spectrum, also known as bispectrum, of a non-Gaussian stationary stochastic signal. First, we introduce the bicovariance function analogous to the covariance function in second order characterization. Bispectrum is simply the Fourier transform of the bicovariance function.

Bicovariance function: Let $f(x_1,x_2,x_3;n_2-n_1,n_3-n_1)$ be a third order probability density function of a zero mean stationary stochastic process, where x_1,x_2 and x_3 are samples of *x(t)* at time instants n_1,n_2 and n_3, respectively. The third moment, known as bicovariance, is defined as follows:

$$\begin{aligned} B_c(s,h) &= E\{x(n)x(n+s)x(n+h)\} \\ &= \iiint x_1x_2x_3 f(x_1,x_2,x_3;s,h)dx_1dx_2dx_3 \end{aligned} \tag{3.29}$$

We shall assume *x(n)* is a real process. The bicovariance function has following properties:

$$\begin{aligned} &B_c(s,h) = B_c(h,s) = B_c(s-h,-h) = \\ &B_c(-h,s-h) = B_c(h-s,-s) = B_c(-s,h-s) \end{aligned} \tag{3.30}$$

Note that Bc(0,0)=E{$x^3(n)$}, the third moment, is equal to zero for any symmetrically distributed random variable, including the Gaussian random variable. However, only for a Gaussian process the bicovariance function is equal to zero for all s and h.
The bicovariance function defined on a line passing through the center of coordinates with a slope = +1.0, that is, $s=h=\tau$ has a particularly interesting form:

$$Bc_r(\tau) = E\{x(n)x^2(n+\tau)\}$$

Using the symmetry properties (3.30) we obtain that $Bc_r(\tau)=Bc(0,\tau)=Bc(\tau,0)$. Similarly the bicovariance function defined on a line passing through the center of coordinates with a slope = − 1.0 has the following form:

$$Bc_l(\tau) = E\{x(n)x(n+\tau)x(n-\tau)\}$$

where $s=-h=\tau$. Note that $Bc_r(\tau) \neq Bc_r(-\tau)$ but $Bc_l(\tau) = Bc_l(-\tau)$.
The bicovariance function of a non-Gaussian white noise process is $Bc_\eta(s,h) = \beta_\eta^3\delta_{s,h}$. where $\beta_\eta^3 = E\{\eta^3\}$ is the third moment of the white noise $\eta(n)$ (assumed to be zero mean) and $\delta_{s,h}=1$ for s=h=0, otherwise $\delta_{s,h}=0$.

Bispectrum: Bispectrum is a two-dimensional Fourier transform of the bicovariance function,

$$Bs_x(\omega_1,\omega_2) = \sum_s \sum_h Bc_x(s,h)\exp(-j(s\omega_1 + h\omega_2)) \tag{3.31a}$$

where ω_1 and ω_2 are continuous frequencies ranging from $-\pi$ to $+\pi$ and its inverse is given by

$$Bc_x(s,h) = \frac{1}{4\pi^2}\int_{-\pi}^{+\pi}\int Bs_x(\omega_1,\omega_2)\exp(j(s\omega_1 + h\omega_2))d\omega_1 d\omega_2 \tag{3.31b}$$

On account of the symmetry properties of the bicovariance function (3.30) the bispectrum will have the following symmetries:

$$\begin{aligned} Bs_x(\omega_1,\omega_2) &= Bs_x(\omega_2,\omega_1) \\ &= Bs_x(-\omega_1-\omega_2,\omega_2) \\ &= Bs_x(\omega_1,-\omega_1-\omega_2) \end{aligned} \tag{3.32a}$$

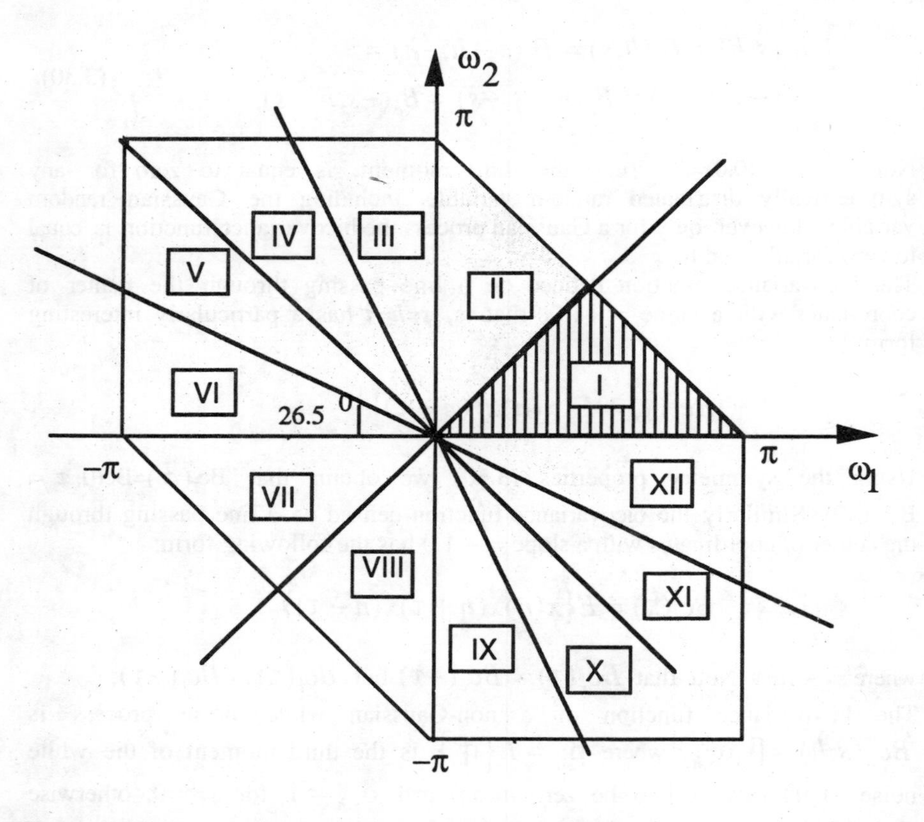

Figure 3.13: Symmetry of digital bispectrum of a real stationary stochastic process is shown above. It is enough if the bispectrum is computed in any one of the regions, for example, in the first region.

When *x(n)* is real we have the following two additional symmetries:

$$Bs_x(\omega_1,\omega_2) = Bs_x^*(-\omega_1,-\omega_2)$$
$$Bs_x(\omega_1,-\omega_2) = Bs_x^*(-\omega_1,\omega_2)) \tag{3.32b}$$

The bispectrum becomes periodic outside $|\omega_1| \le \pi$, $|\omega_2| \le \pi$ and $|\omega_1 + \omega_2| \le \pi$. The first two inequalities are satisfied inside a square of size $2\pi \times 2\pi$ while the last inequality is satisfied in a region bounded by $\omega_1 + \omega_2 = \pi$ and $\omega_1 + \omega_2 = -\pi$. The intersection of these two regions is a hexagon as shown in fig. 3.13. This is the principal domain for the discrete bispectrum. In this domain, on account of symmetries (3.32), it is enough if we compute the bispectrum in any one of the twelve regions, for example, the triangular hatched domain in fig. 3.13 (see [6] for more details). Bispecrum of non-Gaussian white noise is constant equal to third moment, β_η^3. Note that the bispectrum of Gaussian noise is zero for all s and h.

The bispectrum, unlike the spectrum, is a complex function with real and imaginary parts,

$$Bs_x(\omega_1,\omega_2) = \alpha(\omega_1,\omega_2) + j\beta(\omega_1,\omega_2)$$

Bicoherence is defined as

$$bs(\omega_1,\omega_2) = \frac{Bs(\omega_1,\omega_2)}{\sqrt{S_x(\omega_1)S_x(\omega_2)S_x(\omega_1+\omega_2)}} \tag{3.33}$$

It represents the fraction of the power at sum frequency $(\omega_1 + \omega_2)$ due to interaction of power at ω_1 and ω_2. It is equal to 1.0 when all the power is due to interaction. An example of this is given in Example 3.5.

Example 3.5: A test signal is generated as follows:

$$x(n) = \cos(0.22\pi n + \theta_k) + \cos(0.375\pi n + \theta_l) + \frac{1}{2}\cos(0.595\pi n + \theta_a)$$
$$+ \cos(0.22\pi n + \theta_k)\cos(0.375\pi n + \theta_l) + \eta(n)$$

The phases θ_k, θ_l and θ_a are taken from random numbers in the range of $-\pi$ to $+\pi$. $\eta(t)$ is the background Gaussian white noise (SNR=20 dB). The power spectrum of the above test signal showed four peaks at frequencies 0.155π, 0.22π, 0.375π, 0.595π. The fourth term will give rise to two sinusoids at the sum frequency, 0.595π, and difference frequency, 0.155π. The bicoherence at ω_1=0.22π and ω_2=0.375π was found to be 0.48 and that at ω_1=-0.22π and ω_2=0.375π was found to be 0.99. The higher bicoherence is due to the fact that the power at the difference frequency, that is, at $\pm 0.155\pi$, is entirely due to the interaction. This theoretical model formed the basis for understanding the non linearly coupled waves in plasma reported in [19] from where the above example is taken ●.

Interpretation of Bispectrum: We would now like to provide a physical meaning to the bispectrum. Let us consider three ideal bandpass filters, $h_k(n)$, $h_l(n)$ and $h_m(n)$, centered at frequencies $\pm\omega_k$, $\pm\omega_l$, and $\pm\omega_m$, respectively. Further, the center frequencies are related as $\omega_m = -(\omega_k + \omega_l)$. The bandwidth is Δ $(\Delta \approx 0)$. A signal, $x(n)$, is passed through each filter (see fig. 3.14). Let the outputs be $x_k(n)$, $x_l(n)$ and $x_m(n)$ where

$$x_k(n) = h_k(n) * x(n)$$

$$x_l(n) = h_l(n) * x(n)$$

$$x_m(n) = h_m(n) * x(n)$$

where * denotes convolution. We would like to evaluate the expectation of the triple product,

$$E\{x_k x_l x_m\}$$

$$= \sum\sum\sum h_k(n')h_l(n'')h_m(n''')E\{x(n-n')x(n-n'')x(n-n''')\}$$

$$= \sum\sum\sum h_k(n')h_l(n'')h_m(n''')Bc_x(n''-n', n'''-n') \tag{3.34}$$

Let us replace the filter impulse response functions in (3.34) by their Fourier representation.

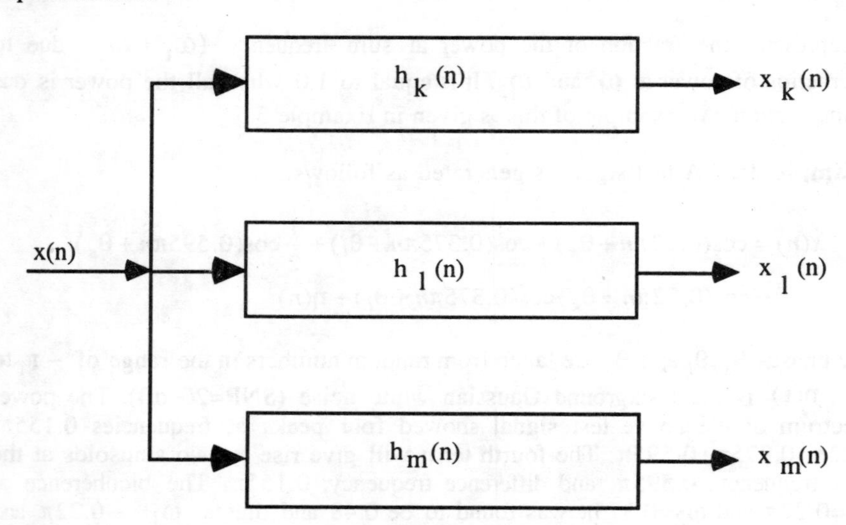

Figure 3.14: A signal is passed through three ideal bandpass filters.

$h_k(n) = \frac{1}{2\pi} \int_{-\pi}^{\pi} H_k(\omega) \exp(j\omega n) d\omega$ and so on. We obtain

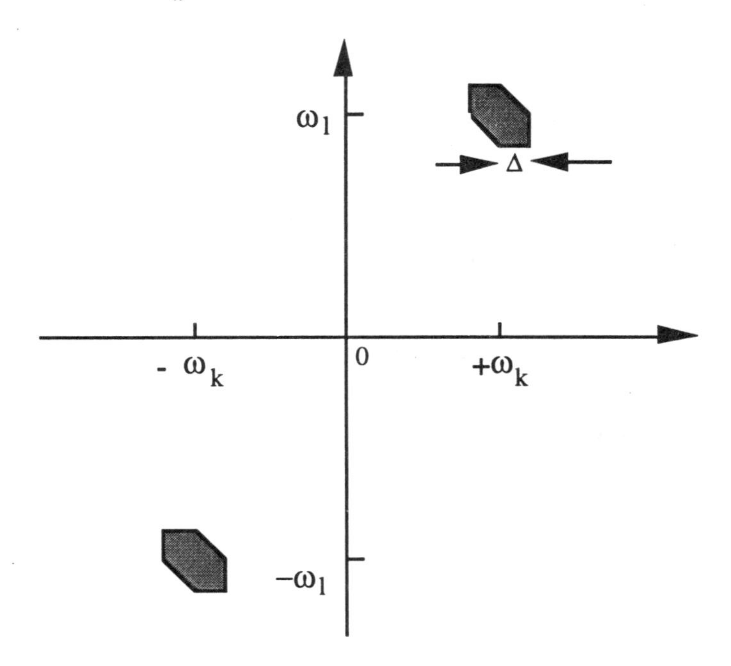

Figure 3.15: The pass band of the product $H_k(\omega_1)H_l(\omega_2)H_m(-\omega_1 - \omega_2)$.

$$E\{x_k x_l x_m\}$$

$$= \frac{1}{(2\pi)^3} \int\int\int \begin{bmatrix} H_k(\omega_1)H_l(\omega_2)H_m(\omega_3) \\ Bs_x(\omega_1,\omega_2)\delta(\omega_1 + \omega_2 + \omega_3) \end{bmatrix} d\omega_1 d\omega_2 d\omega_3 \tag{3.35}$$

$$= \frac{1}{(2\pi)^2} \int\int H_k(\omega_1)H_l(\omega_2)H_m(-\omega_1 - \omega_2)Bs_x(\omega_1,\omega_2)d\omega_1 d\omega_2$$

The filter pass band of the product $H_k(\omega_1)H_l(\omega_2)H_m(-\omega_1 - \omega_2)$ in (3.35) is shown in fig. 3.15. Note that the area under the pass region is equal to $\frac{3}{4}\Delta^2$ where Δ is the pass bandwidth of the filter and the shape of the pass region is a hexagon instead of the usual square shape. To show this, consider a point with coordinates $\omega_1 + \delta\omega_1$ and $\omega_2 + \delta\omega_2$ lying inside the pass region. Further, this point must also lie within the pass region of individual filters, therefore, the coordinates must satisfy the following inequalities:

$$-\frac{\Delta}{2} \le \delta\omega_1 \le \frac{\Delta}{2}$$

$$-\frac{\Delta}{2} \le \delta\omega_2 \le \frac{\Delta}{2}$$

$$-\frac{\Delta}{2} \le \delta\omega_1 + \delta\omega_2 \le \frac{\Delta}{2}$$

This requires that the shape of the pass region be as shown in fig. 3.15. We shall assume that the bispectrum is almost constant within the pass region. Using the symmetry properties of the bispectrum of a real process (3.33) we obtain the following result:

$$E\{x_k x_l x_m\} = \frac{3}{2}\Delta^2 \operatorname{Re}\{Bs_x(\omega_k, \omega_l)\} \tag{3.36}$$

Estimation of Bispectrum: A method analogous to the Welch method for spectrum estimation may be devised to estimate the bispectrum. The first step is to split the signal into smaller overlapping segments each of length L as illustrated in fig.3.5. Each segment is first multiplied with a suitable window, then Fourier transformed (FDFT). Let $X_p(k),\ p = 1,\ldots,P$ be the Fourier transforms of the segments. An estimate of the bispectrum is given by

$$\hat{B}_s(k,l) = \frac{1}{LP}\sum_p^P X_p(k) X_p(l) X_p(-k-l) \tag{3.37}$$

The expected value of the bispectrum estimate for large L is given by

$$E\{\hat{B}_s(k,l)\}$$

$$= \sum_{s=-L+1}^{L-1} \sum_{h=-L+1}^{L-1} \frac{(L-1-\min(s,h))}{L} Bc(s,h) e^{-js\frac{2\pi k}{L}} e^{-jhl\frac{2\pi}{L}}$$

$$\approx \frac{1}{4\pi^2}\int_{-\pi}^{\pi}\int_{-\pi}^{\pi} Bs(\omega_1,\omega_2) \sum_{s=-L+1}^{L-1} \sum_{h=-L+1}^{L-1} e^{j(\omega_1-\frac{2\pi k}{L})s} e^{j(\omega_2-\frac{2\pi l}{L})h} d\omega_1 d\omega_2$$

$$= \frac{1}{4\pi^2}\int_{-\pi}^{\pi}\int_{-\pi}^{\pi} \left[Bs(\omega_1,\omega_2) \frac{\sin((\omega_1 - \frac{2\pi k}{L})\frac{2L-1}{2})}{\sin((\omega_1 - \frac{2\pi k}{L})\frac{1}{2})} \frac{\sin((\omega_2 - \frac{2\pi l}{L})\frac{2L-1}{2})}{\sin((\omega_2 - \frac{2\pi l}{L})\frac{1}{2})} \right] d\omega_1 d\omega_2 \tag{3.38)}$$

§3.5 Time Varying Spectrum

The assumptions of stationarity and time invariance, while mathematically attractive, are impractical in most cases. For example, a speech signal clearly does not remain stationary beyond a few seconds and so is the seismic signal from an earthquake. Consider a frequency-hopping signal, often used in radar and communication. The transition from one frequency to another can be determined by means of time varying spectrum analysis. We shall introduce two simple methods of estimating the time varying spectrum, namely, a method suggested by Bendat and Piersol [20] and Short Term Fourier Transform method.

Linear Time Varying System: The input-output relation of a linear time varying (LTV) system is given by

$$x(t) = \int_0^\infty h(t,\tau)\varepsilon(t-\tau)d\tau \qquad (3.39)$$

$h(t,\tau)$ is the time varying impulse response function. Recall that for LTI system the impulse response function is independent of t. $\varepsilon(t)$ is stationary stochastic input to LTV system. The Fourier representation of the time varying impulse response function is given by

$$h(t,\tau) = \frac{1}{2\pi}\int_{-\infty}^{\infty} H(t,\omega)e^{j\omega\tau}d\omega$$

where $H(t,\omega)$ is a time varying transfer function of a filter. We use this representation in (3.39) and obtain an input-output relation in the frequency domain,

$$x(t) = \frac{1}{2\pi}\int_{-\infty}^{\infty} H(t,\omega)\mathrm{E}(\omega)e^{j\omega t}d\omega \qquad (3.40)$$

where $\mathrm{E}(\omega)$ is Fourier transform of $\varepsilon(t)$. This is schematically shown in fig. 3.16. When the input is a discrete stochastic signal the output will also be a discrete stochastic signal. In place of (3.40) we have

$$x(n) = \frac{1}{2\pi}\int_{-\pi}^{\pi} H(n,\omega)\mathrm{E}(\omega)e^{j\omega n}d\omega \qquad (3.41)$$

which we shall call a Fourier representation of discrete time LTV system. A component of *x(n)* at some frequency ω_0, $x(n,\omega_0) = H(n,\omega_0)\mathrm{E}(\omega_0)e^{j\omega_0 n}$, may be considered as a discrete complex exponential signal of frequency ω_0 with time varying amplitude and phase. On squaring $x(n,\omega_0)$ and evaluating its expected value we obtain the time varying spectrum, also known as evolutionary spectrum.

$$S_x(n,\omega) = |H(n,\omega)|^2 S_\varepsilon(\omega) \tag{3.42}$$

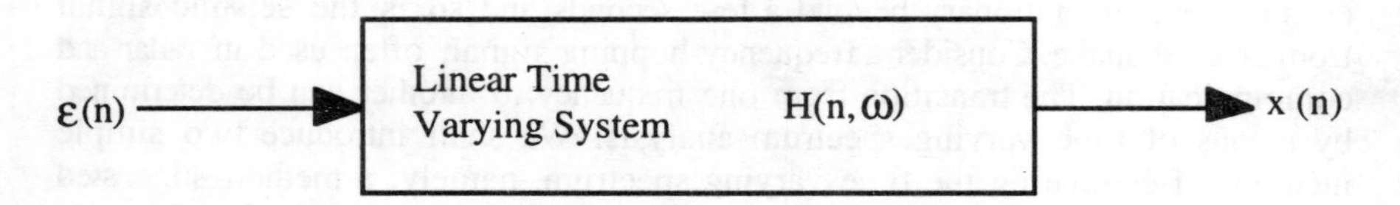

Figure 3.16: A linear time varying system driven by stationary white noise processes.

We define a time varying covariance function as follows:

$$c_x(n,\tau) = E\left\{x(n+\frac{\tau}{2})x(n-\frac{\tau}{2})\right\} \tag{3.43}$$

From the above equation we note that $c_x(n,\tau)$ is real and symmetric,

$$c_x(n,\tau) = c_x(n,-\tau)$$

Now, using (3.41) in (3.43) we obtain

$$c_x(n,\tau) = \frac{1}{2\pi}\int_{-\pi}^{\pi} H(n+\frac{\tau}{2},\omega)H^*(n-\frac{\tau}{2},\omega)S_\varepsilon(\omega)e^{j\omega\tau}d\omega \tag{3.44}$$

which may be considered as a spectral representation of the time varying covariance function. However, there is no unique definition of a time varying spectrum. Bendat and Piersol [20] proposed the following definition of a time varying spectrum

$$\begin{aligned} S_x(n,\omega) &= \sum_{\tau=-\infty}^{\infty} c_x(n,\tau)e^{-j\omega\tau} \\ &= \sum_{\tau=-\infty}^{\infty} E\left\{x(n+\frac{\tau}{2})x(n-\frac{\tau}{2})\right\}e^{-j\omega\tau} \end{aligned} \tag{3.45}$$

Example 3.6: In order to illustrate the significance of time varying spectrum we consider a frequency hopping signal where there are four random sinusoids with radian frequencies, 0.6π, 0.7π, 0.4π and 0.5π. as illustrated in fig. 3.17a. The time varying spectrum, given by (3.45), was programmed in Matlab. For the purpose of evaluating the expected operation in (3.45) eight independent samples of the signal were generated. The maximum value of the lag was fixed at 16. In order to avoid interpolation we have used only even lags. The resulting time varying spectrum is shown in fig. 3.17b.

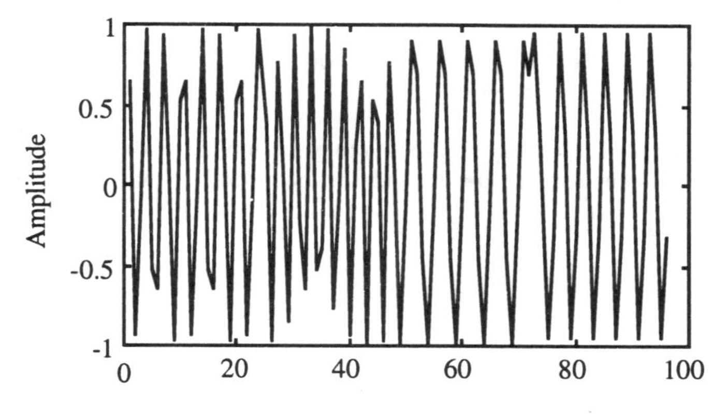

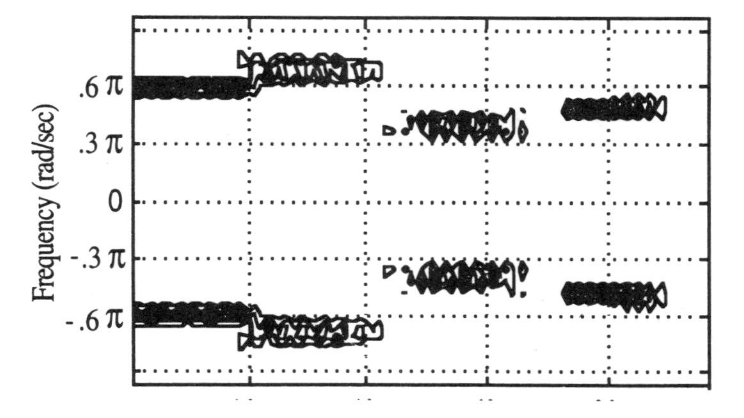

Figure 3.17: An illustration of time varying spectrum of a frequency-hopping signal .

A complete statistical characterization of a non-stationary signal is difficult when just one sample of the stochastic process available. It may be more appropriate to treat the single time series as a deterministic signal that is then mapped into the time-frequency plane. The map shows us how the various frequency components are distributed in the time-frequency plane. The physical quantity measured as a function of time and frequency is related to the energy but this relationship is often involved. Yet, a considerable insight into the structure of the signal can be gained from the time-frequency distributions. One such example of time-frequency distribution is spectrogram, which we shall briefly outline below. Other examples of time-frequency distributions are WV distribution, Wavelet transform, and Gabor transform [6].

Short Time Fourier Transform (STFT): When the time varying filter, $H(t, \omega)$, is a slowly varying function of time it may be approximated by the Fourier transform of a short segment of data assuming that the filter transfer function does not vary much over the duration of the segment. The short time Fourier transform (STFT) is defined as

$$X(n,\omega) = \sum_{\tau=-\infty}^{\infty} x(\tau)h(n-\tau)e^{-j\omega\tau} \tag{3.46}$$

where $h(n)$ is a lowpass filter. Equation (3.46) may be looked upon as an output of a lowpass filter (centered at $\omega = 0$ and the bandwidth $= \Delta\omega$) with input $x(n)e^{-j\omega n}$ (see fig. 3.18).

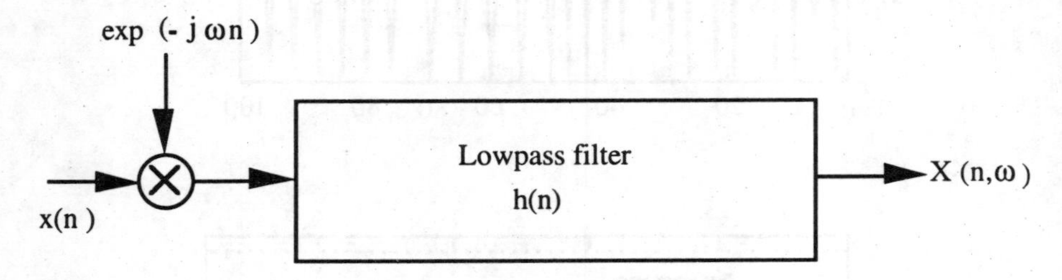

Figure 3.18: Short time Fourier transform is shown as an output of a lowpass filter whose input is a frequency shifted signal.

The STFT $X(n,\omega)$ is a complex function even though the actual signal is real. Further, $X(n,\omega)$, is discrete with its time and frequency intervals determined by the effective bandwidth and the effective length of the lowpass filter. There is, however, no unique definition of an effective length or an effective bandwidth. We shall define these quantities as follows: Let $h(n)$ be an infinite length lowpass filter and $H(\omega)$ be its Fourier transform. The effective length of a signal and its effective bandwidth may be defined as

$$\Delta n_{eff} = 2\sqrt{\frac{\sum_{n=-\infty}^{\infty} n^2 h^2(n)}{\sum_{n=-\infty}^{\infty} h^2(n)}}$$

and $$\Delta\omega_{eff} = 2\sqrt{\frac{\int_{-\pi}^{\pi} \omega^2 |H(\omega)|^2 d\omega}{\int_{-\pi}^{\pi} |H(\omega)|^2 d\omega}}$$

respectively. It is known that

$$\Delta n_{eff}\Delta\omega_{eff} \geq 2\sqrt{2} \qquad \text{(Uncertainty principle)}$$

where the sign of equality holds for a Gaussian window. The STFT with Gaussian window is known as a Gabor transform [6]. The STFT will be sampled in time with interval equal to Δn_{eff} and in frequency with interval equal to $\Delta\omega_{eff}$..

Next we show a relationship between squared STFT and time varying spectrum. In (3.46) substitute for *x(n)* using (3.41). We get

$$X(n,\omega)=\frac{1}{2\pi}\int_{-\pi}^{\pi}\mathrm{E}(\omega')d\omega'\sum_{\tau=-\infty}^{\infty}h(n-\tau)H(\tau,\omega')e^{j(\omega'-\omega)\tau}$$

$$=\frac{1}{2\pi}\int_{-\pi}^{\pi}\mathrm{E}(\omega')e^{j(\omega'-\omega)n}d\omega'\sum_{\tau'=-\infty}^{\infty}h(\tau')H(n-\tau',\omega')e^{-j(\omega'-\omega)\tau'} \tag{3.47}$$

We assume that

$$H(n-\tau',\omega')\approx H(n,\omega)$$

over the effective width of the lowpass filter, then (3.47) reduces to

$$X(n,\omega)\approx H(n,\omega)e^{-j\omega n}\frac{1}{2\pi}\int_{\omega-\frac{\Delta\omega}{2}}^{\omega+\frac{\Delta\omega}{2}}\mathrm{E}(\omega')e^{j\omega'n}d\omega' \tag{3.48}$$

It is now possible to relate the STFT to the time varying spectrum defined in (3.42). For this find the expected value of the magnitude square on both sides of (3.48)

$$E\left\{|X(n,\omega)|^2\right\}\approx|H(n,\omega)|^2\frac{\sigma_\varepsilon^2}{2\pi}\Delta\omega$$
$$=S_x(n,\omega)\Delta\omega \tag{3.49}$$

where σ_ε^2 is the variance of the input to time varying system. Equation (3.49) forms the basis for estimation of the time varying spectrum via STFT. The simplest approach, therefore, is to use a bank of narrowbandpass filters whose outputs are squared and averaged over a moving window. The result is often known as a spectrogram. The method works well when the time variations are slow.

§3.6 Exercises

Problems:

1. Consider the following arrangement (fig. 3.19) where two LTI systems are connected in parallel. Compute coherence between $y_1(n)$ and $y_2(n)$.

2. Consider a signal, which is a sum of p random sinusoids, $x(n) =$

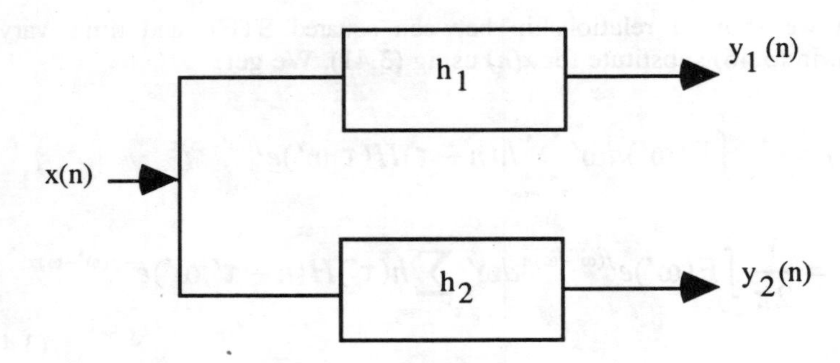

Figure 3.19: Two LTI systems connected in parallel.

$\sum_{i=1}^{p} \alpha_i \cos(\omega_i n) + \beta_i \sin(\omega_i n)$ where α_i and β_i are zero mean uncorrelated random variables with variance σ^2. Compute the covariance function of *x(n)* defined via the expected operation, $c_x(\tau) = E\{x(n)x(n+\tau)\}$ and via the arithmetic averaging,

$$\hat{c}_x(\tau) = 1/(N-\tau)\sum_{n=0}^{N-\tau} x(n)x(n+\tau) \;\; N \to \infty.$$

What is the difference between $c_x(\tau)$ and $\hat{c}_x(\tau)$?.

3. Evaluate the factor due to a window in equation (3.38c). Assume that a uniform window is used. [Hint: Apply the Parseval's theorem to autocorrelation of the window].

4. Consider a triangular (quadratic) window given by $w_1(n) = 1 - |n|/L$. Compute its Fourier transform and its 3dB width, the first null and the height of the first sidelobe. When this window is used in BT spectrum estimation show that the estimate is guaranteed to be positive.

5. Given two random sinusoids with distinct frequencies it is possible to minimize power leakage by means of a suitable choice of the length of a window. This may be achieved by positioning the peak of the second sinusoid just above the null of the first sinusoid. Devise a numerical example to illustrate this possibility. Assume Hanning window.

6. We often subtract the mean from the signal before computing its spectrum. As a result, the estimated spectrum at $\omega = 0$ is always zero even though the true spectrum has a finite value. How would you then estimate the spectrum at the zero frequency?

7. We have two sinusoids at frequencies $\omega = 2\pi \times 0.20$ and $\omega = 2\pi \times 0.27$. The amplitude of the first sinusoid is 1.0 and that of the second is 0.12 units. It is desired to estimate the magnitude of the second sinusoid. For simplicity we use a uniform window. Compute the minimum length of the signal duration required for minimum bias.

8. The BT spectrum is likely to be negative which is against the basic requirement that the power spectrum, being a measure of power, must be always positive. Explain what leads to negative spectral estimate.

9. Let $w_1(n) = \frac{1}{N_1} a^{\left| n - \frac{N_1 - 1}{2} \right|}$, $n = 0,1,\ldots N_1 - 1$ where N_1 is odd and $0 \le a \le 1$. For smoothly varying spectrum show that the BT spectrum estimate is unbiased and its variance is given by,

$$\mathrm{var}\left\{\hat{S}_x(\omega)\right\} = \frac{2}{N_1}\left[\frac{a^{N_1+1} - 1}{a^2 - 1} - 0.5\right] S_x^2(\omega)$$

10. Referring to fig. 3.2 let the signal-to-noise ratio at the output be 10 dB. Compute the magnitude of the coherence between the input and output of LTI. Assume that the LTI is a simple delay unit. What is the phase of the coherence.

11. Let $x(n) = 0.4\cos(0.2\pi n + \varphi) + \eta(n)$ where $\eta(n)$ is white noise and φ is a random variable, which is uniformly distributed over $\pm\theta$. Compute the theoretical covariance function of *x(n)*. For what values of θ will *x(n)* becomes stationary.

12. What is the effect an asymmetric window on the BT spectrum? Does it also affect the Welch spectrum?

13. Consider use of a window whose all sidelobes are of equal magnitude. Such a window may be an equiripple FIR filter, which we shall cover in the next chapter. Assume that the spectrum of the signal has several peaks spread over the entire frequency range. How does such a window compare with a common window such as Hanning or Hamming.

14. A linear time invariant (LTI) system is being driven by a stationary non-Gaussian stochastic signal, *x(n)*. Let *h(n)* be its impulse response function. Compute the bicovariance function of the output. Taking 2D DFT of the output covariance function evaluate the bispectrum of the output.

15. In problem 14 assume that the input function is a non-Gaussian white noise process Relate the phase of the output bispectrum to that of the transfer function. In contrast, the spectrum of the output is independent of the phase of the transfer function (see Example 3.1).

16. Integrating on both sides of (3.45) show that

$$\frac{1}{2\pi}\int_{-\pi}^{\pi} S_x(n,\omega)d\omega = \sigma_x^2(n)$$

where $\sigma_x^2(n)$ is time varying variance.

17. A sum of two random sinusoids differing in frequency by 0.01 Hz is given. We wish to estimate the power of each sinusoid from the spectrum. What is the minimum data length required when Hanning is used. It is required to meet the Rayleigh resolution criterion. Repeat this exercise for other windows, namely, uniform, Hamming, optimum.

18. It is often recommended that each segment in the Welch spectrum estimation method be padded with equal number of zeros. Apart from giving a smoother estimate of the spectrum, is there any other purpose?

19. Periodogram is equal to magnitude square of FDFT coefficients, that is,

$$Periodogram = \left| \sum_{n=0}^{N-1} x(n) e^{-j\frac{2\pi kn}{N}} \right|^2$$

Show that the periodogram can also be obtained by squaring the output of a filter whose impulse response function is $e^{-j\frac{2\pi kn}{N}} u(n)$ at n=N. Note that u(n) is a step function.

20 Which of the following functions (fig. 3.20) is acceptable as a covariance function and why?

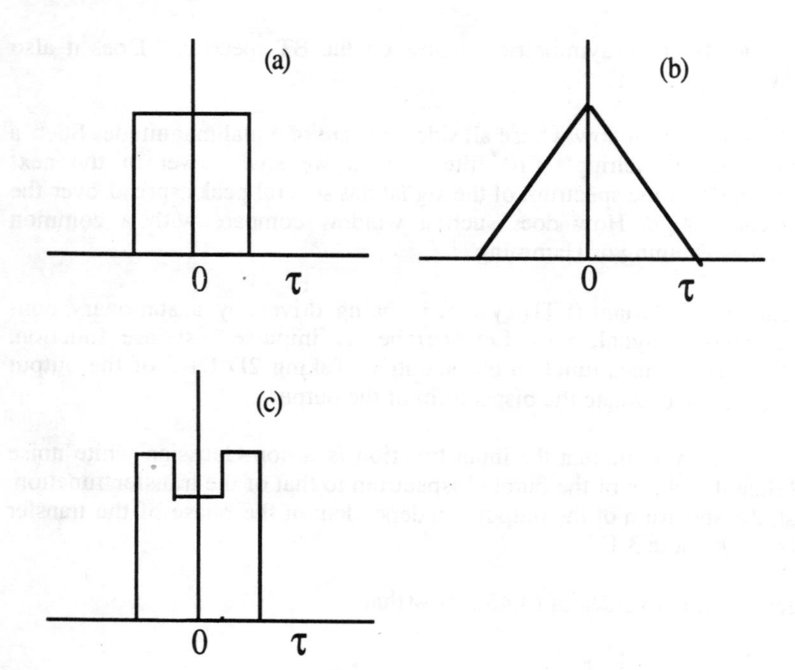

Figure 3.20: Which of above functions is acceptable as a covariance function?

21. In fig. 3.18 let the input signal be a linear FM signal which is illustrated in fig. 1.4a. Assume that the linear frequency increase lasts for 10 seconds, after

which time the signal continues at a constant frequency. Sketch the output of the filter shown in fig. 3.18

22. Referring to fig. 3.13, given a point ($\omega_1 = 3\pi/4, \omega_2 = \pi/8$) lying inside the region I, map this point into all other remaining eleven regions. Use the symmetry properties of the bispectrum given in (3.32).

23.Consider the following (fig. 3.21) highly simplified imaging problem:

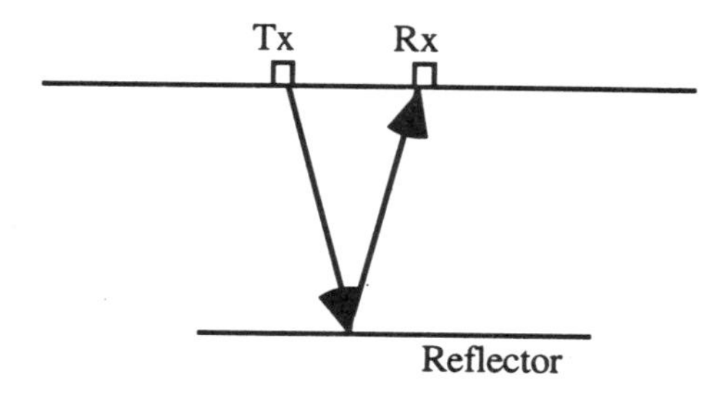

Figure 3.21: A simple arrangement to map a reflector.

The transmitter Tx sends out a BPSK (fig. 1.28) signal that is reflected back to the receiver, Rx, on the surface. From the received signal and the known transmitted signal estimate the depth to the reflector and its reflection coefficient. Assume that the propagation speed is known.

Computer Projects:

1. Any window can be used either as a linear window in Welch method or as a quadratic window in BT method. But the window characteristics in terms of the width of the main lobe and the magnitude of the first sidelobe differ. Illustrate these differences by considering any one of the windows described in the text.

2. The definition of the effective length (page158) of a window was intentionally left out. There is no single definition of the effective length. Consider the following possibility: Replace a given window by a uniform window of length L_{eff} such the areas under both curves are equal. Compute the effective lengths of the three types of windows, which we have considered in this chapter.

3. Generate a second order moving average signal given by

$$x(n) = \varepsilon(n) + 1.6\varepsilon(n-1) + 0.48\varepsilon(n-2)$$

where $\varepsilon(n)$ is pseudo-random Gaussian numbers with zero mean and unit variance (Use Matlab mfile "randn"). Compute the autocorrelation function of 1024 points of the moving average signal. Verify the property of the covariance function of a moving average signal given in eq.(1.30b).

4. In Example 3.6 let us assume that the frequency switching times are random time instants. Repeat this example to find out how well the transition times can be estimated from the time varying spectrum.

5. Compute the effective length and effective bandwidth, as defined on page158 of an FIR filter whose coefficients are listed in table 4.5 (Chapter Four).

4 Finite Impulse Response (FIR) Filters

Digital filtering is at the core of digital signal processing. The present chapter and the next two chapters are devoted to digital filters. In this chapter we describe finite impulse response (FIR) filters and in the subsequent two chapters we shall cover infinite impulse response (IIR) filters and adaptive filters. After a brief introduction to digital filters we go on to develop the FIR filters; their properties, structure and design which includes simple approach of windowing, optimizing the transition zone and equiripple response. From the point of practical applications FIR filters, on account of their linear phase characteristics and guaranteed stability, are most preferred.

§4.1 Introduction to Digital Filters

The frequency content of a signal is altered by means of a running weighted average of the sampled signal. The weight coefficients are the filter coefficients operating on a discrete signal, therefore the operation is known as digital filtering. The running weighted averaging or filtering is also known as convolution in mathematical parlance. We have covered convolution extensively in Chapter Two. A filter may be one-sided or two-sided sequence of weighting coefficients. It can be either infinite length or finite length sequence. A filter may also be classified based on the passband, such as lowpass, bandpass and highpass. This section is devoted to classification of filters.

Types of Filters: Digital filters are basically of two types depending upon whether the filtering is operating in real-time or on the stored data working off line. A digital filter to work in real-time must be one-sided or causal whereas a filter to work on the stored data can be two-sided or non-causal (see fig. 4.1). A two-sided filter is to be preferred as it can have linear or constant (zero) phase impulse response function. The direct implementation involves computation of a convolution sum as in (4.1)

$$y(n) = \sum_{m=0}^{\infty} h(m)x(n-m) \qquad \text{One-sided} \qquad (4.1a)$$

$$y(n) = \sum_{m=-\infty}^{\infty} h(m)x(n-m) \qquad \text{Two-sided} \qquad (4.1b)$$

Note that the impulse response function, $h(m)$ is either semi-infinite (0 to $+\infty$) or infinite ($\pm\infty$) duration. Such a filter is known as an infinite impulse response (IIR) filter. It requires infinite past signal or infinite stored signal. In either case practical implementation of IIR filter becomes extremely expensive except when it can be expressed as a recursive structure, which we shall describe later. Only a two-sided filter can be a zero or linear phase filter.

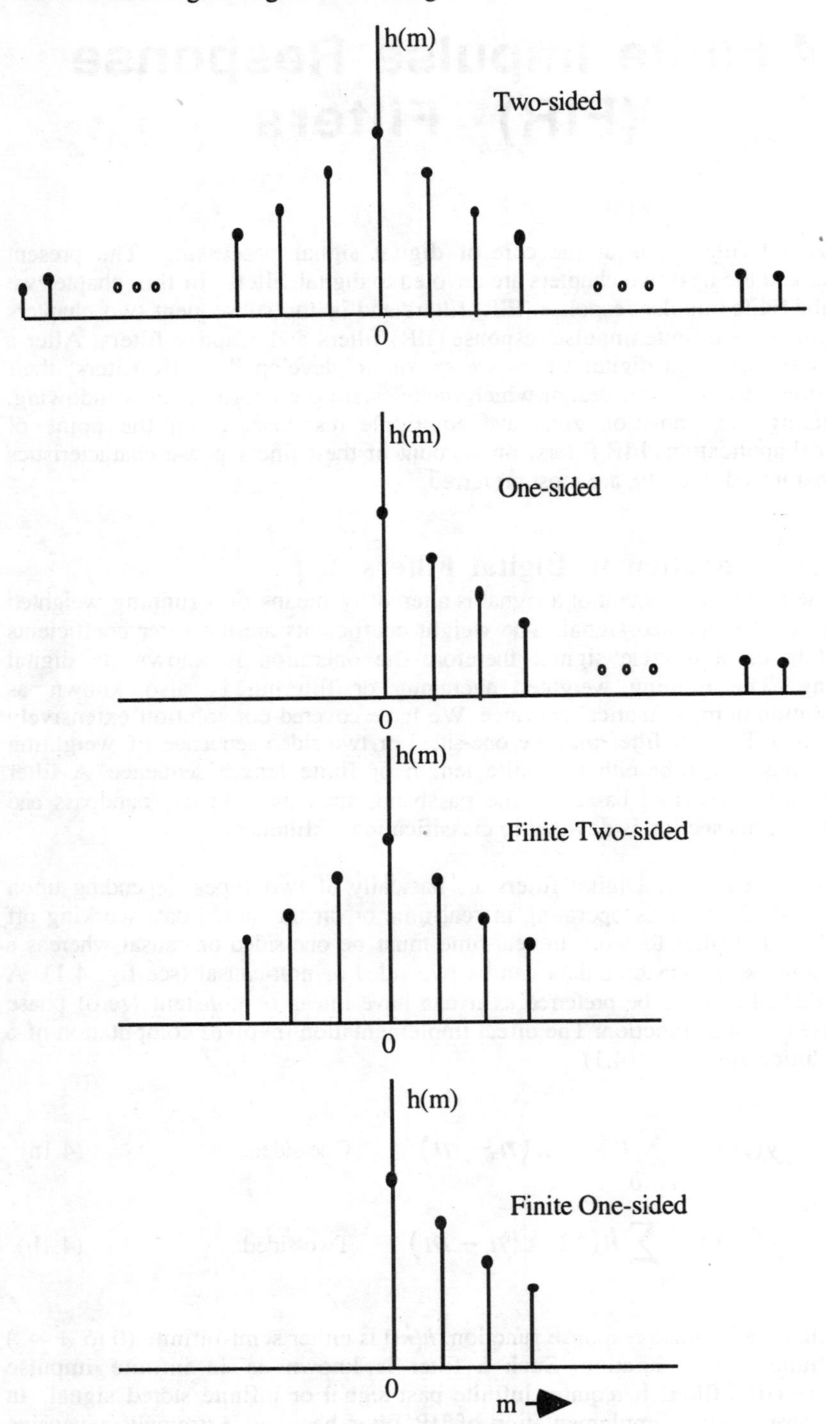

Figure 4.1: Four different types of digital filters

A two-sided filter may be written as a sum of two one-sided filters, taking +ve indexed and -ve indexed halves of the impulse response function as two one-sided filters. Eq. (4.1b) may be expressed as

$$\begin{aligned} y(n) &= \sum_{m=-\infty}^{\infty} h(m)x(n-m) \\ &= \sum_{m=0}^{\infty} h(m)x(n-m) + \sum_{m=-\infty}^{-1} h(m)x(n-m) \\ &= \sum_{m=0}^{\infty} h^{+}(m)x(n-m) + \sum_{m=0}^{\infty} h^{-}(m)x(n+m) \\ &= \sum_{m=0}^{\infty} h^{+}(m)x(n-m) + \sum_{m=0}^{\infty} h^{-}(m)x(n+m) \end{aligned} \qquad (4.2a)$$

where we have assumed that $h^{+}(0) = h^{-}(0) = 0.5h(0)$. Note that for a fixed *n*, *x(n-m)* refers to the past input and *x(n+m)* refers to the future input. Since the data is stored we have an access to the past as well as to the future data. If we assume that the filter is symmetric, that is, $h^{+}(m) = h^{-}(m), m = 0,1,\ldots,\infty$ (4.2a) may be expressed as a on-sided filter.

$$y(n) = \sum_{m=0}^{\infty} h^{+}(m)[x(n-m) + x(n+m)] \qquad (4.2b)$$

The advantage in the above representation of a two-sided filter as a one-sided filters lies in the possibility that, since one-sided filter can be implemented recursively, a two-sided filter may be implemented recursively.

A causal filter will be required to satisfy some additional constraints as those in causal systems (see Chapter Two, page 94). For example, its phase response is closely connected to its amplitude response. Further, it must satisfy the Paley-Wiener criterion (Chapter Two, page 95) which requires that the amplitude response cannot be zero over any finite frequency range. No filter of practical interest, however, will satisfy this criterion.

Recursive Implementation: A filter may include a feedback path, that is, the past outputs are used along with the past and present inputs. Such a filter is said to be *recursive*. The mathematical form of a recursive filter is be given by

$$y(n) = \sum_{k=1}^{p-1} a_k y(n-k) + \sum_{i=0}^{q-1} b_i x(n-i) \qquad (4.3)$$

where $a_k, k = 1,2,\ldots,p$ and $b_i, i = 0,1,\ldots,q$ are filter coefficients and p and q refer to the order of the recursive filter. Note that the present output depends upon the present and past inputs as well as on the past outputs. The presence of the past outputs in (4.3) results into an infinite impulse response. By suitably selecting the filter coefficients it is possible to approximate any prescribed infinite impulse response function, but one-sided. Thus, an infinite impulse response (IIR) filter of causal type can be recursively expressed as in (4.3).

Example 4.1: A simple example of infinite impulse response filter is an exponentially decreasing filter sequence,

$$h(m) = a^m, \quad m = 0,1,2,\ldots\infty \; and \; |a| < 1$$

whose z-transform is $H(z) = \dfrac{1}{1 - az^{-1}}$. The output of the filter in z-domain is given by

$$(1 - az^{-1})Y(z) = X(z)$$

Upon computing the inverse z-transform on both sides of the output equation we obtain a recursive filter

$$y(n) = x(n) + ay(n-1)$$

Thus, we have been able to get a recursive filter equivalent of very low order of an otherwise infinitely long IIR filter. An IIR filter implemented recursively is indeed a very powerful digital filter ◆.

Finite Impulse Response (FIR): Further classification of filters is based on the length of the filter response function. Ideally an impulse response function has infinite length but it is impractical to implement such a filter. We have either to truncate the infinite length impulse response function (see fig. 4.1) or devise a method obtaining a finite length filter whose transfer function is as close as possible to the true theoretical transfer function. Such a finite impulse response (FIR) filter is indeed an approximation to an infinite one-sided or two-sided filter. A two-sided FIR filter may be used as one-sided filter but the output will be delayed. Let *h(m)*, m=0,1,...,M-1 be a finite approximation of an infinite two-sided filter with zero phase response. Therefore, the FIR filter will have a symmetry, for M even, *h(m)=h(M-m-1)*, m=0, 1,2,..., $M/2$-1 and for odd M, *h(m)= h(M-m-1)*, *m*=0,1,..., $(M-3)/2$. The filter output is centered at $(M-1)/2$ hence a delay of $(M-1)/2$ time units (non-integer delay in case M is even). The transfer function of a symmetric FIR filter, for M even, is given by

$$\begin{aligned} H(\omega) &= \sum_{m=0}^{M-1} h(m)e^{-j\omega M} \\ &= 2\sum_{m=0}^{\frac{M}{2}-1} h(m)\cos(\omega(\frac{M-1}{2} - m))e^{-j\omega\frac{M-1}{2}} = H_r(\omega)e^{-j\omega\frac{M-1}{2}} \end{aligned} \tag{4.4a}$$

and for M odd it is given by

$$H(\omega)=\left[2\sum_{m=0}^{\frac{M-3}{2}}h(m)\cos(\omega(\frac{M-1}{2}-m))+h(\frac{M-1}{2})\right]e^{-j\omega\frac{M-1}{2}} \tag{4.4b}$$

$$=H_r(\omega)e^{-j\omega\frac{M-1}{2}}$$

An FIR filter may also be anti symmetric, that is, for M even,

$$h(m)=-h(M-m-1),\ m=0,\ 1,2,\ldots,\frac{M}{2}-1$$

and its transfer function is given by

$$H(\omega)=\sum_{m=0}^{M-1}h(m)e^{-j\omega M}$$

$$=2j\sum_{m=0}^{\frac{M}{2}-1}h(m)\sin(\omega(\frac{M-1}{2}-m))e^{j(-\omega\frac{M-1}{2})}=H_r(\omega)e^{j(-\omega\frac{M-1}{2}+\frac{\pi}{2})} \tag{4.5a}$$

and for M odd it is given by

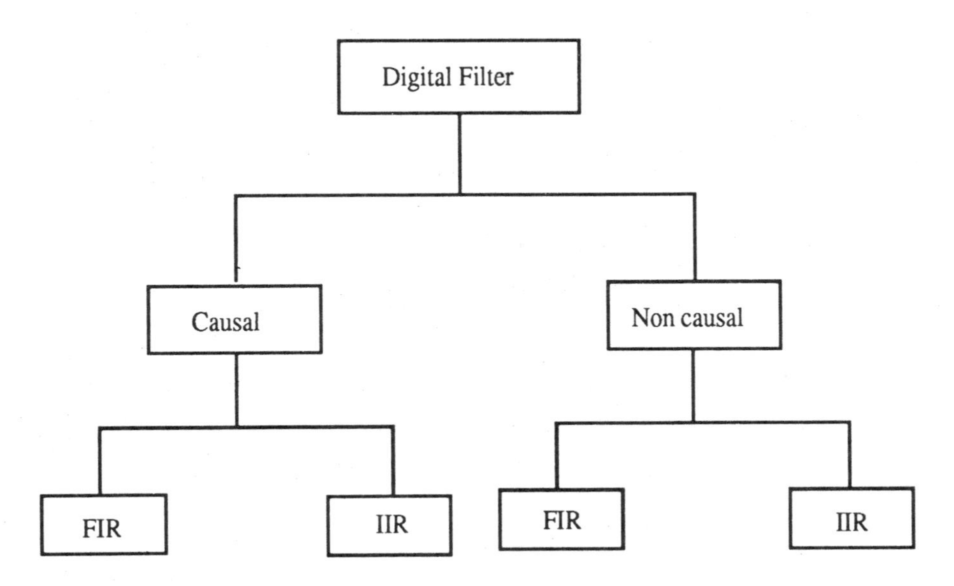

Figure 4.2: Classification of digital filters.

$$H(\omega) = \left[2j \sum_{m=0}^{\frac{M-3}{2}} h(m) \sin(\omega(\frac{M-1}{2} - m)) \right] e^{j(-\omega \frac{M-1}{2})} \tag{4.5b}$$

$$= H_r(\omega) e^{j(-\omega \frac{M-1}{2} + \frac{\pi}{2})}$$

Note that $H_r(\omega)$ is a real quantity and it represents the magnitude of the filter. The phase is given by $(-\omega \frac{M-1}{2})$ for symmetric filter and by $(-\omega \frac{M-1}{2}) + \frac{\pi}{2}$ for anti symmetric filter. The phase is linearly decreasing, hence such filters are known as linear phase filters. Finally, in fig.4.2 we relate the different types of digital filteis.

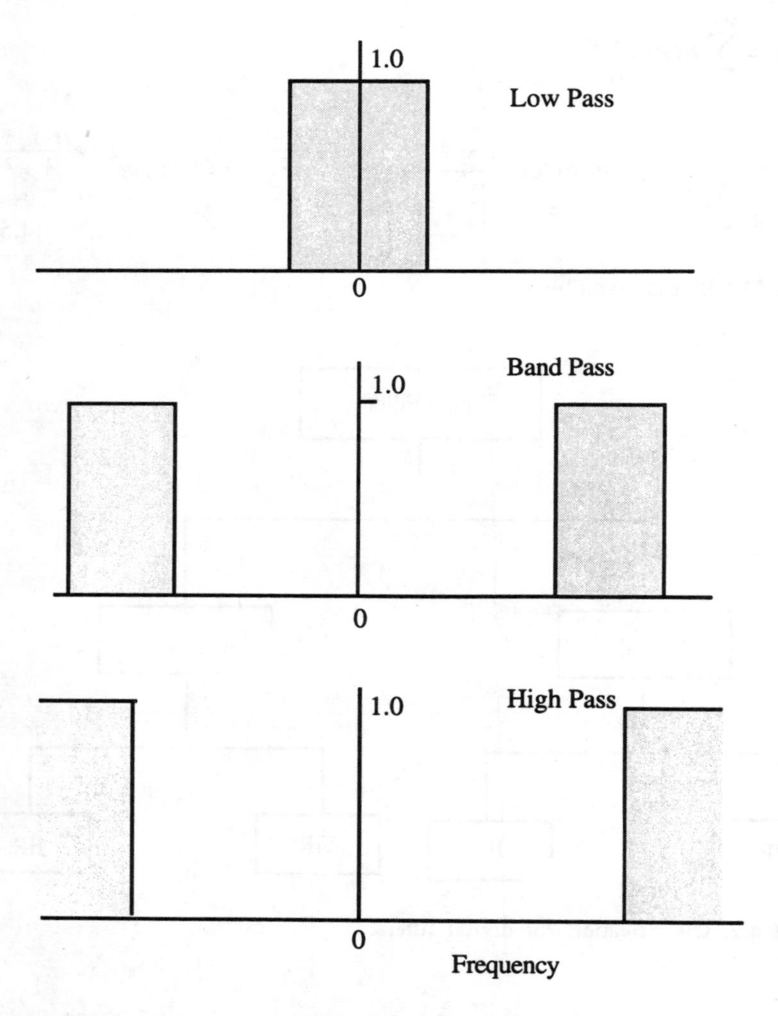

Figure 4.3: Ideal transfer characteristics of lowpass, bandpass and highpass filters.

The digital filters are also classified based on their transfer characteristics, namely, lowpass, bandpass and highpass, having ideally unit response in the pass region and zero outside. The phase is, however, zero or linearly varying. The filters are illustrated in fig. 4.3. The lowpass filter is the basic filter from which the other two can be derived by simple transformation.

Example 4.2: Consider a lowpass filter. The coefficients are listed in table 4.1 and its magnitude square response is shown in fig. 4.4(a). We would like to show how to relocate its passband at any desired center frequency and thus transform the basic lowpass filter into a bandpass and a highpass filters. To locate the center of the passband at ω_c the lowpass filter coefficients are modulated by $\cos(\omega_c(m-(M-1)/2))$. The filter response for $\omega_c = \pi/2$ (bandpass) is shown in fig. 4.4(b) and for $\omega_c = \pi$ (highpass) in fig. 4.4(c). Note that for highpass filter the modulation function is simply $\cos(\pi(m-(M-1)/2)) = (-1)^{m-(M-1)/2}$, that is, the sign of lowpass filter coefficients is alternatively reversed.

Table 4.1: Digital filter coefficients of a Lowpass filter. Only one half of the coefficients (m=14 to 27) are shown. The remaining are obtained by mirror imaging.

m	h(m)	m	h(m)
0	0.4	7	0.0157769
1	0.299692	8	-0.0115598
2	0.0898359	9	-0.0134373
3	-0.0569018	10	0.0
4	-0.0642052	11	0.00623505
5	0.0	12	0.00239574
6	0.0345003	13	-0.00132685

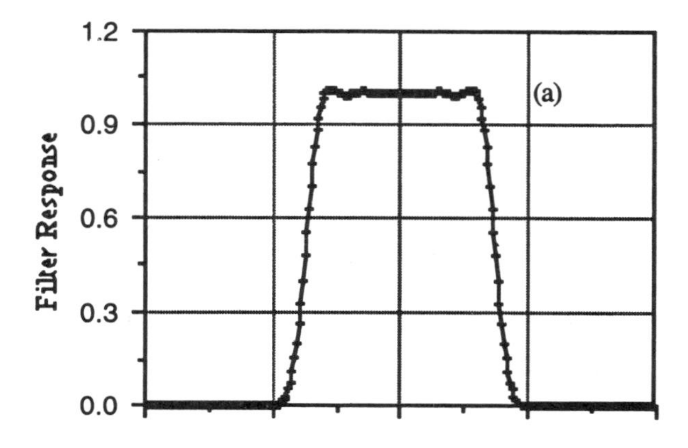

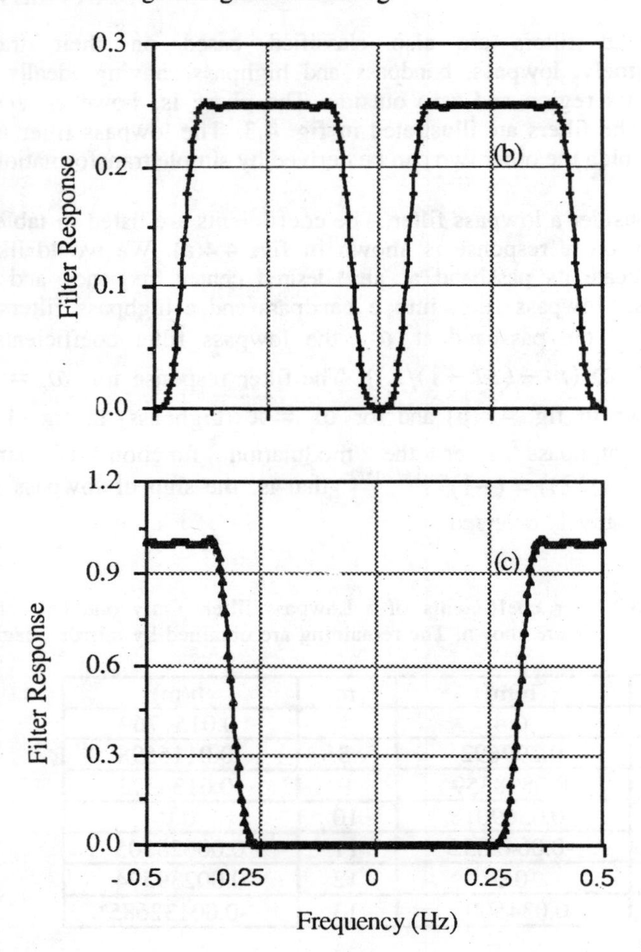

Figure 4.4: The response (magnitude square) of (a) lowpass (b) and (c) highpass filters. The bandpass filter and highpass filter were obtained by relocating the center frequency at $\pi/2$ and π.

§4.2 FIR Filter Realization

A finite impulse response filter may be realized in more than one way. The simplest form is the so-called taped delay line structure shown in fig.4.5. Another approach is to factor the transfer function into two of more smaller factors or filters those may be combined in parallel or series form. There are some special structures available for implementation of FIR filters. These are lattice filter and FDFT coefficient based recursive filter. This section is devoted to the study of different types FIR filter realization.

Direct Form: The simplest realization of an FIR filter is the direct form structure which follows the basic convolution sum,

$$y(n) = \sum_{m=0}^{M-1} h(m)x(n-m) \tag{4.6a}$$

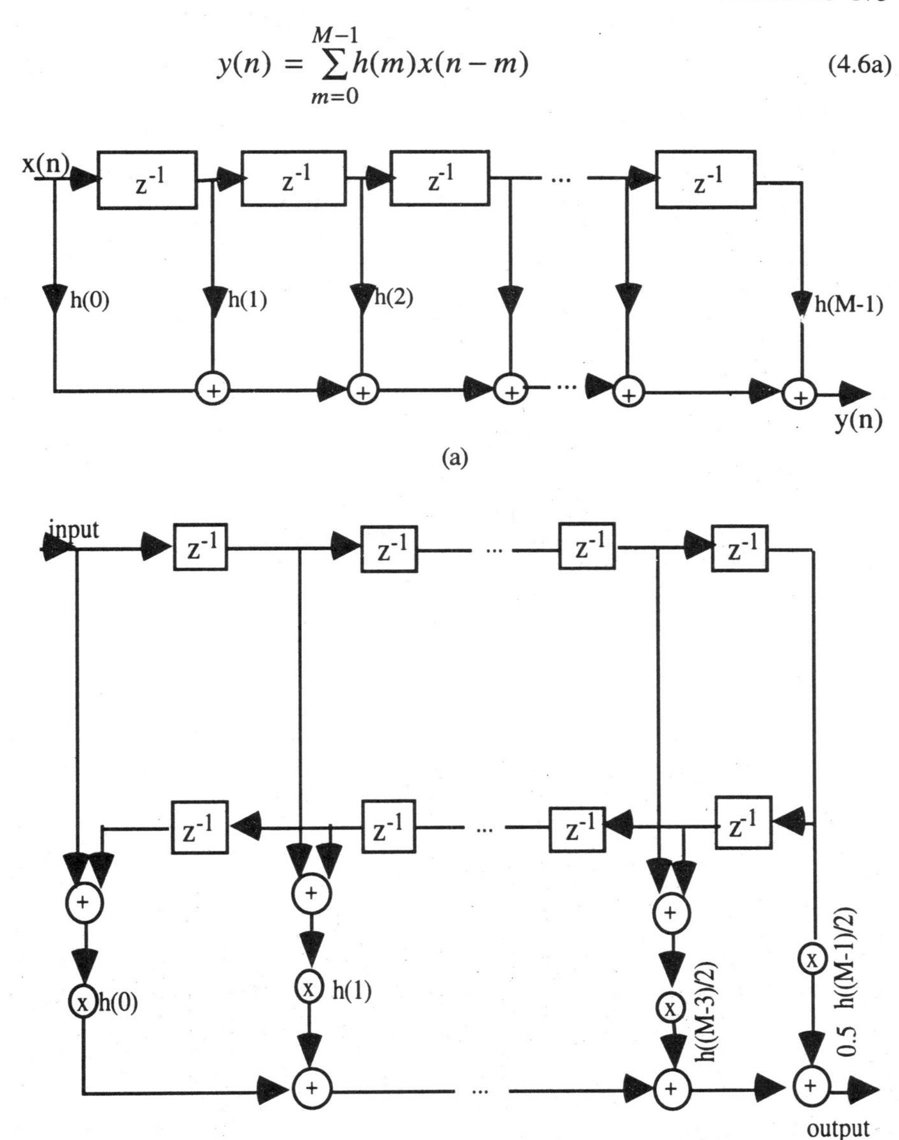

Figure 4.5: Tapped delay line structure for realization of an FIR filter. (a) any FIR filter (b) linear phase FIR filter. M is odd.

For linear phase FIR filter, since the filter coefficients are symmetric, eq.(4.6a) may be written for M odd as

$$y(n) = \sum_{m=0}^{\frac{M-1}{2}} a(m)h(m)\{x(n-m) + x(n-M+1-m)\} \tag{4.6b}$$

where $a(m) = 1, m = 0,1,...,(M-3)/2$ and $a((M-1)/2) = 0.5$. The structure of the filter is a tapped delay line or transversal system as shown in figure 4.5a. For linear phase filter the structure exploits the symmetry in the FIR filter coefficients. The tapped delay line structure for linear phase filter for M odd is shown in figure 4.5b.

Cascade Form: Another type of FIR filter realization is the cascade form. The z-transform of a FIR filter is expressed as a product of basic quadratic factors,

$$H(z) = \prod_{q=1}^{Q} H_q(z) \tag{4.7}$$

where $H_q(z) = b_{q0} + b_{q1}z^{-1} + b_{q2}z^{-2}$ is a quadratic factor, Q= integer part of {(M+1)/2}. The roots of a quadratic factor must occur in complex conjugate pairs so that the coefficients b_{q0}, b_{q1}, b_{q2} are real. The roots of a quadratic equation are

$$r_{q1} = \frac{-b_{q1} + \sqrt{b_{q1}^2 - 4b_{q0}b_{q2}}}{2b_{q0}}$$

$$r_{q2} = \frac{-b_{q1} - \sqrt{b_{q1}^2 - 4b_{q0}b_{q2}}}{2b_{q0}}$$

For roots to be in complex conjugate position we must have $b_{q1}^2 < 4b_{q0}b_{q2}$. Further, the magnitude of the roots be equal,

$$|r_{q1}| = |r_{q2}| = \sqrt{\frac{b_{q1}^2 - 2b_{q0}b_{q2}}{2b_{q0}^2}}$$

The roots could lie either inside the unit circle or outside the unit circle. Let us express $H_q(z)$ as a product of two linear factors,

$$H_q(z) = (1 - r_1 z^{-1})(1 - r_2 z^{-1})$$

where $|r_1| = |r_2|$. Consider another factor with roots at reciprocal location

$$H_{q'}(z) = (1 - \frac{1}{r_1} z^{-1})(1 - \frac{1}{r_2} z^{-1})$$

whose roots are at $1/r_1$ and $1/r_2$. The factors $H_{q'}(z)$ and $H_q(z^{-1})$ are related as

$$H_{q'}(z) = (1 - \frac{1}{r_1} z^{-1})(1 - \frac{1}{r_2} z^{-1})$$
$$= \frac{1}{|r_1|^2} z^{-2} H_q(z^{-1})$$

We shall call them as complementary factors. Let us now form an FIR filter as a product of two complementary factors. It may be noted that since $\left|H_q(\omega)\right|^2 = H_q(z) H_q(z^{-1})\Big|_{z=e^{j\omega}}$ the phases of $H_q(z)$ and $H_q(z^{-1})$ are in opposite sign.

$$H(z) = H_q(z) H_{q'}(z)$$
$$= \frac{1}{|r_1|^2} z^{-2} \left|H_q(z)\right|^2$$

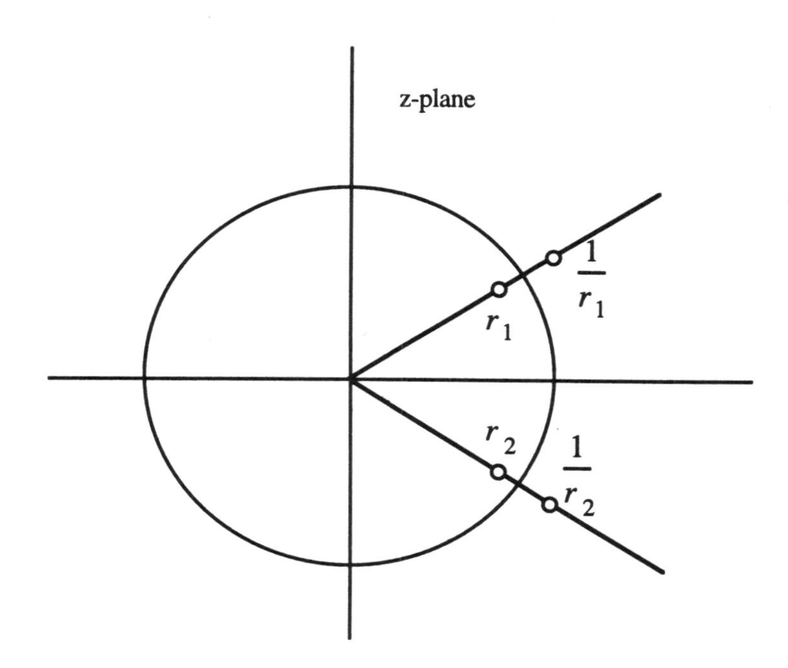

Figure 4.6: Position of the zeros of a quadratic factor and its complement.

This is a two-sided filter shifted by two units and it is a linear phase FIR filter. Conversely, a linear phase filter may be factored into pairs of complementary factors. The roots of a linear phase filter will lie both inside and outside the unit circle at reciprocal positions. Hence, the roots of a two-sided filter will lie both inside and outside the unit circle. Some of the factors in (4.7) can have zeros outside the unit circle. This will result into a mixed phase filter. We shall elaborate more on the concept of minimum phase and maximum phase when we deal with IIR filters in Chapter Five.

Each quadratic factor may be realized through a tapped delay structure shown in fig. 4.7. All quadratic factors are arranged in a series form as in fig. 4.8. Such a realization is known as cascade form.

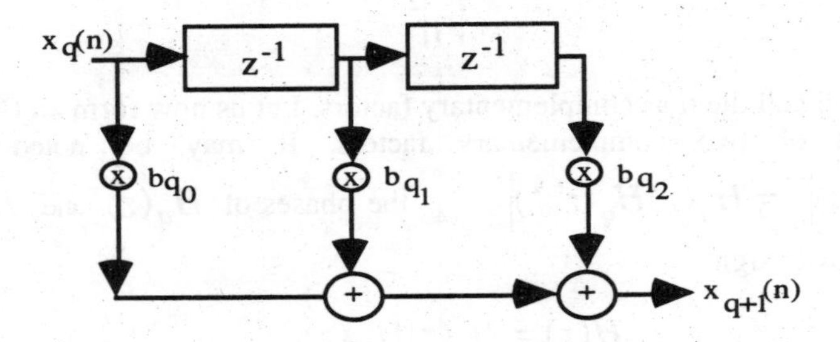

Figure 4.7: Tapped delay realization of a single quadratic factor in (4.7).

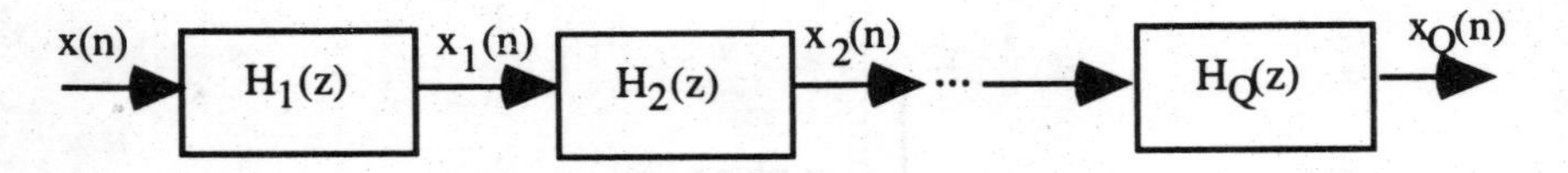

Figure 4.8: Cascade structure of FIR filter. FIR filter response in z domain is factored into Q quadratic factors. Each quadratic factor is realized through a tapped delay structure as shown in fig. 4.7. All quadratic filters are arranged in a series connection.

Polyphase Realization: Another type of realization which allows parallel implementation of an FIR filter is polyphase realization. We decompose the FIR filter coefficients as follows:

$$e_l(n) = h(Ln+l) \quad 0 \le n \le \left[\frac{M}{L}\right]_{Int}, \quad 0 \le l \le L-1 \qquad (4.8a)$$

where L is an integer, known as order of decomposition. $e_l(n)$ is the lth polyphase component of *h(n)*. The z-transform of the lth component is given by

$$E_l(z) = \sum_{n=0}^{\left[\frac{M}{L}\right]_{Int}} h(Ln+l)z^{-n} \qquad 0 \le l \le L-1 \qquad (4.8b)$$

The transfer function of an FIR filter may be expressed in terms of the polyphase components

$$H(z)=\sum_{l=0}^{L-1} z^{-l}E_l(z^L) \tag{4.8c}$$

which is known as polyphase decomposition. An FIR filter may be realized as a parallel combination of L polyphase components as shown in fig.4.9 for L equal to 2, 3, and 4. Such a realization is known as a polyphase realization, which is often used in multirate digital signal processing.

Example 4.3: Consider a nine point FIR filter, that is, M=9. Let L=2. The polyphase decomposition of the filter is given by

$$e_0=\{h(0), h(2), h(4), h(6), h(8)\}$$
$$e_1=\{h(1), h(3), h(5), h(7)\}$$

and the z-transforms

$$E_0(z)=h(0)+h(2)z^{-1}+h(4)z^{-2}+h(6)z^{-3}+h(8)z^{-4}$$
$$E_1(z)=h(1)+h(3)z^{-1}+h(5)z^{-2}+h(7)z^{-3}$$

The polyphase decomposition of H(z) is given by $H(z)=E_0(z^2)+z^{-1}E_1(z^2)$ and the corresponding realization is shown in fig.4.9a. Next, let L=3. The polyphase components are

$$e_0=\{h(0), h(3), h(6)\}$$
$$e_1=\{h(1), h(4), h(7)\}$$
$$e_2=\{h(2), h(5), h(8)\}$$

and their z-transforms are

$$E_0(z)=h(0)+h(3)z^{-1}+h(6)z^{-2}$$
$$E_1(z)=h(1)+h(4)z^{-1}+h(7)z^{-2}$$
$$E_2(z)=h(2)+h(5)z^{-1}+h(8)z^{-2}$$

The polyphase decomposition of the transfer function is given by

$$H(z)=E_0(z^3)+z^{-1}E_1(z^3)+z^{-2}E_2(z^3)$$

and the corresponding filter realization is shown in fig.4.9b. Finally, let L=4. The polyphase components of the filter are given by

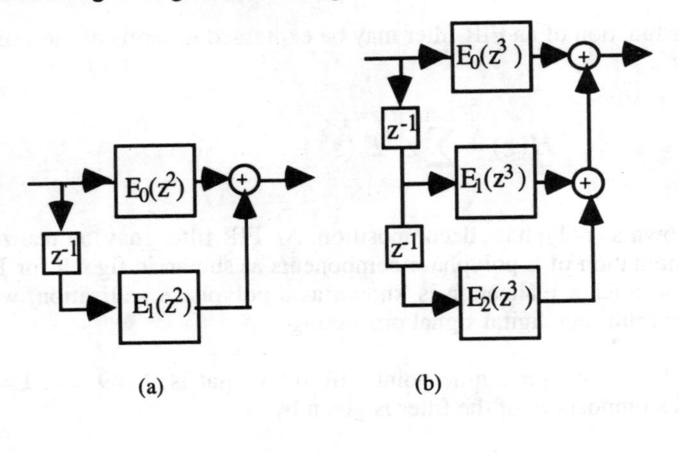

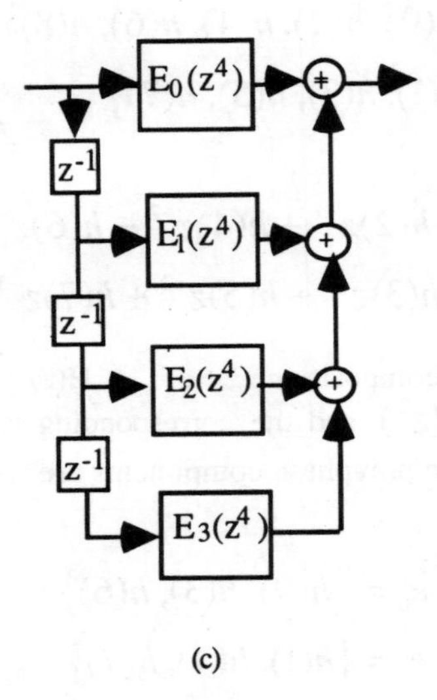

Figure 4.9: Polyphase realization of a nine point FIR filter transfer function. (a) Two component realization (b) Three component realization and (c) Four component realization. Each box in the above scheme may be realized either through direct form or cascade form.

$$e_0 = \{h(0), h(4), h(8)\}$$
$$e_1 = \{h(1), h(5)\}$$
$$e_2 = \{h(2), h(6)\}$$
$$e_3 = \{h(3), h(7)\}$$

and their z-transforms are

$$E_0(z) = h(0) + h(4)z^{-1} + h(8)z^{-2}$$
$$E_1(z) = h(1) + h(5)z^{-1}$$
$$E_2(z) = h(2), + h(6)z^{-1}$$
$$E_3(z) = h(3) + h(7)z^{-1}$$

The polyphase decomposition of the transfer function is given by

$$H(z) = E_0(z^4) + z^{-1}E_1(z^4) + z^{-2}E_2(z^4) + z^{-3}E_3(z^4)$$

and the corresponding filter realization is shown in fig.4.9c .

Recursive Realization: Given the FDFT coefficients of an FIR filter we can compute the filter coefficients using inverse FDFT and use them for filtering employing the standard discrete convolution sum (4.1). It is also possible to use the FDFT coefficients directly in a recursive implementation. Let $H(k), k = 0,1,\ldots M-1$ be the FDFT coefficients and $h(m),\quad m = 0,1,\ldots M-1$ be filter coefficients. The z-transform of the filter coefficients is given by.

$$\begin{aligned} H(z) &= \sum_{m=0}^{M-1} h(m) z^{-m} \\ &= \frac{1}{M} \sum_{m=0}^{M-1} \sum_{k=0}^{M-1} H(k) e^{j\frac{2\pi}{M}km} z^{-m} \\ &= \frac{1-z^{-M}}{M} \sum_{k=0}^{M-1} \frac{H(k)}{1-z^{-1}e^{j\frac{2\pi}{M}k}} \end{aligned} \tag{4.9}$$

For linear phase FIR filter the FDFT coefficients, using (4.4), may be expressed as

$$H(k) = H_r(\frac{2\pi}{M}k)e^{-j\pi k\frac{M-1}{M}}$$

which we use in (4.9) and write as

$$H(z) = \frac{1-z^{-M}}{M} \sum_{k=0}^{M-1} \frac{H_r(\frac{2\pi}{M}k)}{1-z^{-1}e^{j\frac{2\pi}{M}k}} e^{-j\pi k\frac{M-1}{M}} \tag{4.10}$$

Note the symmetry property,

$$H_r(\frac{2\pi}{M}k) = H_r(\frac{2\pi}{M}(M-k)), k = 1,\ldots,\frac{M-1}{2} \qquad \text{for M odd.}$$

$$H(z) = \frac{1-z^{-M}}{M}\left[\sum_{k=1}^{\frac{M-1}{2}} \frac{2H_r(\frac{2\pi}{M}k)\left\{\cos(\frac{2\pi}{M}k\frac{M-1}{2}) - z^{-1}\cos(-\frac{2\pi}{M}k\frac{M+1}{2})\right\}}{1-2z^{-1}\cos(\frac{2\pi}{M}k)+z^{-2}} + \frac{H_r(0)}{1-z^{-1}}\right]$$

$$= \frac{1-z^{-M}}{M}\left[\sum_{k=1}^{\frac{M-1}{2}} \frac{2H_r(\frac{2\pi}{M}k)\left\{(1-z^{-1})\cos(\frac{\pi}{M}k)\right\}(-1)^k}{1-2z^{-1}\cos(\frac{2\pi}{M}k)+z^{-2}} + \frac{H_r(0)}{1-z^{-1}}\right] \tag{4.11a}$$

The filter transfer function has zeros on the unit circle equally spaced around the circle at

$$o_k = e^{-j\frac{2\pi}{M}k}, k = 0,1,\dots,M-1$$

and also poles at

$$z_k = e^{\pm j\frac{2\pi}{M}k}, k = 1,\dots,\frac{M-1}{2}$$

and on the real axis, $z_0 = 1.0$. Notice that the poles and zeros occupy the same locations resulting in a cancellation. To avoid instability problems it is recommended that poles and zeros be pushed marginally inside the unit circle. The transfer function then takes a form

$$H(z) = \frac{1-r^M z^{-M}}{M}\left[(1-rz^{-1})\sum_{k=1}^{\frac{M-1}{2}} \frac{2H_r(\frac{2\pi}{M}k)\cos(\frac{\pi}{M}k)(-1)^k}{1-2rz^{-1}\cos(\frac{2\pi}{M}k)+r^2z^{-2}} + \frac{H_r(0)}{1-rz^{-1}}\right] \tag{4.11b}$$

where r is a constant close to one but less than one. The recursive realization of above transfer function is shown in fig. 4.10. The input (u_n) and output (v_n) of a resonator (fig. 4.10b) are given by

$$v_n = u_n + 2r\cos(\frac{2\pi}{M}k)v_{n-1} - r^2 v_{n-2} \tag{4.12}$$

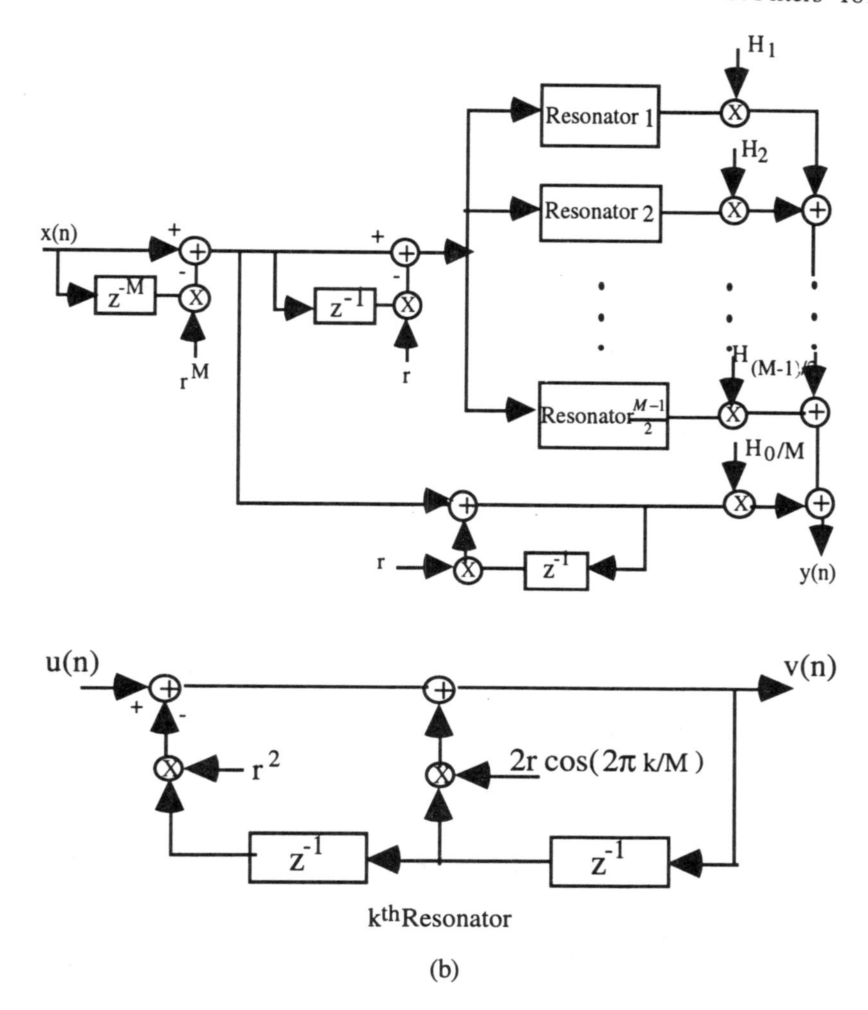

Figure 4.10: Recursive realization of FIR filter whose samples of the transfer function are given. (a) Complete signal flow diagram where $H_k = (-1)^k 2/M \cos(\pi k/M) H_r(2\pi k/M)$, k=1,...,$(M-1)/2$ and (b) kth Resonator.

The use of FDFT coefficients in place of the impulse response function reduces the computational load particularly when the passband of the filter is narrow; then, only a few FDFT coefficients will be non-zero and therefore only a few resonators (in fig. 4.10) will be required. On the other hand, direct form realization of a narrowband filter would require many impulse response coefficients.

Lattice Filter: The main advantage of lattice filters is its modularity. A new stage may be added without having to recompute the previous stages. Thus, a filter may be continuously improved upon until a desired design criterion is met. Consider a simple two coefficient FIR filter, *h(0)* and *h(1)*. Without any loss of generality we can set $h(0)=1$ or normalize all filter coefficients with respect to the zeroth coefficient which we shall assume to be non-zero. We introduce a single stage lattice (fig. 4.11) structure to implement above two-coefficient filter:

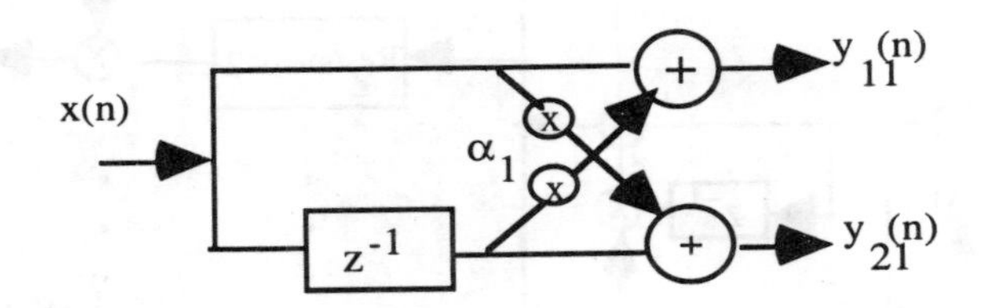

Figure 4.11: A single stage lattice filter to realize a simple two-coefficient FIR filter

$$\begin{aligned} y_{11}(n) &= x(n) + \alpha_1 x(n-1) \\ y_{21}(n) &= \alpha_1 x(n) + x(n-1) \end{aligned} \tag{4.13}$$

where $\alpha_1 = \dfrac{h(1)}{h(0)}$. There are two outputs, namely, $y_1(n)$ and $y_2(n)$. The first output is the normal filter output. The second output corresponds to reverse order filter output. Next, we consider two-stage lattice filter (fig. 4.12)

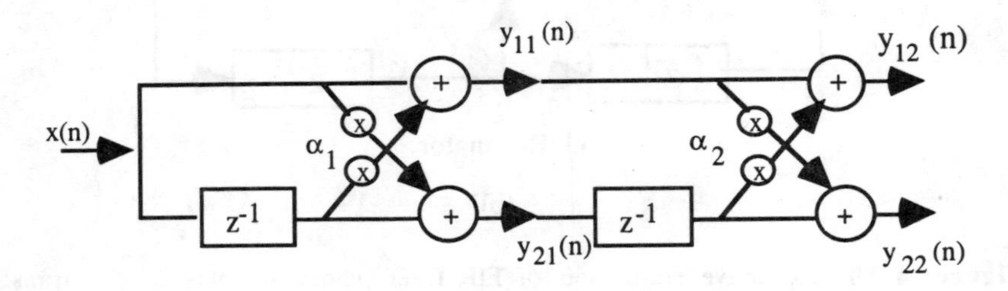

Figure 4.12: Two-stage lattice filter

$$\begin{aligned} y_{12}(n) &= y_{11}(n) + \alpha_2 y_{21}(n-1) \\ &= x(n) + \alpha_1 x(n-1) + \alpha_2\{\alpha_1 x(n-1) + x(n-2)\} \\ &= x(n) + \alpha_1(1+\alpha_2)x(n-1) + \alpha_2 x(n-2) \end{aligned} \tag{4.14a}$$

$$\begin{aligned} y_{22}(n) &= \alpha_2 y_{11}(n) + y_{21}(n-1) \\ &= \alpha_2\{x(n) + \alpha_1 x(n-1)\} + \alpha_1 x(n-1) + x(n-2) \\ &= \alpha_2 x(n) + \alpha_1(\alpha_2+1)x(n-1) + x(n-2) \end{aligned} \tag{4.14b}$$

$y_{12}(n)$ is the desired output of FIR filter with filter coefficients $h(0)=1$, $h(1)=\alpha_1(\alpha_2+1)$ and $h(2)=\alpha_2$ or the lattice coefficients may be expressed in terms of the filter coefficients,

$$\alpha_1 = \frac{h(1)}{1+h(2)} \qquad \alpha_2 = h(2) \tag{4.15}$$

The second output, $y_{22}(n)$, corresponds to the reverse order filter output. We shall consider a three stage lattice filter by appending (on the right hand side) one more single stage lattice filter to fig. 4.11. The outputs are

$$y_{13}(n) = x(n) + (\alpha_2\alpha_3 + \alpha_1(1+\alpha_2))x(n-1) + (\alpha_2 + \alpha_1\alpha_3(1+\alpha_2))x(n-2) + \alpha_3 x(n-3)$$

$$y_{23}(n) = \alpha_3 x(n) + (\alpha_2 + \alpha_1\alpha_3(1+\alpha_2))x(n-1) + (\alpha_2\alpha_3 + \alpha_1(1+\alpha_2))x(n-2) + x(n-3)$$

$y_{13}(n)$ is the desired output of a four coefficient FIR filter with coefficients,

$$h(0) = 1.0, \quad h(1) = (\alpha_2\alpha_3 + \alpha_1(1+\alpha_2)) \qquad h(2) = (\alpha_2 + \alpha_1\alpha_3(1+\alpha_2)), \quad h(3) = \alpha_3 \tag{4.16}$$

or the lattice coefficients may be expressed in terms of the filter coefficients,

$$\alpha_1 = \frac{h(1) - h(3)\dfrac{h(3)h(1)-h(2)}{h^2(3)-1}}{1+\dfrac{h(3)h(1)-h(2)}{h^2(3)-1}} \qquad \alpha_2 = \frac{h(3)h(1)-h(2)}{h^2(3)-1} \qquad \alpha_3 = h(3) \tag{4.17}$$

We shall now consider M-1 lattice filters in tandem. The input/output relation of ht lattice filter is given by

$$y_{1m}(n) = y_{1,m-1}(n) + \alpha_m y_{2,m-1}(n-1) \qquad y_{2m}(n) = \alpha_m y_{1,m-1}(n) + y_{2,m-1}(n-1) \tag{4.18}$$

In z-transform domain (4.18) may be written as

$$\begin{aligned} Y_{1m}(z) &= Y_{1,m-1}(z) + \alpha_m z^{-1} Y_{2,m-1}(z) \\ Y_{2m}(z) &= \alpha_m Y_{1,m-1}(z) + z^{-1} Y_{2,m-1}(z) \end{aligned} \tag{4.19}$$

with initial condition $Y_{10}(z) = Y_{20}(z) = X(z)$. Dividing (4.19) through out by $X(z)$ and noting that $\dfrac{Y_{1m}(z)}{X(z)} = H_{1m}(z)$ and $\dfrac{Y_{2m}(z)}{X(z)} = H_{2m}(z)$ we obtain

$$\begin{aligned} H_{1m}(z) &= H_{1,m-1}(z) + \alpha_m z^{-1} H_{2,m-1}(z) \\ H_{2m}(z) &= \alpha_m H_{1,m-1}(z) + z^{-1} H_{2,m-1}(z) \end{aligned} \tag{4.20a}$$

which may be expressed in a matrix form

$$\begin{bmatrix} H_{1m}(z) \\ H_{2m}(z) \end{bmatrix} = \begin{bmatrix} 1 & \alpha_m z^{-1} \\ \alpha_m z^{-1} & 1 \end{bmatrix} \begin{bmatrix} H_{1,m-1}(z) \\ H_{2,m-1}(z) \end{bmatrix} \tag{4.20b}$$

Note that $H_{1m}(z)$ and $H_{2m}(z)$ are transfer functions of m lattice filters connected in tandem. There are M-1 lattice filters. The overall transfer functions are

$$\begin{aligned} H_{1,M-1}(z) &= H(z) \\ H_{2,M-1}(z) &= z^{-M+1} H(z^{-1}) \end{aligned} \tag{4.21}$$

Also, it may be noted that $H_{10}(z) = H_{20}(z) = 1$. The second equation in (4.21) follows from the observation that the filter coefficients in $H_{2m}(z)$ are in the reverse order as illustrated in (4.14) and (4.16) for M=2 and M=3, respectively. Further, it is observed that the first coefficient in $H_{1m}(z)$ is always unity and the last coefficient is equal to α_m. The above mentioned constraints allow us to compute $H_{1m}(z)$ and $H_{2m}(z)$, starting from mm-1 downward. To obtain α_m we just need to look for the last filter coefficient in $H_{1m}(z)$. Simplifying (4.20) we obtain

$$H_{1m}(z) = H_{1,m-1}(z) + \alpha_m \left\{ H_{2m}(z) - \alpha_m H_{1,m-1}(z) \right\}$$

$$\frac{H_{1m}(z) - \alpha_m H_{2m}(z)}{1 - \alpha_m^2} = H_{1,m-1}(z) \tag{4.22a}$$

and using (4.21) in (4.22) we obtain an equation for $H_{1,m-1}(z)$

$$H_{1,m-1}(z) = \frac{H_{1m}(z) - \alpha_m z^{-m} H_{1m}(z^{-1})}{1 - \alpha_m^2} \tag{4.22b}$$

It may be pointed out that whenever $\alpha_m^2 = 1$ the above method fails as the denominator in (4.22b) goes to zero. A natural consequence of this condition is that a linear phase FIR filter cannot be converted into a lattice filter.

Example 4.4: Consider a four coefficient FIR filter (M=4):

$$h(0) = 1.0$$
$$h(1) = 0.8$$
$$h(2) = 0.45$$
$$h(3) = 0.6$$

$\alpha_3 = h(3) = 0.6$ and $H_{13}(z) = 1.0 + 0.8z^{-1} + 0.45z^{-2} + 0.6z^{-3}$.First, using (4.22b) we shall compute $H_{12}(z)$. All intermediate steps are shown below:

$$H_{12}(z) = \frac{1.0 + 0.8z^{-1} + 0.45z^{-2} + 0.6z^{-3} - 0.6z^{-3}\begin{Bmatrix}1.0 + 0.8z \\ +0.45z^2 + 0.6z^3\end{Bmatrix}}{0.64}$$

$$= \frac{1.0 + 0.8z^{-1} + 0.45z^{-2} + 0.6z^{-3} - 0.6\begin{Bmatrix}1.0z^{-3} + 0.8z^{-2} \\ +0.45z^{-1} + 0.6\end{Bmatrix}}{0.64}$$

$$= \frac{0.64 + .53z^{-1} - 0.03z^{-2}}{0.64}$$

$$= 1.0 + 0.8281z^{-1} - 0.04687z^{-2}$$

$$H_{12}(z) = \frac{1.0 + 0.8z^{-1} + 0.45z^{-2} + 0.6z^{-3} - 0.6z^{-3}\begin{Bmatrix}1.0 + 0.8z \\ +0.45z^2 + 0.6z^3\end{Bmatrix}}{0.64}$$

$$= \frac{1.0 + 0.8z^{-1} + 0.45z^{-2} + 0.6z^{-3} - 0.6\begin{Bmatrix}1.0z^{-3} + 0.8z^{-2} \\ +0.45z^{-1} + 0.6\end{Bmatrix}}{0.64}$$

$$= \frac{0.64 + .53z^{-1} - 0.03z^{-2}}{0.64}$$

$$= 1.0 + 0.8281z^{-1} - 0.04687z^{-2}$$

The lattice coefficient is equal to the last coefficient of $H_{12}(z)$, therefore, $\alpha_2 = -.04687$. Finally we shall compute $H_{11}(z)$ using the same procedure. We give the final result

$$H_{11}(z) = 1 + 0.8687z^{-1}$$
$$\alpha_1 = 0.8687$$

The lattice coefficients thus obtained may be verified against the ones obtained by direct computation using the equations in (4.17).

$$\alpha_1 = 0.8688$$
$$\alpha_2 = -0.4687$$
$$\alpha_3 = 0.6$$

§4.3 **Design if FIR Filters:** Windowing

An ideal lowpass filter, on account of sharp transition from unit response to zero response, would necessarily possess an infinite impulse response. Consider a two-sided lowpass filter whose frequency response is unity in the frequency range $\pm\omega_c$ and zero outside (see fig.4.13). The theoretical impulse response function is given by

$$h(m) = \frac{\sin(\omega_c m)}{\pi m} \qquad m = 0, \pm 1, \pm 2, \ldots, \pm\infty$$

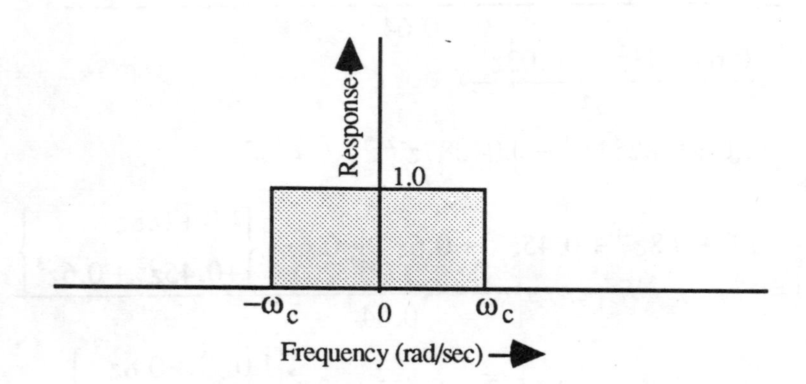

Figure 4.13: An ideal lowpass filter frequency response. ω_c is cut-off frequency.

which is shown in fig. 4.14. It is of infinite length. For practical implementation the impulse response function must be truncated, which results into large sidelobes. To suppress these large sidelobes there are three possibilities: 1. Use a window and smoothly truncate the response. 2. Provide smooth transition from passband to stopband. 3. Distribute the power in the sidelobes uniformly (equiripple filter). In this section we describe the first two possibilities. The equiripple filter design will be taken up in §4.4.

Truncation: A simple approach to obtain an FIR filter is to truncate the infinite impulse response filter. We shall truncate the theoretical response function to create an M length FIR filter,

$$h_{FIR}(m)=\begin{cases} h(m) & m=-\dfrac{M-1}{2},-\dfrac{M-3}{2},...,0,...,\dfrac{M-3}{2},\dfrac{M-1}{2} \\ 0 & otherwise \end{cases}$$

where we have assumed for simplicity that M is odd. The frequency response function of $h_{FIR}(m)$ is no longer the same as the theoretical lowpass filter. The frequency response of the truncated filters is shown in fig.4.15 for different lengths. From this figure we observe that the truncated filter frequency response has oscillatory behaviour as it wiggles around the theoretical frequency response. Interestingly, the amplitude of the ripples does not decrease with increasing filter length. Instead the ripples move closer to the discontinuity. This phenomenon is known as Gibb's oscillations. Further it is observed that the sharpness of the transition is increased by increasing the filter length.

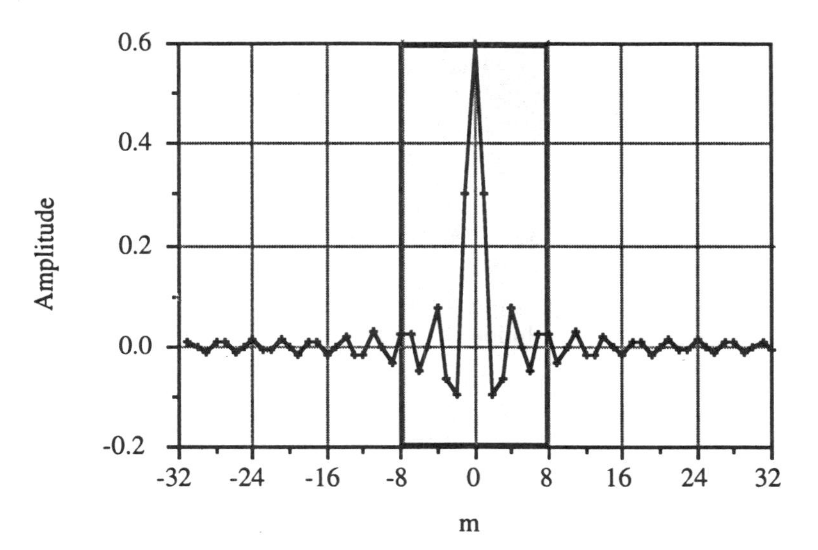

Figure 4.14: The impulse response function of a lowpass filter ($\omega_c = 0.6\pi$) truncated to a finite 16 coefficient filter. The impulse response lying outside the hatched band is lost creating a discontinuous termination.

Transition Zone: The remedy against Gibb's oscillation is simply to replace a sharp discontinuity with a gentle discontinuity. An example of such a

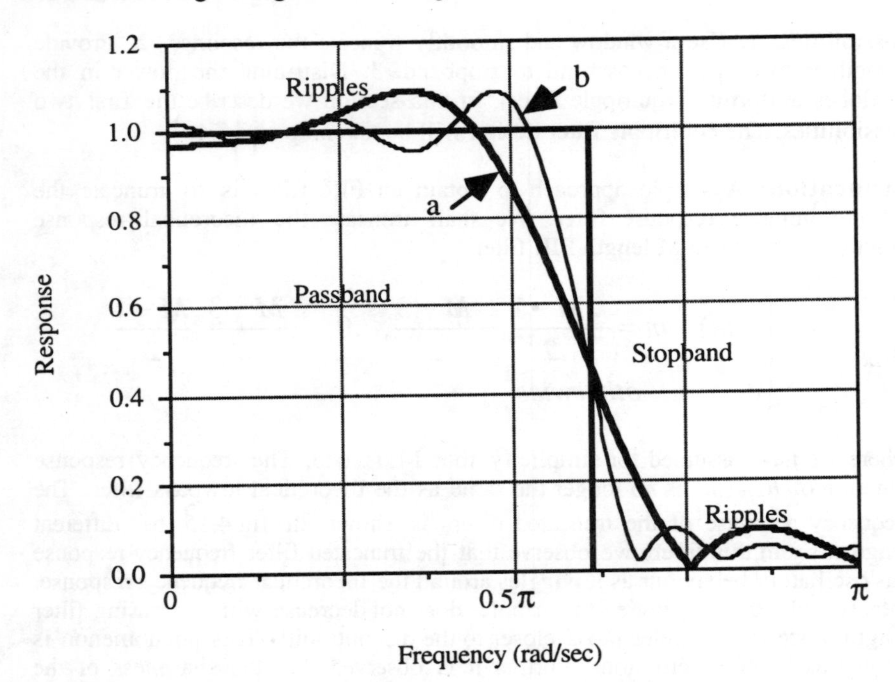

Figure 4.15: The phenomenon of Gibb's oscillations is caused by a sharp discontinuity in the filter transfer function $\omega_c = 0.6\pi$. By increasing the size of the filter the amplitude of the ripples cannot be reduced, however, the sharpness of the transition can be increased. The response function of the filters of different lengths are shown: (a) M=8 and (b) M=16.

replacement is shown in fig. 4.16. The sharp step change in the lowpass transfer function is now replaced by a smooth linear change from 1 to 0 over a frequency interval $\Delta\omega$. We call this as a transition zone centered at $\omega = \omega_c$, the maximum frequency of the lowpass filter. The transfer function of such a tapered lowpass filter is given by a convolution of two rectangular functions,

$$H(\omega) = rect(\frac{\omega}{\omega_c}) * rect(\frac{\omega}{\frac{\Delta\omega}{2}}) \tag{4.23a}$$

where

$$rect(\frac{\omega}{\omega_c}) = \begin{cases} 1 & for \quad |\omega| \le \omega_c \\ 0 & otherwise \end{cases}$$

and

$$rect(\frac{\omega}{\frac{\Delta\omega}{2}}) = \begin{cases} \dfrac{1}{\Delta\omega} & for \quad |\omega| \le \dfrac{\Delta\omega}{2} \\ 0 & otherwise \end{cases}$$

The impulse response function of the filter is obtained by inverse Fourier transformation of $H(\omega)$. The result is

$$h(m) = \frac{\sin(\omega_c m)}{\pi m} \frac{\sin(\frac{\Delta\omega}{2} m)}{\frac{\Delta\omega}{2} m} \quad (4.23b)$$

Notice that the ideal impulse response is multiplied by a window function given by

$$w(m) = \frac{\sin(\frac{\Delta\omega}{2} m)}{\frac{\Delta\omega}{2} m}$$

Thus the effect of a transition zone is same as that of a window function.

Example 4.5: Let us consider a numerical example. Take the case of lowpass filter shown in fig. 4.13. The sharp step change in the lowpass transfer function is at $\omega_c = 0.6\pi$ which is replaced by a smooth linear change from 1 to 0 over a frequency interval, from $\omega = 0.5\pi$ to $\omega = 0.7\pi$. (see fig. 4.16) The transition width is $\Delta\omega = 0.2\pi$. The transfer function of such a tapered lowpass filter is given by a convolution of two rectangular function of width 1.2π and 0.2π respectively. The frequency response of the FIR filter (M=16) is shown in fig. 4.17. Compare this result with curve b in fig. 4.15.

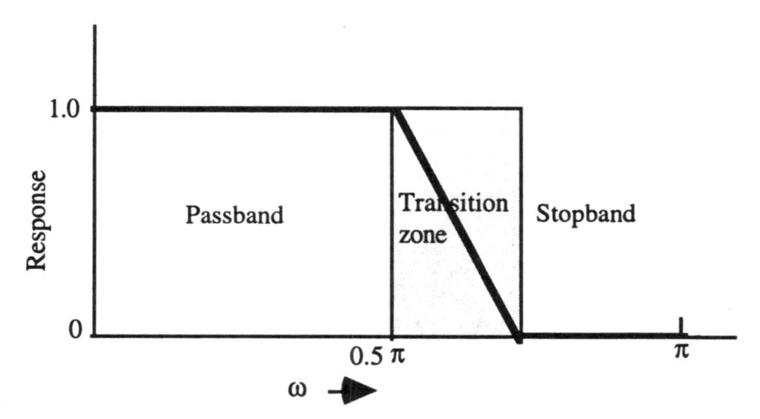

Figure 4.16: A transition zone is created between the passband and the stopband in the hope that the transition will be smooth which will result in reduced ripple amplitude in FIR filter response. $\omega_c = 0.6\pi$ and $\Delta\omega = 0.2\pi$.

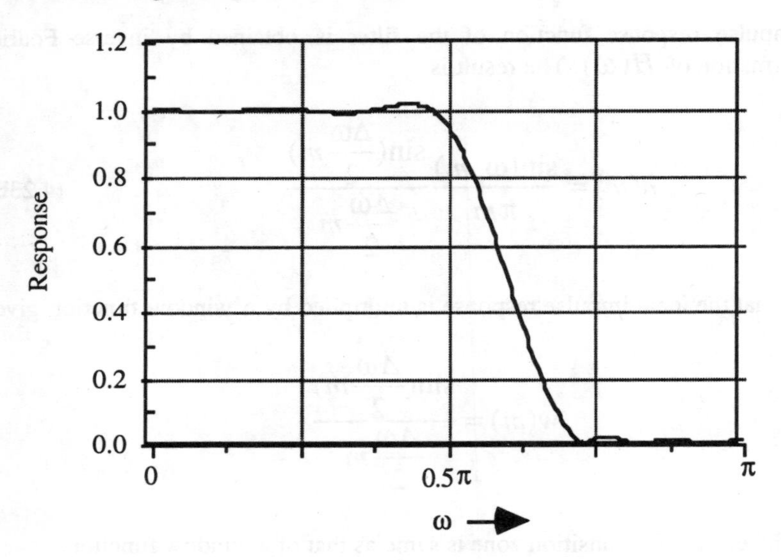

Figure 4.17: The magnitude of the Gibb's oscillations has been reduced by a providing a smooth transition from the passband to the stopband, as shown in fig. 4.16. The transition width is $\Delta\omega = 0.2\pi$. Filter size is M=16.

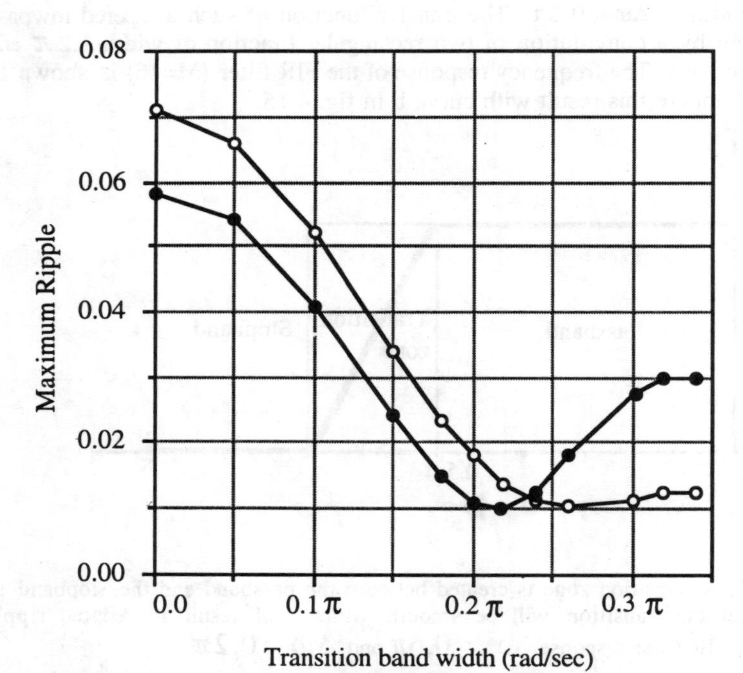

Figure 4.18: The attenuation of the maximum ripple in the passband (●) and in stop-band (○) as a function of the width of the transition zone.

The attenuation of the maximum ripple evidently depends upon the transition width as shown in fig. 4.18. We have assumed that the center of the transition band is held fixed at $\omega_c = 0.6\pi$.

Windowing: Another equivalent remedy to reduce the Gibb's oscillations is to smoothen the discontinuity in the impulse response function caused by truncation. Let us take the earlier example of the lowpass filter and truncate its impulse response (shown in fig 4.14) to 16 points. We shall use four different windows (see Chapter Three for discussion on windows) and compute the filter transfer function in each case. The first window is the default uniform window, that is, all filter coefficients are equally weighed, $w(m) = 1.0$. The filter coefficients are weighted with the window, $h(m)w(m)$, m=0,1,..., M-1. The resulting filter response function is shown by thick curve in fig. 4.19. The minimum attenuation (in stopband) is -19.5 dB. We define the attenuation as 20log(δ) where δ stands for ripple magnitude. The second window we have used is the Hanning window,

$$w(m) = 0.5 - 0.5\cos(\frac{2\pi m}{M})$$

The transfer function of the weighted filter coefficients is shown by the thin

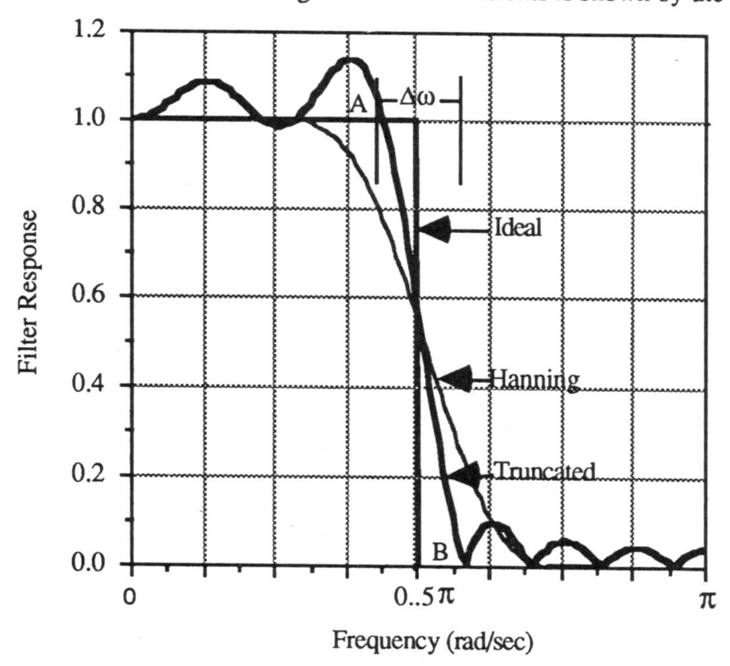

Figure 4.19: The transfer function of the weighted (Hanning window) filter coefficients is shown by the thin curve and that of the truncated filter coefficients (before weigthing) is shown by thick curve. The ideal response is also shown.

curve in fig. 4.19. The minimum attenuation (in the stopband) is -44 dB. We may note that the amplitude of the ripples in both passband and stopband are considerably reduced. The third window we have considered is so-called optimum window, which is obtained by maximizing the relative power in the main lobe. It turns out to be an eigenvector corresponding to the largest eigenvalue of certain matrix as described in Chapter Three. The minimum attenuation in the stopband is -35 dB.

Kaiser Window: The optimum window appears to be the best choice considering the ripple height and the transition width but it involves eigenvector decomposition, which is computationally expensive. An approximation to optimum window is the so-called Kaiser window defined in terms of modified Bessel function of the first type, $I_0(x)$. It is given by

$$w(m) = \frac{I_0\left(\alpha\sqrt{1-\left(\frac{2m}{M-1}\right)^2}\right)}{I_0(\alpha)} \qquad |m| \le \frac{M-1}{2} \tag{4.24}$$

for M odd. By varying α we can get different windows with different ripple characteristics. In fact we can control the stopband ripple magnitude in the windowed filter response through α. A plot of α as a function of stopband ripple is shown in fig.4.20. Note that when α=0 the Kaiser window reduces to a uniform window.

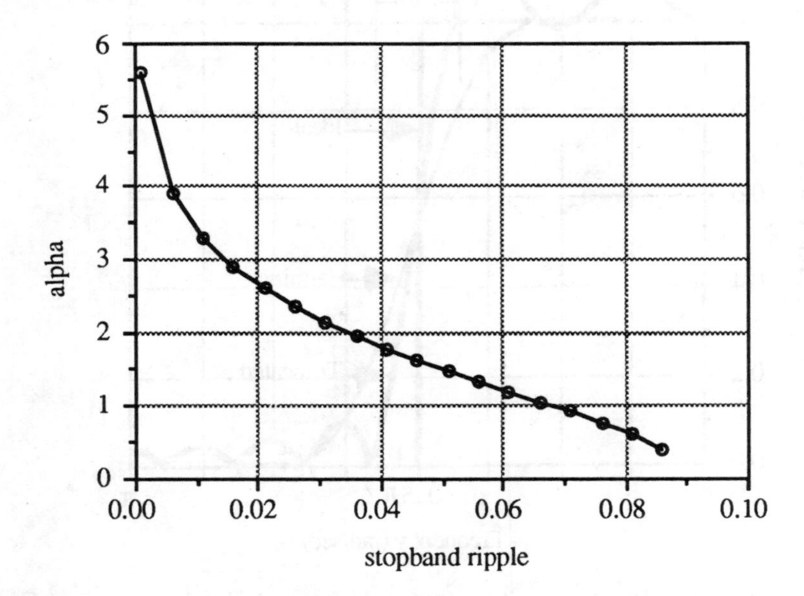

Figure 4.20: α in eq.(4.24) as a function of stopband ripple.

Example 4.6: As an illustration of how the parameter α influences the stopband ripple magnitude we have created two Kaiser windows of length 17 points with α=1.5 and α =3.2, respectively. The corresponding stopband ripples, predicted from fig. 4.20, are 0.051 (-26dB) and 0.011 (-39 dB), respectively. An impulse response of an ideal lowpass filter ($\omega_c = \pi/2$) was windowed using above Kaiser windows. The frequency response of the windowed lowpass filter is shown in fig. 4.21. The minimum attenuation in the stopband is 40 dB (for α =3.2.)

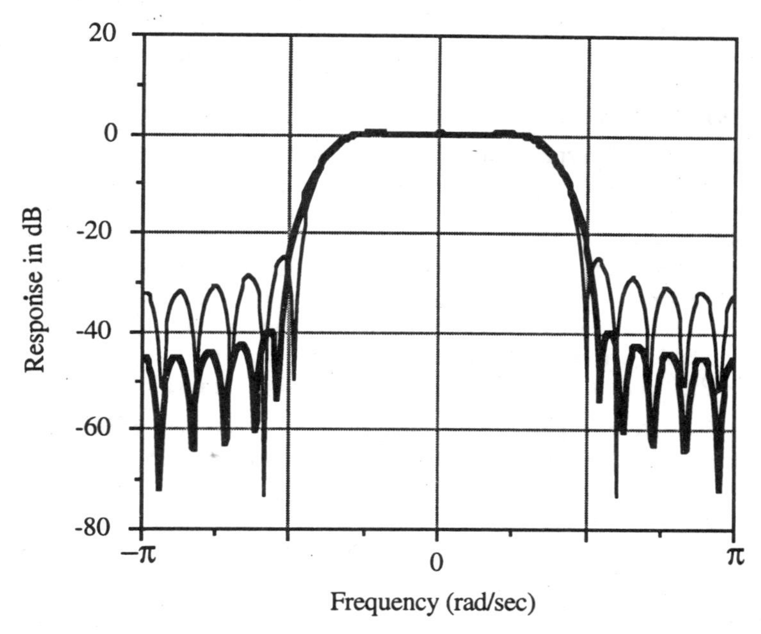

Figure 4.21: Frequency response of a lowpass filter using two different Kaiser windows Thin curve α=1.5 and thick curve α=3.2

Transition Width: The role of a window is basically to introduce a transition zone between the pass and stopbands. The effective width, measured as a distance

Table 4.2: The transition width and the minimum stopband attenuation of a lowpass filter frequency response using different types of windows

Type of window	Attenuation of the first sidelobe	Transition width, $\Delta\omega$
Uniform	-19.5 dB	$1.26\frac{2\pi}{M}$
Hanning	-44 dB	$3.25\frac{2\pi}{M}$
Optimum	-35 dB	$2.08\frac{2\pi}{M}$
Kaiser	-40dB	$2.52\frac{2\pi}{M}$

between the points A and B in fig 4.19, depends upon the type of window used and its size. For example, the transition widths and the maximum ripple height are shown for four different windows in table 4.2.

Design Procedure: A lowpass FIR filter may be specified in the frequency domain in terms of the passband edge, stopband edge and the minimum attenuation (in dB) in the stopband. The frequency information is given in terms normalized frequency. Note that the maximum normalized frequency is 0.5Hz. The sampling rate needs to be specified in order to convert the normalized frequencies into practical frequencies. For example, let the sampling rate, $f_s = 1000$Hz and the passband edge be $0.12 \text{x} \, 2\pi$ (in radian/sec) which translates into 120Hz. The specification of the maximum ripple enables us to select a right window from table 4.2, which for simplicity includes just four windows but we can include quite a few more. For example, if the desired maximum ripple is -40 dB or less we have a choice between Hanning and Kaiser window. The next step is to select the length of the filter, which fortunately does not influence the ripple magnitude. The length is controlled by the desired transition width, the difference between the passband edge and the stopband edge, both are available from the filter specification. For example let $\omega_p = 0.25 \times 2\pi$ and $\omega_s = 0.40 \times 2\pi$, the desired transition width is $\Delta\omega = 0.15 \times 2\pi$. Referring to column three in table 4.2 we can readily obtain the required length of the window (to the nearest odd integer) as 21 for Hanning window and 17 for Kaiser window. Finally, we still need to compute the theoretical two-sided IIR lowpass filter. This requires knowledge of the cut-off frequency, which is simply given as an average of the passband edge and stopband edge.

§4.4 **Design of FIR Filter**: Optimization

In optimization approach the filter coefficients are selected such that the maximum ripple in passband and/or in stopband is the minimum possible. The optimization can be carried out either in the Fourier domain or in the time domain. We shall briefly describe both approaches for the design of a lowpass filter. In frequency domain the frequency response in the transition zone is optimally selected such that the maximum ripple is minimum possible (minimax criterion). In time domain an impulse response is sought such that its frequency response oscillates around the desired frequency response (Chebyshev approximation).

Frequency Sampling: In the previous section we have shown how the ripple magnitude can be reduced by simply introducing a transition zone which separates the passband from the stopband; wider the transition zone smaller is the ripple magnitude. What is desirable is both narrow transition band and low ripples. Such a conflicting requirement cannot be satisfied hence a need for optimization. For a given transition zone we like to find transition coefficients which will minimize the maximum ripple in the passband and/or stopband. The linear transition is our starting first order approximation.

The frequency response is sampled at uniform interval, M samples $H((2\pi/M)k)$, $k = 0,1,\ldots,M-1$ in the interval $\pm\pi$. It is assumed to be periodic outside this interval. From these frequency samples we obtain, through inverse FDFT, M point FIR filter.

$$h(m) = \frac{1}{M}\sum_{k=0}^{M-1} H(\frac{2\pi}{M}k)e^{j\frac{2\pi}{M}mk} \qquad (4.25a)$$

The FIR coefficients thus obtained can be related to the theoretical impulse response coefficients, $h_d(m)$.

$$H(\omega = \frac{2\pi}{M}k) = \sum_{m=-\infty}^{\infty} h_d(m)e^{-j\frac{2\pi}{M}mk} \qquad (4.25b)$$

Substituting (4.25b) in (4.25a) we obtain

$$\begin{aligned} h(m) &= \frac{1}{M}\sum_{k=0}^{M-1}\sum_{m'=-\infty}^{\infty} h_d(m')e^{-j\frac{2\pi}{M}m'k}e^{j\frac{2\pi}{M}mk} \\ &= \sum_{m'=-\infty}^{\infty} h_d(m')\frac{1}{M}\sum_{k=0}^{M-1} e^{j\frac{2\pi}{M}k(m-m')} \end{aligned} \qquad (4.25c)$$

Note that

$$\frac{1}{M}\sum_{k=0}^{M-1} e^{j\frac{2\pi}{M}k(m-m')} = \begin{cases} 1 & for\ m' = m + pM \\ 0 & otherwise \end{cases}$$

where p is an integer. Using the above result in (4.25c) we obtain

$$h(m) = \sum_{p=-\infty}^{\infty} h_d(m + pM) \qquad (4.26)$$

The FIR filter is now obtained not by truncating an IIR filter but by folding it, a phenomenon, which we have seen in the Fourier transform of a discrete signal (Chapter Two). The different folds contribute to the aliasing error but now in the time domain. A numerical example is shown in Example 4.7.

Example 4.7: Consider a lowpass filter with cut off frequency $\omega_c = 0.5\pi$ and a linear transition band of width $\Delta\omega = 0.2\pi$. The impulse response function is computed using equation (4.23b). Let us form an FIR filter of length 17. The first fold will be at ± 8 $(\pm(M-1)/2)$. In fig. 4.22 the impulse response function (0 to 8) and the first segment due to fold are shown. The maximum amplitude of the folded segment is only 0.0047. Other folds are not shown, as the impulse response function becomes negligible. The actual FIR filter is then a sum of two segments shown in the figure. The aliased filter is tabulated in table 4.3.

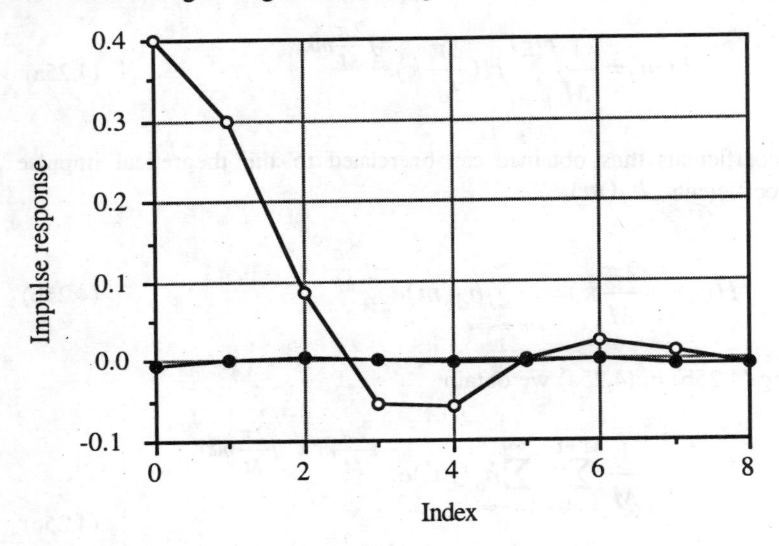

Figure 4.22: Impulse response of a lowpass filter ($\omega_c = 0.5\pi$) with a linear transition band $\Delta\omega = 0.2\pi$ is shown by empty circles and the first fold at m=8 is shown by filled circles. Other folds are not shown, as their magnitude is negligible.

Table 4.3: Non aliased and aliased FIR filter coefficients. Only half of the coefficients are shown (M=17).

Index	Non-aliased	Aliased
0	0.4000	0.3970
1	0.2978	0.2978
2	0.0087	0.0134
3	-0.053	-0.0501
4	-0.057	-.0.0594
5	0.0	-0.0025
6	0.0256	0.00256
7	0.0098	0.0061
8	-0.0055	-0.0110

Non-linear Transition: We have previously studied linear transition from the passband to the stopband. Theoretical impulse response function was derived. The FIR filter derived by truncating the theoretical impulse response function was found about 20 dB below that when the transition band had zero width. Further improvement is possible by optimally selecting the frequency response in the transition band and preselecting in rest of the region, that is, setting the response equal to one in the passband and zero in the stopband.(fig. 4.23). The width of the transition band in terms the number of frequency samples is first fixed. If $P = \Delta\omega M/2\pi$ is the width of the transition band, there will be P-1 transition coefficients.

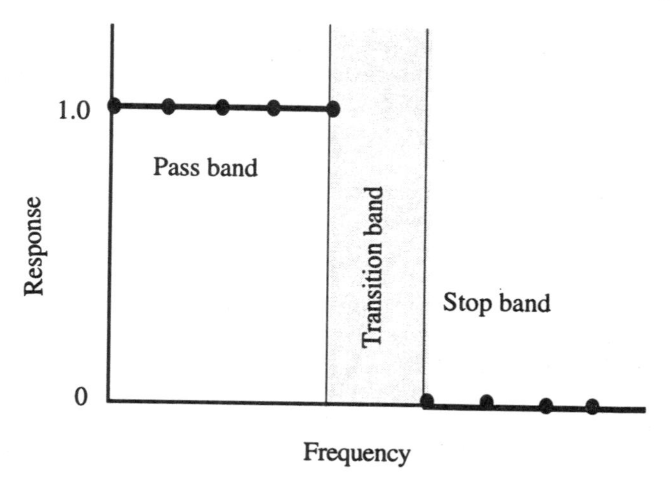

Figure 4.23: Frequency samples are preset in the passband and in the stopband but they are to be estimated such that the maximum ripple magnitude is minimum possible (minimax criterion) The sampling interval is $2\pi/M$.

Given the frequency samples we can obtain the response function everywhere, by means of interpolation. The interpolation is given by (see Exercise #5)

$$H(\omega)=\frac{e^{-j\frac{M\omega}{2}}}{M}\sin\frac{M\omega}{2}\sum_{k=0}^{M-1}\frac{H_r e^{-j\pi k\frac{M-1}{M}}}{e^{-j\frac{1}{2}(\omega-\frac{2\pi k}{M})}\sin\frac{1}{2}(\omega-\frac{2\pi k}{M})}$$

$$=\frac{e^{-j\frac{M\omega}{2}}}{M}\sin\frac{M\omega}{2}\left[\sum_{pass\ band}\frac{1*e^{-j\pi k\frac{M-1}{M}}}{e^{-j\frac{1}{2}(\omega-\frac{2\pi k}{M})}\sin\frac{1}{2}(\omega-\frac{2\pi k}{M})}+\sum_{p=1}^{P-1}\frac{t_p e^{-j\pi k\frac{M-1}{M}}}{e^{-j\frac{1}{2}(\omega-\frac{2\pi k}{M})}\sin\frac{1}{2}(\omega-\frac{2\pi k}{M})}\right] \tag{4.27}$$

where t_p, $p=1,\ldots,P-1$ are the unknown transition coefficients. Observe that the transition coefficients in (4.27) occur linearly. These coefficients are estimated by minimizing the largest ripple present in the passband or in stopband or in both. It has been established by Rabiner et al [21] that there is a unique solution to above minimax problem. They have also provided a steepest descent algorithm for obtaining these coefficients. Here we shall describe a simple minded approach where the minimax search is carried out over a dense grid of transition coefficients. In one dimension the search is over a line segment but in two dimensions the search is over a plane segment. The initial or the starting point is the one determined by linear transition.

Example 4.8: We shall illustrate the simple-minded approach by considering a design of lowpass filter with the following specifications:

M=17
Passband: -3 to +3
Transition band: +3 to +5 (one transition coefficient)
-3 to -5 (one transition coefficient)
Stopband: +5 to +8 and -5 to -8
Initial value of the transition coefficient = 0.5 (linear transition)

The minimax search was carried out in the stopband by allowing the transition coefficient to vary in the range, 0.3 to 0.6 with spacing 0.005. The minimum was encountered at around 0.40, which was further improved upon by scanning over a range 0.39 to 0.41 with a spacing 0.0005. The result is shown in fig. 4.24.

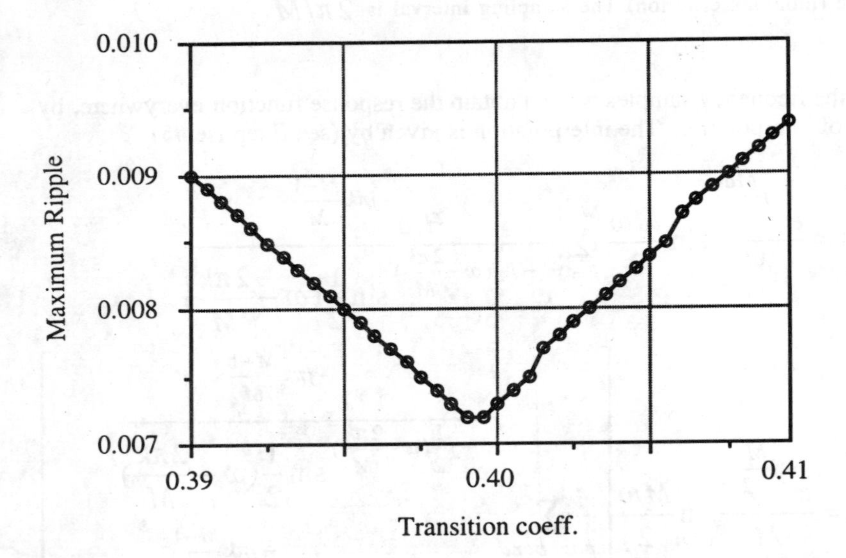

Figure 4.24: Maximum ripple in the stopband. One transition coefficient and M=17. The minimax occurs at 0.39925 and the minimax ripple magnitude is -42.85 dB.

The minimax ripple was found in stopband for transition coefficient = 0.39925 and its value was 0.0072 (-42.85 dB). Compare this with the linear transition (t=0.5) when the maximum ripple in the stopband is 0.0341 (-29.34 dB). The non-linear transition has considerably reduced the ripple in the stopband.

The search procedure may be extended to two dimensions but at greater computational load. We have carried out the optimization for two transition coefficients. The result is shown in fig. 4.25 where a contour plot of minimax ripple magnitude is plotted as a function of t_1 and t_2, the two transition coefficients. After searching in the neighbourhood of linear transition coefficient value with three different intervals, 0.01, 0.001 and 0.0001 we have estimated the optimum transition coefficients as (0.104, 0.585). The minimax ripple magnitude in the stopband is 0.000367 (-68.68 dB). A plot of the frequency response using the optimum transition coefficients is shown in fig. 4.26a. The maximum ripple in the passband however is quite high, 0.027 (-31.37dB). There is about 25 dB decrease in the minimax ripple magnitude over single transition coefficient. If we include the passband also in the minimax search it is found that the position of the minimum as well as the value of the minimum change. A minimax search carried out both in passband and stopband has yielded the following result: transition coefficients are (0.24, 0.783) and the value of the minimum is 0.0037 (-48.63 dB). A plot of the frequency response using the optimum transition coefficients is shown in fig. 4.26b. The maximum ripples in the passband and in the stopband are of same magnitude. The minimum is about 20 dB above that when the search is confined to the stopband only.

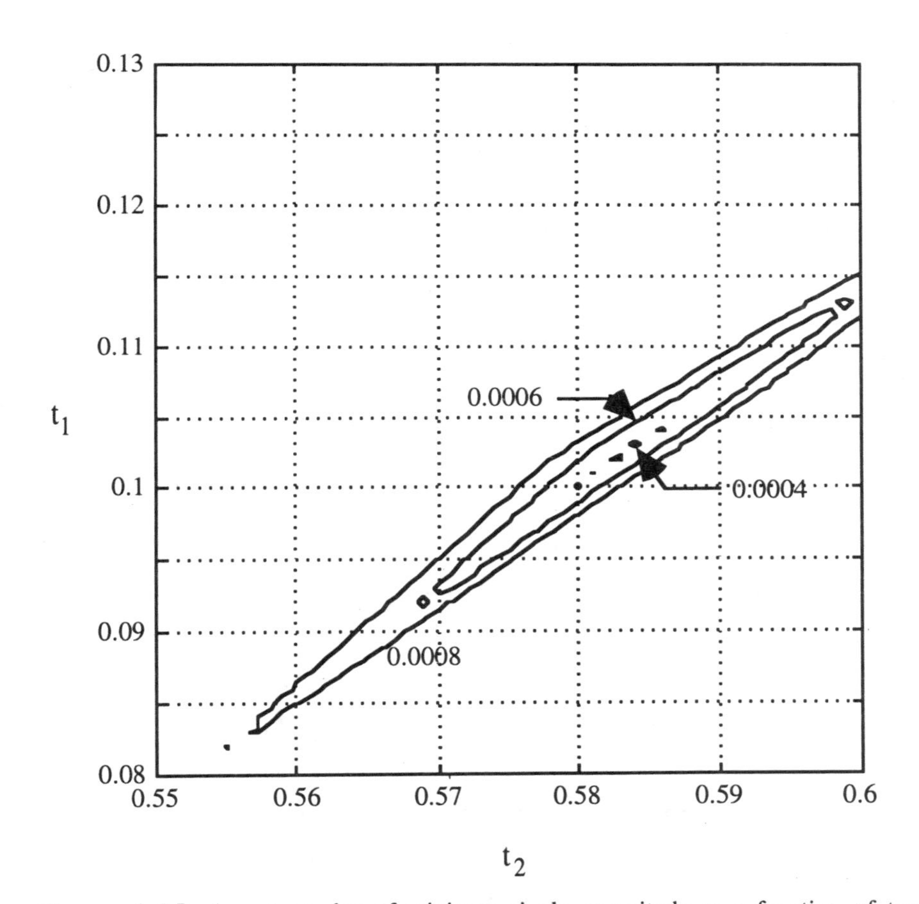

Figure 4.25: A contour plot of minimax ripple magnitude as a function of two transition coefficients. The contour values are as labeled. The minimum is at (0.104, 0.585) and the value is 0.000367 (-68.68 dB).

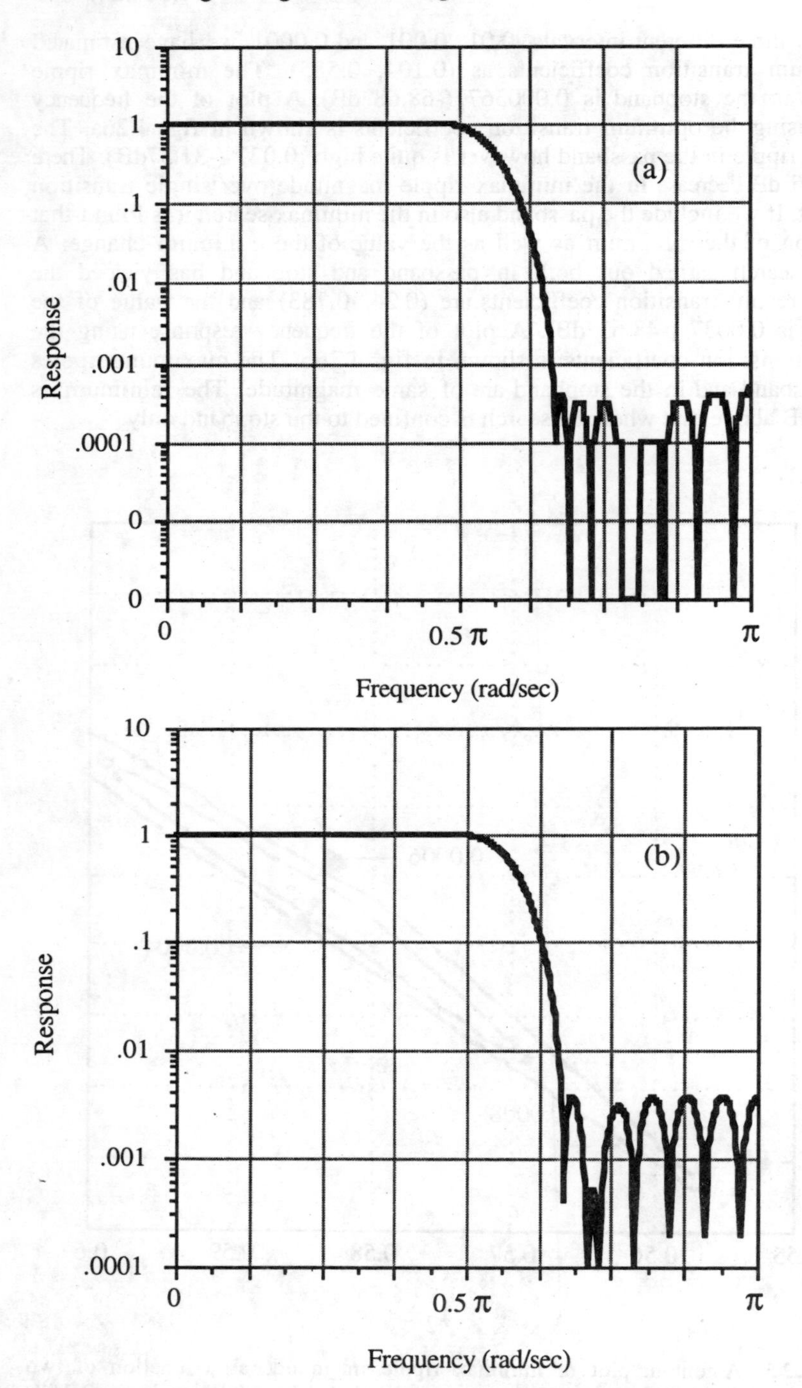

Figure 4.26: The frequency response function of a lowpass filter with two transition coefficients M=33. (a) Minimax search was carried out only in the stopband. Transition coefficients are (0.104, 0.585). The maximum ripple is 0.000367 (-68.68 dB) (b) Minimax search was carried out both in passband and

stopband. Transition coefficients are (0.24, 0.783). The maximum ripple is 0.0037 (-48.63 dB) .

Simply increasing the filter size can reduce the transition bandwidth. For example, if we increase the filter size from 17 to 33 the transition bandwidth will be reduced from 0.3529π to 0.1818π. Keeping the transition width approximately fixed as in table 4.4 the effect of increasing the number of transition coefficients is shown in table 4.4. From this table it is apparent that not much is gained by going beyond two transition coefficients. The transition coefficients for different passband widths have already been computed by Rabiner et al [21]. They are reprinted in [1].

Table 4.4: Minimum attenuation for different number of transition coefficients. The transition width is approximately kept fixed. $\omega_c \approx 0.5\pi$. There is not much to gain beyond two transition coefficients.

Filter size M	Minmax attenuation (Stopband)	Minmax attenuation (Stop & Passbands)	No. of transition coefficients	Transition bandwidth
23	-38.49	-29.27	1	0.174π
33	-68.68	-48.63	2	0.182π
43	-62.07	-35.8	3	0.186π

Equiripple FIR Filters: The central idea in the equiripple FIR filter design is to evenly distribute the approximation error over the entire passband(s) and stopband(s). There is a theorem known as Chebyshev equi-oscillation theorem [22], which states that there exists a unique polynomial of degree L or less for which the maximum absolute error is equal to δ_L. The polynomial has at least L+2 points in the range of approximation, (a, b) where the approximation error is given by $(-1)^m \delta_L$, m=0,1,2,...,L+1. We shall first show that the FIR filter transfer function can indeed be expressed as a polynomial in x (=cos(ω)). For this we go back to (4.4b) for M odd and consider the magnitude term.

$$H_r(\omega) = \left[2\sum_{m=0}^{\frac{M-3}{2}} h(m)\cos(\omega(\frac{M-1}{2} - m)) + h(\frac{M-1}{2}) \right] \quad (4.28)$$

Substitute q for $(\frac{M-1}{2} - m)$ in (4.28)

$$H_r(\omega) = \left[2\sum_{q=1}^{\frac{M-3}{2}} h(\frac{M-1}{2} - q)\cos(\omega q) + h(\frac{M-1}{2}) \right] \tag{4.29}$$

$$= \sum_{q=0}^{\frac{M-1}{2}} a(q)\cos(\omega q)$$

where

$$a(0) = h(\frac{M-1}{2})$$

and

$$a(q) = 2h(\frac{M-1}{2} - q).$$

Let us express

$$\cos(\omega q) = \sum_{i=0}^{q} b_i^q \cos(\omega)^i \tag{4.30a}$$

where b_i^q are listed in [27, p.795] in (4.29) Using (4.30a) in (4.29) and after simplification we obtain the required polynomial form for the transfer function,

$$H_r(\omega) = \sum_{i=0}^{\frac{M-1}{2}} \beta(i)\cos(\omega)^i \tag{4.30b}$$

where

$$\beta(0) = \sum_{l=0}^{\frac{M-1}{2}} b_0^l a(l), \quad \beta(1) = \sum_{l=1}^{\frac{M-1}{2}} b_1^l a(l), \quad \ldots, \quad \beta(i) = \sum_{l=i}^{\frac{M-1}{2}} b_i^l a(l), \quad \ldots$$

$$\beta(\frac{M-1}{2}) = b_{\frac{M-1}{2}}^{\frac{M-1}{2}} a(\frac{M-1}{2})$$

Let $H_d(\omega)$ represent the desired response, one in the passband(s) and zero in the stopband(s) but undefined in the transition band(s). We seek a polynomial approximation of the type (4.30b). Express the error function as

$$E(\omega) = W(\omega)[H_d(\omega) - \sum_{i=0}^{\frac{M-1}{2}} \beta(i)\cos(\omega)^i] \tag{4.31}$$

where $W(\omega)$ is a piecewise constant weight function. Let the extrema of $E(\omega)$ be at $0 \le \omega_0 < \omega_1 < \ldots < \omega_{L+1} \le \pi$. Following the Chebyshev oscillation theorem we have

$$E(\omega_m) = (-1)^m \delta_L, \quad m=0,1,2,...,L+1.$$

We use this result in (4.31), which leads us to the following system of linear equations

$$\sum_{q=0}^{L} a(q)\cos(\omega_m q) + \frac{(-1)^m \delta_L}{W(\omega_m)} = H_d(\omega_m), \quad m = 0,1,\ldots,L+1 \tag{4.32a}$$

The equation (4.32a) can be expressed in a matrix form,

$$\begin{bmatrix} 1 \; \cos(\omega_0) \; \cos(\omega_0 2) & \ldots & \cos(\omega_0 L) & \frac{1}{W(\omega_0)} \\ 1 \; \cos(\omega_1) \; \cos(\omega_1 2) & \ldots & \cos(\omega_1 L) & \frac{-1}{W(\omega_1)} \\ \ldots & & & \\ \ldots & & & \\ \ldots & & & \\ 1 \; \cos(\omega_{L+1}) \; \cos(\omega_{L+1} 2) & \ldots & \cos(\omega_{L+1} L) & \frac{(-1)^{L+1}}{W(\omega_{L+1})} \end{bmatrix} \begin{bmatrix} a_0 \\ a_1 \\ a_2 \\ . \\ . \\ . \\ a_L \\ \delta_L \end{bmatrix} = \begin{bmatrix} H_d(\omega_0) \\ H_d(\omega_1) \\ H_d(\omega_2) \\ . \\ . \\ . \\ H_d(\omega_{L+1}) \end{bmatrix} \tag{4.32b}$$

Remez Exchange Algorithm: Given the extremal points we can clearly solve (4.32b) for the filter coefficients as well as the maximum absolute error but the extremal points are not known. However, a recursive approach is possible. We start with some initial position of extremal points and solve for the filter coefficients. Using these coefficients we compute the frequency response of the filter over a fine grid and obtain fresh estimates of the extremal points, which are then used for computing the filter coefficients once again. The process is continued until we encounter stable extremal points, that is, there is no appreciable change in the position of the extremal point between two iterations. Notice that during each iteration we need to invert a (L+2)x(L+2) matrix. This can be computationally quite expensive for large L. Fortunately, we have a fast algorithm which uses the Lagrange interpolation scheme for obtaining fresh estimates of the extremal points. The Remez Exchange m-file in signal processing toolbox of Matlab is a very powerful filter design program. It can design a variety of FIR filters including Hilbert transform and differentiator. In the next example we illustrate a design of a lowpass filter.

Note that the ripple magnitude is not a design parameter but the width of the transition band is. The ripple magnitude is a part of the output of the algorithm along with the filter coefficients. The ripple magnitude depends upon the filter length. We have plotted the dependence of the minimum stopband attenuation (in dB) as a function of the filter length as in fig.4.28.

Example 4.9: We like to design an FIR filter of length 25 coefficients. The passband is 0 to 0.55π and stopband from 0.65π to π. The band between 0.55π and 0.65π is transition band or also known as "don't care band'. The filter coefficients (table 4.5) and its frequency response (fig. 4.27) are shown below. Observe that all ripples in fig. 4.27 are of same height, as desired. The minimax stopband ripple attenuation is -27.5 dB.

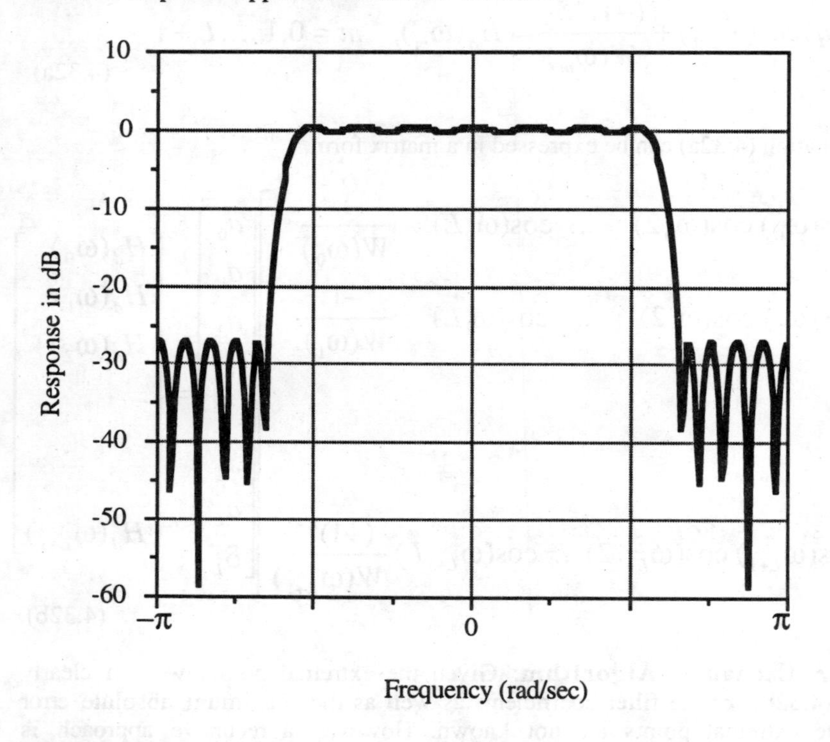

Figure 4.27: The frequency response of a 25 coefficient equiripple lowpass filter.

Table 4.5: FIR filter coefficients of the filter whose response is shown in fig.4.27

h(0)	-0.0067	*h(24)*
h(1)	0.0318	*h(23)*
h(2)	0.0033	*h(22)*
h(3)	-0.0235	*h(21)*
h(4)	0.0181	*h(20)*
h(5)	0.0220	*h(19)*
h(6)	-0.0437	*h(18)*
h(7)	-0.0001	*h(17)*
h(8)	0.0711	*h(16)*
h(9)	-0.0601	*h(15)*
h(10)	-0.0921	*h(14)*
h(11)	0.3015	*h(13)*
h(12)	0.6001	*h(12)*

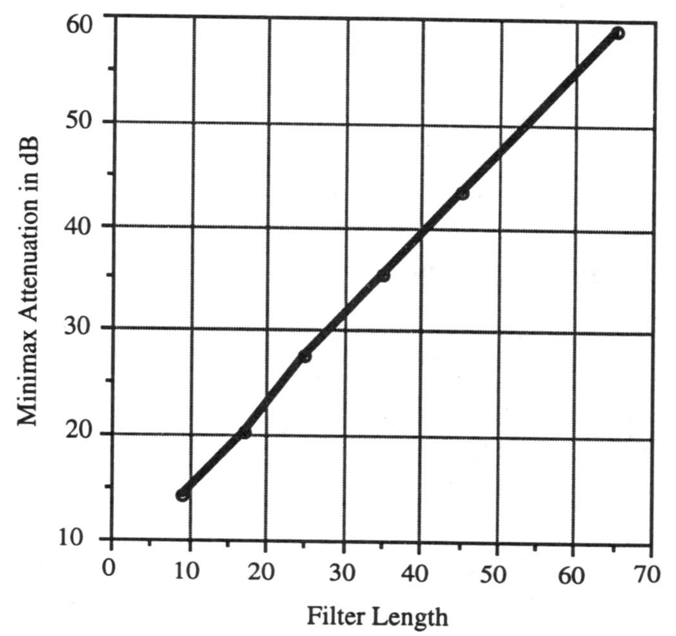

Figure 4.28: Minimax stopband attenuation (in dB) as a function of the filter length. The transition bandwidth was fixed at 0.1π.

The most important advantage of the equiripple filter design method over the frequency sampling method is the relative ease and precision with which one can specify the passband and stopband edges or the transition band width in terms of the radian frequency. On the other hand in the frequency sampling method, the filter specification has to be in units of the sampling interval which is proportional to inverse of the filter length.

§4.5 Exercises

Problems:

1. We have considered a linear transition from the passband edge to stopband edge. The transition can also be non-linear which may be achieved by convolving a lowpass filter by a triangular function shown in fig. 4.29:

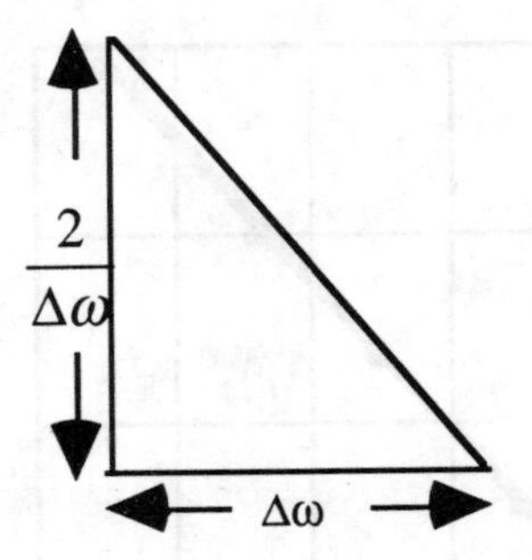

Figure 4.29: A triangular function when convolved with a lowpass filter will produce a non-linear transition zone.

show that the non-linear variation is indeed a quadratic variation. Calculate the impulse response function of the filter

2. A triangular window (also known as Bartlett window) of length (2L+1) is given by

$$w(m) = 1 - \frac{|m|}{L}, \quad 0 \le |m| \le L$$

Show that it can be obtained by linear convolution of two scaled rectangular windows. Determine the scale factor and the length of the rectangular window. From this relation derive the frequency response of the triangular window.

3. As an extension of Exercise #2 define a window as a linear convolution of p identical rectangular windows. Obtain the transfer function of this window.

4. Consider the following frequency response (fig. 4.30) of a filter

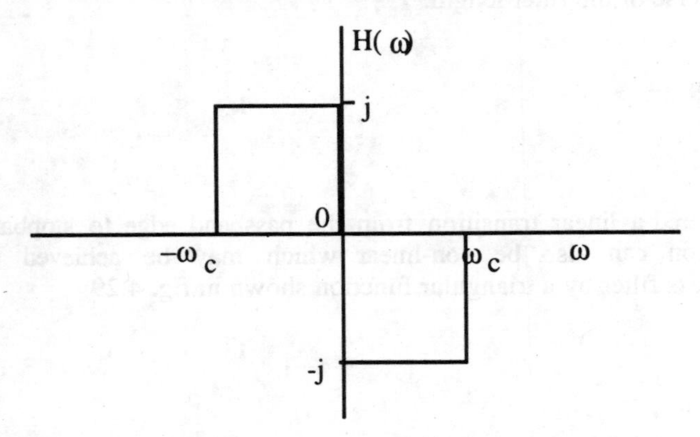

Figure 4.30: Frequency response of the desired filter.

Derive the impulse response function. Now, if we let $\omega_c \to \pi$ the frequency response tends to Hilbert transform. Thus, obtain the impulse response function of the Hilbert transform.

5. Given the frequency response of filter at uniform intervals, $2\pi k/M$, k=0.1,...,M-1 we can interpolate over the entire frequency range $\pi \leq \omega \leq -\pi$. Using eq. (4.10) on page 179 derive the interpolation formula,

$$H(\omega)=\frac{e^{-j\frac{M\omega}{2}}}{M}\sin\frac{M\omega}{2}\sum_{k'=0}^{M-1}\frac{H_r e^{-j\pi k'\frac{M-1}{M}}}{e^{-j\frac{1}{2}(\omega-\frac{2\pi k'}{M})}\sin\frac{1}{2}(\omega-\frac{2\pi k'}{M})}$$

Show that $H(\omega)\big|_{\omega=\frac{2\pi k}{M}} = H_r e^{-j\pi k\frac{M-1}{M}}$.

6. The zeros of a filter are shown in the z-plane in fig. 4.31 The unit circle is shown by thick circle. Compute the impulse response function of the filter. Is this a linear phase FIR filter?

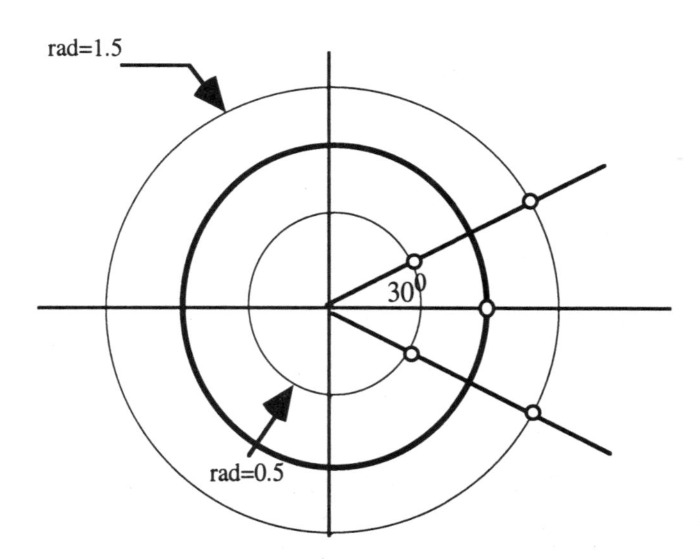

Figure 4.31: The position of the zeros in the z-plane.

7. Let $H(z)=(1-1.6z^{-1}+.873z^{-2})$ Obtain roots of H(z) and then define its complement., H'(z). Show that H(z)H'(z) is a linear phase FIR filter.

8. The frequency response of a linear phase FIR filter satisfies the following symmetry. The magnitude response is symmetric about $\omega=\pi/2$ such that

$$H(\omega+\frac{\pi}{4})+H(\frac{\pi}{4}-\omega)=1 \quad 0\leq\omega\leq\frac{\pi}{2}$$

Show that $h(0) = 0.5$, $h(\frac{\pi}{2}) = 0.5$ and $h(2n) = 0$, $n \neq 0$. Such a filter is known as a half band filter. It has only $(M+1)/2$ (M: odd) non-zero coefficients. (See [3, p408]).

9. Compute the impulse response function of a differentiator whose frequency response is given by $H(\omega) = j\omega$. Assume that the signal is bandlimited to $\pm\omega_c$.

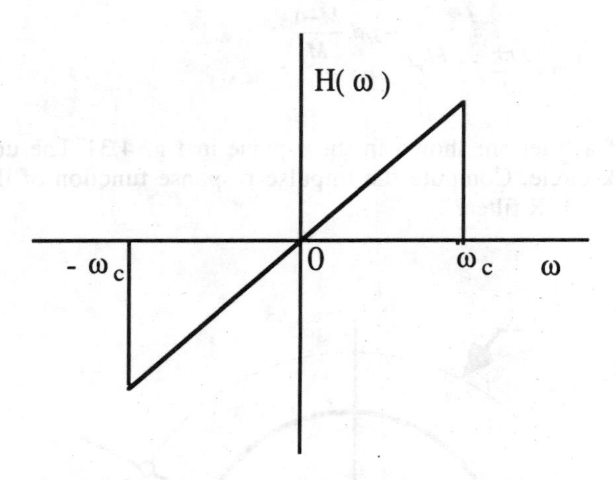

Figure 4.32: Frequency response of a differentiator.

Simplify the result for $\omega_c = \frac{\pi}{2}$. [Ans: $h(n) = \frac{\omega_c \cos(\omega_c n)}{\pi n} - \frac{\sin(\omega_c n)}{\pi n^2}$ and *h(0)*=0.0] .

10. Rewrite the interpolation formula (1.2) as a digital filter. It is useful for converting the sampling rate. For example, a bandlimited signal, which was sampled at a rate $2f_b$, but is now required to be resampled at a rate of $4f_b$ where f_b is the maximum frequency in the signal.

11. Obtain the impulse response function of a lowpass filter with tapered corners as shown in fig. 4.33

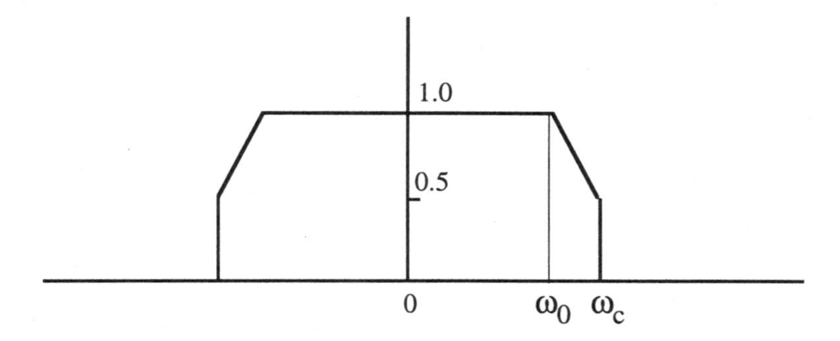

Figure 4.33: A lowpass filter with tapered corners.

12. Compute the frequency response of the following FIR filters: h_1=[0.5, 0.5] and
h_2=[-0.5, +0.5]. These are simple examples of lowpass and highpass filters, respectively.

13. Let $E_0(z)$ and $E_1(z)$ be second order polyphase components of an FIR filter transfer function, H(z). Define two other transfer functions

$$H_0(z^2) = \frac{1}{2}(E_0(z^2) + E_1(z^2))$$
$$H_1(z^2) = \frac{1}{2}(E_0(z^2) - E_1(z^2))$$

Show the following result, known as a generalized polyphase decomposition

$$H(z) = (1 + z^{-1})H_0(z^2) + (1 - z^{-1})H_1(z^2)$$

14. What is the transfer function of the kth resonator in fig.4.10b. Find the poles of the transfer function. The resonators act as a bank of tuned filters.

15. Convert the following FIR filter with four coefficients into a lattice filter

$$h(0) = 1.0 \qquad h(2) = 0.445$$
$$h(1) = 0.75 \qquad h(3) = 0.24$$

Compute the lattice coefficients using the method illustrated in Example 4.4 and verify the results through direct computation using eq.4.17.

16. Given the lattice coefficients $\alpha_1 = 1/3, \alpha_2 = 1/2, \alpha_3 = 1/8$ compute the filter coefficients *h(0)*, *h(1)*, *h(2)*, and *h(3)* (See eq.4.16). Is this a linear phase FIR filter?

17. A lowpass FIR filter has a passband ± 0.06 Hz (normalized frequency). The filter length is 15 coefficients long What is the relative computational advantage in using the FDFT coefficients as in the recursive realization against the direct form realization?

18. It is desired to design a lowpass filter with a passband edge at 160 Hz and a stopband edge at 200 Hz. The maximum ripple in the stopband should be better than -40 dB. The sampling frequency is 8 kHz. Choose an appropriate window and the corresponding length.

19. Design a lowpass filter using the frequency sampling method. Following are the desired specifications:

Passband edge: 0.35π
Stopband edge: 0.47π

Use a single transition coefficient. Choose an appropriate size of the filter consistent with the desired specification as closely as possible. What is the expected stopband attenuation. You may use the transition coefficients listed in [1].

20. Design an equiripple FIR filter of the shortest length such that the stopband minimum attenuation is 40 dB. Assume that the transition band is 0.1π. wide. The passband edge is at 0.55π (see Example 4.8).

21. Let $h=\{h(0), h(1),...,h(M-1)\}$ be an FIR filter (non-linear phase). Define another filter as follows:

$$h_1=0.5\{h(0)+h(M-1)\}, h(1)+h(M-2),... ,h(M-1)\}+h(0)\}$$

Show that h_1 is a linear phase filter. Compute its transfer function.

22. The tapped delay line structure of an FIR filter without and with symmetry is shown in fig 4.5. Compute the number of multiplications and additions required per output sample in both cases.

Computer Projects:

1.Truncate the lowpass filter shown in table 4.1 to the central nine points. Carry out a second order polyphase decomposition. Compute the amplitude and phase responses of the polyphase components, $E_0(z^2)$ and $E_1(z^2)$.

2. The transfer functions of a lattice filter are governed by a matrix recursive equation (4.20) which is reproduced here for convenience. Let the lattice coefficients be $\alpha_1 = 1/3, \alpha_2 = 1/2, \alpha_3 = 1/8, \alpha_4 = 1/5$. Calculate $H_{14}(z)$ and $H_{24}(z)$. Assume that $H_{10}(z) = 1$ and $H_{20}(z) = 1$ (why?)

$$\begin{bmatrix} H_{1m}(z) \\ H_{2m}(z) \end{bmatrix} = \begin{bmatrix} 1 & \alpha_m z^{-1} \\ \alpha_m z^{-1} & 1 \end{bmatrix} \begin{bmatrix} H_{1m-1}(z) \\ H_{2m-1}(z) \end{bmatrix} \qquad (4.20)$$

Verify the relationship given in equation (4.21). Also study the amplitude and phase characteristics of the transfer functions, $H_{1m}(z)$, $m = 1,...,4$.

3. Design a lowpass FIR filter using windowing method. The filter specifications are as follows:

Passband edge, $\omega_p = 0.3 \times 2\pi$

Stopband edge, $\omega_s = 0.4 \times 2\pi$

Stopband maximum ripple= -40 dB.

Find the required length of Hanning and Kaiser windows meeting above specifications. Obtain the frequency responses of the filter and verify whether they meet the desired specifications. Measure the maximum ripple in the passband.

4. Design an equiripple bandstop filter using the remez m-file from the signal processing toolbox of Matlab. The filter specifications are as follows:

Filter length:	25
Passbands	0 to 0.3π & 0.7π to 1.0π
Stopband	0.4π to 0.6π

What is the minimax stopband attenuation?. Compute the frequency response of the filter and verify whether the design specifications have been achieved.

5.Design an equiripple two passband FIR filter. The filter specifications are as follows:

			Response
Filter length:		33	
Passband	I	0 to 0.3π	1.0
Stopband		0.4π to 0.5π	0.0
Passband	II	0.6π to 0.8π	0.5
Stopband		0.9π to 1.0π	0.0

What is the minimax stopband attenuation?. Compute the frequency response of the filter and verify whether the design specifications have been achieved.

5 Infinite Impulse Response (IIR) Filters

A causal infinite impulse response filter is appropriate where the signal to be filtered is a causal signal, which may be an output, for example, of a dynamic system. In §2.6 we have studied the properties of a causal system. It may be recalled that the magnitude response completely determines the phase response. This is true of IIR filter also. In contrast, for FIR filter we were able to select the magnitude and phase responses independently and then, by choice, we could set the phase a linear function. The IIR filters are non-linear phase filters. Further, IIR filters are not guaranteed to be stable. Also, the coefficient quantization effects can be serious. Yet, IIR filters are commonly used in real-time applications as fewer coefficients are required to achieve the same filter (magnitude) response than in FIR filter. Another important property of an IIR filter is that it may be implemented recursively as in (4.3) where an infinite impulse response is condensed to p+q-1 coefficients. The presence of poles and associated feedback paths play an important role in determining the filter response function. There are fewer coefficients in IIR filter, hence it is faster to implement on a digital signal processor (see Chapter Eight) but its phase characteristics are far from desirable.

In this chapter we shall study the infinite impulse response (IIR) filters. The IIR filters are often derived from the well-known analog filters such as Butterworth, Chebyshev and Elliptic filters through sampling. Bilinear transformation plays central role in such a design procedure. It is also possible to design an IIR filter directly by placing pole and zeros in the z-plane. The notch filter is an example of such a filter. Given an impulse response function of a filter one can obtain an equivalent IIR filter. Finally, we take up transformation of the standard lowpass filter into another lowpass, bandpass or highpass filter.

§5.1 IIR Filter

The attractive feature of an IIR filter is its recursive form. The transfer function may be expressed as a ratio of two polynomials. Alternatively, an IIR filter may be characterized by the poles and zeros of the polynomials.

Recursive Form: Taking z-transform on both sides of (4.3) we obtain

$$Y(z) = B(z)X(z) + A(z)Y(z)$$

where

$$A(z) = a_1 z^{-1} + a_2 z^{-2} + \dots a_p z^{-p}$$

$$B(z) = b_0 + b_1 z^{-1} + b_2 z^{-2} + \dots b_q z^{-q}$$

$$p \geq q$$

The transfer function is defined as a ratio of the z-transforms of the output to input.

$$\boxed{H(z)=\frac{Y(z)}{X(z)}=\frac{B(z)}{1-A(z)}} \tag{5.1}$$

For a recursive filter to be realizable, the transfer function of the filter must satisfy the following conditions:

(1) It must be a rational function of z (i.e. a ratio of two polynomials) with real coefficients.

(2) The degree of the numerator polynomial must be equal to or less than that of the denominator polynomial ($p \geq q$)

(3) All poles must lie within the unit circle in the z-plane.

When the degree of the numerator polynomial is greater than the degree of the denominator polynomial the resulting impulse response function becomes non-causal. This may be shown as follows. Let q>p, the transfer function can be expressed as

$$H(z)=\frac{\sum_{i=0}^{q} b_i z^{-i}}{1-\sum_{i=1}^{p} a_i z^{-i}}=\sum_{k=1}^{q-p} c_k z^k+\frac{\sum_{i=0}^{q} b_i' z^{-i}}{\sum_{i=0}^{p} a_i' z^{-i}}$$

where we note that $\sum_{k=1}^{q-p} c_k z^k$ represents the non-causal part which acts on the future inputs.

Example 5.1: A simple example of infinite impulse response filter is an exponentially decreasing filter sequence (Example 1.1),

$$h(m)=r^m, \quad m=0,1,2,\ldots\infty \text{ and } |r|<1$$

The filter output is given by

$$y(n)=\sum_{m=0}^{\infty} h(m)x(n-m)=\sum_{m=0}^{\infty} r^m x(n-m)$$

Taking z-transform on both sides of above equation we obtain using the convolution theorem

$$Y(z)=\sum_{m=0}^{\infty} r^m z^{-m} X(z)=\frac{1}{1-rz^{-1}}X(z)$$

We rewrite above equation after cross multiplication

$$(1-rz^{-1})Y(z)=X(z)$$

Now compute the inverse z-transform on both sides of above equation. We obtain,

$$y(n) = x(n) + ry(n-1)$$

which is a recursive filter equivalent of an infinite impulse response filter. Just one coefficient is required in place of infinite response ■.

A higher order difference equation can often be split into interconnected first or second order difference equations. To achieve this we must first express the transfer function as a product of first and second order factors and transform each factor into a difference equation. We demonstrate this through an example.

Example 5.2: Consider the following third order difference equation:

$$y(n) = x(n) + \frac{9}{4}y(n-1) - \frac{77}{64}y(n-2) + \tfrac{49}{64}y(n-3)$$

Taking z-transform on both sides of above equation we obtain

$$Y(z) = \frac{1}{1 - \frac{9}{4}z^{-1} + \frac{77}{64}z^{-2} - \frac{49}{64}z^{-3}} X(z)$$

The transfer function may be factored as below

$$H(z) = \frac{1}{1 - \frac{9}{4}z^{-1} + \frac{77}{64}z^{-2} - \frac{49}{64}z^{-3}}$$

$$= \frac{1}{(1 - \frac{7}{4}z^{-1} + \frac{49}{32}z^{-2})(1 - \frac{1}{2}z^{-1})}$$

Each factor translates into a difference equation

$$y(n) = y_1(n) + \frac{7}{4}y(n-1) - \tfrac{49}{32}y(n-2)$$

$$y_1(n) = x(n) + \frac{1}{2}y_1(n-1)$$

We have thus been able to express a third order difference equatior. as two interconnected difference equations ■.

Poles and Zeros: A digital filter is fully characterized by its poles and zeros which may be estimated from the roots of the denominator and numerator polynomials (see (5.1)). The frequency response of a digital filter, which is fundamental in the understanding of the functioning of a filter, can be computed from the knowledge of the location of the poles and zeros. For this note that

$$H(z) = \frac{B(z)}{1 - A(z)} = \frac{(z-\beta_1)(z-\beta_2)...(z-\beta_q)}{(z-\alpha_1)(z-\alpha_2)...(z-\alpha_p)} \tag{5.2}$$

where $(\alpha_1, \alpha_2, ... \alpha_q)$ and $(\beta_1, \beta_2, ... \beta_p)$ are the positions of poles and zeros, respectively. The poles and zeros must occur with a complex conjugate symmetry so that the coefficients of the numerator and denominator polynomials in (5.1) are real.
To compute the frequency response of a digital filter we must evaluate its z-transform on the unit circle, that is, for

$$z = e^{j\omega}$$

Referring to fig. 5.1 we note that

$$(z - \alpha_i)\big|_{z=e^{j\omega}} = d_{\alpha_i} e^{j\theta_{\alpha_i}}$$

z-plane

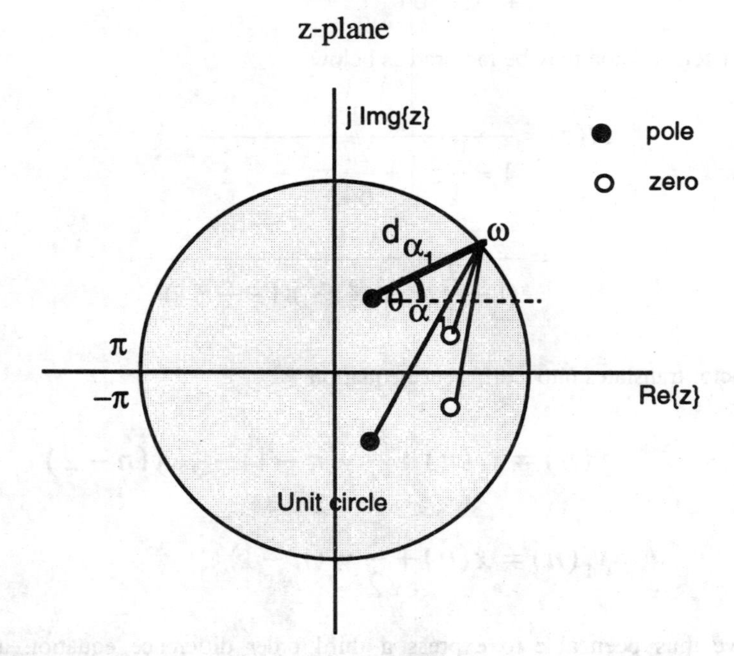

Figure 5.1: The frequency response of a digital filter may be obtained by evaluating its z-transform on the unit circle. Geometrically, the frequency response is equal to the ratio of the distance to zeros and to poles (5.3).

Hence, the frequency response function may be expressed as

$$H(z)\big|_{z=e^{j\omega}} = \frac{\prod_{i=0}^{q} d_{\beta_i} e^{j\theta_{\beta_i}}}{\prod_{i=0}^{p} d_{\alpha_i} e^{j\theta_{\alpha_i}}} \tag{5.3}$$

This result is useful in the design of a filter given its poles and zeros. We shall elaborate this approach in § 5.5.

§5.2 IIR Filter Structure

The IIR filter structure has a special feature in the form of a feedback network, which is not present in FIR filter. The denominator polynomial represents the feedback component. A digital filter may be realized through different filter networks. Some of these networks are

1. Direct
2. Cascade
3. Parallel

Direct Realization: Consider a general transfer function such as

$$H(z) = \frac{N(z)}{D(z)} = \frac{N(z)}{1 - D'(z)} \tag{5.4}$$

Let X(z) and Y(z) be the z transforms of the input and output, respectively.

$$Y(z)(1 - D'(z)) = N(z)X(z)$$

or, after rearrangement,

$$Y(z) = N(z)X(z) + D'(z)Y(z) \tag{5.5}$$

Filter equation (5.5) may be realized in terms of two simpler transfer functions, namely, $N(z)$ (forward) and $D'(z)$, feedback (see fig. 5.2).

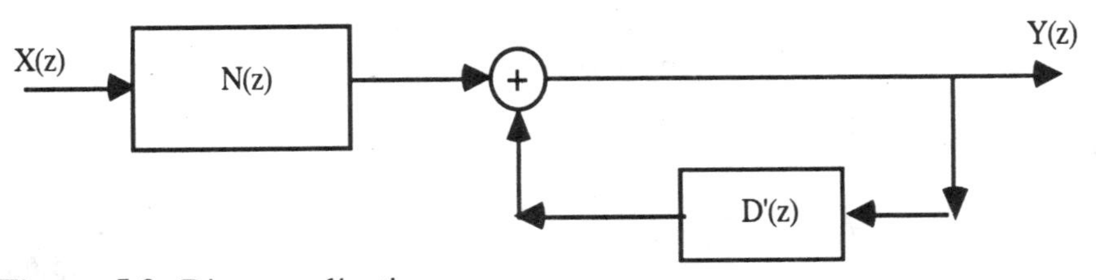

Figure 5.2: Direct realization.

Now consider realization of $N(z)$ which we express as follows:

$$N(z) = \sum_{i=0}^{q} b_i z^{-i}$$
$$= b_0 + z^{-1} \sum_{i=1}^{q} b_i z^{-i+1} = b_0 + z^{-1} N_1(z) \tag{5.6a}$$

Equation (5.6a) may be realized in two different ways as shown in fig. 5.3.

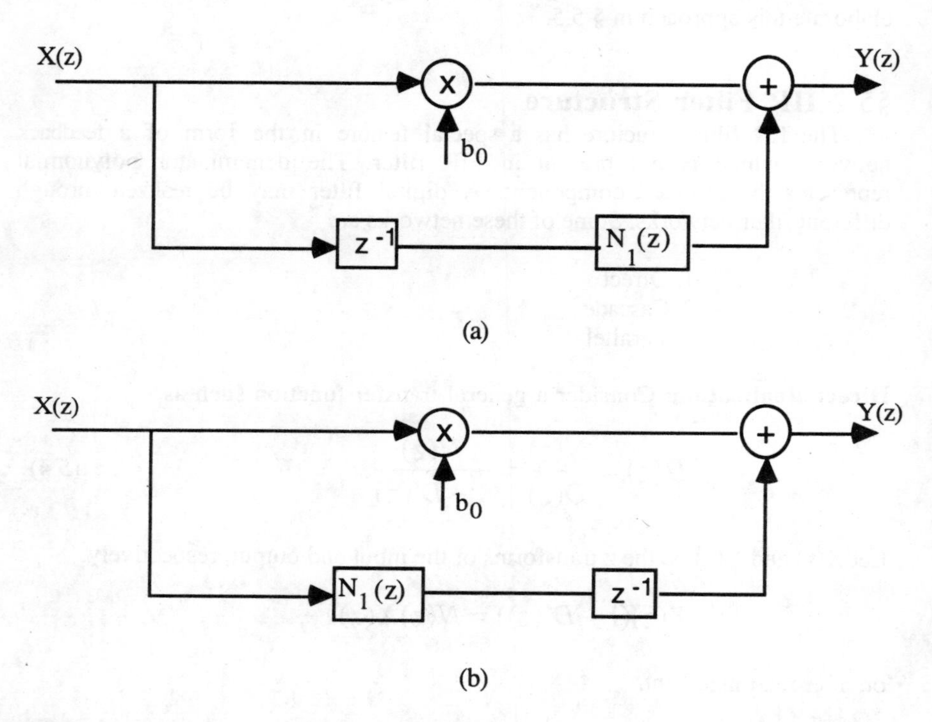

Figure 5.3: Two different ways of realizing eq. (5.6a). The essential difference is in the position of the delay element.

Next we replace $N_1(z)$ by

$$N_1(z) = b_1 + z^{-1} N_2(z)$$

as shown in (5.6b)

$$N(z) = b_0 + z^{-1}\{b_1 + z^{-1} N_2(z)\} \tag{5.6b}$$

There are four possible realizations of (5.6b), one of which is shown in fig. 5.4.

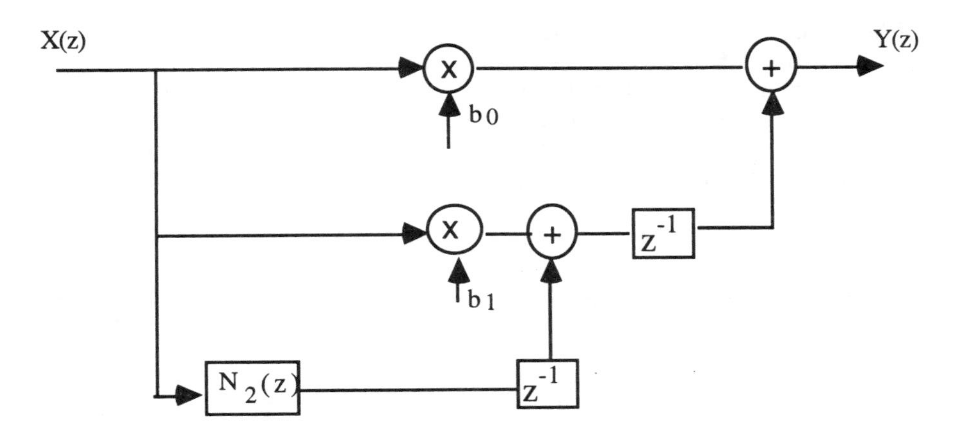

Figure 5.4: One of the four possible ways of realizing (5.6).

Thus, the simplification may be carried on until the last term contains just a constant, namely, b_q. Since at each stage there are two possible ways of realization, there will be in all 2^q realizations. One such realization is shown in fig. 5.5a

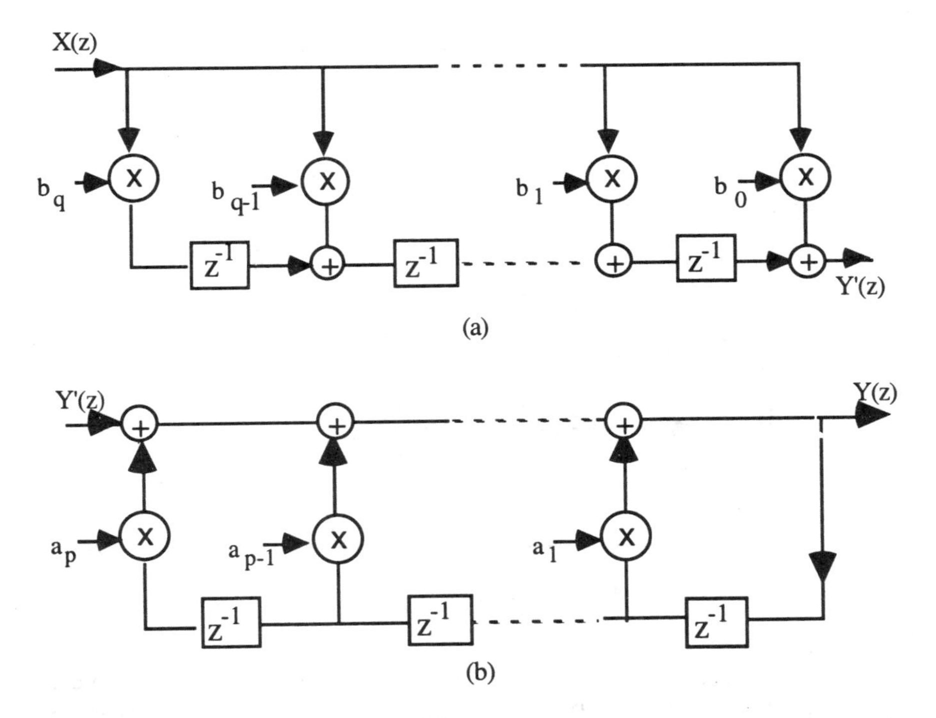

Figure 5.5: (a) One of the possible 2^q realizations of the numerator (b) One of the possible 2^p realizations of the denominator.

The output of the numerator network (fig. 5.5a) is fed into the denominator network. (fig. 5.5b). This completes direct form realization of IIR filter. This is often referred to as direct form I realization.

A digital network is said to be canonical if the number of unit delays employed is equal to the order of the transfer function. In this respect, direct form I is not a canonical network. Let us consider a slightly different arrangement of (5.4) as in (5.7a).

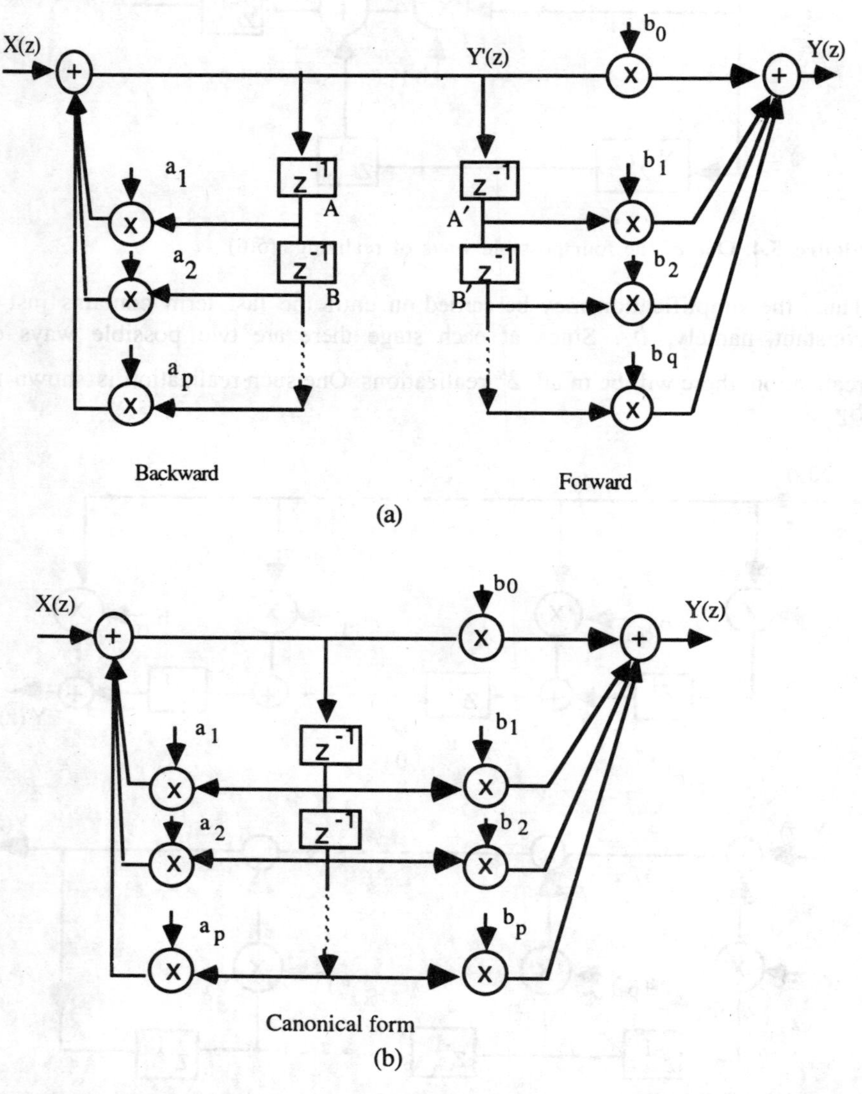

Figure 5.6: Realization of the filter shown in equation (5.7). (a) Forward and backward networks realized separately (b) the two are combined into a single network (canonical form) for q=p.

$$Y(z) = \frac{N(z)}{(1 - D'(z))} X(z) \tag{5.7a}$$
$$= N(z)Y'(z)$$

where

$$Y'(z) = \frac{X(z)}{(1 - D'(z))}$$
$$Y'(z) = X(z) + D'(z)Y'(z) \tag{5.7b}$$

The filtering operation shown in (5.7) may be realized through the network shown in fig. 5.6. The signals at nodes A', B' are equal to the respective signals at nodes A, B. Hence unit delays in the path $A'B'$ can be eliminated to yield a canonical realization of *H(z)*. An example of canonical realization is shown in fig. 5.6(b). This is often known as direct form II realization.

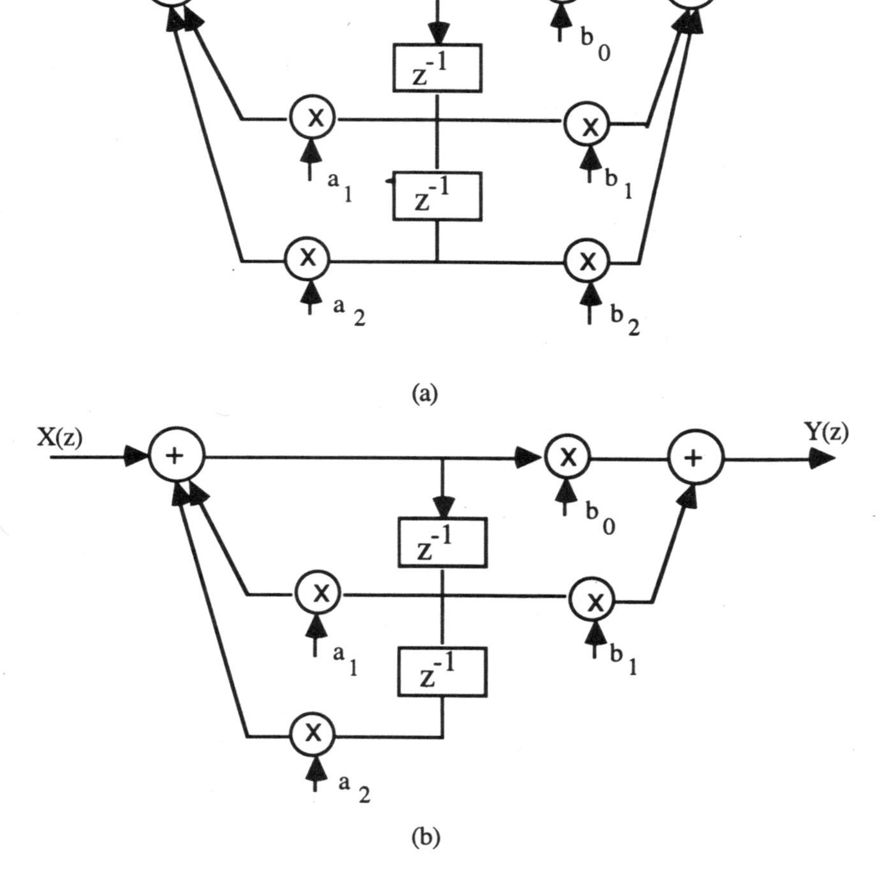

Figure 5.7: Canonical realization of each factor in cascade realization(a) and of each factor in parallel realization (b).

Cascade Realization: A transfer function can be factored into a product of second order transfer functions,

$$\boxed{H(z) = \prod_{i=1}^{r} H_i(z)} \tag{5.8a}$$

where

$$H_i(z) = \frac{b_{0i} + b_{1i}z^{-1} + b_{2i}z^{-2}}{1 - a_{1i}z^{-1} - a_{2i}z^{-2}} \tag{5.8b}$$

The second order filters are arranged in series form as shown in fig. 5.8. Each second order filter may be realized in the direct form as shown in fig. 5.7a. Note that, in practical applications, the coefficients b_{0i}, b_{1i}, b_{2i} and a_{1i}, a_{2i} must be real. This is the only condition one needs to impose while deriving the second order factors. Often there is more than one way to achieve the desired factorization.

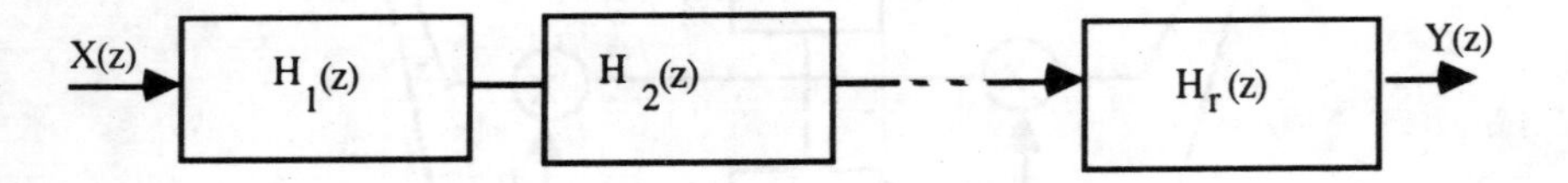

Figure 5.8: Cascade realization

Parallel Realization: Alternatively, the transfer function may be expanded into partial fractions as follows:

$$H(z) = \sum_{i=1}^{r} H_i(z) \tag{5.9a}$$

where

$$H_i(z) = \frac{b_{0i} + b_{1i}z^{-1}}{1 - a_{1i}z^{-1} - a_{2i}z^{-2}} \tag{5.9b}$$

The partial fractions are then arranged in a parallel form as shown below (fig. 5.9). Each partial fraction factor may be realized as in fig. 5.7b. Note that, since the numerator in (5.9b) has no term in z^{-2}, that is, $b_2 = 0$ the bottom line in the forward part in fig. 5.7b is absent.

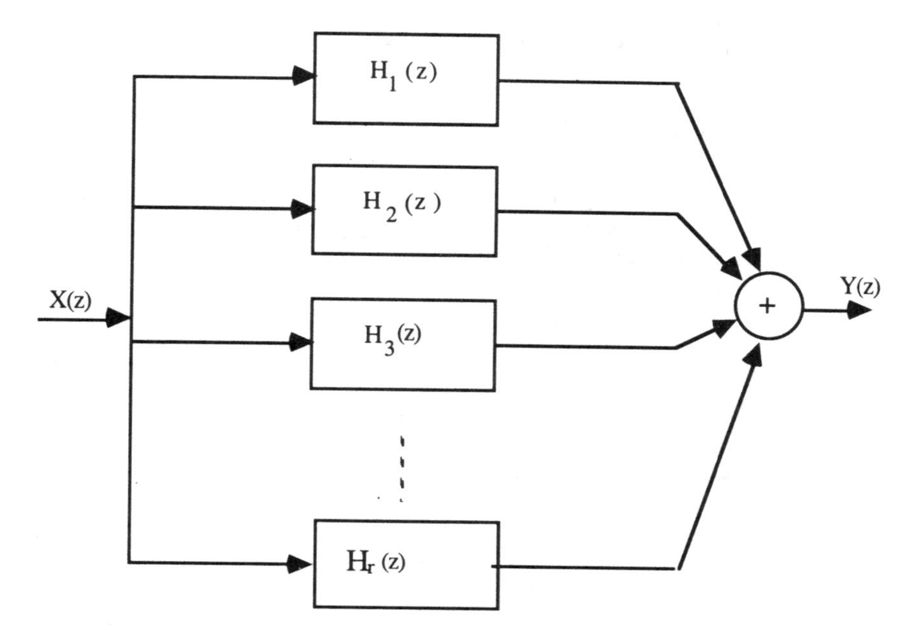

Figure 5.9: Parallel realization of the transfer function after its partial factor expansion has been done.

§5.3 Analog Filters

The digital filter design in its simplest form, with which we are concerned here, deals with discrete approximation of some well-known analog filters, namely, Butterworth, Chebyshev, Elliptic, Bessel filters, etc. We shall study three simple analog filters of great practical importance. They are Butterworth, Chebyshev and Elliptic filters. Analog filters are characterized by a loss function, $L(\omega)$ or by its logarithm, 10log($L(\omega)$). The loss function is related to the transfer function of the filter, H(s).

$$L(\omega) = \left. \frac{1}{|H(s)|^2} \right|_{s=j\omega} \tag{5.10}$$

Given the analytic form for $L(\omega)$ we need to find *H(s)* satisfying (5.10) and that all its poles are to the left of the imaginary axis of s-plane. Recall that for causal filter only the amplitude response or its loss function can be independently prescribed. The phase is determined by calculating the Hilbert transform of the log amplitude (see Chapter Two) or in the present case from the poles lying to the left of the imaginary axis.
The filter of interest is a lowpass filter from which other filters (bandpass, highpass, etc) can be derived. An ideal lowpass filter has unit loss (or zero dB) in the range 0 to ω_c and infinite loss in the range $>\omega_c$ to ∞. The range 0 to ω_c is the passband and the range $>\omega_c$ to ∞ is the stopband. In practice the ideal loss function cannot be achieved but it can only be approximated. We shall study

three important approximations, namely, Butterworth, Chebyshev and Elliptic approximations.

Butterworth Approximation: The loss function in Butterworth approximation is given by $L(\omega) = 1 + \omega^{2n}$ where n is the order of the filter. The loss in the range $\omega = 0$ to 1 monotonically increases from 0 dB to 3dB approximately. There are no ripples in the passband or in the stopband. The loss function increases smoothly with increasing frequency. In fig. 5.10 a few examples of Butterworth filters are shown.

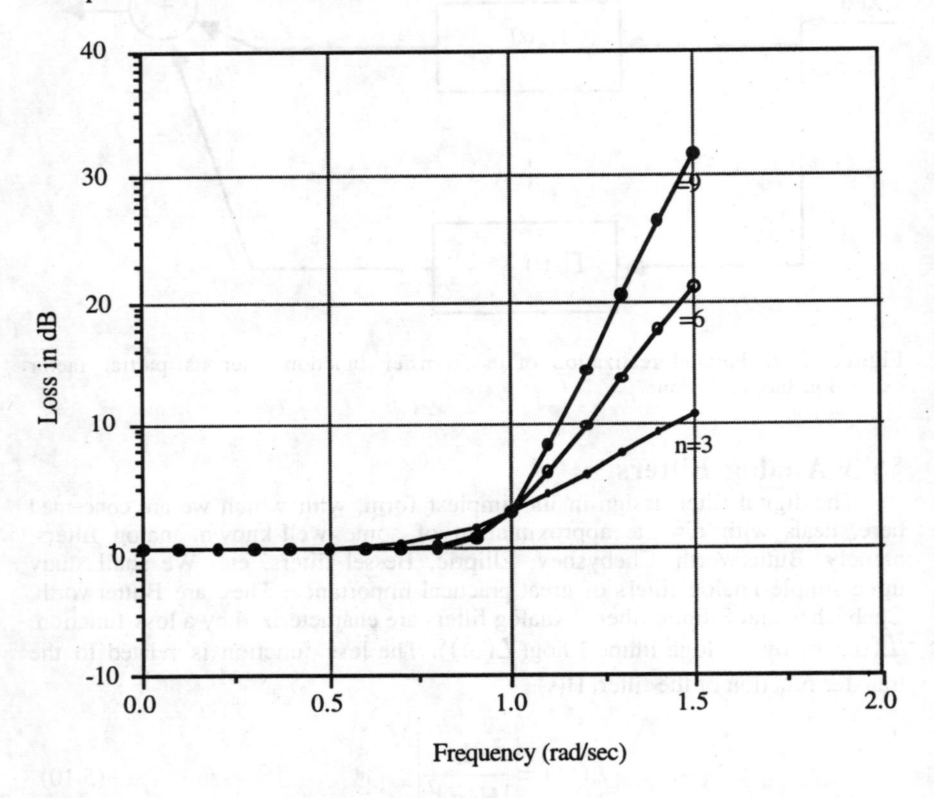

Figure 5.10: A few examples of Butterworth filter. The loss as a function of frequency is plotted for three different orders.

Transfer function: We shall first map the loss function onto s plane where $s = \sigma + j\omega$. For this purpose replace ω by $\frac{s}{j}$.

$$L(-s^2) = 1 + (-s^2)^n$$

$$= \prod_{k=1}^{2n} (s - s_k) \tag{5.11a}$$

where

$$s_k = \begin{cases} \exp(j(2k-1)\dfrac{\pi}{2n} & \textit{for n even} \\ \exp(j(k-1)\dfrac{\pi}{n} & \textit{for n odd} \end{cases}$$

The transfer function is easily obtained by collecting all factors in (5.11a) which correspond to the zeros lying to the left of jω axis of the s-plane. The transfer function may be written as

$$H_n(s) = \frac{1}{\prod_{k=1}^{n}(s - s_k)} \tag{5.11b}$$

where s_k, k=1, 2, ..., n are the poles of the transfer function which lie to the left of the imaginary axis in s-plane.

Example 5.3: In this example we show how to find the poles of a 2nd order Butterworth filter and then find its transfer function. The zeros of the loss function are as follows:

$$s_1 = \exp(j\frac{\pi}{4})$$

$$s_2 = \exp(j\frac{3\pi}{4})$$

$$s_3 = \exp(j\frac{5\pi}{4})$$

$$s_4 = \exp(j\frac{7\pi}{4})$$

The zeros of the loss function are shown in fig. 5.11. The zeros that lie to the left of jω axis are the poles of the transfer function. The transfer function may now be expressed as follows:

$$H_2(s) = \frac{1}{\prod_{i=1}^{2}(s - p_i)}$$

where $p_1 = e^{j\frac{3\pi}{4}}$ and $p_2 = e^{j\frac{5\pi}{4}}$

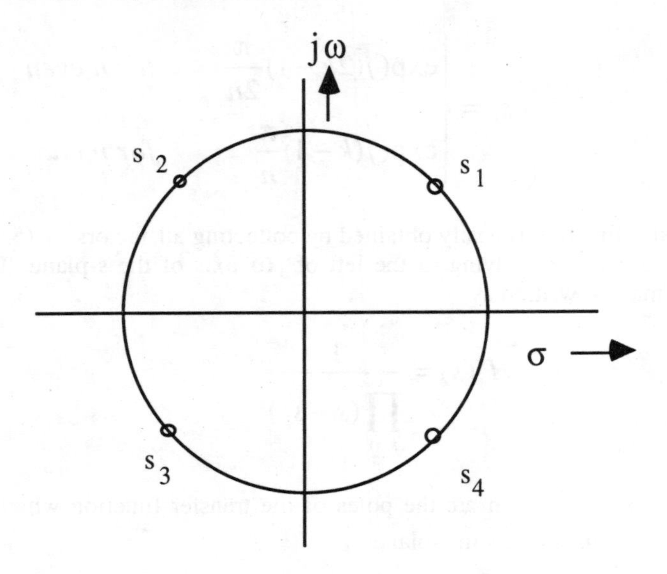

Figure 5.11: Zeros of a 2nd order Butterworth loss function. Zeros s_2 and s_3 become poles of the transfer function.

Chebyshev Approximation: The loss function is defined as

$$L(\omega) = 1.0 + \varepsilon^2 T_n^2(\omega)$$

where $T_n(\omega)$ is a Chebyshev polynomial, which is defined as

$$\begin{aligned} T_n(\omega) &= \cos(n\cos^{-1}(\omega)) & |\omega| \le 1 \\ &= \cosh(n\cosh^{-1}(\omega)) & |\omega| > 1 \end{aligned}$$

and ε is related to the ripple magnitude in the passband

$$\varepsilon^2 = 10^{0.1A_p} - 1$$

where A_p is ripple amplitude in dB. Note that the Chebyshev polynomial of order n has n zeroes and (n-1) maxima in the range $\omega = \pm 1$ and outside this range the function grows monotonically. A sketch of the loss function of a 4th order Chebyschev filter is shown in fig. 5.12.

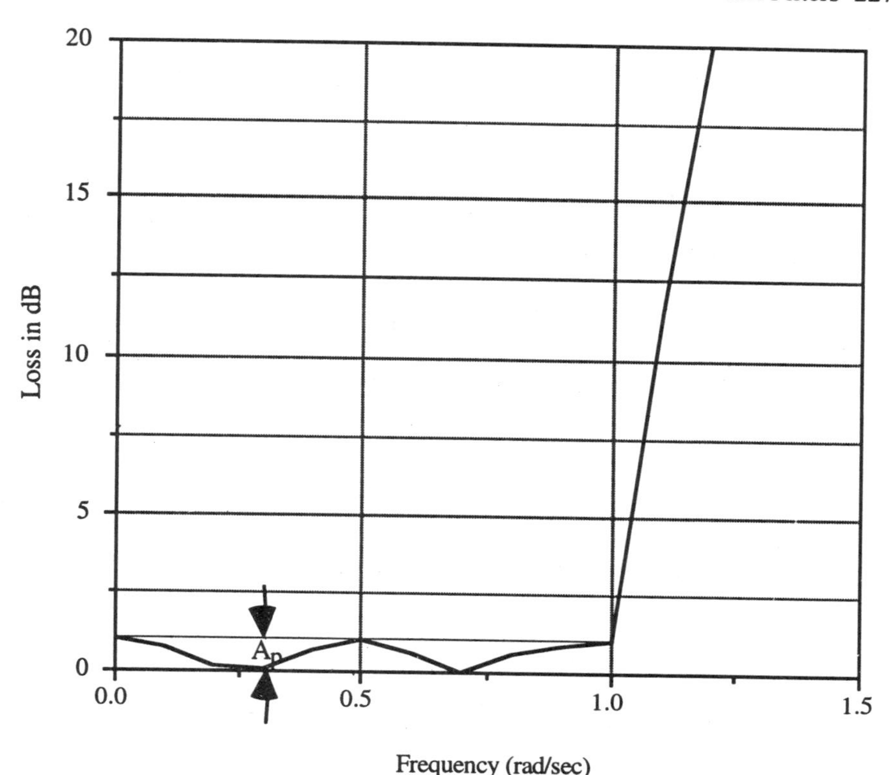

Figure 5.12: Loss function of 4th order Chebyshev filter. Notice the ripples of 1 dB magnitude (A_p=1 dB) in the passband. By selecting $\varepsilon^2 = 10^{0.1A_p} - 1$ we can alter the ripple magnitude $A_p = 10\log(1+\varepsilon^2)$. Remember that Butterworth filter had no such ripples in the passband.

The normalized transfer function of order n is given by

$$H_n(s) = \frac{H_0}{\prod_{i=1}^{n}(s - p_i)} \tag{5.12}$$

where p_i, i=1,2,...n are the poles in the left half s-plane and H_0 is a constant,

$$H_0 = \prod_{i=1}^{n}(-p_i) \qquad \text{for n odd}$$

$$= 10^{-0.05A_p}\prod_{i=1}^{n}(-p_i) \qquad \text{for n even}$$

The poles are given by

$$p_i = (\sigma_i + j\omega_i), \quad i = 1,2,\ldots,n$$

where

$$\sigma_i = \pm\sinh\left(\frac{1}{n}\sinh^{-1}\left(\frac{1}{\sqrt{10^{0.1A_p}-1}}\right)\right)\sin\frac{(2i-1)\pi}{2n}$$
$$\omega_i = \cosh\left(\frac{1}{n}\sinh^{-1}\left(\frac{1}{\sqrt{10^{0.1A_p}-1}}\right)\right)\cos\frac{(2i-1)\pi}{2n} \tag{5.13}$$

Example 5.4: We consider a 4th order Chebyshev filter, that is, n=4. Let A_p=1.0 dB.

$$\sigma_i = \pm 0.364625\sin\frac{(2i-1)\pi}{8}$$
$$\omega_i = 1.064402\cos\frac{(2i-1)\pi}{8}, \quad \text{i=1.2,...,4}$$

Poles lying in the left half of the s-plane are

$$-0.364625\sin\frac{\pi}{8} + j1.064402\cos\frac{\pi}{8}$$
$$-0.364625\sin\frac{3\pi}{8} + j1.064402\cos\frac{3\pi}{8}$$
$$-0.364625\sin\frac{5\pi}{8} + j1.064402\cos\frac{5\pi}{8}$$
$$-0.364625\sin\frac{7\pi}{8} + j1.064402\cos\frac{7\pi}{8}$$

Using the symmetry properties of the sine and cosine functions the pole position may be reduced to

$$-0.364625\sin\frac{\pi}{8} \pm j1.064402\cos\frac{\pi}{8}$$
$$-0.364625\sin\frac{3\pi}{8} \pm j1.064402\cos\frac{3\pi}{8}$$

Finally, the poles and the constant H_0 are as follows:

$$p_1, p_1^* = -0.139536 \pm j0.983379$$
$$p_2, p_2^* = -0.336870 \pm j0.407329$$
$$H_0 = 0.245653$$

as shown in fig. 5.13. The transfer function is formed by taking into account the zeros lying in the left half plane,

$$H_4(s) = \frac{0.245653}{\prod_{i=1}^{2}(s-p_i)(s-p_i^*)}$$

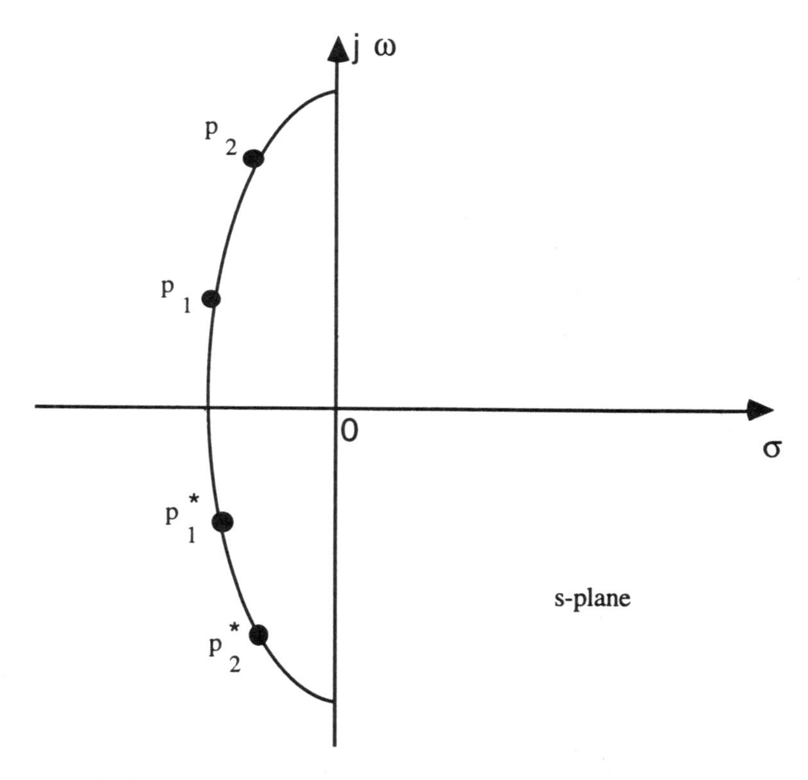

Figure 5.13: Location of the poles of a 4th order Chebyschev filter in s-plane.

Design of Chebyshev Filter: The Chebyshev filter is completely specified by selecting any three parameters from the following list:

1. n: order of the filter.
2. ε: parameter related to the passband ripple $(10\log(1+\varepsilon^2) = A_p)$.
3. ω_s: frequency at which the stopband loss attains a prescribed attenuation.
4. A_S: prescribed attenuation in dB.

Example 5.5: Let A_P=1 dB, A_S=40dB at ω_s=1.2. The order of the Chebyshev filter may be obtained as follows: First we evaluate D as

$$D = \frac{10^{4.0} - 1}{10^{0.1} - 1} = 38617.30$$

$$\frac{\ln(D + \sqrt{D^2 - 1})}{\ln(\omega_s + \sqrt{\omega_s^2 - 1})} = 11.2546/0.6224 = 18.08$$

Hence the order of the filter is 18 or above. The frequency response of the filter is shown in fig. 5.14.

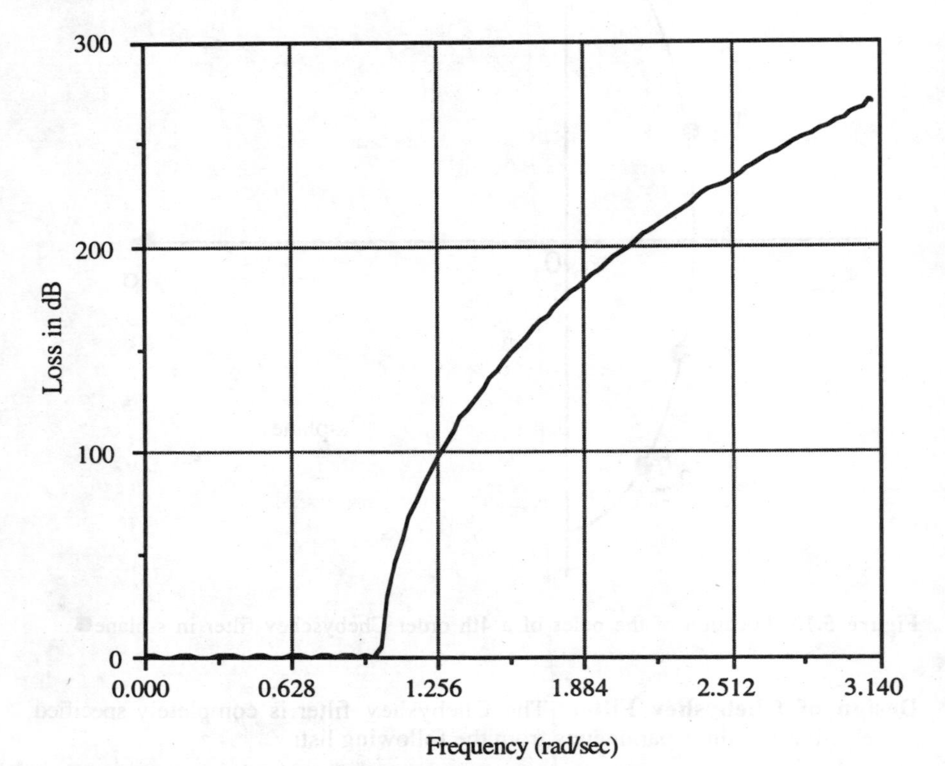

Figure 5.14: 18th order Chebyshev filter with A_P=1.0 dB, A_S=40 dB and ω_s=1.2 ✱.

All four parameters are related through the following equation:

$$n \geq \frac{\ln(D + \sqrt{D^2 - 1})}{\ln(\omega_s + \sqrt{\omega_s^2 - 1})} \tag{5.14}$$

where

$$D = \frac{10^{0.1A_s} - 1}{10^{0.1A_p} - 1}$$

There also exists a Chebyshev filter where the ripples are in the stopband and the passband is maximally flat. This is often known as Chebyshev approximation of type II or also as inverse Chebyshev approximation.

Elliptic Approximation: It may be noted that in Butterworth and Chebyshev approximations the transfer function has only poles (no zeros), hence they are often known as all-pole filters. We shall now consider a filter with both poles and zeros. One such filter is the Elliptic filter. The loss function of an Elliptic filter is given by

$$L(z) = 1 + \varepsilon^2 sn^2 (n \frac{K_1}{K} z, k_1) \qquad (5.15)$$

where z is related to the frequency by $\omega = \sqrt{k}\ sn(z,k)$. $sn(\ )$ is an elliptic sine function.

$k = \dfrac{\omega_p}{\omega_s}$ is selectivity factor and $k_1 = \sqrt{\dfrac{10^{0.1A_p} - 1}{10^{0.1A_s} - 1}}$ is discrimination factor.

Further K and K_1 are defined as

$$K = \int_0^{\frac{\pi}{2}} \frac{d\theta}{\sqrt{1 - k^2 \sin^2\theta}} \quad \text{and} \quad K_1 = \int_0^{\frac{\pi}{2}} \frac{d\theta}{\sqrt{1 - k_1^2 \sin^2\theta}}$$

In the elliptic approximation the ripples are permitted both in the passband and in the stopband. This results into a much sharper transition. The loss function is given now in terms of the elliptic function, which is conveniently expressed in a series form. In the derivation of what follows such a series expansion of an elliptic function was employed. Here we shall note only the final result leaving out all details, which may be found in [23]. A typical 4th order elliptic approximation is shown in fig. 5.15. Note the abbreviations used in above figure:

$\omega_p = \sqrt{k}$: Passband edge

$\omega_s = \dfrac{1}{\sqrt{k}}$: Stopband edge

$\omega_c = \sqrt{\omega_p \omega_s} = 1$: Cut-off frequency

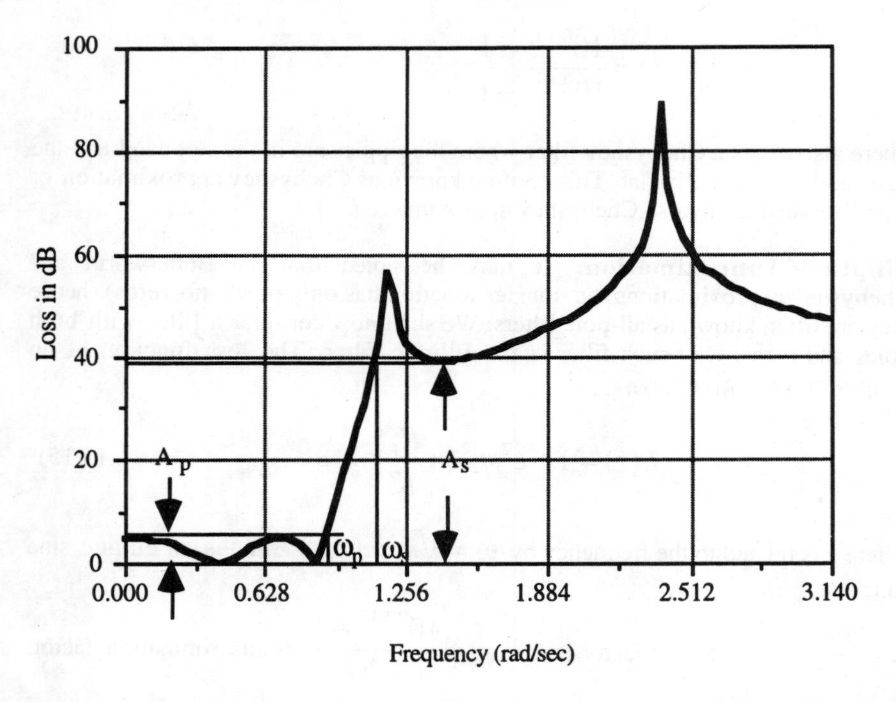

Figure 5.15: Loss characteristics of a 4th order elliptic filter. Note the definitions of various parameters used in the design of elliptic filters.

The normalized transfer function of an elliptic filter with selectivity factor k ($k = \omega_p/\omega_s$), passband ripple A_p and minimum stopband loss equal to or in excess of A_s is of the form given below:

$$H_n(s) = \frac{H_0}{s+\sigma_0}\prod_{i=1}^{r}\frac{s^2+A_{0i}}{s^2+B_{1i}s+B_{0i}} \qquad r=\frac{n-1}{2} \qquad n\text{: odd} \tag{5.16}$$

where

$$A_{0i} = \frac{1}{\Omega_i^2}$$

$$B_{0i} = \frac{(\sigma_0 v_i)^2 + (\Omega_i w)^2}{(1+\sigma_0^2\Omega_i^2)^2}$$

$$B_{1i} = \frac{2\sigma_0 v_i}{(1+\sigma_0^2\Omega_i^2)}$$

$$H_0 = \sigma_0\prod_{i=1}^{r}\frac{B_{0i}}{A_{0i}}$$

An elliptic filter is specified in terms of three parameters, namely, A_p (passband ripple), A_s (minimum stopband loss) and k selectivity factor. The fourth parameter is the order of the filter, which may be computed given the other three parameters. The procedure is as follows:

$$k' = \sqrt{1-k^2}$$

$$q_0 = 0.5\frac{1-\sqrt{k'}}{1+\sqrt{k'}}$$

$$q = q_0 + 2q_0^5 + 15q_0^9 + 150q_0^{13}$$

$$D = \frac{10^{0.1A_s}-1}{10^{0.1A_p}-1}$$

$$n \geq \frac{\log(16D)}{\log(1/q)}$$

Other parameters in (5.16) are given in terms of above quantities as

$$\sigma_0 = \left| \frac{2q^{1/4}\sum_{m=0}^{\infty}(-1)^m q^{m(m+1)} \sinh((2m+1)\Lambda)}{1+2\sum_{m=1}^{\infty}(-1)^m q^{m^2} \cosh(2m\Lambda)} \right|$$

where

$$\Lambda = \frac{1}{2n}\ln\frac{10^{0.05A_p}+1}{10^{0.05A_p}-1}$$

$$w = \sqrt{(1+k\sigma_0^2)(1+\frac{\sigma_0^2}{k})}$$

$$\Omega_i = \frac{2q^{1/4}\sum_{m=0}^{\infty}(-1)^m q^{m(m+1)} \sin(\frac{(2m+1)}{n}i\pi)}{1+2\sum_{m=1}^{\infty}(-1)^m q^{m^2} \cos(\frac{2m}{n}i\pi)}$$

$$v_i = \sqrt{(1-k\Omega_i^2)(1-\frac{\Omega_i^2}{k})}$$

For n= even, the transfer function is given by

$$H_n(s) = H_0 \prod_{i=1}^{r} \frac{s^2 + A_{0i}}{s^2 + B_{1i}s + B_{0i}} \qquad r = \frac{n}{2} \qquad n{:}even \tag{5.17}$$

The parameters in (5.17) remain unchanged except the following

$$\Omega_i = \frac{2q^{1/4}\sum_{m=0}^{\infty}(-1)^m q^{m(m+1)} \sin(\frac{(2m+1)}{n}(i-0.5)\pi)}{1+2\sum_{m=1}^{\infty}(-1)^m q^{m^2} \cos(\frac{2m}{n}(i-0.5)\pi)}$$

$$H_0 = 10^{-0.05A_p} \prod_{i=1}^{r} \frac{B_{0i}}{A_{0i}}$$

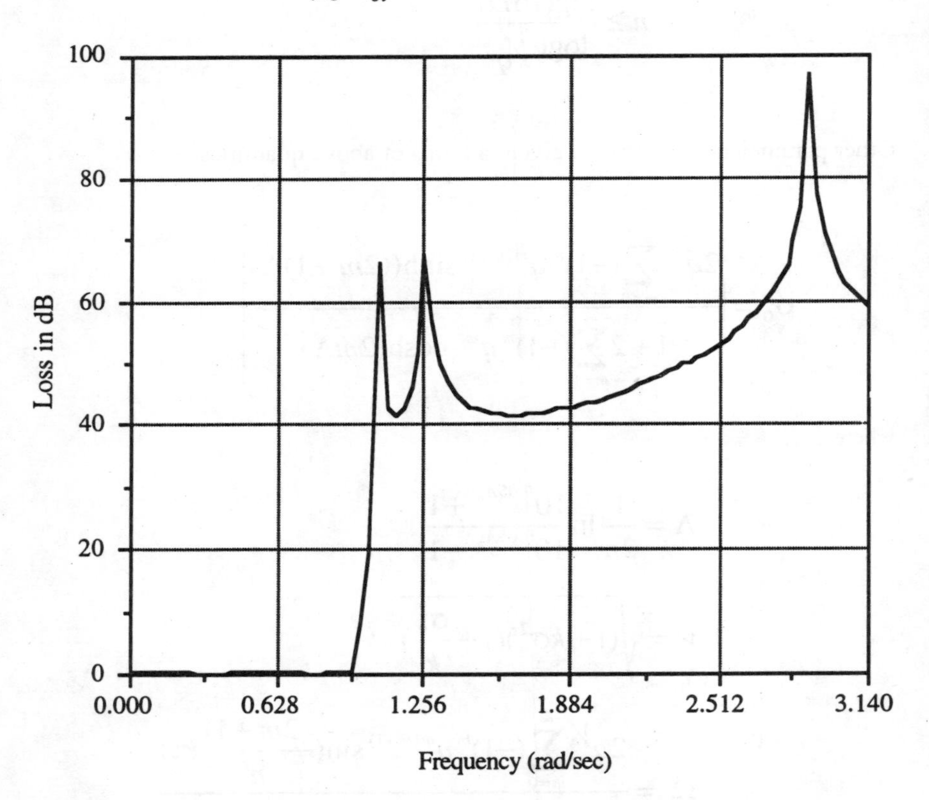

Figure 5.16: Loss characteristics of a 6th order elliptic filter. A_p=1 dB, A_s=40dB, k=0.9.

The transfer function has zeros at $\pm j1/\Omega_i$, $i=1,2,\ldots r$ and poles at $(-B_{1i} \pm \sqrt{B_{1i}^2 - 4B_{0i}})/2$, $i=1,2,\ldots r$.

Example 5.6: Consider an example, let $k \approx 1$ (but <1). Then, $k' \approx 0$, $q_0 \approx q = 0.5$. Further let $A_p = 1$ dB and A_s=40 dB. The order of the filter is $n \geq 20$. Let us consider yet another example, k=0.9, A_p=1.0 dB and A_s= 40 dB. The order of the required filter is 5.85. The loss characteristics of the 6th order elliptic filter are shown in fig. 5.16 There are six zeros at 0.0 ± 2.8931j, 0.0 ± 1.2631j, 0.0 ± 1.0693j and six poles at -0.2952 ± 0.3806j, -0.1154 ±+ 0.8240j -0.0240 ± 0.9480j ♣.

Of the four parameters characterizing an elliptic filter given any three the fourth can be found. Given any two we can study the mutual dependence of the remaining two parameters, for example, the dependence of A_s on k for n=4 and A_p=1 dB is shown in fig. 5.17.

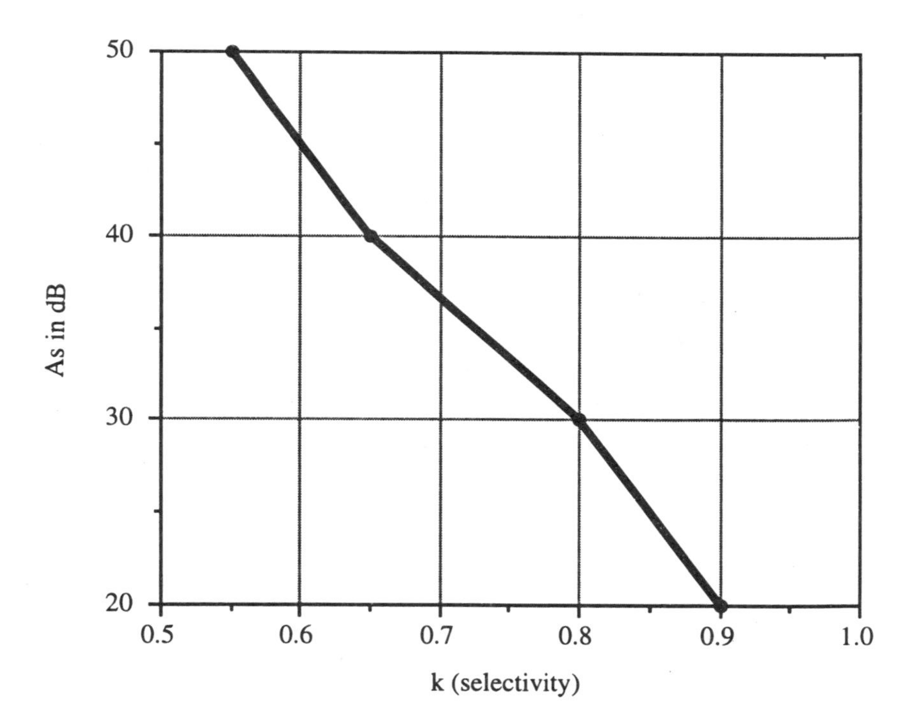

Figure 5.17: Minimum loss (in dB) in stopband as a function of k, the selectivity factor for fixed ripple magnitude (A_p=1 dB) and filter order (n=4).

§5.4 Sampling of Analog Filters

In this section we see how to convert an analog filter into a digital filter preserving as far as possible its transfer characteristics. We shall study three types of transformations, namely, impulse inverse, bilinear and matched-z transformations.

Impulse Invariance Transformation: Let h(t) be the impulse response function of an analog filter. The corresponding transfer function, $H(\omega)$ is related to its impulse response function through Fourier transform.

$$H(\omega) = \int_0^\infty h(t)\exp(-j\omega t)dt \tag{5.18}$$

Let us approximate (5.18) by its discrete version

$$H(\omega) \approx \sum_0^\infty h(n\Delta t)\exp(-jn\omega\Delta t) \tag{5.19}$$

where Δt is the sampling interval. We shall introduce the z transform in (5.19) (see Chapter Two for more details on z transform). Let $z = \exp(j\omega\Delta t)$,

$$\sum_0^\infty h(n\Delta t)\exp(-jn\omega\Delta t) = \sum_0^\infty h(n\Delta t)z^{-n} = H_d(z)$$

$H_d(z)$ is the transfer function of the equivalent digital filter, $h(n\Delta t)$, n=0,1,2, ... obtained by sampling a continuous filter. The sampling rate must be greater than twice the highest frequency present in the transfer function. Evidently, error free sampling is possible only if the transfer function is bandlimited. It may be noted that the digital approximation will be equal to the analog filter transfer function, that is, $H_d(z)\big|_{z=e^{j\omega\Delta t}} = H(\omega)$, only in the range $|\omega| \le \frac{\pi}{\Delta t}$. We shall now use above approach for mapping an analog filter transfer function into digital filter transfer function. Let an arbitrary analog transfer function be given by

$$H(s) = \sum_{i=1}^{n} \frac{A_i}{s - p_i} \tag{5.20}$$

We shall first find out its impulse response function by computing its inverse Laplace transform, which in our case is a well-known result, $h(t) = \sum_{i=1}^{p} A_i e^{p_i t}$ and its sampled version is

$$h(\Delta t\, n) = \sum_{i=1}^{p} A_i e^{p_i \Delta t\, n}, \quad n = 0,1,2,\ldots\infty$$

Its z-transform is easily computed as follows:

$$H_d(z) = \sum_{0}^{\infty} h(n\,\Delta t) z^{-n}$$
$$= \sum_{i=1}^{p} A_i e^{p_i \Delta t\, n} z^{-n} = \sum_{i=1}^{p} A_i \frac{z}{z - e^{p_i \Delta t}} \tag{5.21}$$

Observe that $H_d(z)$ has poles at $z_i = e^{p_i \Delta t}$ where $p_i = \sigma_i + j\omega_i, \sigma_i < 0$. Thus, all poles of the digital approximation will lie inside the unit circle in the z-plane.

Example 5.7: Design a digital filter as an approximation of an analog filter whose transfer function (Bessel filter function) is

$$H(s) = \frac{1}{1 + s + \frac{3}{7}s^2 + \frac{2}{21}s^3 + \frac{1}{105}s^4}$$
$$= \sum_{i=1}^{2} [\frac{A_i}{s - p_i} + \frac{A_i^\bullet}{s - p_i^\bullet}]$$

where

$$A_1, A_1^\bullet = 1.663392 \pm (-j)8.396299$$
$$A_2, A_2^\bullet = -1.663392 \pm j2.244076$$
$$p_1, p_1^\bullet = -2.896211 \pm j0.8672341$$
$$p_2, p_2^\bullet = -2.103789 \pm j2.657418$$

The transfer function is in a form assumed in (5.20). Hence, to obtain the digital approximation to the given analog filter we make use of (5.21) and obtain the following result ($\Delta t = \pi/4$ sec)

$$H_d(z) = \sum_{i=1}^{2} \frac{A_{1i} z + A_{2i} z^2}{B_{0i} + B_{1i} z + z^2}$$

where

$$A_{1i} = 0.64523, -0.83452$$
$$A_{2i} = 2.61851, -2.61851$$
$$B_{0i} = 0.01057, \ 0.03673$$
$$B_{1i} = -0.15977, 0.18919$$

The transfer function (magnitude) of the digital filter are compared in fig.5.17 with the actual transfer of the analog filter for two different sampling intervals, namely, $\pi/4$ and $\pi/8$ sec. Since the analog transfer function is not bandlimited there will always be some aliasing error, depending upon the sampling interval. This is clearly brought out in fig. 5.18.

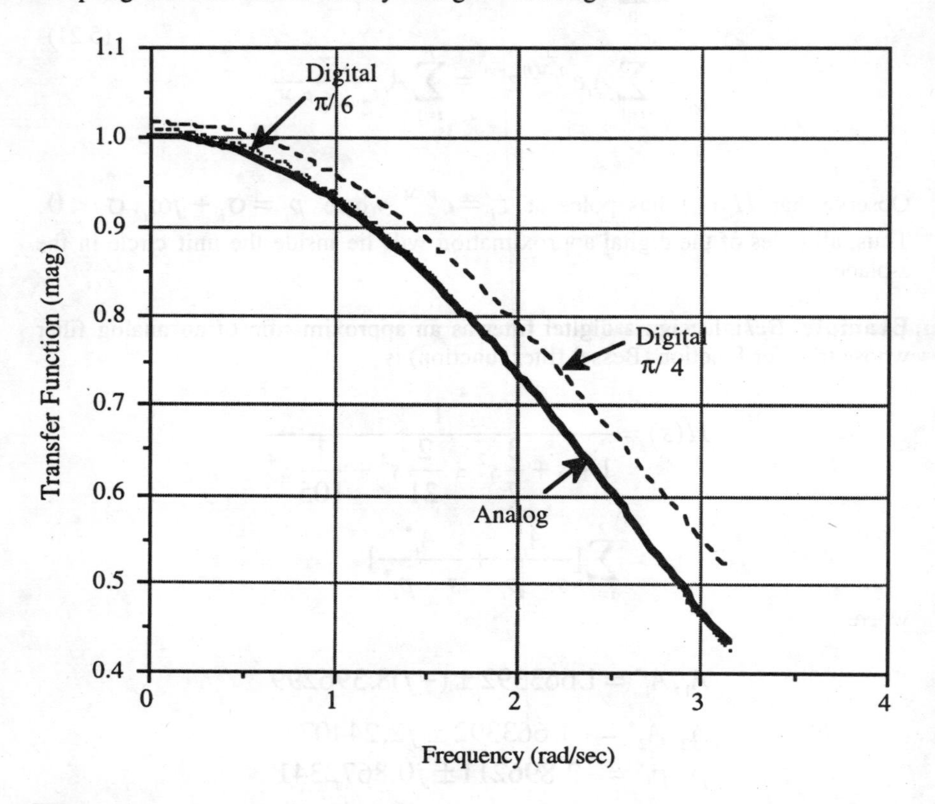

Figure 5.18: Transfer function of the Bessel filter mapped into digital domain by sampling the impulse response function at two different sampling intervals, $\pi/4$ and $\pi/6$ sec. The aliasing error is reduced drastically when sampled at higher rate ♠.

Bilinear Transformation: Consider an analog filter

$$H(s)=\frac{\sum_{i=0}^{p} a_i s^{p-i}}{s^p-\sum_{i=1}^{p} b_i s^{p-i}}=\frac{\sum_{i=0}^{p} a_i s^{-i}}{1-\sum_{i=1}^{p} b_i s^{-i}} \tag{5.22}$$

which may be realized using a circuit configuration as shown fig. 5.19.

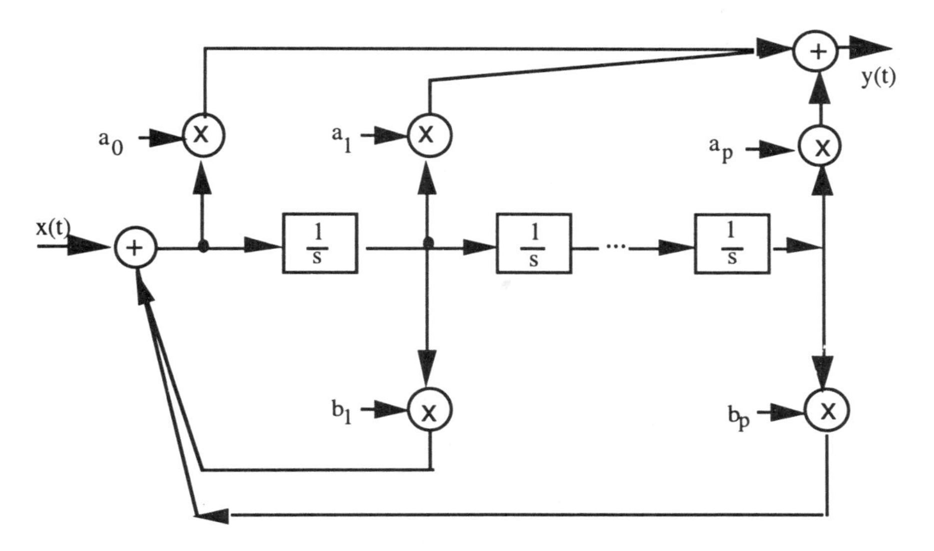

Figure 5.19: Realization of an analog filter shown in eq. (5.22)

If each analog element in fig. 5.19 is replaced by the corresponding digital element we obtain the digital approximation of the analog filter. The digital adders and multipliers are straightforward but the digital integrator with transfer function $1/s$ needs some clarification. Let us express the integrator as

$$y = \int_{o}^{t} x(t')h(t-t')dt'$$

where h(t) is an impulse response function of the integrator,

$$\begin{aligned} h(t) &= 1 \quad t \geq o^{+} \\ &= 0 \quad t \leq o^{-} \end{aligned}$$

From this it is easy to obtain

$$\begin{aligned} y(t_2) - y(t_1) &= \int_{o}^{t_2} x(t')h(t-t')dt' - \int_{o}^{t_1} x(t')h(t-t')dt' \\ &\approx \frac{(t_2 - t_1)}{2}(x(t_2) + x(t_1)) \end{aligned}$$

Left hand side is area under $x(t)$ between t_1 and t_2 as shown in fig. 5.20.

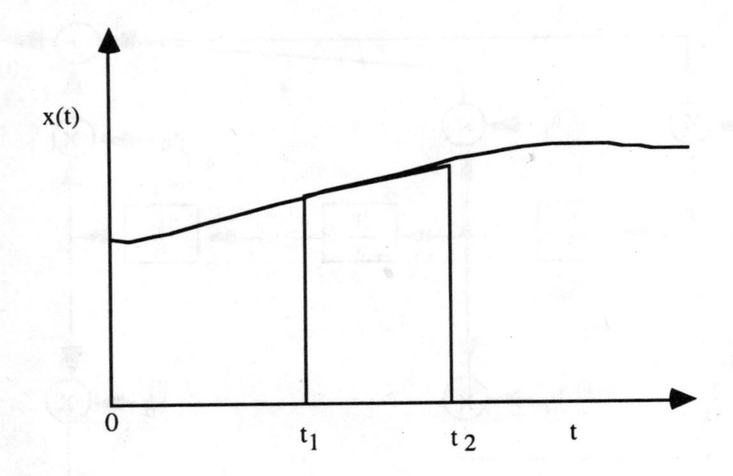

Figure 5.20: Area under the curve is equal to $y(t_2) - y(t_1)$.

Let $t_2 = n\Delta t$ and $t_1 = (n-1)\Delta t$

$$y(n\Delta t) - y((n-1)\Delta t) \approx \frac{\Delta t}{2}[x(n\Delta t) + x((n-1)\Delta t)] \qquad (5.23)$$

Taking z-transform on both sides of (41) we obtain

$$Y(z) - z^{-1}Y(z) \approx \frac{\Delta t}{2}[X(z) + z^{-1}X(z)]$$
$$\frac{Y(z)}{X(z)} = \frac{\Delta t}{2}\frac{z+1}{z-1} \qquad (5.24)$$

Equation (5.24) is the transfer function of a digital integrator. The analog integrator in fig. 5.20 may be replaced by the digital integrator. The transformation involved is replacement of s by $\frac{2}{\Delta t}\frac{z-1}{z+1}$, that is,

$$\boxed{s = \frac{2}{\Delta t}\frac{z-1}{z+1}} \qquad (5.25a)$$

which is known as a bilinear transformation. Let us examine some of the characteristics of the bilinear transformation. Equation (5.25a) may be reversed as shown below:

$$\boxed{z = \frac{\frac{2}{\Delta t} + s}{\frac{2}{\Delta t} - s}} \qquad (5.25b)$$

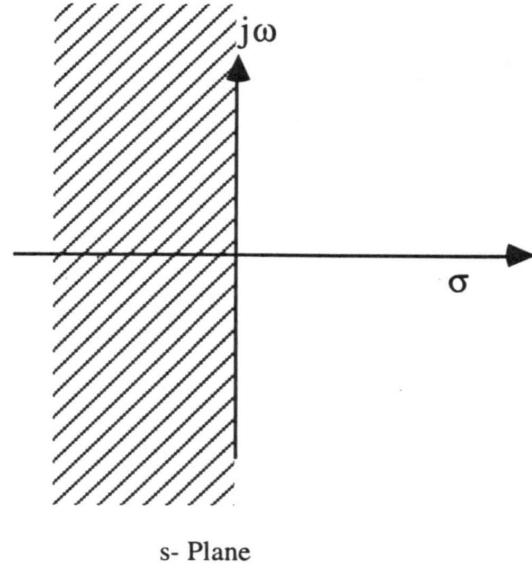

s- Plane

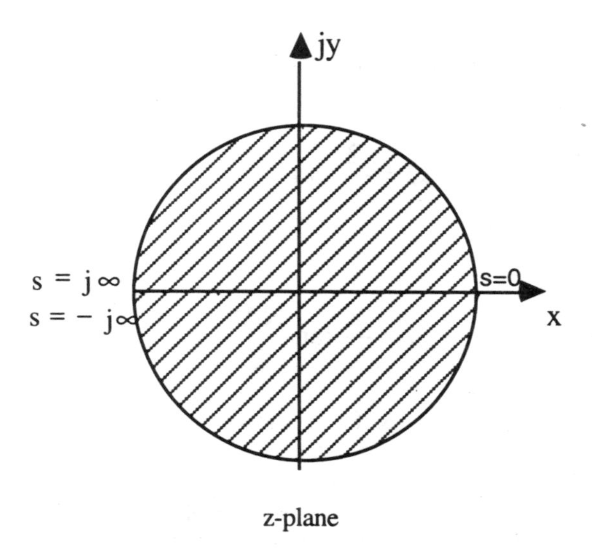

z-plane

Figure 5.21: Mapping of s-plane into z-plane using (5.25).

Let $z = re^{j\theta}$ and $s = \sigma + j\omega$, then

$$r = \sqrt{\frac{(\frac{2}{\Delta t} + \sigma)^2 + \omega^2}{(\frac{2}{\Delta t} - \sigma)^2 + \omega^2}} \tag{5.26a}$$

and

$$\theta = \tan^{-1}\frac{\omega}{(\frac{2}{\Delta t}+\sigma)} + \tan^{-1}\frac{\omega}{(\frac{2}{\Delta t}-\sigma)} \tag{5.26b}$$

The left half of the s-plane is mapped into a unit circle. The imaginary frequency axis, that is, $j\omega$ is mapped onto the circumference of the unit circle (see fig. 5.21). Since the range of ω is infinity (i.e. $\pm\infty$) the frequency axis is compressed as it goes over the circumference of the unit circle. The mapping rule is

$$\theta = 2\tan^{-1}\frac{\omega\Delta t}{2}$$

or

$$\omega = \frac{2}{\Delta t}\tan\frac{\theta}{2} \tag{5.27}$$

We shall rewrite (5.27) by replacing θ by $\Omega\Delta t$ where Ω is frequency variable in z-plane. We obtain a mapping rule between two frequency axes.

$$\Omega = \frac{2}{\Delta t}\tan^{-1}\frac{\omega\Delta t}{2}$$

or

$$\omega = \frac{2}{\Delta t}\tan\frac{\Omega\Delta t}{2} \tag{5.28}$$

The mapping rule given by (5.28) is shown in fig. 5.22. From the plot we notice that a passband in analog domain get compressed (also known as warped) when mapped into the digital domain. Therefore, the bilinear transformation would distort the transfer function when mapped to the digital domain. Nevertheless, a bandpass response remains a bandpass response though with a reduced width of the passband. The bilinear transformation is thus not appropriate for highpass filter.

Matched-z Transform: Another simple approach is to map all poles and zeros from the s-plane to the z-plane. Let $p_k, k=1,2,\ldots p$ be the poles (re{ p_k }<0) and $q_k, k=1,2,\ldots q$ be the zeros. The transfer function in the s-plane is given by

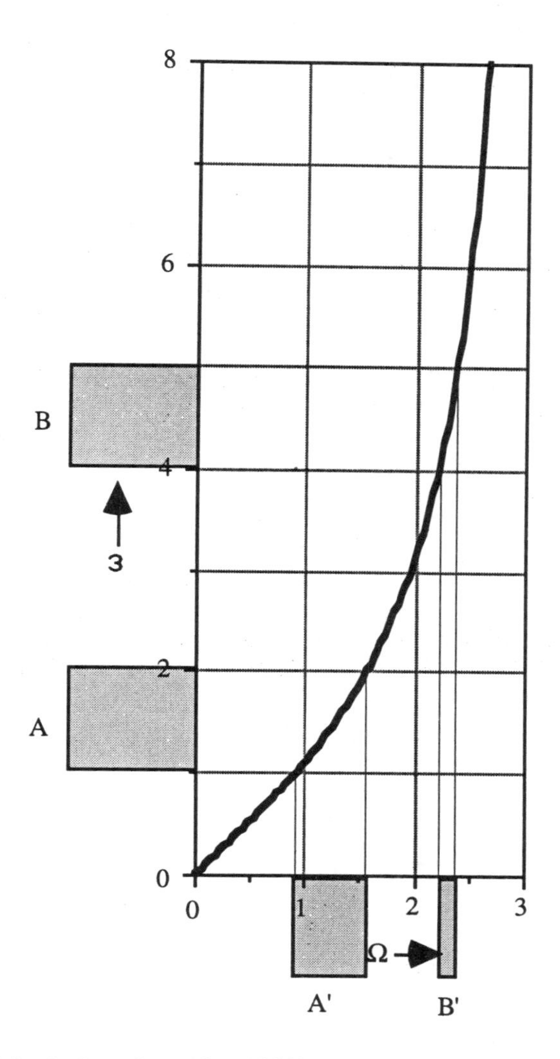

Figure 5.22: A plot of equation (5.28). A passband A in analog domain (ω axis) gets compressed into band A' when it is mapped to digital domain (Ω axis). The compression increases at higher frequencies or higher sampling interval. Note the compression of passband B is much greater than that of A.

$$H(s) = \frac{\prod_{k=1}^{q}(s - q_k)}{\prod_{k=1}^{p}(s - p_k)} \tag{5.29a}$$

The transfer function of the corresponding digital filter is

$$H(z) = \frac{\prod_{k=1}^{q}(1 - e^{q_k \Delta t} z^{-1})}{\prod_{k=1}^{p}(1 - e^{p_k \Delta t} z^{-1})} \tag{5.29b}$$

where Δt is sampling interval. This is known as matched-z transformation. The impulse invariance method and matched-z transformation result in digital filters having identical poles but differing zeros. This will be clear in the next example.

Example 5.8: We shall revisit the previous example of Elliptic approximation (Example 5.6) whose transfer function is given by

$$H_n(s) = H_0 \prod_{i=1}^{3} \frac{s^2 + A_{0i}}{s^2 + B_{1i}s + B_{0i}}$$

where

$$\begin{aligned} A_{01}, A_{02}, A_{03} &= (8.3702,\ 1.5954,\ 1.1434) \\ B_{01}, B_{02}, B_{03} &= (0.2320,\ 0.6923,\ 0.8993) \\ B_{11}, B_{12}, B_{13} &= (0.5903,\ 0.2307,\ 0.0479) \\ H_0 &= 0.0084 \end{aligned}$$

The partial fraction expansion of the transfer function is

$$H_n(s) = c_0 + \sum_{i=1}^{3} [\frac{c_{1i}}{s - p_{1i}} + \frac{c_{2i}}{s - p_{2i}}]$$

where

$$c_0 = H_0$$

$$c_{1i} = H_0 \left\{ (s - p_{1i}) \prod_{k=1}^{3} \frac{(s^2 + A_{0i})}{s^2 + B_{1k}s + B_{0k}} \right\} \Bigg|_{s=p_{1i}}$$

$$c_{2i} = H_0 \left\{ (s - p_{2i}) \prod_{k=1}^{3} \frac{(s^2 + A_{0i})}{s^2 + B_{1k}s + B_{0k}} \right\} \Bigg|_{s=p_{2i}}$$

and $p_{1i}, p_{2i}, i = 1,2,3$ are roots of the denominator polynomial, [-0.2952 ± 0.3806j, -0.1154 ± 0.8240j -0.0240 ± 0.9480j]. The zeros are [0.0 ± 2.8931j, 0.0 ± 1.2631j, 0,0 ± 1.0693j].

We shall now transform the analog filter into digital filter using all three different transformations described earlier:

1. Impulse Invariance:

$$H_D(z) = c_0 \delta(z) + \sum_{i=1}^{3} \left[c_{1i} \frac{z}{z - e^{p_{1i}\Delta t}} + c_{2i} \frac{z}{z - e^{p_{2i}\Delta t}} \right]$$

The poles of above transformation are the same as those in the s-plane but stand mapped into the z-plane. However, the zeros are placed at entirely new locations.
2. Bilinear Transformation:

$$H_n(z) = H_0 \prod_{i=1}^{3} \frac{((\frac{2}{\Delta t}\frac{z-1}{z+1})^2 + A_{0i})(z+1)^2}{[(\frac{2}{\Delta t} - p_{1i})z - p_{1i} - \frac{2}{\Delta t})][(\frac{2}{\Delta t} - p_{2i})z - p_{2i} - \frac{2}{\Delta t})]}$$

It may be noted that the poles which are now at $\frac{p_{1i} + \frac{2}{\Delta t}}{(\frac{2}{\Delta t} - p_{1i})}, \frac{p_{2i} + \frac{2}{\Delta t}}{(\frac{2}{\Delta t} - p_{2i})}$,

i=1,2,3, correspond to the poles in the s-plane but mapped into z plane. This is

also true of the zeros which are now at $\frac{\frac{\Delta t}{2} \pm j\sqrt{A_{0i}}}{\frac{\Delta t}{2} \pm (-j)\sqrt{A_{0i}}}$, $i = 1,2,3$ and -1.

3. Matched-z Transformation:

$$H_D(z) = \prod_{i=1}^{3} \frac{(z - e^{q_i \Delta t})(z - e^{q_i^* \Delta t})}{(z - e^{p_i \Delta t})(z - e^{p_i^* \Delta t})}$$

The poles and zeros of above transformation are the same as those in the s-plane but stand mapped into the z-plane ♠.

§5.5 Design by Placement of Poles and Zeros

The poles and zeros play an important role in characterizing a digital filter. This we have demonstrated for some well-known analog filters transformed into digital domain. In fact, a more direct approach would be to design a digital filter directly in the z-plane by optimally placing the poles and zeros so as to achieve the desired transfer function. This problem is much more involved, hence outside the scope of this introductory book. However, there are some simple arrangements of poles and zeros, often done intuitively but with good results. Some examples of filters designed by placement of poles and zeros are Notch filter, Comb filter, All-pass filter, etc.

Notch Filter: An IIR notch filter has a zero on the unit circle and a pole close to the unit circle in order to make the notch sharper as shown in fig. 5.23. First, we consider two zeros on the unit circle as in fig. 5.23a. We get an FIR filter whose transfer function is

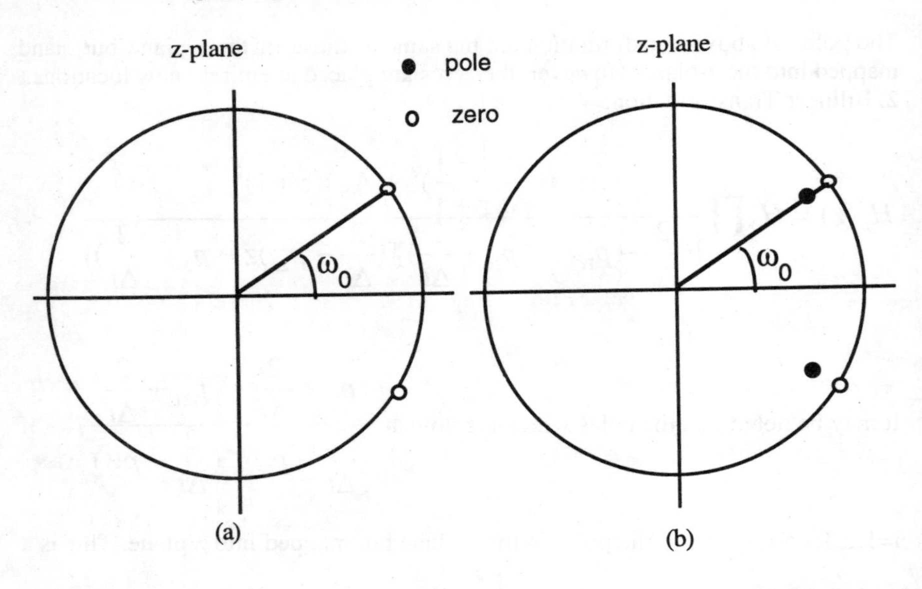

Figure 5.23: Notch filter. (a) placement of zeros and (b) placement of poles and zeros

given by

$$H(z) = (1 - e^{j\omega} z^{-1})(1 - e^{-j\omega} z^{-1})$$
$$= (1 - 2\cos\omega_0 z^{-1} + z^{-2})$$

The frequency response of such a filter has a null at $\omega = \omega_0$ but the null is very wide. The frequency response is shown in fig. 5.24. This may be easily rectified, though in an ad hoc manner, by placing a pole next to the zero but inside the unit circle as shown in the fig. 5.23(b). The transfer function is now given by

$$\boxed{\begin{aligned} H(z) &= \frac{(1 - e^{j\omega} z^{-1})(1 - e^{-j\omega} z^{-1})}{(1 - re^{j\omega} z^{-1})(1 - re^{-j\omega} z^{-1})} \\ &= \frac{(1 - 2\cos\omega_0 z^{-1} + z^{-2})}{(1 - 2r\cos\omega_0 z^{-1} + r^2 z^{-2})} \end{aligned}}$$

where $r < 1$ is the radial position of the pole. Depending upon r the null in the frequency response can be made sharp (see fig. 5.24). With the introduction of a pole the notch filter now becomes an IIR filter which may be implemented recursively.

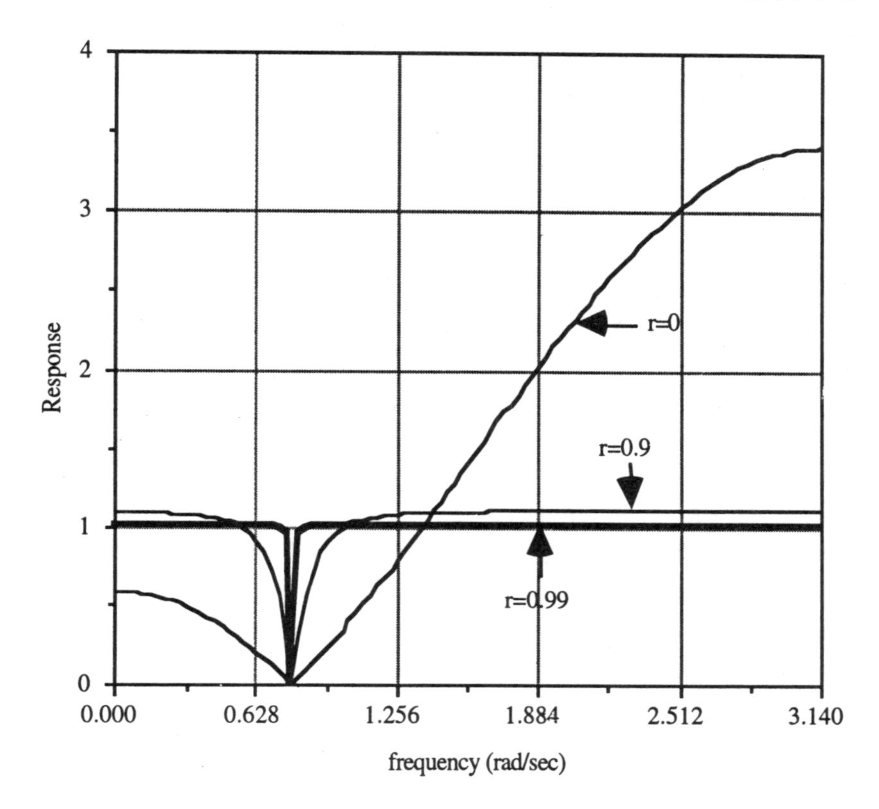

Figure 5.24: The frequency response of a notch filter for different values of r, the pole position. The curve labeled as r=0 corresponds to FIR filter.

Comb Filter: It is straightforward to build a filter, which has more than one notch, say at frequencies $\omega_i, i = 1,2,\ldots,p$. The filter transfer function is given by

$$H(z^{-1}) = \prod_{i=1}^{p} \frac{(1 - 2\cos\omega_i z^{-1} + z^{-2})}{(1 - 2r\cos\omega_i z^{-1} + r^2 z^{-2})} \qquad (5.30a)$$

Often the different frequencies present in a signal are harmonically related, that is, $\omega_i = i\omega_0$ where ω_0 is the fundamental frequency. It is possible to rewrite the comb filter as a function of a single parameter. Let $b_i = -2\cos\omega_i$.

$$H(z^{-1}) = \prod_{i=1}^{p} \frac{(1 + b_i z^{-1} + z^{-2})}{(1 + rb_i z^{-1} + r^2 z^{-2})} \qquad (5.30b)$$

It is easy to show that all coefficients may be written in terms b_1 where $b_1 = -2\cos\omega_0$ as below:

$$b_2 = 2 - b_1^2$$
$$b_3 = b_1^3 - 3b_1 \qquad (5.30c)$$
$$b_4 = -b_1^4 + 4b_1^2 - 2$$

A special case arises when $\omega_i, i = 0,1,2,\ldots, p-1$ are equally spaced over the frequency range $\pm\pi$. We shall place p equally spaced zeros on the unit circle and p equally spaced poles on a circle of radius r (r<1). The transfer function is now given by

$$\boxed{H(z) = \frac{(1 - z^{-p})}{(1 - rz^{-p})}}$$

The transfer function will have p nulls and poles that are equally spaced in the frequency range $\pm\pi$. Such a filter is known as a comb filter. A comb filter is useful in extracting/eliminating a sinusoidal signal/interference (e.g. power line interference) which repeats at many harmonics.

Example 5.9: Here is an example of comb filter with 8 notches (p=8) located at $\omega_i = \frac{2\pi}{8} i, i = 0,1,2,\ldots,7$. Let r-=0.99. The filter transfer function (magnitude) is shown in fig. 5.25.

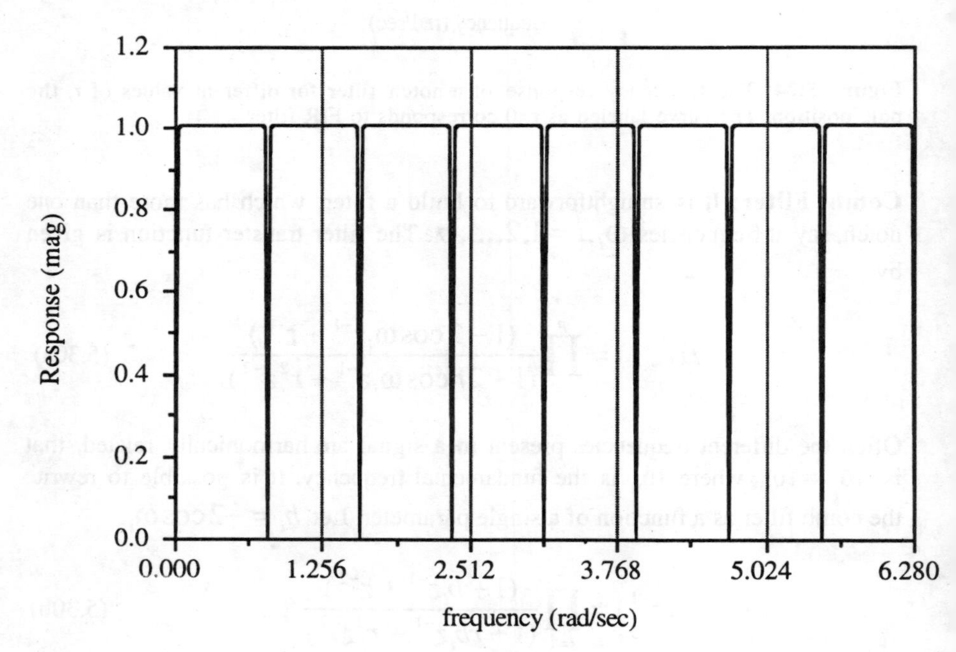

Figure 5.25: A comb filter with 8 notches in the frequency range 0-2π. The notch at frequency 0 and 2π is one and the same.

All-Pass Filter: An all-pass filter is one whose frequency response (magnitude) is one over the entire frequency band but its phase is variable. The simplest all-pass filter is a delay filter $H(z) = z^{-\tau}$ where τ is delay. A more general all-pass filter has following structure:

$$H(z) = \frac{a_p + a_{p-1}z^{-1} + \ldots + a_1 z^{-p+1} + z^{-p}}{1 + a_1 z^{-1} + \ldots + a_{p-1} z^{-p+1} + a_p z^{-p}} = \frac{\sum_{k=0}^{p} a_k z^{-p+k}}{\sum_{k=0}^{p} a_k z^{-k}}, \quad a_0 = 1.0$$

$$= z^{-p} \frac{A(z^{-1})}{A(z)} \tag{5.31}$$

Note that, since $|H(\omega)|^2 = H(z)H(z^{-1})\big|_{z=e^{j\omega}} = 1.$, $H(z)$ is an all-pass filter. The structure of an all-pass filter is such that if z_0 is a pole of $H(z)$ then $1/z_0$ is a zero of $H(z)$. Thus for all poles present in $H(z)$ there must be zeros at reciprocal position, that is, outside the unit circle. This is shown in fig. 5.26.

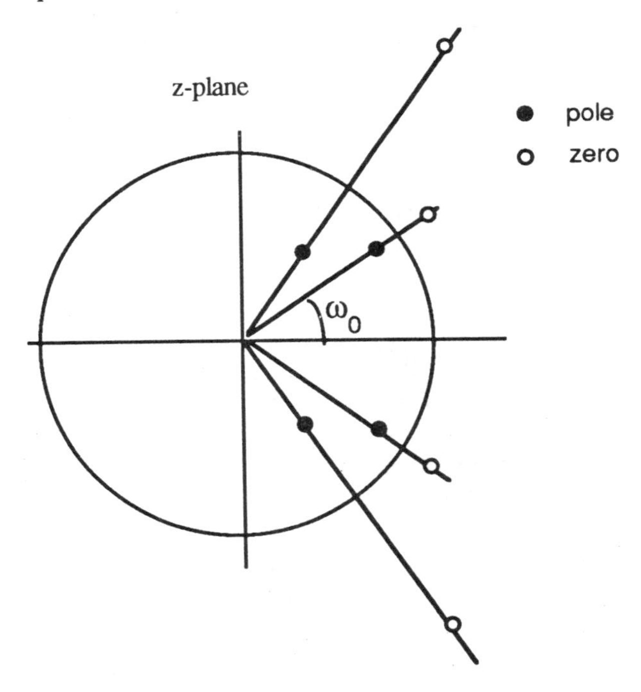

Figure 5.26: Distribution of poles and zeros of an all-pass filter. Note that the zeros must necessarily lie outside the unit circle. There are no zeros inside the unit circle (why?).

Example 5.10: Consider an all-pass filter having one pole inside the unit circle and one zero at reciprocal position outside the unit circle (see fig. 5.27a)

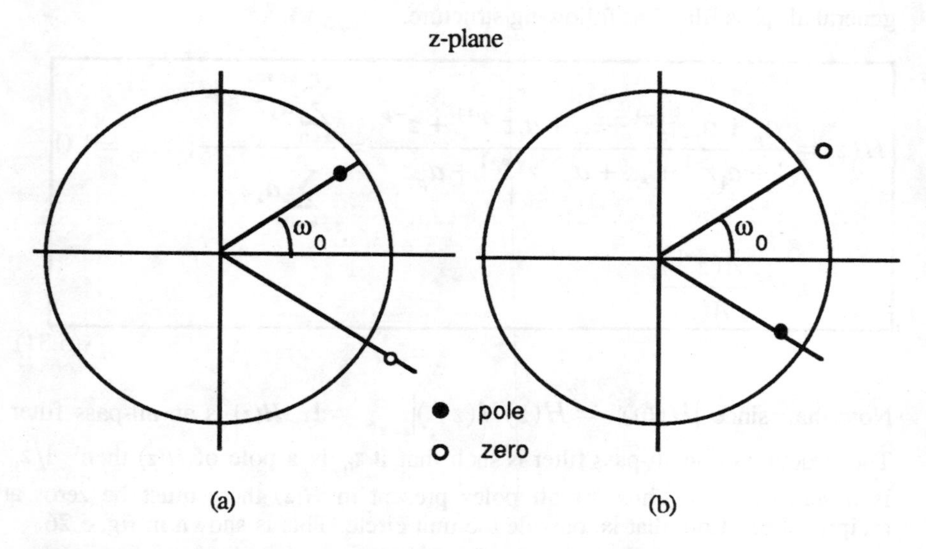

Figure 5.27: (a) Single pole and zero at reciprocal position outside the unit circle (b) Conjugate pole and zero.

The transfer function is given by

$$H(z) = \frac{(z^{-1} - re^{j\omega_0})}{(1 - re^{j\omega_0} z^{-1})}$$

The phase function is given by

$$\phi(\omega) = -\omega + \tan^{-1}\frac{r\sin(\omega_0 - \omega)}{1 - r\cos(\omega_0 - \omega)} - \tan^{-1}\frac{r\sin(\omega + \omega_0)}{1 - r\cos(\omega + \omega_0)}$$

With pole and zero at complex conjugate position (see fig. 5.27b) the phase of the all-pass filter remains unchanged, $\phi_{conj}(\omega) = \phi(\omega)$.

The filter response function with two poles and two zeros satisfying reciprocal and complex conjugate symmetry is given by

$$H(z) = \frac{(z^{-1} - re^{j\omega_0})}{(1 - re^{j\omega_0} z^{-1})} \frac{(z^{-1} - re^{-j\omega_0})}{(1 - re^{-j\omega_0} z^{-1})} \tag{5.32}$$

The response, magnitude and phase, of the all-pass filter in (5.32) for r=0.9 is shown in fig. 5.28.

All-pass filter is useful in compensating the phase distortion introduced by, say, a low- pass digital filter (IIR). It may be necessary to use more than one all-pass filters in tandem for complete compensation over a wide frequency band.

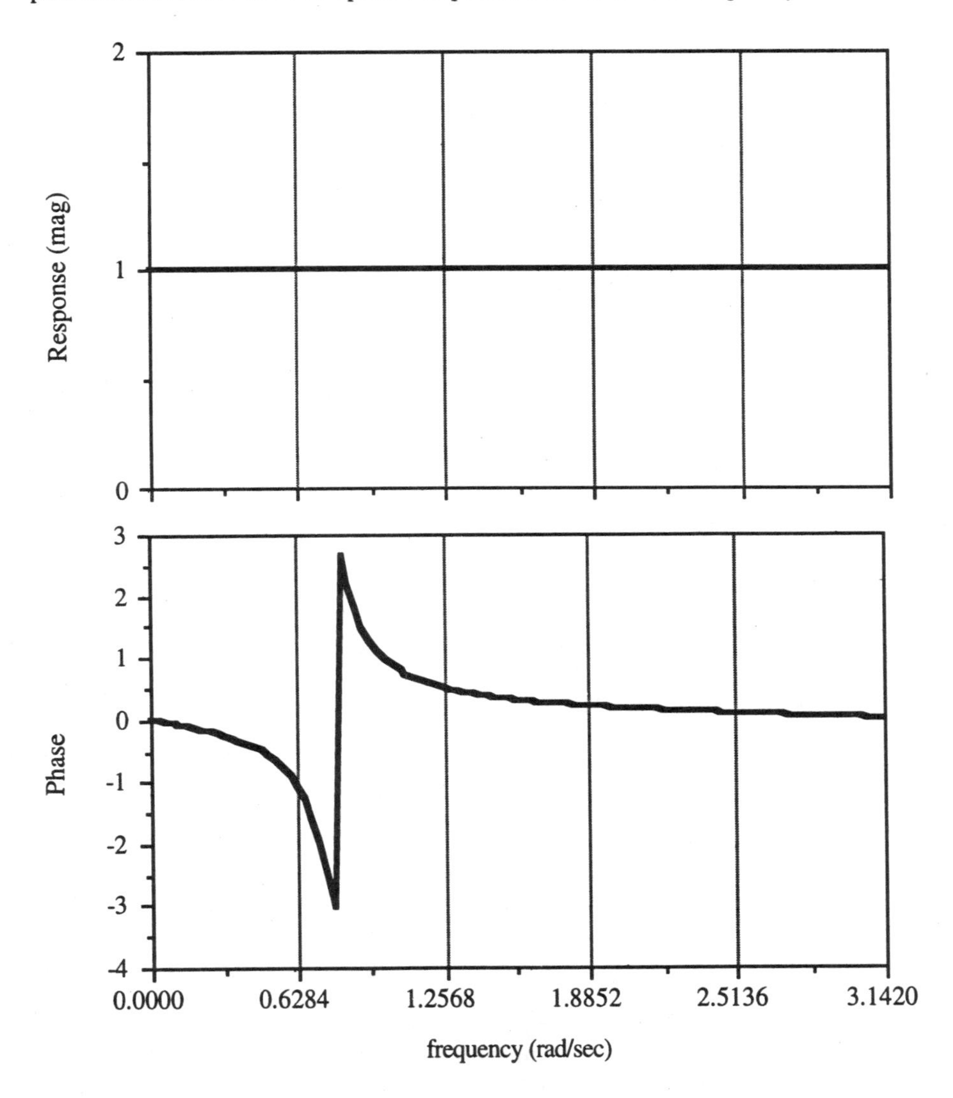

Figure 5.28: All-pass filter with poles at $0.9e^{\pm j\frac{\pi}{4}}$ and zeros at $1.1111e^{\pm j\frac{\pi}{4}}$.

Digital Sinusoidal Generator: In digital frequency synthesizer a sinusoidal function may be generated as an output of a digital filter which has a pole and its complex conjugate on the unit circle (see fig. 5.29).

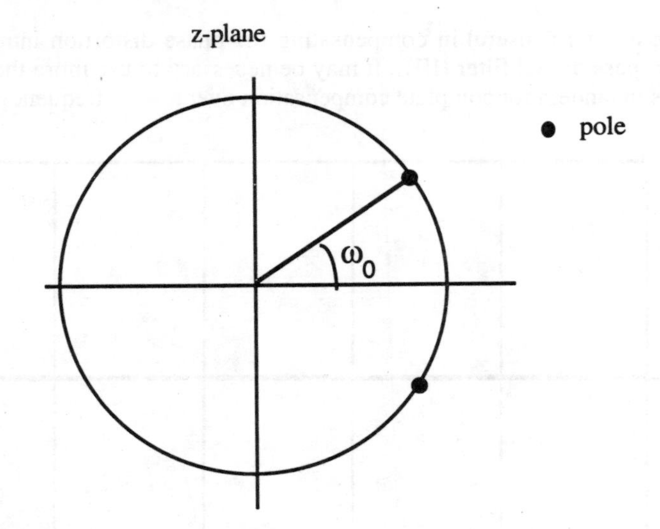

Figure 5.29: Poles of a digital filter to generate digital sinusoidal function.

The transfer function is given by

$$H(z) = \frac{b}{1 - 2r\cos(\omega_0)z^{-1} + r^2 z^{-2}}$$

The filter output to an impulse input may be expressed as

$$\boxed{y(n) = 2r\cos(\omega_0)y(n-1) - r^2 y(n-2) + b\delta(n)} \tag{5.33}$$

To solve above difference equation we must set the initial conditions, namely, $y(-1) = z(-2) = 0$ and the scale factor b =A $\sin(\omega_0)$ where A is the amplitude of the sinusoid (see Exercise #8).

Minimum Phase Filter: A linear time invariant system is said to be minimum phase when all its zeros are inside the unit circle (poles, if any, will have to be inside the unit circle anyway for the sake of stability). If some of the zeros are outside the unit circle (a mixed phase filter) it is possible to convert it into a minimum phase filter by simply relocating the zeros into the unit circle at their reciprocal location without altering the filter magnitude response. Let a mixed phase filter be expressed as

$$H_{mix}(z) = \frac{B(z)}{A(z)} = \frac{B_{in}(z)B_{out}(z)}{A(z)} \tag{5.34}$$

where $B_{in}(z)$ has all its zeros inside the unit circle and $B_{out}(z)$ has all its zeros outside the unit circle. Next we rewrite (5.34) as follows:

$$H_{mix}(z) = \frac{B_{in}(z)B_{out}(z^{-1})}{A(z)} \frac{B_{out}(z)}{B_{out}(z^{-1})} = H_{\min}(z)\frac{B_{out}(z)}{B_{out}(z^{-1})} \tag{5.35a}$$

$$\boxed{H_{\min}(z) = H_{mix}(z)\frac{B_{out}(z^{-1})}{B_{out}(z)}} \tag{5.35b}$$

Note that the zeros of $B_{out}(z^{-1})$ are inside the unit circle at reciprocal position as those of $B_{out}(z)$. $B_{out}(z)/B_{out}(z^{-1})$ is an all-pass filter as its poles and zeros are at reciprocal position as shown in fig 5.26. The magnitude square of the response of $H_{\min}(z)$ may be obtained from (5.35),

$$\left|H_{mix}(z)\right|^2\Big|_{z=e^{j\omega}} = \left|H_{\min}(z)\right|^2\left|\frac{B_{out}(z)}{B_{out}(z^{-1})}\right|^2\Bigg|_{z=e^{j\omega}} = \left|H_{\min}(z)\right|^2\Big|_{z=e^{j\omega}} \tag{5.36}$$

The result is valid for all-zero FIR filter or pole-zero IIR filter.

Now consider the phase response of $H_{mix}(z)$. It is easy to show that

$$\theta_{mix}(\omega) = \theta_{\min}(\omega) + \theta_{all}(\omega) \tag{5.37a}$$

and the corresponding group delays

$$\tau_{mix}(\omega) = \tau_{\min}(\omega) + \tau_{all}(\omega) \tag{5.37b}$$

Since the group delays are positive, the minimum of $\tau_{mix}(\omega)$ is $\tau_{\min}(\omega)$.

§5.6 IIR Filter Satisfying Desired Impulse Response

Let $h(\tau), \tau = 0,1,2,\ldots$ be the impulse response of an IIR filter and its z-transform is given by

$$H(z) = \sum_{\tau=0}^{\infty} h(\tau)z^{-\tau}$$

We can relate the impulse response to the coefficient of IIR filter. Equating the z transform of the filter impulse response function and its parametric representation, (5.1),

$$\sum_{\tau=0}^{\infty} h(\tau) z^{-\tau} = \frac{\sum_{k=0}^{q} b_k z^{-k}}{1 - \sum_{k=1}^{p} a_k z^{-k}} \tag{5.38}$$

By cross multiplying and taking the inverse z-transform on both sides of (5.38) we obtain

$$\begin{aligned} h(k) - \sum_{i=1}^{p} h(k-i) a_i &= b_k \qquad 0 \le k \le q \\ &= 0 \qquad k > q \end{aligned} \tag{5.39}$$

where a_0=1.0. We can also express (5.39) in a matrix form for $0 \le k \le p$:

$$\begin{bmatrix} h(0) & & & \\ h(1) & h(0) & & \mathbf{0} \\ h(2) & h(1) & h(0) & \\ \vdots & & \ddots & \\ h(p) & h(p-1) & \cdots & h(0) \end{bmatrix} \begin{bmatrix} 1 \\ -a_1 \\ -a_2 \\ \vdots \\ -a_p \end{bmatrix} = \begin{bmatrix} b_0 \\ b_1 \\ \vdots \\ b_q \\ 0 \\ 0 \end{bmatrix} \tag{5.40a}$$

and for $q \le k \le p+q-1$:

$$\begin{bmatrix} h(q) & \cdots & h(q-p+1) \\ h(q+1) & \cdots & h(q-p+2) \\ \vdots & & \\ h(p+q-1) & \cdots & h(q) \end{bmatrix} \begin{bmatrix} a_1 \\ a_2 \\ \vdots \\ a_p \end{bmatrix} = \begin{bmatrix} h(q+1) \\ h(q+2) \\ \vdots \\ h(p+q) \end{bmatrix} \tag{5.40b}$$

where we have assumed p>q. The denominator coefficient of the IIR filter transfer function, $a_1, a_2, \ldots, a_p$ may be obtained from (5.40b). Using these coefficients we can next solve (5.40a) for the numerator coefficients, $b_0, b_1, \ldots, b_q$. Note that the impulse response function is known. Thus, given p+q impulse response coefficients we have shown how to obtain the denominator and the numerator coefficients of an IIR filter. Note that we are not employing the entire available impulse response coefficients unless a very high order filter model is required to be estimated. When p and q are known apriori, the excess impulse response coefficients may be used to compute a least squares estimate of $a_1, a_2, \ldots, a_p$.

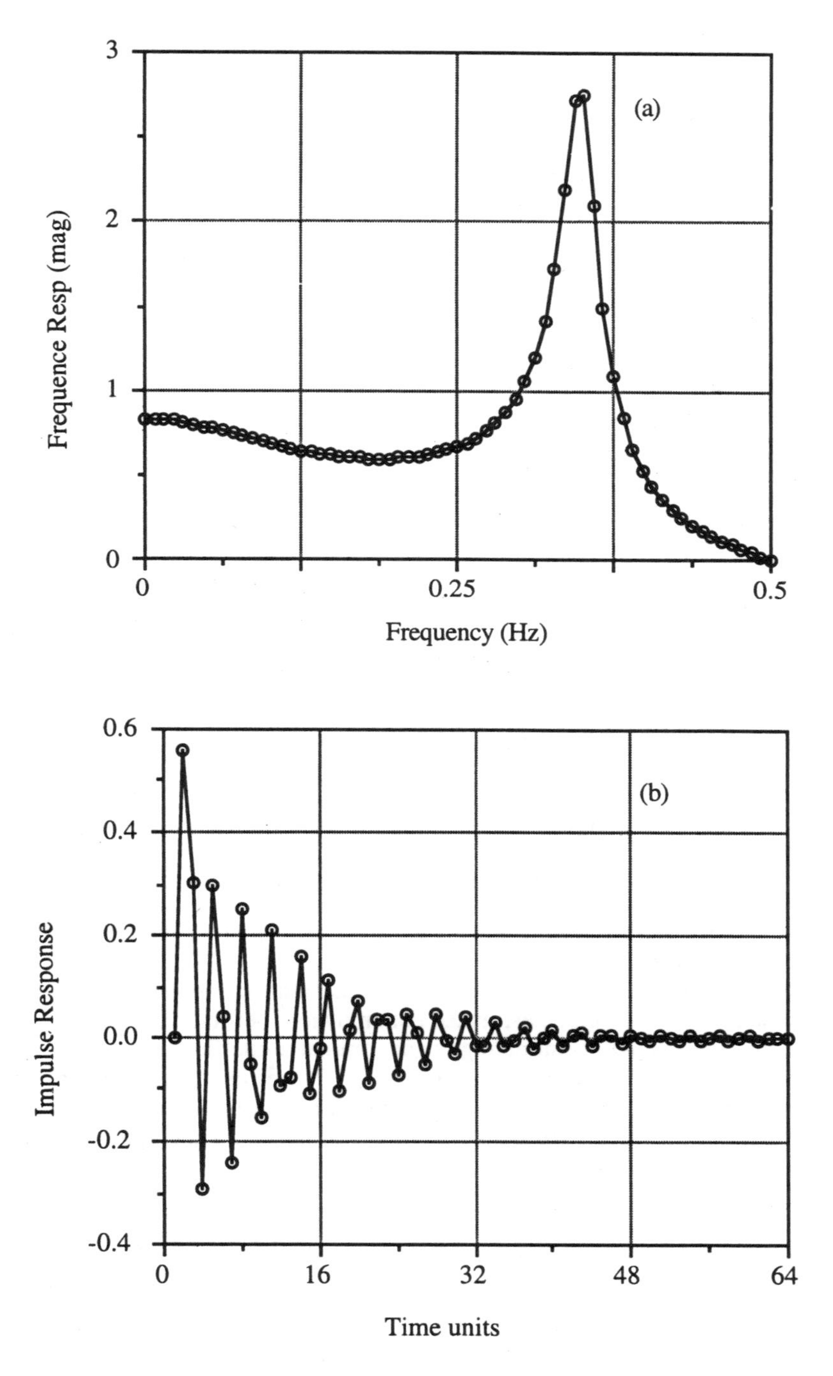

Figure 5.30: The frequency response (magnitude) and the impulse response function of IIR filter used in Example 5.11 (a) Frequency response and (b) Impulse response.

Example 5.11: First we consider a simple impulse response function considered in Example 5.1, $h(m) = r^m, \quad m = 0,1,2,\ldots\infty$ *and* $|r| < 1$. Here p=1 and q=0. Eq. (5.40b) reduces to

$$ah(0) = h(1)$$

or

$$a = \frac{h(1)}{h(0)} = r$$

To obtain b_0 we go to eq. (5.40a), which reduces to

$$\begin{bmatrix} 1 & 0 \\ r & 1 \end{bmatrix} \begin{bmatrix} 1 \\ -r \end{bmatrix} = \begin{bmatrix} b_0 \\ 0 \end{bmatrix}$$

Evidently, b_0=1 and $b_1 = b_2 = b_3 \cdots = 0$.
Let us consider another example. The transfer function of tan IIR filter is given below:

$$H(z) = \frac{o.56z^{-1} + 0.638z^{-2} + 0.08z^{-3}}{1 + 0.6z^{-1} + 0.34z^{-2} + -0.4z^{-3}}$$

The frequency response is shown in fig. 5.30(a). It was computed in the range 0 t0 0.5 Hz with a sampling interval of $\frac{1}{128}$. The sampled frequency response was inverse Fourier transformed to obtain the impulse response function, which is shown in fig. 5.30(b).
The denominator coefficients were first computed using the computed impulse response function in (5.40b). The estimated denominator coefficients were : -0.6000 -0.3400 0.4000. After estimating the denominator coefficients the numerator coefficients were computed using (5.40a). The estimated numerator coefficients were 0.0000, 0.5600, 0.6380, 0.0800. It is no surprise that the results are exact as the impulse response function was noise free and the filter order was known precisely .

§5.7 Filter Transformation

Given a lowpass normalized digital filter we like to transform it into another lowpass, or a bandpass or a band stop or a highpass filter. To achieve this we consider the following transformation:

$$\boxed{z = \pm \prod_{i=1}^{m} \frac{\bar{z} - a_i^*}{1 - a_i \bar{z}}} \tag{5.41}$$

where $a_i, i = 1,2,\ldots,m$ are complex constants and m is an integer constant. Note that the transformed filter is in the $\overline{z}$ plane. It may be noted the unit circle in the z plane is mapped into a unit circle in the $\overline{z}$ plane, that is, $|z| = 1$,

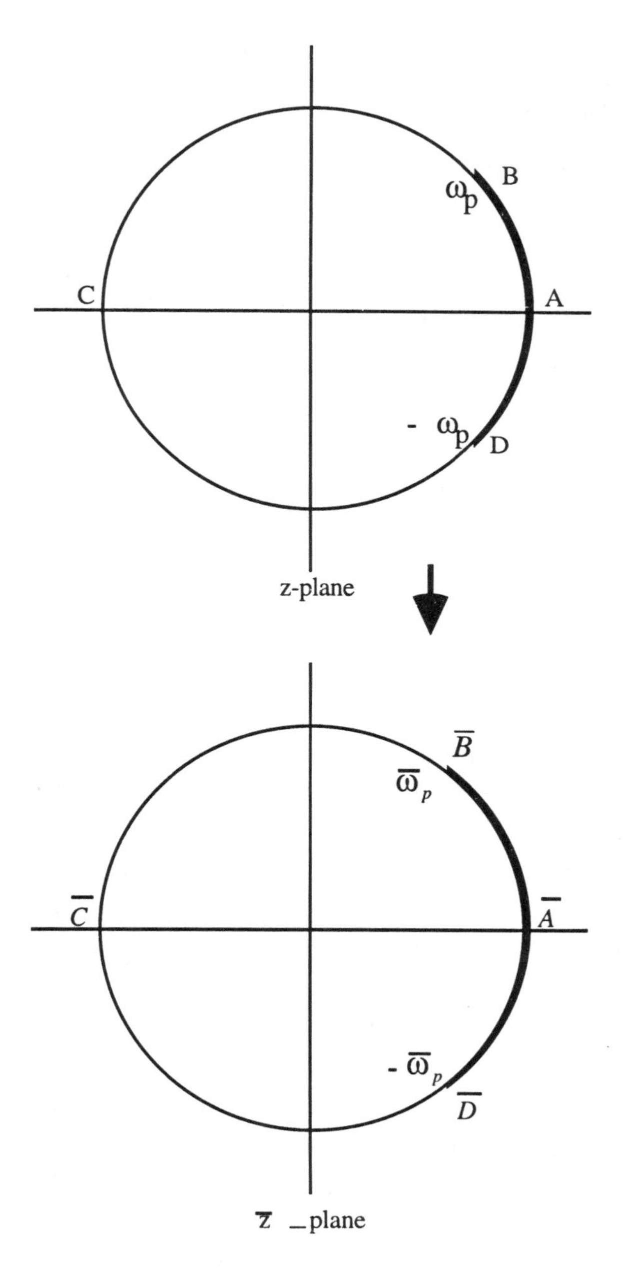

Figure 5.31: Lowpass to Lowpass transformation.

implies $|\bar{z}| = 1$. Further, the interior of the unit circle in the z plane is mapped into the interior of the unit circle in the $\bar{z}$ plane. Therefore, all poles of a filter lying inside the unit circle remain inside even after mapping. This ensures that a stable filter remains stable. The integer m stands for multiplicity of mapping, that is, a lowpass band is mapped over m different passbands, for example, m=1 is for lowpass to lowpass and m=2 is for lowpass to bandpass. As an example, we shall describe lowpass to lowpass transformation but merely mention other types of transformations (see table 5.1).

Lowpass to Lowpass Transformation: A lowpass filter with a passband, BAD, is desired to be mapped into another lowpass filter with a different passband, $\overline{BAD}$. Let the passband edges be $\pm\omega_p$ and $\pm\bar{\omega}_p$ respectively (see fig. 5.31). The transformation is of multiplicity one; hence m=1. Further, the points A and C must remain undisturbed. The lowpass to lowpass transformation is given by

$$\boxed{z = \frac{\bar{z} - \alpha}{1 - \alpha\bar{z}}}$$

where α is real constant which may be obtained by comparing the passband edges before and after transformation. This gives us

$$\alpha = \frac{\sin[(\omega_p - \bar{\omega}_p)\frac{\Delta t}{2}]}{\sin[(\omega_p + \bar{\omega}_p)\frac{\Delta t}{2}]}$$

Note that when $\alpha > 0$ we have contraction and $\alpha < 0$ we have expansion. One to one mapping is possible when $\alpha = 0$. The transformation from lowpass to highpass and bandpass are listed in table 5.1.

Example 5.12: Here we give an example of mapping a standard lowpass filter (ω_p=1) into another lowpass filter with the passband edge at $\bar{\omega}_p = \pi/6$. Assume that $\Delta t = 1$. We substitute these values in the expression for α above.

$$\alpha = \frac{\sin[0.2382]}{\sin[0.7618]} = 0.3419$$

The transformation is given by $z = (\bar{z} - 0.3419)/(1 - 0.3419\bar{z})$. As an illustration consider the 4th order normalized analog Chebychev filter derived in Example 5.4. It will be transformed into a lowpass digital filter with cutoff frequency at $\pi/6$. First let us map the normalized analog filter

$$H_4(s) = \frac{0.245653}{\prod_{i=1}^{2}(s - p_i)(s - p_i^*)}$$

into a normalized digital filter using the bilinear transformation, $s = 2(z-1)/(z+1)$. We obtain

$$H(z) = \frac{0.245653}{\prod_{i=1}^{2}(2\frac{z-1}{z+1} - p_i)(2\frac{z-1}{z+1} - p_i^*)}$$

The normalized lowpass filter $\omega_p = 1$ is mapped to a lowpass filter with cutoff frequency $\frac{\pi}{6}$. The mapping function is $z = (\bar{z} - 0.3419)/(1 - 0.3419\bar{z})$ or $(z-1)/(z+1) = 2.0391(\bar{z}-1)/(\bar{z}+1)$.

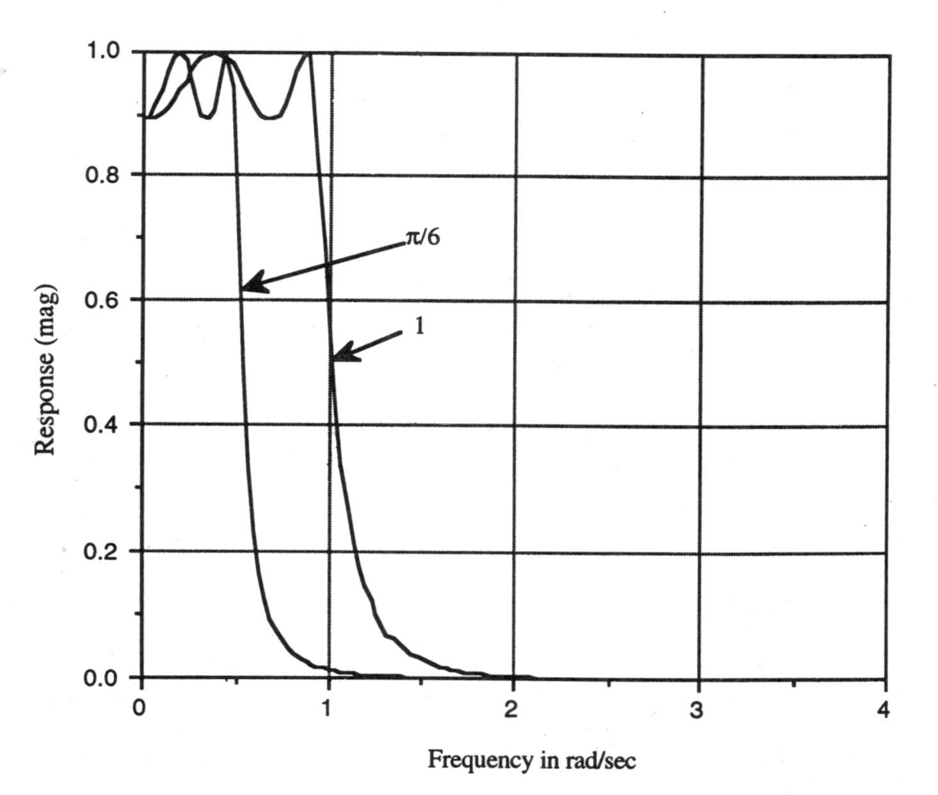

Figure 5.32: A lowpass Chebyshev filter with unit cut-off frequency is mapped into another lowpass Chebyshev filter with cut-off frequency $\pi/6$.

$$H(\bar{z}) = \frac{0.245653}{\prod_{i=1}^{2}(4.0781\frac{\bar{z}-1}{\bar{z}+1} - p_i)(4.0781\frac{\bar{z}-1}{\bar{z}+1} - p_i^*)}$$

The lowpass filter before and after mapping is shown in fig. 5.32 ⚫.

Table 5.1: Some of the commonly used transformations are listed above.

Type	Digital filter transformation	Parameters
LP to LP	$z=\frac{\bar{z}-\alpha}{1-\alpha\bar{z}}$	$\alpha=\frac{\sin((\omega_p-\bar{\omega}_p)\frac{\Delta t}{2})}{\sin((\omega_p+\bar{\omega}_p)\frac{\Delta t}{2})}$
LP to HP	$z=-\frac{\bar{z}-\alpha}{1-\alpha\bar{z}}$	$\alpha=\frac{\cos((\omega_p-\bar{\omega}_p)\frac{\Delta t}{2})}{\cos((\omega_p+\bar{\omega}_p)\frac{\Delta t}{2})}$
LP to BP	$z=-\frac{\bar{z}^2-\frac{2\alpha k}{1+k}\bar{z}-\frac{1-k}{1+k}}{1-\frac{2\alpha k}{1+k}\bar{z}-\frac{1-k}{1+k}\bar{z}^2}$	$\alpha=\frac{\cos((\bar{\omega}_u+\bar{\omega}_l)\frac{\Delta t}{2})}{\cos((\bar{\omega}_u-\bar{\omega}_l)\frac{\Delta t}{2})}$ $k=\cot\frac{(\bar{\omega}_u-\bar{\omega}_l)\Delta t}{2}\tan\frac{\omega_p\Delta t}{2}$ $\bar{\omega}_u$: Upper cutoff frequency $\bar{\omega}_l$: Lower cutoff frequency

Finally it is instructive to compare the FIR and IIR filters. The important differences between the two filters are listed in table 5.2 below

Table 5.2: Shows the important differences between the FIR and IIR filters.

FIR Filter	**IIR Filter**
Linear phase	Non-linear phase
Two-sided	One-sided (causal)
Non-recursive	Recursive
All-zero model	Pole-zero model
Always stable	Possibility of being unstable
More coefficients are required to achieve a desired response	Fewer coefficients are required for the same result
Coefficient are symmetric	No symmetry exits

§5.8 Exercises

Problems:

1. A recursive filter is characterized by following equations

$$y(n) = y_1(n) + \frac{7}{4}y(n-1) - \frac{49}{32}y(n-2)$$

$$y_1(n) = x(n) + \frac{1}{2}y_1(n-1)$$

(a) Obtain the transfer function of the filter.
(b) Find all poles and zeros of the transfer function.

2.The poles and zeros of an IIR filter are given as under. Show, by actual multiplication, that the coefficients of the numerator and denominator polynomials are real.

Poles: -0.7880+j5812, -0.7880-j5812, 0.812+j0.4812, 0.812-j0.4812
Zeros: 0.4661+j0.8514, 0.4661-j0.8514, 0.8843

Compute the value of the transfer function at z=1.0.

3. Draw a digital network similar to the one in fig. 5.6 or 5.7 to realize a filter whose transfer function is given in problem 2 in terms of its poles and zeros.

4.One of the four possible realizations of equation 5.5 is shown in fig. 5.4. Draw the remaining three realizations. What is the essential difference?

5. Consider a pole-zero combination on the real axis as shown in fig. 5.33

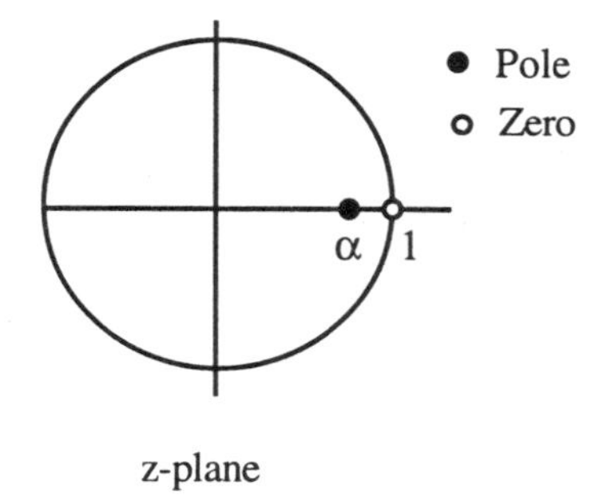

Figure 5.33: Pole and zero on the real axis. The resulting filter is an example of a simple lowpass filter.

Derive the response of the filter. Observe that the filter is an example of a simple highpass filter. Show that the 3dB cutoff frequency of the highpass band is given by

$$\omega_c = \cos^{-1}\left(\frac{2\alpha}{1+\alpha^2}\right)$$

6. Let H(z) represent an ideal lowpass filter whose magnitude response is shown in fig. 5.34(a). Sketch the magnitude response of $H(z^2)$ and $H(z^3)$. Do the same with fig. 5.34 (b)

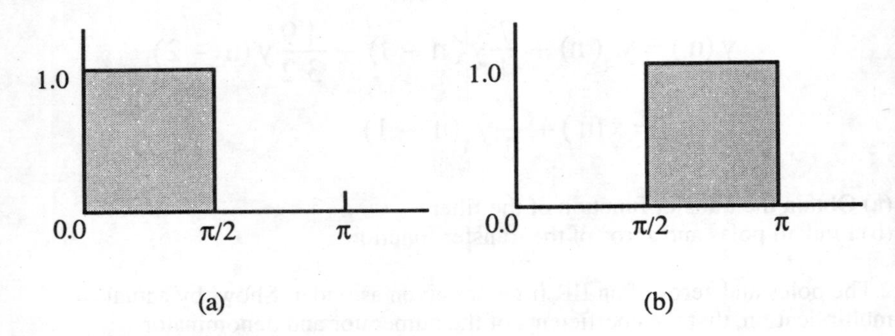

Figure 5.34: Magnitude response of an ideal (a) lowpass and (b) highpass filters

7. An analog passband filter with passband from 80 to 120 Hz is sampled with sampling interval $\Delta t = 10/3$ms. The transfer function of the digital filter is derived from the analog filter transfer function through a bilinear transformation. What is the passband of the digital filter? What happens if we were to halve the sampling interval?

8. Solve the difference equation (5.33) with initial conditions $y(-1) = y(-2) = 0$. By letting $b = A\sin(\omega_0)$, show that the solution of the difference equation is $y(n) = r^n A\sin(\omega_0(n+1))$, n=0,1,2,... As the pole tends to the unit circle, that is, $r \to 1$ the solution turns into a pure sinusoid

9. With b equal to zero in eq. (5.32) (i.e. zero input) and the initial conditions as $y(-1) = 0$, $y(-2) = -A\sin(\omega_0)$, show that the solution of the difference equation is $y(n) = r^{n+2} A\sin(\omega_0(n+1)$, n=0,1,2,...

10. Derive transfer and impulse response functions of the following network (fig. 5.35):

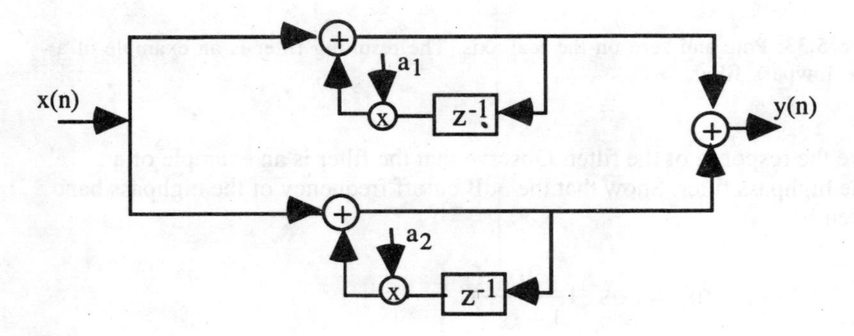

Figure 5.35: Filter network whose impulse response is desired.

11. Another type of direct form realization of IIR filter is shown in fig. 5.36. Show that the transfer function remains the same as in fig.5.6. Fig. 5.36 is known as the transpose of fig. 5.6.(2, p360)

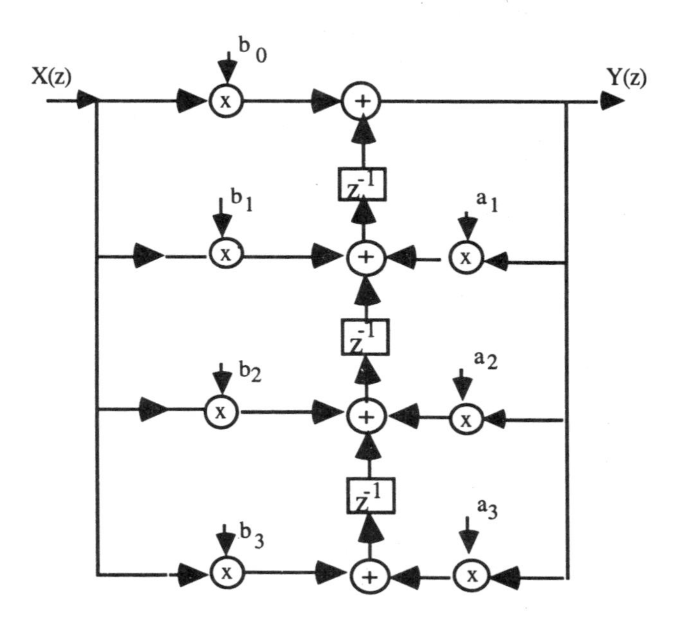

Figure 5.36: Transpose of IIR filter

12. Let

$$H(z) = \frac{z^2 - 1.6971z + 1.44}{z^2 - 1.6914z + 0.81}$$

(a) Find its poles and zeros
(b) Compute the minimum phase transfer function, $H_{\min}(z)$, having the same magnitude response as that of *H(z)*.
(c) Find an all-pass filter which can convert $H_{\min}(z)$ into *H(z)*.

13. The transformation from lowpass to highpass is achieved through an expression, which differs from that for lowpass to lowpass transformation in sign (see table 5.1). That is, we choose a -ve sign in (5.41). How do you explain this choice?

14. A comb filter for harmonically related frequencies (5.30) may be expressed in terms of a single parameter, $b_1 = -2\cos\omega_0$, where ω_0 is the fundamental frequency. Prove the relation between b_1 and b_2, b_3, b_4,... given by eq. (5.30c).

15. Given the numerator and denominator coefficients of an IIR filter one can compute the impulse response function directly from (5.39b). Solve this equation for p=2 and q=1 in terms of a1, a2 and b0, b1.

16. Find all poles of a third order Butterworth filter and derive its transfer function. Compute the transfer function on the imaginary axis (frequency axis).

17. Find all poles of a third order (A_p=1.0 dB) Chebyshev filter and derive its transfer function. Compute the transfer function on the imaginary axis (frequency axis). What is the minimum stopband attenuation assuming that the stopband starts at $\omega = 1.2$.

18. A pole in the s-plane when mapped into the z-plane using the bilinear transformation. The pole in the s-plane is located at -1.0+j10. Assume that $\Delta t = 1$. What is the path followed by the pole in the z-plane when the pole in the s-plane is moved along a line, $\sigma = -1$.

19. We wish to design a digital lowpass filter by mapping an analog lowpass filter using the bilinear transformation. The desired cut-off frequency of the digital filter is 0.3 Hz ($\Delta t = 1$). What should be the cut-off frequency of the analog lowpass filter.

20. Consider the following filter with three interconnected difference equations:

$$y(n) = y_2(n) + \frac{7}{4}y(n-1) - \frac{49}{32}y(n-2)$$

$$y_2(n) = 0.7y_2(n-1) + \frac{1}{2}y_1(n)$$

$$y_1(n) = x(n) + 0.5x(n-1)$$

Compute the transfer function of the filter. Obtain the poles and zeros of the filter. Is this a stable filter?

21. Let

$$H(z) = \frac{(1-0.9z^{-1})}{(1-1.455z^{-1}+0.9025z^{-2})(1-0.94z^{-1})}$$

Express the filter equation as a system of two interconnected difference equations.

22. Show that a digital filter transformer (Eq. 5.41) is indeed an all-pass filter for any value of m.

23. An FIR filter has zeros at $0.9e^{\pm j28^0}$ and $1.24e^{\pm j40^0}$. Design a minimum phase filter that has the same magnitude response as the one defined above. Show, by actual computation, the magnitude responses are indeed same.

24. Consider the following IIR filter:

$$y(n) = 0.95y(n-4) + x(n) - x(n-4)$$

For some inputs, the output is zero. What are these inputs? Explain this property in terms of the transfer function of the filter.

25. A maximum phase filter is one whose all zeros are outside the unit circle. A mixed phase filter (with zeros both inside and outside the unit circle) may be converted into a maximum phase filter by mapping the zeros which are inside the unit circle to their reciprocal position, reverse of minimum phase filter, as defined in eq.(5.35). Construct a maximum phase filter from

$$H_{\max}(z) = \frac{z^2 - 1.6z^1 + 0.89}{z^2 - 1.69z^1 + 0.81}$$

Show by actual multiplication $|H_{\max}(z)| = |H_{mix}(z)|$.

Computer Projects:

1. Generate a digital sinusoid $y(n) = A\sin(\omega_0 n)$ without evaluating the sine function. It involves numerically solving the difference equation (5.32) with r close to 1.
Next, we like to generate a digital sinusoid with a prescribed phase, i.e. $y(n) = A\sin(\omega_0 n + \phi_0)$. This may be achieved by generating at the same time $A\sin(\omega_0 n)$ and $A\cos(\omega_0 n)$ and forming a linear combination, $A\cos(\phi_0)\sin(\omega_0 n) \;\; + A\sin(\phi_0)\cos(\omega_0 n) = A\sin(\omega_0 n + \phi_0)$. Show that the digital sine and cosine functions may be generating by solving a coupled difference equation [1]

$$y_c(n) = \cos(\omega_0) y_c(n-1) - \sin(\omega_0) y_s(n-1)$$
$$y_s(n) = \sin(\omega_0) y_c(n-1) + \cos(\omega_0) y_s(n-1)$$

with initial conditions $y_c(-1) = A\cos(\omega_0)$ and $y_s(-1) = -A\sin(\omega_0)$.

2. Consider a digital filter with a frequency response given by

$$H(z) = \frac{1-\alpha}{2} \frac{1 - z^{-2}}{1 - \beta(1+\alpha)z^{-1} + \alpha z^{-2}}$$

where α and β are constants. Compute the magnitude response of H(z) for a fixed β but variable α. The response has a peak at $\omega_c = \cos^{-1}(\beta)$ that is independent of α but its 3-dB width depends upon α. Verify these statements. Note that H(z) has zeros at ± 1 (that is, at $\omega = 0$ and π) and has poles which tend to the unit circle at $e^{\pm j\omega_c}$.This is an example of a simple passband filter whose passband is centered at $\cos^{-1}(\beta)$ and 3dB width is given by $\cos^{-1}(2\alpha/(1+\alpha))$ [2,].

3. Consider the filter response function used in problem #10. Evaluate the group delay of H(z) and $H_{\min}(z)$. Use numerical differentiation routine from the Matlab Mfiles. Verify the relation (5.37b) between the group delays of H(z) and $H_{\min}(z)$.
4. Convert the 4th order normalized analog Chebyshev filter derived in Example 5.4. into a digital filter using bilinear transformation (This has already been derived in Example 5.12). Now we like to map this normalized lowpass filter into a bandpass filter with upper cut-off at 1.3 rad/sec and the lower cut-off at 0.9 rad/sec. Write a Matlab program to carry out the above transformation. Assume the sampling interval as one second.

5. Map a standard lowpass filter (ω_p=1) into a bandpass filter with its upper cutoff frequency at $\overline{\omega}_u = \pi/3$ and the lower cutoff frequency at $\overline{\omega}_u = \pi/6$. Assume that the sampling interval is one second. Consider the 4th order normalized analog Chebychev filter derived in Example 5.4 and used in Example 5.12 for mapping into a lowpass filter. As in Example 5.12 use a bilinear transformation to go into the digital domain. Write a Matlab program to carry out the required mapping and plot the response function.

6. In Example 5.11 we have preferred to compute the impulse response function of an IIR filter as inverse DFT of its frequency response. However, as suggested in problem 14, the impulse response function may be computed directly from (5.39b) given the numerator and denominator coefficients. Write a Matlab program to compute the impulse response function of the IIR filter used in Example 5.11 and compare it with that shown in fig. 5.29b.

6 Adaptive Filters

It is often required to find a filter that minimizes the mean square error between the observed signal, which is corrupted with noise and a reference signal. In the strict statistical sense Wiener filter is the right choice but this requires knowledge of the statistical quantities such as the auto and cross correlation functions which are not known apriori. This difficulty is bypassed in a clever manner. The error is progressively minimized by going down the error surface in the direction of the steepest gradient which must be computed from the observation - a task not easily accomplished in real-time. We need to look for alternate possibilities. One such possibility is the least mean square (LMS) algorithm where the steepest gradient is replaced by an instantaneous gradient. It is now possible to update the filter coefficients as new data arrives in real-time. There are important issues related to speed of convergence and step size. In another approach (recursive least-squares (RLS) algorithm) the statistical model is itself completely replaced by a deterministic model where one minimizes the sum of squares of the errors. The least-squares solution requires data matrix, but no correlation matrix. Fortunately, the data matrix and its inverse can be updated as new data arrives. This results into a neat recursive algorithm for the solution of the desired filter. In some applications the signal model is known in place of the required reference signal. It is then possible to exploit this information for recursive estimation of the unknown signal. It is a simplified version of the celebrated Kalman filter.

One of the important applications of the adaptive algorithm, given the input and output of a system, is to determine the impulse response function of the system, which is either a FIR or an IIR filter. This is often known as system identification. We shall give a few examples of identification of FIR and IIR filters. Having identified a system, it is of interest to estimate the input from the observed output. This is known as equalization in communication or inverse filtering in physics. Finally, we give a few simple practical examples of adaptive filtering such as for echo cancellation, channel equalization in shallow water communication, source-receiver coupling.

§6.1 Optimum Filters

A typical adaptive filter consists of a tapped delay line with unit delays and adjustable weights. The weighted outputs are summed to form the filter output that is then compared with the desired or reference signal. The difference between the reference signal and the filter output, that is, error is then used to adjust the weights so that the error is reduced further. This is schematically shown in fig. 6.1. The input signal vector is defined as

$$\mathbf{x}_n = [x_n, x_{n-1}, \ldots, x_{n-M+1}]^T$$

at time instant n and the weight vector of length M is defined as

$$\mathbf{h} = [h_0, h_1, \ldots, h_{M-1}]^T.$$

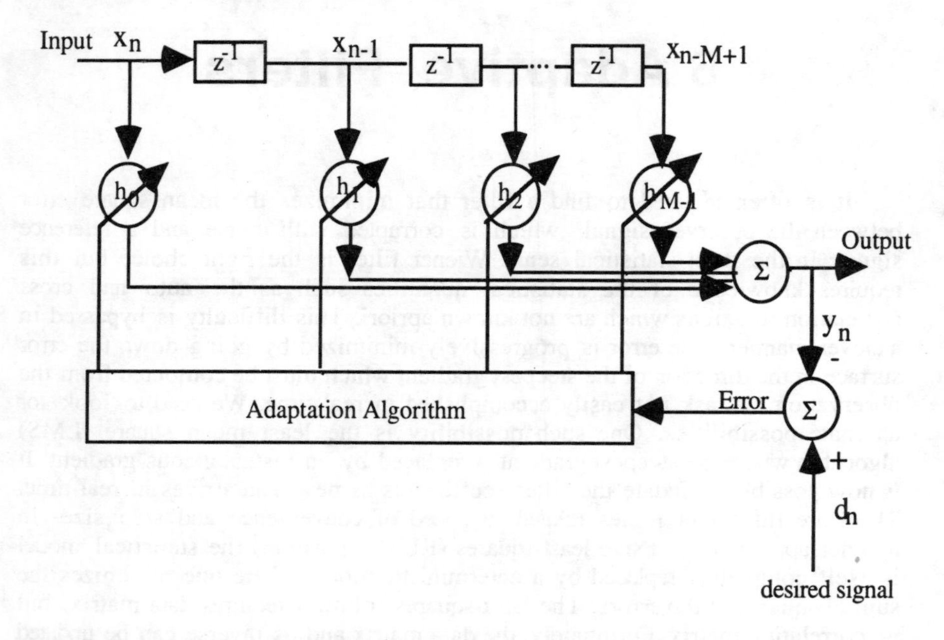

Figure 6.1: A typical adaptive filter consists of a tapped delay line with unit delays and adjustable weights. The desired signal is known and depends upon application.

The weighted summed output is given by

$$y_n = \mathbf{h}^T \mathbf{x}_n = \mathbf{x}_n^T \mathbf{h} \tag{6.1}$$

The error is defined as

$$e_n = d_n - y_n = d_n - \mathbf{h}^T \mathbf{x}_n \tag{6.2}$$

where d_n is reference signal. The mean square error may be obtained by squaring (6.2) and computing the ensemble average

$$\xi = E\{e_n^2\} = E\{d_n^2\} - 2E\{d_n \mathbf{x}_n^T\}\mathbf{h} + \mathbf{h}^T E\{\mathbf{x}_n \mathbf{x}_n^T\}\mathbf{h} \tag{6.3a}$$

We will assume that both the input signal and desired signals are stationary time series. Then, $E\{\mathbf{x}_n \mathbf{x}_n^T\}$ and $E\{d_n \mathbf{x}_n^T\}$ are covariance and cross-covariance matrices respectively,

$$E\{\mathbf{x}_n\mathbf{x}_n^T\} = \begin{bmatrix} E\{x_n x_n\} & E\{x_n x_{n-1}\} & \cdots & E\{x_n x_{n-M+1}\} \\ E\{x_{n-1} x_n\} & E\{x_{n-1} x_{n-1}\} & \cdots & E\{x_{n-1} x_{n-M+1}\} \\ & & \cdots & \\ & & \cdots & \\ E\{x_{n-M+1} x_n\} & E\{x_{n-M+1} x_{n-1}\} & \cdots & E\{x_{n-M+1} x_{n-M+1}\} \end{bmatrix} = \mathbf{c}_x$$

$$E\{d_n\mathbf{x}_n^T\} = \begin{bmatrix} E\{d_n x_n\} \\ E\{d_n x_{n-1}\} \\ \\ \\ E\{d_n x_{n-M+1}\} \end{bmatrix} = \mathbf{c}_{dx}$$

Using the above results in (6.3) we obtain

$$\xi = \sigma_d^2 - 2\mathbf{c}_{dx}^T\mathbf{h} + \mathbf{h}^T\mathbf{c}_x\mathbf{h} \tag{6.3b}$$

where σ_d^2 is variance of the desired signal. The mean square error as a function of the weights is a concave hyperparaboloidal surface or a bowl shaped surface (see fig. 6.2). It never goes negative. It has a unique minimum, the bottom of the bowl.

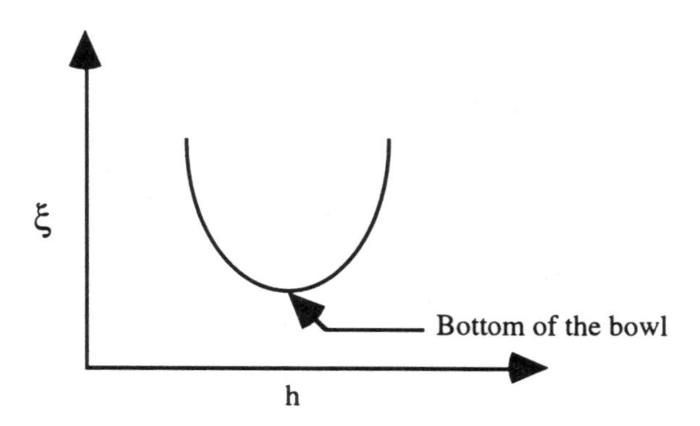

Figure 6.2: Bowl shaped error surface. The bottom of the bowl represents minimum mean square error (mmse) point.

Wiener Filter: The minimum may be obtained by differentiating (6.3b) with respect to the weights and setting the derivative to zero. The derivative of ξ is given by

$$\nabla\xi = -2\mathbf{c}_{dx} + 2\mathbf{c}_x\mathbf{h} \tag{6.4}$$

From this we get a solution for the weight vector.

$$\boxed{\mathbf{h}_W = \mathbf{c}_x^{-1}\mathbf{c}_{dx}} \tag{6.5a}$$

This is the well-known Wiener-Hopf equation. The minimum mean square error (mmse) may be obtained by substituting (6.5a) into (6.3b)

$$\xi_{\min} = \sigma_d^2 - \mathbf{c}_{dx}^T\mathbf{h}_W \tag{6.5b}$$

It may be noted that computation of the weight vector for mmse requires inversion of a covariance matrix of the size MxM. This is computationally expensive in practical applications. We now describe an alternate approach that does not require matrix inversion.

Steepest Gradient: The goal is to reach the bottom of the bowl. The simplest strategy would be to go down the slope, preferably down the steepest gradient. After a finite number of steps we can get to the bottom of the bowl. At k^th step let the mse be ξ_k and the corresponding weight vector be $\mathbf{h}_k$

$$\xi_k = \sigma_d^2 - 2\mathbf{c}_{dx}^T\mathbf{h}_k + \mathbf{h}_k^T\mathbf{c}_x\mathbf{h}_k \tag{6.6a}$$

Adding and subtracting $\mathbf{c}_{dx}^T\mathbf{h}_W$ on the right hand side of (6.6a) we can write

$$\xi_k = \xi_{\min} - 2\mathbf{c}_{dx}^T\mathbf{h}_k + \mathbf{h}_k^T\mathbf{c}_x\mathbf{h}_k + \mathbf{c}_{dx}^T\mathbf{h}_W \tag{6.6b}$$

Further after some algebraic simplification we can show that

$$\begin{aligned} -2\mathbf{c}_{dx}^T\mathbf{h}_k + \mathbf{h}_k^T\mathbf{c}_x\mathbf{h}_k + \mathbf{c}_{dx}^T\mathbf{h}_W &= (\mathbf{h}_k - \mathbf{h}_W)^T\mathbf{c}_x(\mathbf{h}_k - \mathbf{h}_W) \\ &= \Delta\mathbf{h}_k^T\mathbf{c}_x\Delta\mathbf{h}_k \end{aligned} \tag{6.7}$$

Using (6.7) in (6.6b) we obtain

$$\xi_k = \xi_{\min} + \Delta\mathbf{h}_k^T\mathbf{c}_x\Delta\mathbf{h}_k \tag{6.8}$$

The gradient of ξ_k at the kth step from (6.4) and using (6.5a)

$$\begin{aligned} \nabla\xi_k &= -2\mathbf{c}_{dx} + 2\mathbf{c}_x\mathbf{h}_k \\ &= -2\mathbf{c}_x\mathbf{h}_W + 2\mathbf{c}_x\mathbf{h}_k \\ &= 2\mathbf{c}_x\Delta\mathbf{h}_k \end{aligned}$$

To reach the bottom of the bowl (i.e. error surface) we go down the steepest gradient in small steps. After a sufficient number of steps we are guaranteed to reach the minimum. Starting from some initial point on the error surface the weight vector is changed in the direction of the maximum gradient. The weight vector at the kth step $\mathbf{h}_k$ is altered in accordance with the negative of the gradient of ξ_k. The updated weight vector is given by

$$\begin{aligned}\mathbf{h}_{k+1} &= \mathbf{h}_k + \mu(-\nabla\xi_k) \\ &= \mathbf{h}_k - 2\mu\mathbf{c}_x\Delta\mathbf{h}_k\end{aligned} \tag{6.9}$$

where μ is the step size whose magnitude controls the convergence of the iterative process. Subtracting $\mathbf{h}_W$ from both sides of (6.9) we obtain

$$\Delta\mathbf{h}_{k+1} = (\mathbf{I} - 2\mu\mathbf{c}_x)\Delta\mathbf{h}_k \tag{6.10}$$

where $\Delta\mathbf{h}_k = \mathbf{h}_k - \mathbf{h}_W$. Eq. (6.10) is a vector difference equation whose solution is given by

$$\Delta\mathbf{h}_k = (\mathbf{I} - 2\mu\mathbf{c}_x)^k\,\Delta\mathbf{h}_0 \tag{6.11a}$$

where $\Delta\mathbf{h}_0$ stands for the weight vector difference at some starting point, k=0. The solution is convergent if $(\mathbf{I} - 2\mu\mathbf{c}_x)^k \to 0$ as $k \to \infty$. To further understand the convergence properties we introduce the eigenvalue eigenvector decomposition of the covariance matrix. As noted in Chapter One the covariance matrix is toeplitz positive definite and hence its eigenvalues are real and positive. $\mathbf{c}_x = \mathbf{u}\Lambda\mathbf{u}^H$ where the columns of $\mathbf{u}$ are the eigenvectors and Λ is a diagonal matrix with eigenvalues of $\mathbf{c}_x$ on its diagonal. Using the eigenvalue eigenvector decomposition of $\mathbf{c}_x$ in (6.11a) we obtain

$$\Delta\mathbf{h}_k = \mathbf{u}(\mathbf{I} - 2\mu\Lambda)^k\mathbf{u}^H\Delta\mathbf{h}_0 \tag{6.11b}$$

The condition for convergence may now be expressed as

$$\begin{aligned}&(\mathbf{I} - 2\mu\Lambda)^k \to 0 \\ &k \to \infty\end{aligned} \tag{6.12}$$

which will be satisfied if $|(1 - 2\mu\lambda_m)| < 1, \quad m = 1,2,\ldots,M$ where $\lambda_m, m = 1,2,\ldots,M$ are the eigenvalues of $\mathbf{c}_x$. Since all eigenvalues are positive,

$$\frac{1}{\lambda_{\max}} > \mu > 0 \tag{6.13}$$

where λ_{max} is the largest eigenvalue of $\mathbf{c}_x$. The mth term or mode in (6.12) decays as $(1-2\mu\lambda_m)^k$, as a geometric progression. Let us fit an exponential decay model of the type $\exp(-k/\tau_m)$ where τ_m is a decay constant, which is numerically equal to the number of steps required to let the error decay to $1/e$(=0.367879)th fraction of the initial value. We obtain

$$(1-2\mu\lambda_m)=\exp(-\frac{1}{\tau_m})\approx 1-\frac{1}{\tau_m}$$
$$\tau_m=\frac{1}{2\mu\lambda_m} \tag{6.14}$$

Using (6.13) in (6.14) we obtain a lower bound for the decay constant,

$$\tau_m>\frac{\lambda_{max}}{2\lambda_m}.$$

Let λ_{min} be the minimum eigenvalue, then the lower bound on the decay constant is given by $\frac{\lambda_{max}}{2\lambda_{min}}$, which is known as eigenvalue spread of a matrix. Thus, the covergence will be slow whenever the eigenvalue spread of the covariance matrix is large. The eigenvalue spread of a covariance matrix of white noise is one, therefore the convergence of the weight vector shall be rapid for white noise input. In contrast, the eigenvalue spread of highly correlated signal such as a sinusoid plus noise will be large, depending upon the signal-to-noise ratio (SNR).

Learning Curve: A plot of ξ_k as a function of number of iterations is known as a learning curve. We shall now derive an expression for learning curve. The decay in the mean square error may be obtained by using (6.11a) in (6.8),

$$\xi_k=\xi_{min}+\Delta\mathbf{h}_0^T(\mathbf{I}-2\mu\mathbf{c}_x)^k\mathbf{c}_x(\mathbf{I}-2\mu\mathbf{c}_x)^k\Delta\mathbf{h}_0 \tag{6.15a}$$

which may be further simplified by using the eigenvalue eigenvector decomposition of the covariance matrix,

$$\xi_k=\xi_{min}+\Delta\mathbf{h}_0^T\mathbf{u}(\mathbf{I}-2\mu\Lambda)^{2k}\Lambda\mathbf{u}^H\Delta\mathbf{h}_0 \tag{6.15b}$$

As long as (6.13) is satisfied the mse of the adaptive process will converge to the minimum point of mse surface, that is, $\xi_k\rightarrow\xi_{min}$ as $k\rightarrow\infty$. The decay constant τ_m is equal to $1/4\mu\lambda_m$. Let

$$\Delta\mathbf{h}_0'=\mathbf{u}^H\Delta\mathbf{h}_0$$

and note that $(\mathbf{I}-2\mu\Lambda)^{2k}\Lambda$ is a diagonal matrix whose mth element on the diagonal is $(1-2\mu\lambda_m)^{2k}\lambda_m$. Eq (6.15) may be expressed as a sum of M decaying modes

$$\begin{aligned}\xi_k &= \xi_{\min} + \Delta\mathbf{h}_0'^T(\mathbf{I}-2\mu\Lambda)^{2k}\Lambda\Delta\mathbf{h}_0' \\ &= \xi_{\min} + \sum_{m=1}^{M}(1-2\mu\lambda_m)^{2k}\lambda_m\left|\Delta\mathbf{h}_0'^T\mathbf{u}_m\right|^2\end{aligned} \quad (6.16)$$

where $\mathbf{u}_m$ is mth column of $\mathbf{U}$. Note that different modes decay with different speeds, the fastest is the one with $\lambda_{\max}$ and the slowest is the one with $\lambda_{\min}$.

§6.2 Least Mean Square (LMS) Algorithm

In the method of steepest descent, an update of weight vector (6.9) requires computation of the steepest gradient that in turn requires a prior knowledge of the covariance matrix, which is often not available. This obstacle is overcome by replacing the statistical gradient by an instantaneous gradient

Instantaneous gradient: The least mean square (LMS) algorithm is an implementation of the steepest descent method where the statistical gradient estimate is replaced by an instantaneous gradient which is defined as follows:

$$\hat{\nabla}_k\xi = \begin{bmatrix}\frac{\partial e_k^2}{\partial h_1} \\ \frac{\partial e_k^2}{\partial h_2} \\ . \\ . \\ . \\ \frac{\partial e_k^2}{\partial h_M}\end{bmatrix} = 2e_k\begin{bmatrix}\frac{\partial e_k}{\partial h_1} \\ \frac{\partial e_k}{\partial h_2} \\ . \\ . \\ . \\ \frac{\partial e_k}{\partial h_M}\end{bmatrix} = -2e_k\mathbf{x}_k \quad (6.17)$$

Using the instantaneous gradient in (6.9) we get the LMS algorithm,

$$\boxed{\mathbf{h}_{k+1} = \mathbf{h}_k + 2\mu e_k\mathbf{x}_k} \quad (6.18)$$

After the arrival of a new observation we compute a fresh update of the weight vector as in (6.18). Note that we have thus got over the need for covariance and cross-covariance matrices, which require a large number of observations and a large computing power. But the instantaneous gradient is a noisy estimate of the statistical gradient. The quality of the instantaneous gradient is of some interest.

$$E\{\hat{\nabla}_k \xi\} = -2E\{e_k \mathbf{x}_k\} = -2E\{(d_j - \mathbf{x}_k^T \mathbf{h}_k)\mathbf{x}_k\}$$
$$= -2\mathbf{c}_{dx} + 2\mathbf{c}_x \mathbf{h}_k = \nabla \xi \tag{6.19}$$

where we have assumed that $\mathbf{h}_k$ is held fixed. Hence, the instantaneous gradient is unbiased estimate of the statistical gradient. Next, we compute the covariance matrix of $\hat{\nabla}_k \xi$ close to the bottom of the bowl where $\nabla \xi \approx 0$. We obtain

$$E\{(\hat{\nabla}_k \xi)^2\} = E\{(2e_k \mathbf{x}_k)^2\}$$
$$= 4E\{e_k^2\}E\{\mathbf{x}_k \mathbf{x}_k^T\} \tag{6.20}$$
$$= 4\xi_{\min} \mathbf{c}_x$$

where we have assumed that e_k and $\mathbf{x}_k$ are independent. Since the variance of the LMS gradient estimate is always finite (except when the minimum error is zero), the weight vector will tend to perform a random walk around the true minimum (given by Wiener-Hopf equation (6.5a)).

Convergence of Weight Vector: The weight vector given by (6.18) is a stochastic quantity as it is a solution of a difference equation driven by stochastic observed signal. We compute the expected value on eq. (6.18)

$$E\{\mathbf{h}_{k+1}\} = E\{\mathbf{h}_k\} + 2\mu E\{e_k \mathbf{x}_k\}$$
$$= E\{\mathbf{h}_k\} + 2\mu E\{d_k \mathbf{x}_k - \mathbf{x}_k \mathbf{x}_k^T \mathbf{h}_k\}$$
$$= E\{\mathbf{h}_k\} + 2\mu(\mathbf{c}_{dx} - \mathbf{c}_x E\{\mathbf{h}_k\})$$
$$= (\mathbf{I} - 2\mu \mathbf{c}_x)E\{\mathbf{h}_k\} + 2\mu \mathbf{c}_x \mathbf{h}_W$$

Rearranging the terms we get

$$E\{\Delta \mathbf{h}_{k+1}\} = (\mathbf{I} - 2\mu \mathbf{c}_x)E\{\Delta \mathbf{h}_k\} \tag{6.21}$$

where $\Delta \mathbf{h}_k = \mathbf{h}_k - \mathbf{h}_W$. The solution of (6.21) is given by

$$E\{\Delta \mathbf{h}_{k+1}\} = (\mathbf{I} - 2\mu \mathbf{c}_x)^k \Delta \mathbf{h}_0$$
$$= \mathbf{u}(\mathbf{I} - 2\mu \Lambda)^k \mathbf{u}^H \Delta \mathbf{h}_0 \tag{6.22}$$

where $\mathbf{u}$ is matrix with eigenvectors as columns and Λ is eigenvalues of $\mathbf{c}_x$. The expected value of the weight vector approaches the Wiener solution, that is, $\mathbf{h}_k \to \mathbf{h}_W$ as $k \to \infty$ provided the parameter μ satisfies inequality (6.13). Compare (6.22) with (6.11a) for the steepest descent method where there is no error in the estimation of the gradients. In LMS algorithm the estimated weight vector is a noisy estimate, only its mean value approaches the Wiener solution. On the other hand, in the steepest descent method the weight vector estimate itself approaches the Wiener solution.

The computation of the covariance matrix of the weight vector difference is more involved than computing its mean. We shall just give the final asymptotic result taken from [24]

$$E\left\{\Delta\mathbf{h}_k \Delta\mathbf{h}_k^T\right\} \approx \mu\xi_{\min}\mathbf{I} \tag{6.23}$$

It may be observed that the errors in the weight vector are uncorrelated but their asymptotic variance remains constant, proportional to the step size μ. Thus, error free estimate of the weight vector is possible only with $\mu \to 0$.

Misadjustments: The mean square error given by (6.8) is a sum of two terms of which the first term is the minimum error attained when using the Wiener filter as a weight vector, and the second term is the errors in the weight vector. We like to quantify this excess error. Misadjustment is defined as a ratio of the excess error to minimum error,

$$\frac{\xi_k - \xi_{\min}}{\xi_{\min}} = \frac{\Delta\mathbf{h}_k^T \mathbf{c}_x \Delta\mathbf{h}_k}{\xi_{\min}} \tag{6.24a}$$

We will treat $\Delta\mathbf{h}_k$ as a random number, independent of $\mathbf{x}_n$ but dependent on $\mathbf{x}_{n-1}$ (see eq. 6.18). Taking the expected value of (6.24a) we obtain the average misadjustment ratio as

$$\Pi = \frac{E\left\{\Delta\mathbf{h}_k^T \mathbf{c}_x \Delta\mathbf{h}_k\right\}}{\xi_{\min}} \tag{6.24b}$$

Use a relation between a quadratic and trace,

$$\Delta\mathbf{h}_k^T \mathbf{c}_x \Delta\mathbf{h}_k = tr\left\{\mathbf{c}_x \Delta\mathbf{h}_k \Delta\mathbf{h}_k^T\right\}$$

in (6.24b) and obtain

$$\Pi = \frac{tr\left\{\mathbf{c}_x E\left\{\Delta\mathbf{h}_k \Delta\mathbf{h}_k^T\right\}\right\}}{\xi_{\min}} \tag{6.25}$$

For large k , $k \to \infty$, we can simplify (6.25) using (6.23) in (6.25) we get

$$\boxed{\Pi \approx \mu\xi_{\min}\frac{tr\left\{\mathbf{c}_x \mathbf{I}\right\}}{\xi_{\min}} = \mu\, tr\left\{\mathbf{c}_x\right\} = \mu\sum_{m=1}^{M}\lambda_m} \tag{6.26}$$

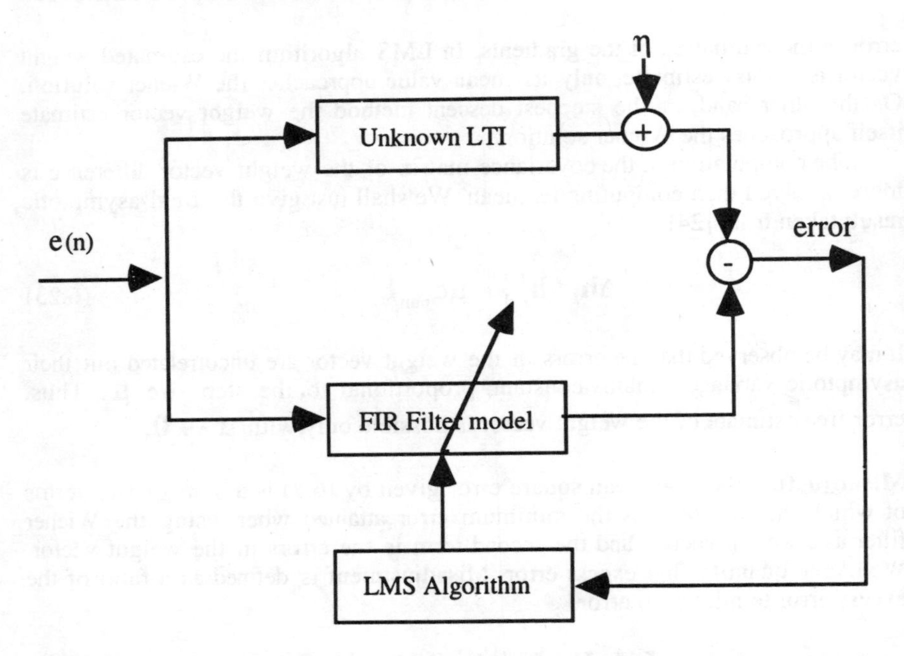

Figure 6.3: Schematic for system identification using LMS algorithm. Both input into and output of the unknown LTI system are known. The coefficients of the FIR filter are altered according to LMS algorithm.

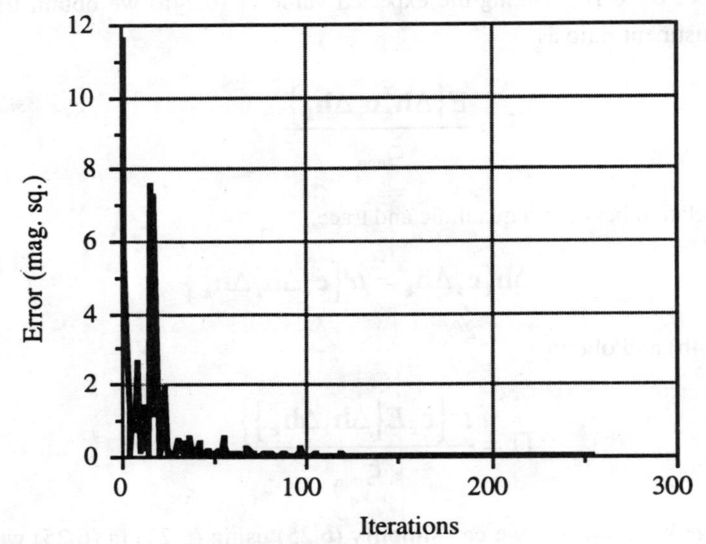

Figure 6.4: Learning curve step size (μ)=0.01.

It is always possible to achieve a small misadjustment by selecting sufficiently small value for μ, which is also the condition for convergence of the weight vector to its optimum value, that is, Wiener filter. But the misadjustment error cannot be reduced to zero however large is the input data. Small step size slows down the convergence as demonstrated in fig. 6.5. But this implies a substantial increase in the computational load.

As for the computational complexity, the LMS algorithm is perhaps the lowest possible requiring o(2M) operations per iteration (one addition + one multiplication); M operations for computing the error and another M operations for the weight vector update.

Example 6.1: As an example of application of LMS algorithm we consider a system identification problem. A linear time invariant (LTI) system is driven by white noise. The output is corrupted by another independent white noise source. The LTI is modeled as a FIR filter. The linear system is a moving average system corrupted by independent noise as given by

$$x(n) = \varepsilon(n) + 1.2\varepsilon(n-1) + 0.9\varepsilon(n-2) + 0.1\eta(n)$$

The input and output of the system are known but the system coefficients are unknown. System identification involves estimating the unknown coefficients. The order of the system, that is, the number of coefficients, is unknown. Since the order is generally more difficult to estimate, we shall assume it to be known.

We devise a FIR filter model with coefficients h_0, h_1, and h_2. Out of the three coefficients we have set $h_0 = 1$ and the other two as unknown. The reference signal is *x(n)*. The output of FIR filter model is compared with the reference signal and the error is used to arrive at improved filter coefficients according to LMS algorithm. A schematic diagram is shown in fig. 6.3 The starting values for $h_1 = h_2 = 0.0$. The learning curve, that is, error square as a function of number of iterations is shown in fig. 6.4. After about fifty iterations the error power becomes very small and it slowly decreases thereafter. The convergence of the unknown coefficients to their true values is shown in fig. 6.5. We have set $\mu = 0.01$. Observe how slow is the convergence to the true values. It took more than 200 iterations to reach the true values. By increasing μ to 0.1 the convergence is improved as seen in fig. 6.6 but there is a lot of scatter around the true values. To make this point clear we have plotted in fig. 6.7 h_1 versus h_2. Observe that the trajectory descends to the bottom of the bowl quite fast but after reaching the bottom it executes a kind of random walk motion around the point of minimum. It does not seem to settle down at the true values. It may be noted that, as the input signal being white noise the eigenvalue spread of the covariance matrix is one, and hence the convergence speed is the fastest possible. In practical problems the convergence issue may become critical, often requiring several hundreds of iterations. This is a serious drawback of the LMS algorithm, though it is simple and easy to implement on a small computer.

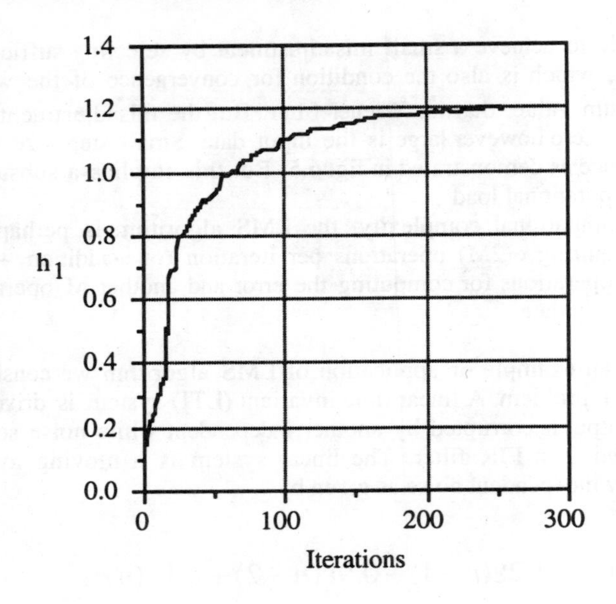

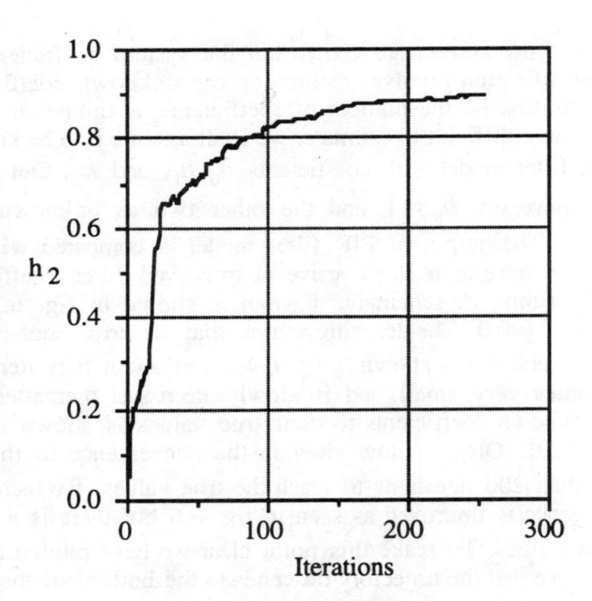

Figure 6.5: Estimates of h_1 and h_2 with increasing number of iterations. Step size (μ)=0.01. Convergence rate is very slow but smooth.

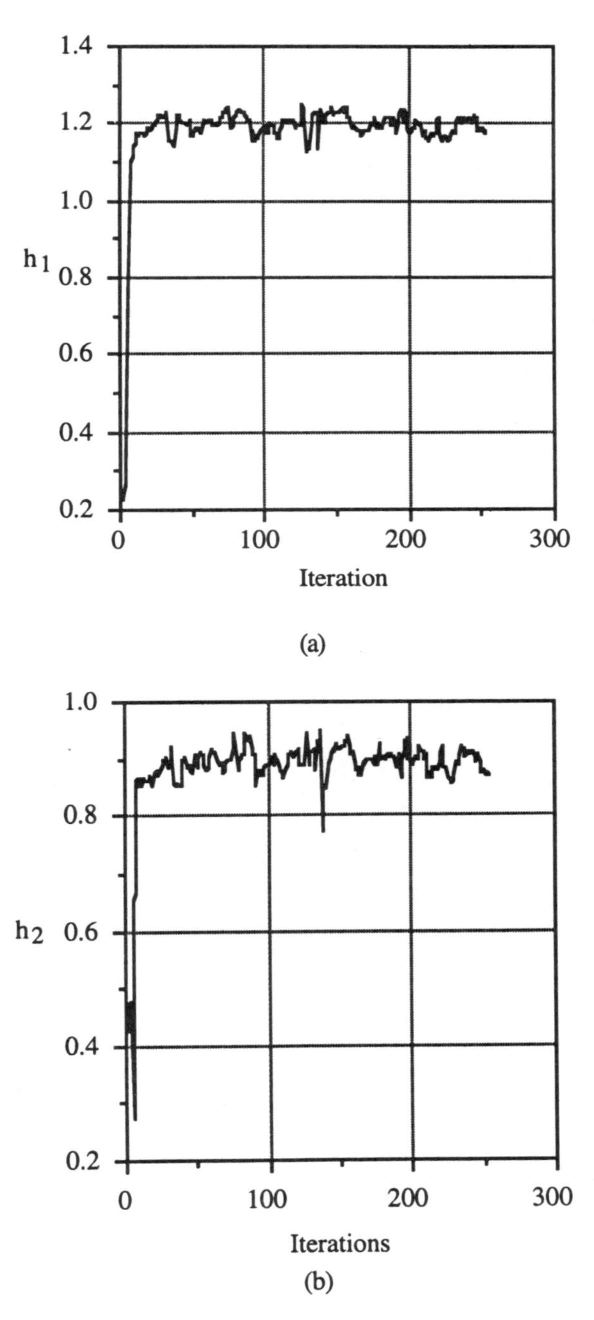

Figure 6.6: Estimates of h_1 and h_2 with increasing number of iterations. Step size (μ)=0.1. While the convergence has improved there is a lot of scatter around the true values, which one may never reach even after many iterations.

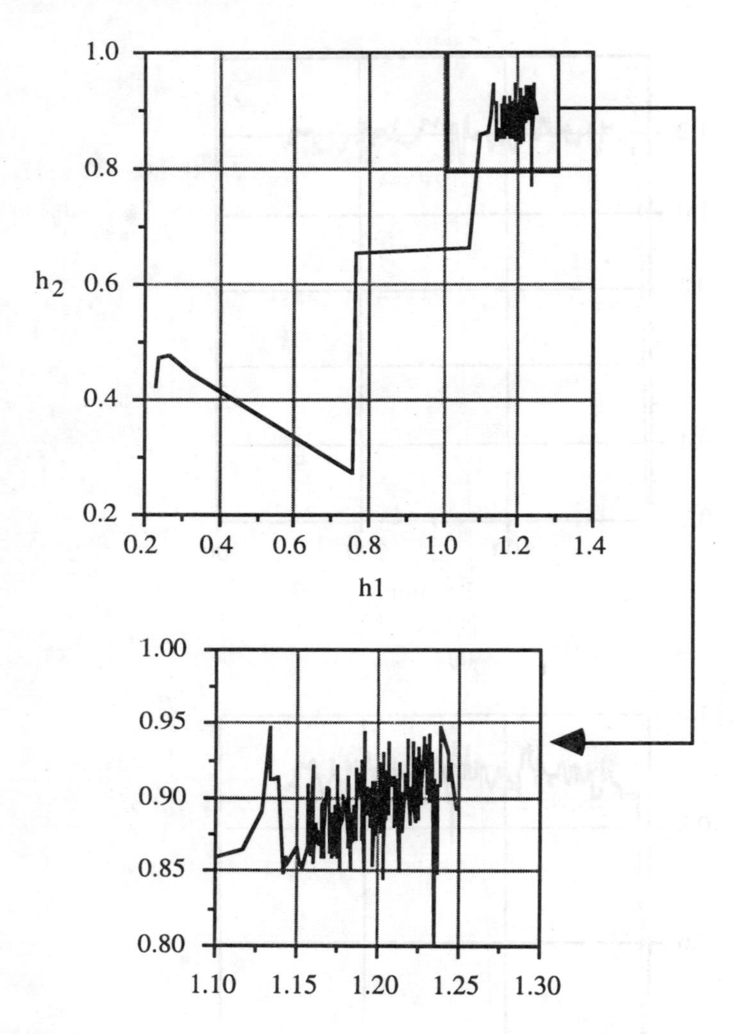

Figure 6.7: A plot of the estimated h_1 versus h_2. Notice that the estimates of h_1 and h_2 converge to their correct values, namely, 1.2 & 0.9. While there is a rapid descent to the bottom of the bowl, it may be observed that in the neighborhood of the minimum there is a kind of random 'walk' motion as is illustrated in the inset .

Variations of LMS Algorithm: The LMS algorithm is computationally simple and robust. It finds applications in many real-time filtering tasks. Among the drawbacks, the convergence of the LMS algorithm is slow when the eigenvalue spread of the input signal is large and the misadjustment error is large unless the step size is made very small, much smaller than the upper bound set by the inverse of the maximum eigenvalue. When the input signal is non-stationary the problem of selecting the correct step size becomes more acute as the maximum eigenvalue may itself be varying with time. The step size will need to be selected for the worst case, which may result in a reduced over all covergence speed. A partial solution to this problem is to make the step size time varying as below:

$$\mathbf{h}_{k+1} = \mathbf{h}_k + \frac{2\mu}{\alpha + \mathbf{x}_k^T \mathbf{x}_k} e_k \mathbf{x}_k$$

where α is a small positive constant. A large increase in the signal power will reduce the step size so that the change in the weight vector is kept small. The LMS algorithm with this modification is known as normalized LMS or nLMS algorithm.

Another variation of LMS algorithm relates to updating of the weight vector after receiving a block of samples. The instantaneous gradient is now averaged over a block length, N_b; thus a better estimate of the statistical gradient is obtained. The weight vector is now given by

$$\mathbf{h}_{k+N_b} = \mathbf{h}_k + \frac{2\mu}{N_b} \sum_{m=k}^{k+N_b} e_m \mathbf{x}_m$$

The LMS algorithm with this modification is known as block LMS or bLMS algorithm.

§6.3 Recursive Least-square (RLS) Algorithm

In LMS algorithm the instantaneous gradient was used as an approximation of the statistical gradient obtained through expected operation. The RLS approach is purely deterministic, involving minimization of the sum of the squares of the errors.

Least-squares Solution: As in §6.2 we define data vectors $\mathbf{x}_n = [x_n, x_{n-1}, \ldots, x_{n-M+1}]^T$, n=0,1,...N-1, filter vector $\mathbf{h} = [h_0, h_1, \ldots, h_{M-1}]^T$ and a reference signal d_n, n=0,1,...N-1. We assume that x_n=0 for n<0. The difference between the reference signal and the filter output that at time instant n is given by

$$e_n = d_n - y_n = d_n - \mathbf{x}_n^T \mathbf{h}, \quad n = 0,1,\ldots N-1 \tag{6.27}$$

Let us express (6.27) in a matrix form:

$$\underbrace{\begin{bmatrix} e_0 \\ e_1 \\ \cdot \\ \cdot \\ \cdot \\ e_{N-1} \end{bmatrix}}_{\mathbf{e}_{N-1}} = \underbrace{\begin{bmatrix} d_0 \\ d_1 \\ \cdot \\ \cdot \\ \cdot \\ d_{N-1} \end{bmatrix}}_{\mathbf{d}_{N-1}} - \underbrace{\begin{bmatrix} \mathbf{x}_0^T \\ \mathbf{x}_1^T \\ \cdot \\ \cdot \\ \cdot \\ \mathbf{x}_{N-1}^T \end{bmatrix}}_{\mathbf{X}_{N-1}} \underbrace{\begin{bmatrix} h_0 \\ h_1 \\ \cdot \\ \cdot \\ \cdot \\ h_{M-1} \end{bmatrix}}_{\mathbf{h}_{M-1}} \tag{6.28a}$$

$$\mathbf{e}_{N-1} = \mathbf{d}_{N-1} - \mathbf{X}_{N-1}\mathbf{h}_{N-1} \tag{6.28b}$$

Note that $\mathbf{x}_0 = [x_0, 0, \ldots, 0]^T$. We like to minimize $\mathbf{e}_{N-1}^T \mathbf{e}_{N-1}$, that is, the sum of the squares of the errors.

$$\begin{aligned}\mathbf{e}_{N-1}^T \mathbf{e}_{N-1} &= (\mathbf{d}_{N-1} - \mathbf{X}_{N-1}\mathbf{h}_{N-1})^T (\mathbf{d}_{N-1} - \mathbf{X}_{N-1}\mathbf{h}_{N-1}) \\ &= \mathbf{d}_{N-1}^T \mathbf{d}_{N-1} - \mathbf{h}_{N-1}^T \mathbf{X}_{N-1}^T \mathbf{d}_{N-1} - \mathbf{d}_{N-1}^T \mathbf{X}_{N-1}\mathbf{h}_{N-1} + \mathbf{h}_{N-1}^T \mathbf{X}_{N-1}^T \mathbf{X}_{N-1}\mathbf{h}_{N-}\end{aligned} \tag{6.29}$$

Differentiating (6.29) with respect $\mathbf{h}_{N-1}$ and setting the derivative to zero we obtain an equation

$$\mathbf{X}_{N-1}^T \mathbf{X}_{N-1}\mathbf{h}_{N-1} = \mathbf{X}_{N-1}^T \mathbf{d}_{N-1} \tag{6.30a}$$

The least-squares solution is given by

$$\boxed{\mathbf{h}_{N-1} = \left(\mathbf{X}_{N-1}^T \mathbf{X}_{N-1}\right)^{-1} \mathbf{X}_{N-1}^T \mathbf{d}_{N-1}} \tag{6.30b}$$

Note that $\mathbf{h}_{N-1}$ is the least-squares solution for N input data points. When a new data arrives, we define a new data matrix incorporating the current data. Then, we can obtain a new up-to-date filter vector, $\mathbf{h}_N$. The process may be carried on as long as desired or till there is no significant change in the filter vector. This approach will require forming a matrix product $\mathbf{X}_{N-1}^T \mathbf{X}_{N-1}$ and inverting it at each step. This indeed is very inefficient and time consuming. We will now consider a recursive approach where the past filter vector is used to get the current filter.

Recursive Least Square: We shall show how to invert $\mathbf{X}_N^T \mathbf{X}_N$ recursively, that is, given the inverse of $\mathbf{X}_{N-1}^T \mathbf{X}_{N-1}$, we like to obtain the inverse of $\mathbf{X}_N^T \mathbf{X}_N$ without actually inverting it. Note that we can write

$$\mathbf{X}_N = \begin{bmatrix} \mathbf{X}_{N-1} \\ \mathbf{x}_N^T \end{bmatrix}$$

Therefore, $\mathbf{X}_N^T \mathbf{X}_N = \mathbf{X}_{N-1}^T \mathbf{X}_{N-1} + \mathbf{x}_N \mathbf{x}_N^T$. We shall use the following matrix identity (Woodbury's identity [25]):

$$(\mathbf{A} + \mathbf{BCD})^{-1} = \mathbf{A}^{-1} - \mathbf{A}^{-1}\mathbf{B}(\mathbf{C}^{-1} + \mathbf{DA}^{-1}\mathbf{B})^{-1}\mathbf{DA}^{-1}$$

where $\mathbf{A}$ is nonsingular matrix and $\mathbf{B}$ $\mathbf{C}$ and $\mathbf{D}$ are matrices of compatible dimensions. In our case $\mathbf{A} = \mathbf{X}_{N-1}^T \mathbf{X}_{N-1}$, $\mathbf{B} = \mathbf{x}_N$, $\mathbf{C} = 1$ and $\mathbf{D} = \mathbf{x}_N^T$. We obtain a recursive relation

$$
\begin{aligned}
\left(\mathbf{X}_N^T\mathbf{X}_N\right)^{-1} &= (\mathbf{X}_{N-1}^T\mathbf{X}_{N-1} + \mathbf{x}_N\mathbf{x}_N^T)^{-1} \\
&= \left(\mathbf{X}_{N-1}^T\mathbf{X}_{N-1}\right)^{-1} \\
&-\left(\mathbf{X}_{N-1}^T\mathbf{X}_{N-1}\right)^{-1}\mathbf{x}_N\left[1+\mathbf{x}_N^T\left(\mathbf{X}_{N-1}^T\mathbf{X}_{N-1}\right)^{-1}\mathbf{x}_N\right]^{-1}\mathbf{x}_N^T\left(\mathbf{X}_{N-1}^T\mathbf{X}_{N-1}\right)^{-1}
\end{aligned}
\tag{6.31}
$$

Using (6.31) in (6.30b) we obtain after some straightforward simplification the following recursive relation between $\mathbf{h}_N$ and $\mathbf{h}_{N-1}$.

$$
\begin{aligned}
\mathbf{h}_N &= \left(\mathbf{X}_N^T\mathbf{X}_N\right)^{-1}\mathbf{X}_N^T\mathbf{d}_N \\
&= \begin{bmatrix} \left(\mathbf{X}_{N-1}^T\mathbf{X}_{N-1}\right)^{-1} \\ -\left(\mathbf{X}_{N-1}^T\mathbf{X}_{N-1}\right)^{-1}\mathbf{x}_N(1+\mathbf{x}_N^T\left(\mathbf{X}_{N-1}^T\mathbf{X}_{N-1}\right)^{-1}\mathbf{x}_N)^{-1}\mathbf{x}_N^T \\ \left(\mathbf{X}_{N-1}^T\mathbf{X}_{N-1}\right)^{-1} \end{bmatrix} \left[\mathbf{X}_{N-1}^T \ \mathbf{x}_N\right]\begin{bmatrix}\mathbf{d}_{N-1} \\ d_N\end{bmatrix} \\
&= \mathbf{h}_{N-1} - \left(\mathbf{X}_{N-1}^T\mathbf{X}_{N-1}\right)^{-1}\mathbf{x}_N(1+\mathbf{x}_N^T\left(\mathbf{X}_{N-1}^T\mathbf{X}_{N-1}\right)^{-1}\mathbf{x}_N)^{-1}\mathbf{x}_N^T\mathbf{h}_{N-1} \\
&+\left(\mathbf{X}_N^T\mathbf{X}_N\right)^{-1}\mathbf{x}_N d_N
\end{aligned}
\tag{6.32}
$$

We can show the following result by post multiplying (6.31) by $\mathbf{x}_N$ and simplifying the resulting expression:

$$
\begin{aligned}
&\left(\mathbf{X}_N^T\mathbf{X}_N\right)^{-1}\mathbf{x}_N \\
&= \left(\mathbf{X}_{N-1}^T\mathbf{X}_{N-1}\right)^{-1}\mathbf{x}_N\left[1-(1+\mathbf{x}_N^T\left(\mathbf{X}_{N-1}^T\mathbf{X}_{N-1}\right)^{-1}\mathbf{x}_N)^{-1}\mathbf{x}_N^T\left(\mathbf{X}_{N-1}^T\mathbf{X}_{N-1}\right)^{-1}\mathbf{x}_N\right] \\
&= \left(\mathbf{X}_{N-1}^T\mathbf{X}_{N-1}\right)^{-1}\mathbf{x}_N\left[\frac{1}{(1+\mathbf{x}_N^T\left(\mathbf{X}_{N-1}^T\mathbf{X}_{N-1}\right)^{-1}\mathbf{x}_N)}\right] \\
&= \mathbf{K}_N
\end{aligned}
\tag{6.33}
$$

Using the above result in (6.32) we obtain a compact recursive relation for the weight vector

$$
\boxed{\begin{aligned}
\mathbf{h}_N &= \mathbf{h}_{N-1} - \mathbf{K}_N\mathbf{x}_N^T\mathbf{h}_{N-1} + \mathbf{K}_N d_N \\
&= \mathbf{h}_{N-1} + \mathbf{K}_N\left(d_N - \mathbf{x}_N^T\mathbf{h}_{N-1}\right)
\end{aligned}}
\tag{6.34}
$$

$\mathbf{K}_N$ is often known as Kalman gain (vector). We have two recursive equations, (6.31) and (6.34) which constitute the RLS algorithm:

$$\mathbf{K}_N = \left(\mathbf{X}_{N-1}^T\mathbf{X}_{N-1}\right)^{-1}\mathbf{x}_N\left[\frac{1}{(1+\mathbf{x}_N^T\left(\mathbf{X}_{N-1}^T\mathbf{X}_{N-1}\right)^{-1}\mathbf{x}_N)}\right]$$

$$\mathbf{h}_N = \mathbf{h}_{N-1} + \mathbf{K}_N\left(d_N - \mathbf{x}_N^T\mathbf{h}_{N-1}\right)$$

$$\left(\mathbf{X}_N^T\mathbf{X}_N\right)^{-1} = \left(\mathbf{X}_{N-1}^T\mathbf{X}_{N-1}\right)^{-1} - \left(\mathbf{X}_{N-1}^T\mathbf{X}_{N-1}\right)^{-1}$$
$$\mathbf{x}_N\left[1+\mathbf{x}_N^T\left(\mathbf{X}_{N-1}^T\mathbf{X}_{N-1}\right)^{-1}\mathbf{x}_N\right]^{-1}\mathbf{x}_N^T\left(\mathbf{X}_{N-1}^T\mathbf{X}_{N-1}\right)^{-1}$$

The starting values for (6.31) and (6.34) may be chosen as $\left(\mathbf{X}_0^T\mathbf{X}_0\right)^{-1} = \sigma\mathbf{I}$ and $\mathbf{h}_0 = 0$, respectively where σ is a large constant.

As for the computational complexity, it takes O(M^2) operations (one multiplication + one addition) per step to obtain the Kalman gain vector. The weight vector update takes just about 2M operations. Thus, most of the computational complexity lies in the evaluation of Kalman gain vector. There exists fast RLS algorithm where the Kalman gain vector requires O(2M) operations rather than O(M^2) operations. Compare this with LMS algorithm which requires only O(2M) per step.

Exponential Window: In a situation where the signal characteristics are changing fast with time as in non-stationary signal model; for example, in speech signal the weight vector should be controlled more by recent signal and less by distant past signal. This can be achieved by introducing an exponentially weighted sum of the squares of the errors. Define a diagonal matrix, $\Lambda_{N-1} = diag\left[\lambda^{N-1},\lambda^{N-2},\ldots,\lambda^0\right]$ where λ $(0<\lambda<1)$ is known as a forgetting factor. The exponentially weighted sum of the squares of the error is now given by $\varepsilon_{N-1}^T\Lambda_{N-1}\varepsilon_{N-1}$. By minimizing this quantity we obtain, similar to (6.30a),

$$\mathbf{X}_{N-1}^T\Lambda_{N-1}\mathbf{X}_{N-1}\mathbf{h}_{N-1} = \mathbf{X}_{N-1}^T\Lambda_{N-1}\mathbf{d}_{N-1} \tag{6.35a}$$

and solving for $\mathbf{h}_{N-1}$ we obtain

$$\mathbf{h}_{N-1} = \left(\mathbf{X}_{N-1}^T\Lambda_{N-1}\mathbf{X}_{N-1}\right)^{-1}\mathbf{X}_{N-1}^T\Lambda_{N-1}\mathbf{d}_{N-1} \tag{6.35b}$$

A recursive solution of (6.35a) is possible in the same manner as that for (6.30). For this note that

$$\mathbf{X}_N^T\Lambda_N\mathbf{X}_N = \lambda\mathbf{X}_{N-1}^T\Lambda_{N-1}\mathbf{X}_{N-1} + \mathbf{x}_N\mathbf{x}_N^T$$
$$= \lambda[\mathbf{X}_{N-1}^T\Lambda_{N-1}\mathbf{X}_{N-1} + \frac{1}{\lambda}\mathbf{x}_N\mathbf{x}_N^T]$$

A recursive relation similar to (6.31) connecting $\left(\mathbf{X}_N^T \Lambda_N \mathbf{X}_N\right)^{-1}$ and $\left(\mathbf{X}_{N-1}^T \Lambda_{N-1} \mathbf{X}_{N-1}\right)^{-1}$ may be derived by using the Woodbury's identity [25]. The result is

$$\left(\mathbf{X}_N^T \Lambda_N \mathbf{X}_N\right)^{-1} = \frac{1}{\lambda}(\mathbf{X}_{N-1}^T \Lambda_{N-1} \mathbf{X}_{N-1} + \frac{1}{\lambda}\mathbf{x}_N \mathbf{x}_N^T)^{-1}$$

$$= \frac{\begin{bmatrix} \left(\mathbf{X}_{N-1}^T \Lambda_{N-1} \mathbf{X}_{N-1}\right)^{-1} - \left(\mathbf{X}_{N-1}^T \Lambda_{N-1} \mathbf{X}_{N-1}\right)^{-1} \\ \mathbf{x}_N \left[\lambda + \mathbf{x}_N^T \left(\mathbf{X}_{N-1}^T \Lambda_{N-1} \mathbf{X}_{N-1}\right)^{-1} \mathbf{x}_N\right]^{-1} \mathbf{x}_N^T \left(\mathbf{X}_{N-1}^T \Lambda_{N-1} \mathbf{X}_{N-1}\right)^{-1} \end{bmatrix}}{\lambda} \quad (6.36)$$

The Kalman gain vector is now given by

$$\mathbf{K}_N = \left(\mathbf{X}_{N-1}^T \Lambda_{N-1} \mathbf{X}_{N-1}\right)^{-1} \mathbf{x}_N \left[\frac{1}{(\lambda + \mathbf{x}_N^T \left(\mathbf{X}_{N-1}^T \Lambda_{N-1} \mathbf{X}_{N-1}\right)^{-1} \mathbf{x}_N)}\right] \quad (6.37)$$

The recursive relation for the weight vector however remains same as (6.34). The forgetting factor controls the effective length of the window beyond which the signal has very little effect on the current weight vector. The effective length of the window is approximately equal to $\frac{1}{1-\lambda}$ (see Exercise # 4), for example, for λ=0.95, the effective length is 20 samples. All input signal samples beyond twenty (with respect to the current input) will have a very little effect on the weight vector. Any change in the signal, within duration of less than twenty samples, cannot be tracked by the filter

.

Misadjustment: We had earlier introduced misadjustment as a ratio of the excess mean square error to the minimum mean square error. The misadjustment ratio Π is given by (6.25), which is reproduced here for convenience

$$\Pi = \frac{tr\left\{\mathbf{c}_x E\left\{\Delta\mathbf{h}_k \Delta\mathbf{h}_k^T\right\}\right\}}{\xi_{\min}}$$

It is shown in [24)] that for large N

$$E\left\{\Delta\mathbf{h}_k \Delta\mathbf{h}_k^T\right\} \approx \frac{\sigma_{\min}^2}{N}\mathbf{c}_x^{-1} \quad (6.38)$$

Using (6.38) in (6.25) we obtain

$$\Pi = \frac{tr\left\{\mathbf{c}_x \frac{\xi_{\min}}{N}\mathbf{c}_x^{-1}\right\}}{\xi_{\min}} = \frac{tr\{\mathbf{I}\}}{N} = \frac{M}{N} \quad (6.39)$$

It is interesting to observe that, unlike in the LMS algorithm, the misadjustment ratio can be made as small as desired by taking sufficiently large data for a finite filter order. For this reason the RLS algorithm is capable of giving superior results compared to the LMS algorithm but at the cost of higher computational load. This is demonstrated in a numerical example included here.

Example 6.2: We re-consider the system identification problem in Example 6.1. A linear time invariant (LTI) system is driven by white noise. The output is corrupted by another independent white noise source. The LTI is modeled as a FIR filter. The linear system is a moving average system corrupted by independent noise as given by

$$x(n) = \varepsilon(n) + 1.2\varepsilon(n-1) + 0.9\varepsilon(n-2) + 0.1\eta(n)$$

The unknown system is modeled as a FIR filter with three coefficients h_0, h_1, h_2. The RLS algorithm was initialized with $\mathbf{P}$=100$\mathbf{I}$, $\mathbf{h}_0 = \mathbf{0}$ and the input is zero prior to the starting of the algorithm. The input signal is a unit variance white Gaussian noise and the output of the system output is corrupted with white Gaussian noise of variance 0.1. The input is passed through the FIR filter model and output is compared with the output of the unknown system. The error is then used to update the weight vector (6.34). The inverse of the covariance matrix is updated according to (6.31).
The results, that is, the evolution of the weight vector with increasing number of iterations, are shown in fig. 6.8. Compare these with the results shown for the LMS algorithm in figures 6.5 and 6.6. Clearly, RLS algorithm provides both rapid and smooth convergence to the correct values of the unknown parameters ■.

Finally, we compare and contrast the LMS and RLS algorithms as in table 6.1:

Table 6.1: A comparison of LMS and RLS algorithms.

LMS Algorithm	**RLS Algorithm**
Stochastic model	Deterministic model
Minimum of mean square error is sought through descent along instantaneous gradient.	Sum of squares of the error is minimized
Convergence is slow and noisy	Convergence is fast and smooth
Computationally very fast	Computationally slow.
Misadjustments error is independent of data length	Misadjustments error decreases with increasing data length

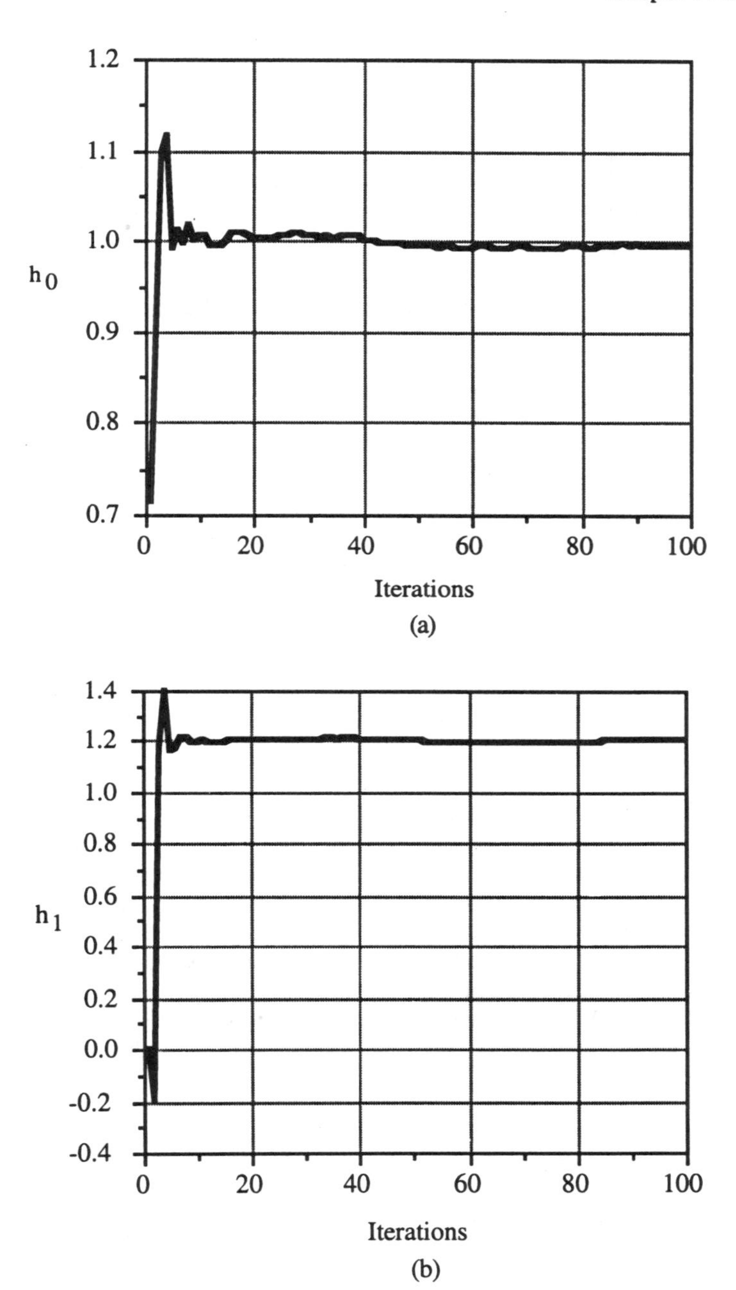
1.2
1.1
1.0
0.9
0.8
0.7
h0
0
20
40
60
80
100
Iterations
(a)
1.4
1.2
1.0
0.8
0.6
0.4
0.2
0.0
-0.2
-0.4
h1
0
20
40
60
80
100
Iterations
(b)

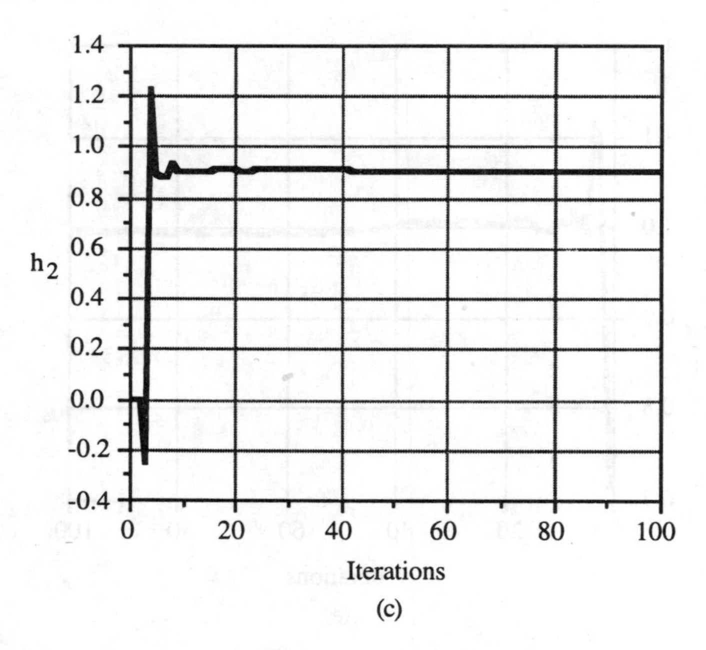

Figure 6.8: FIR filter coefficients as a function of number of iterations. The coefficients converge smoothly to the correct value (within an error <1%) after about 40 iterations.

§6.4 Model Based Least-squares

In Wiener filter or its recursive equivalent (RLS) we need to have a reference signal with respect to which the error is calculated. In practical applications, the reference signal may be a pilot signal, which is often used in communication for channel estimation or a transmitted signal as in radar/sonar. In this section we show how to use the signal model, in particular, its state space representation (Chapter One), for the purpose of error calculation. An update equation similar to (6.34) is obtained. What follows here is a highly simplified version of the celebrated Kalman filter whose full exposition is beyond the scope of this introductory text.

First order AR Signal: The first order AR signal or its state space representation is given by

$$x(n) = ax(n-1) + \varepsilon(n) \tag{6.40a}$$

and it is further assumed that what is measured is a scaled version of *x(n)* plus noise

$$y(n) = cx(n) + g(n) \tag{6.40b}$$

Given *y(n)* and the signal model (including parameters a and c) we like to design a recursive estimate of *x* of the following form:

$$\hat{x}(n) = b(n)\hat{x}(n-1) + k(n)y(n) \tag{6.41}$$

where *b(n)* and *k(n)* are time varying parameters. The mean square error of $\hat{x}(n)$ will be minimized with respect *b(n)* and *k(n)*,

$$\begin{aligned}\xi &= E\{(\hat{x}(n) - x(n))^2\} \\ &= E\{(b(n)\hat{x}(n-1) + k(n)y(n) - x(n))^2\}\end{aligned} \tag{6.42}$$

We differentiate ξ with respect to *b(n)* and *k(n)* and set the derivatives to zero.

$$\begin{aligned}\frac{\partial \xi}{\partial b(n)} &= 2E\{(b(n)\hat{x}(n-1) + k(n)y(n) - x(n))\hat{x}(n-1)\} = 0 \\ \frac{\partial \xi}{\partial k(n)} &= 2E\{(b(n)\hat{x}(n-1) + k(n)y(n) - x(n))y(n\} = 0\end{aligned} \tag{6.43}$$

The first equation in (6.43) provides a relation between *b(n)* and *k(n)*. After some involved simplification we obtain the following result [24],

$$b(n) = a(1 - ck(n)) \tag{6.44}$$

and from the second equation in (6.43) we obtain

$$E\{e(n)y(n)\} = 0 \tag{6.45}$$

where

$$\begin{aligned}e(n) &= \hat{x}(n) - x(n) \\ &= (b(n)\hat{x}(n-1) + k(n)y(n) - x(n))\end{aligned}$$

Recursive Estimator: Using (6.44) in (6.41) we obtain the following recursive estimator or scalar Kalman filter.

$$\hat{x}(n) = a\hat{x}(n-1) + k(n)[y(n) - ac\hat{x}(n-1)] \tag{6.46}$$

The first term in (6.46), $a\hat{x}(n-1)$, is the prediction of the current sample from immediate past sample and the second term, $k(n)[y(n) - ac\hat{x}(n-1)]$, is an adjustment term based on the current error modified by $k(n)$, which is often known as Kalman gain. A block diagram of the recursive estimator is shown in fig. 6.9.

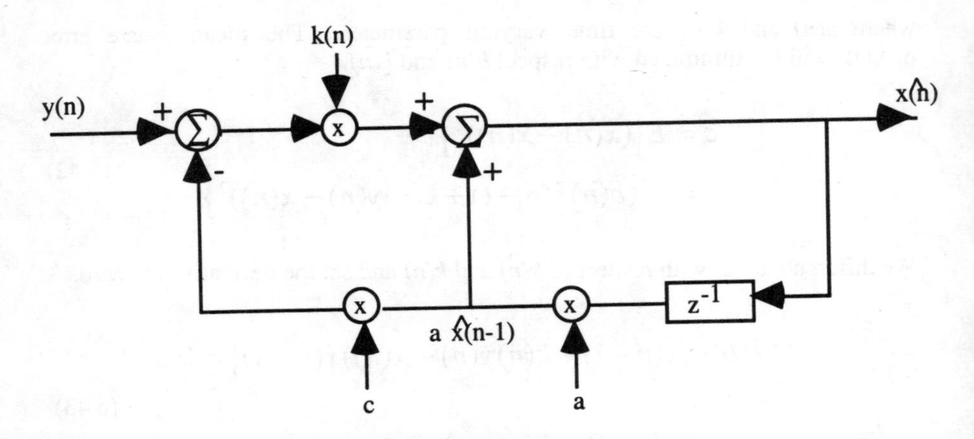

Figure 6.9: A block diagram of the first order AR model based least-squares estimator. The model parameters a and c are assumed to be known. The Kalman gain, k(n), is estimated from (6.53b).

We will now proceed to estimate the Kaman gain. For this we turn our attention to (6.42).

$$\begin{aligned}\xi &= E\{e(n)(b(n)\hat{x}(n-1)+k(n)y(n)-x(n))\} \\ &= b(n)E\{e(n)\hat{x}(n-1)\}+k(n)E\{e(n)y(n)\}-E\{e(n)x(n)\}\end{aligned} \tag{6.47}$$

By virtue of the fact that error is orthogonal to optimum estimator and (6.45), the first two terms in (6.47) vanish, giving

$$\xi = -E\{e(n)x(n)\} \tag{6.48}$$

Substituting (6.40b) in (6.48) and using (6.45) we get

$$\xi = -\frac{1}{c}E\{e(n)(y-g(n))\} = \frac{1}{c}E\{e(n)g(n)\} \tag{6.49}$$

We substitute for *e(n)* in (6.49) and obtain

$$\begin{aligned}\xi &= \frac{1}{c}E\{(b(n)\hat{x}(n-1)+k(n)y(n)-x(n))g(n)\} \\ &= \frac{k(n)}{c}E\{y(n)g(n)\} = \frac{k(n)}{c}E\{g(n)g(n)\} = \frac{k(n)}{c}\sigma_g^2\end{aligned} \tag{6.50}$$

Next we like to obtain a recursive relation connecting $\xi(n)$ to $\xi(n-1)$. Use (6.46) in (6.42) to yield

$$\xi(n) = E\left\{(a\hat{x}(n-1) + k(n)[y(n) - ac\hat{x}(n-1)] - x(n))^2\right\} \tag{6.51}$$

In (6.51) we substitute for *y(n)* and *x(n)* using (6.40) and we obtain after some simplification

$$\begin{aligned}\xi(n) &= E\left\{(a(1-ck(n))e(n-1) - (1-ck(n))\varepsilon(n)) + k(n)g(n))^2\right\} \\ &= (a^2(1-ck(n))^2\xi(n-1) + (1-ck(n))^2\sigma_\varepsilon^2 + k^2(n)\sigma_g^2\end{aligned} \tag{6.52}$$

Using (6.50) in (6.52) we obtain

$$\frac{k(n)}{c}\sigma_g^2 = [a^2\xi(n-1) + \sigma_\varepsilon^2](1-ck(n))^2 + k^2(n)\sigma_g^2$$

which may be arranged in the form of a quadratic equation

$$(c^3Q + c\sigma_g^2)k^2(n) - (2c^2Q + \sigma_g^2)k(n) + cQ = 0 \tag{6.53a}$$

where

$$Q = a^2\xi(n-1) + \sigma_\varepsilon^2$$

The solution of the above quadratic equation is given by

$$k(n) = \frac{cQ}{(c^2Q + \sigma_g^2)} = \frac{c(a^2\xi(n-1) + \sigma_\varepsilon^2)}{c^2(a^2\xi(n-1) + \sigma_\varepsilon^2) + \sigma_g^2} \tag{6.53b}$$

Finally, the model based least-squares algorithm consists of following set of equations:

$$k(n) = \frac{c(a^2\xi(n-1) + \sigma_\varepsilon^2)}{c^2(a^2\xi(n-1) + \sigma_\varepsilon^2) + \sigma_g^2}$$

$$\hat{x}(n) = a\hat{x}(n-1) + k(n)[y(n) - ac\hat{x}(n-1)]$$

$$\xi(n) = \frac{k(n)}{c}\sigma_g^2$$

The first equation provides an estimate of k(n) given $\xi(n-1)$. The second equation is the actual optimum estimator where a and c are assumed to be known. The last equation provides an estimate of ξ (n). The initial conditions are $\xi(0)=1$ and $\hat{x}(0)=0$. Note that when there is no measurement noise, that is, $\sigma_g^2=0$ the Kalman gain equals $1/c$ and the error power, $\xi(n)=0$. Since the measured signal is an exact replica of the actual signal the estimated signal is naturally error free.

Example 6.3: In this example we consider a first order AR signal, $x(n)=0.9x(n-1)+\varepsilon(n)$ where ε(n) is a zero mean unit variance Gaussian white noise. The observed signal is $y(n)=x(n)+2.0g(n)$ where $g(n)$ is a zero mean unit variance Gaussian noise independent of ε(n). The initial conditions are $\xi(0)=1$ and
$\hat{x}(0)=0$. The model based least-squares algorithm, consisting of a set of three equations shown in the box above, was programmed in Matlab. The results are illustrated in fig. 6.10. In fig. 6.10a the estimated or filtered signal is shown by solid curve and the actual signal is shown by dashed curve. The observed signal with noise (SNR ≈ 1.0) is shown in fig 6.10b. The Kalman gain and the error power converged to steady values in less than five iterations.

It may be easily verified that when the measurements are error free the estimated signal is exactly equal to the actual signal, as expected

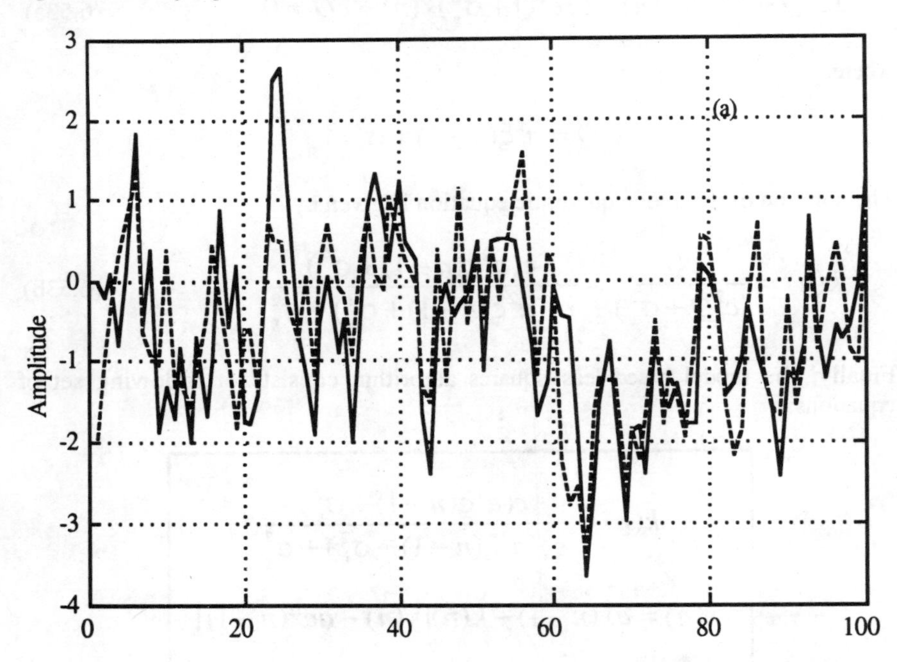

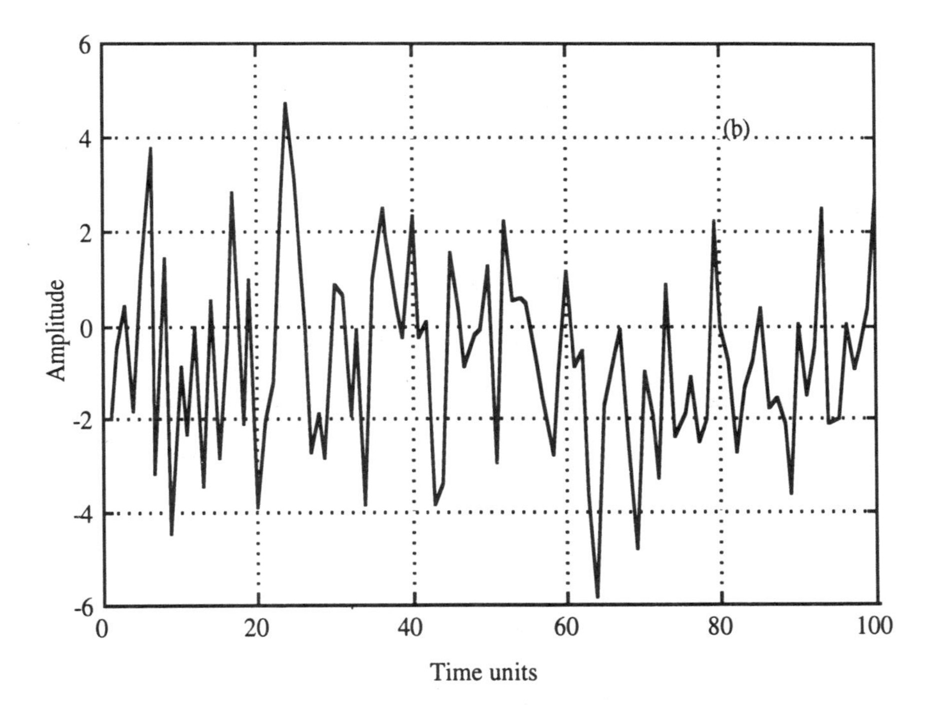

Figure 6.10: An example of application of the model based least-squares method on synthetic data. (a) Estimated signal (solid curve) and the actual signal (dashed curve) (b) measured signal with noise (SNR=1).

§6.5 Adaptive IIR Filters

In the first two sections we have exclusively dealt with the finite impulse response (FIR) filters. Both LMS and RLS algorithms are meant for FIR filters. Adaptive infinite impulse response (IIR) filters may be introduced along the lines of adaptive FIR filter, namely, using the concepts of the steepest descent. There are, however, two major difficulties. Firstly, the error surface is no longer unimodal, that is, there are more than one extremum. Hence, it is possible that one may never be able to reach the global minimum unless one has chosen the right initial values. The second difficulty arises out of the possibility of one or more poles of the IIR filter transfer function wander outside the unit circle during the course of adaptation, which leads to instability. The fundamental problem is in the presence of a feedback path in IIR filters (see fig. 6.11). The IIR filter may be expressed as a difference equation (4.3) from Chapter Four,

$$y(n) = \sum_{k=1}^{p-1} a_k y(n-k) + \sum_{i=0}^{q-1} b_i x(n-i)$$

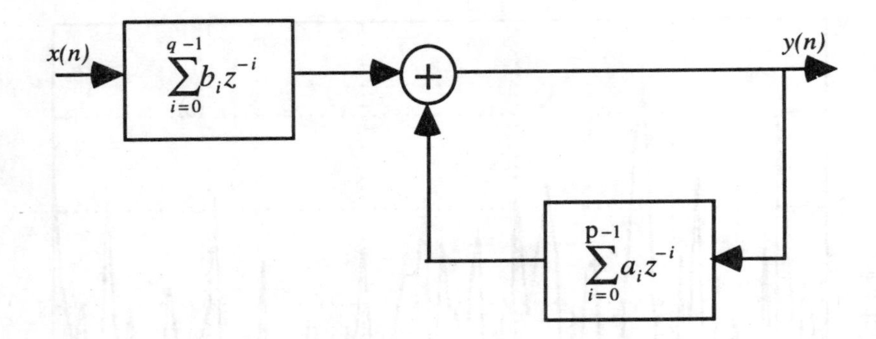

Figure 6.11: Recursive IIR filter consisting of feed forward part and feedback part. There are two sets of coefficients, a's and b's controlling the transfer function of the filter just as in ARMA model of signal referred to in Chapter One.

where *x(n)* is the input signal and *y(n)* is the output signal. The first term on the right hand side of (4.3) represents the feedback component. For a particular input *x(n)*, by adjusting the coefficients *a*'s and *b*'s, the output *y(n)* is made as close as possible to a desired response in the mean square sense. Define a coefficient vector,

$$\theta = \left[a_1, a_2, \ldots, a_{p-1}, b_0, b_1, \ldots, b_{q-1}\right]^T$$

a p+q-1 vector and a data vector,

$$\mathbf{y}_n = \left[y_{n-1}, \ldots, y_{n-p+1}, x_n, \ldots, x_{n-q+1}\right]^T,$$

also a p+q-1 vector. Eq. (4.3) may be expressed as $y_n = \theta^T \mathbf{y}_n = \mathbf{y}_n{}^T\theta$. Let d(n) be the reference signal with which the filter output is compared. The mean square error is now given by

$$\xi = E\{(d_n - \theta^T \mathbf{y}_n)^2\}$$

The most common iterative technique is based on gradient search of the mmse performance surface. The updating equation is given by

$$\theta_{n+1} = \theta_n - \mu \nabla_\theta \xi_n \tag{6.54}$$

where μ is a scalar constant and $\nabla_\theta \xi_n$ is the gradient of ξ with respect to parameters θ. The gradient may be obtained as follows:

$$\nabla_\theta \xi_n(\theta) = -2E\{(d_n - y_n)\nabla_\theta y_n\} \tag{6.55}$$

where $\nabla_\theta y_n$ is gradient of y_n with respect to the parameters. Each component of the gradient is given by

$$\frac{\partial y_n}{\partial a_k} = y(n-k) + \sum_{i=1}^{p-1} a_i \frac{\partial y(n-i)}{\partial a_k}, \quad k = 1,2,\ldots,p-1$$
$$\frac{\partial y_n}{\partial b_k} = x(n-k) + \sum_{i=1}^{p-1} a_i \frac{\partial y(n-i)}{\partial b_k}, \quad k = 0,1,\ldots,q-1 \tag{6.56a}$$

$$\nabla_\theta y_n = \mathbf{y}_n + \sum_{i=1}^{p-1} a_i \nabla_\theta y_{n-i} \tag{6.56b}$$

Using (6.55) in (6.54) we obtain the following update equation

$$\theta_{n+1} = \theta_n + 2\mu E\{(d_n - y_n)\nabla_\theta y_n\} \tag{6.57}$$

As in LMS algorithm we replace the statistical gradient by the instantaneous gradient

$$\theta_{n+1} = \theta_n + 2\mu(d_n - y_n)\nabla_\theta y_n \tag{6.58}$$

The LMS algorithm for IIR filter consists of equations (6.56b) and (6.58) which are shown inside the box below:

$$\nabla_\theta y_n = \mathbf{y}_n + \sum_{i=1}^{p-1} a_i \nabla_\theta y_{n-i}$$

$$\theta_{n+1} = \theta_n + 2\mu(d_n - y_n)\nabla_\theta y_n$$

Note that if we drop the summation term on the right hand side of (6.56b) we indeed obtain the LMS algorithm for FIR filter. Dropping the summation term on the right hand side of (6.56b) implies that the gradient is a linear function of $\mathbf{y}_n$, that is, the error surface is unimodal. But without this approximation the gradient becomes a non-linear function of $\mathbf{y}_n$. Then, the error surface may become multimodal. Starting from some point on the error surface and going down the gradient it is possible that we may get into one of the local minima and never reach the global minimum. In order to avoid such a possibility the adaptive procedure may be started from a number of points and look for the global minimum, though this is an expensive proposition.

Example 6.4: Consider an IIR filter with following coefficients:

a1=-1.2
a2=-0.81
b0=1.0
b1=0.6

and driven by white noise. A schematic of the filter model is shown in fig. 6.12.

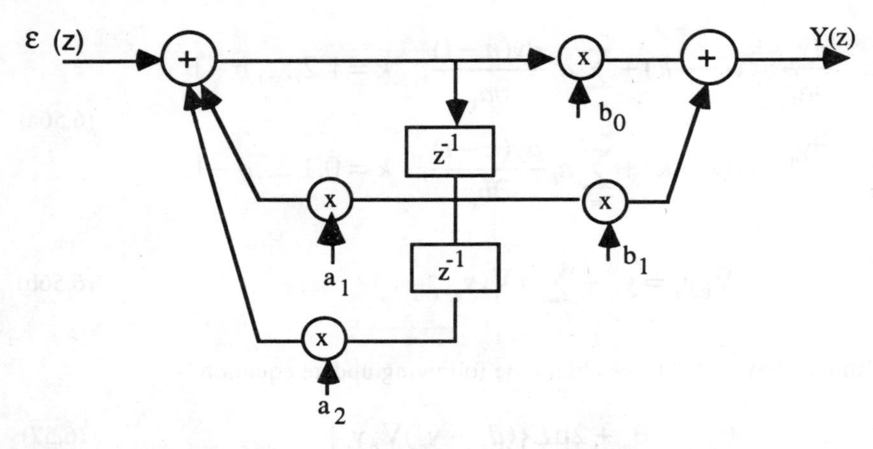

Figure 6.12: A schematic diagram of IIR filter considered in Example (6.4).

Both input and output are known and our aim is to estimate the model, that is, the filter coefficients. We have set μ=0.002 and the initial values of the unknown gradient and the filter coefficients in (6.56b) and (6.57) were also set to zero.

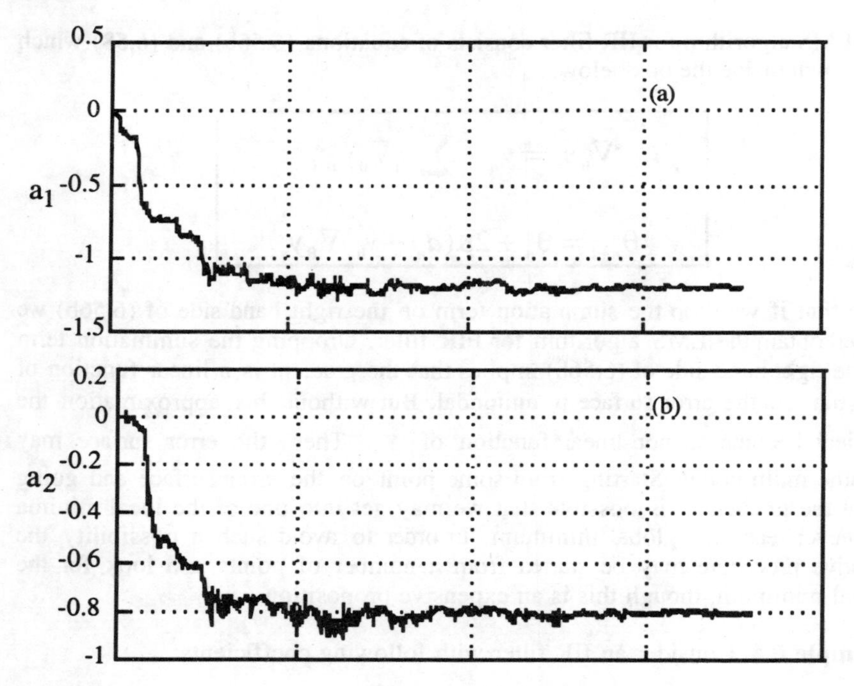

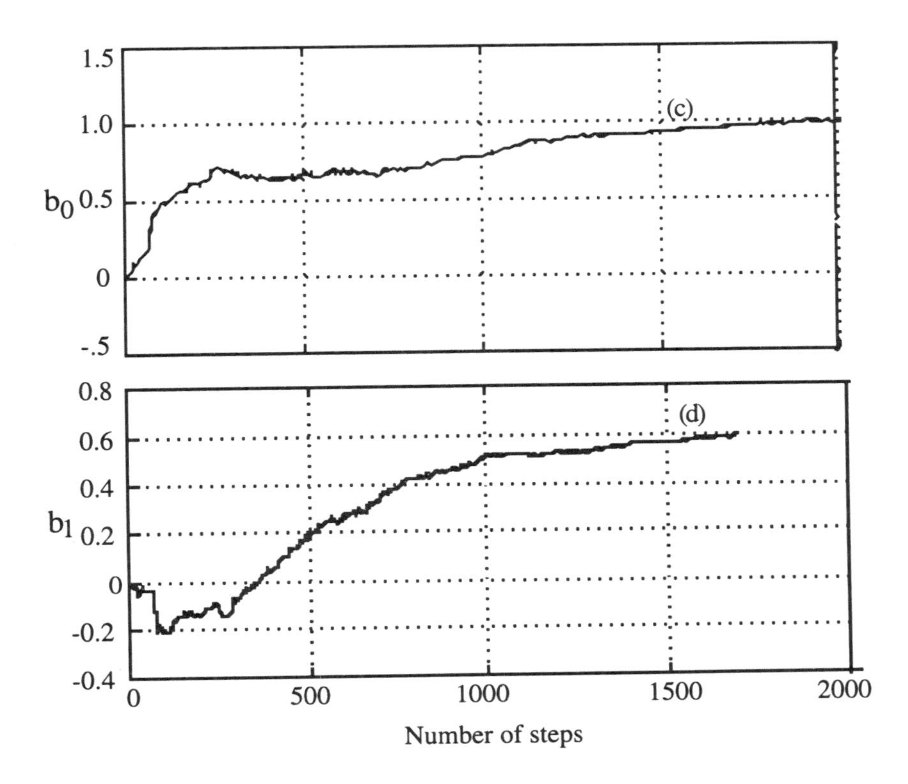

Figure 6.13: The filter coefficients of IIR filter as a function of number of iterations.(a) a_1=-1.2, (b) a_2=-0.81, (c) b_0=1.0, (d) b_1=0.6.

The evolution of the filter coefficients with the number of iterations is shown in fig. 6.13. It required about 2000 iterations before the estimated coefficients came close to the actual values. Compare these results with those in Example 6.1 where we had considered a FIR filter, which required only a few hundred steps to converge to the actual values .

§6.6 Echo Cancellation

A signal is often reflected or scattered at impedance discontinuities. The reflected signal is a scaled down and delayed replica of the incident signal. The direct and reflected signal will reach a sensor or an antenna at different time instants and possibly from different directions. A mathematical model of a signal with echo in its simplest form is given by

$$y(n) = x(n) + \alpha x(n - \tau) \tag{6.59}$$

where α ($|\alpha| < 1$) is the reflection coefficient and τ is the propagation delay with respect to the direct arrival. A schematic of echo signal model is shown in fig. 6.14. In the language of signal processing (6.59) is a FIR filter equation where $x(n)$ is an input to a FIR filter with coefficients $[1,0,0,...,\alpha]$ (τ -1 zeros) and y(n) is the output.

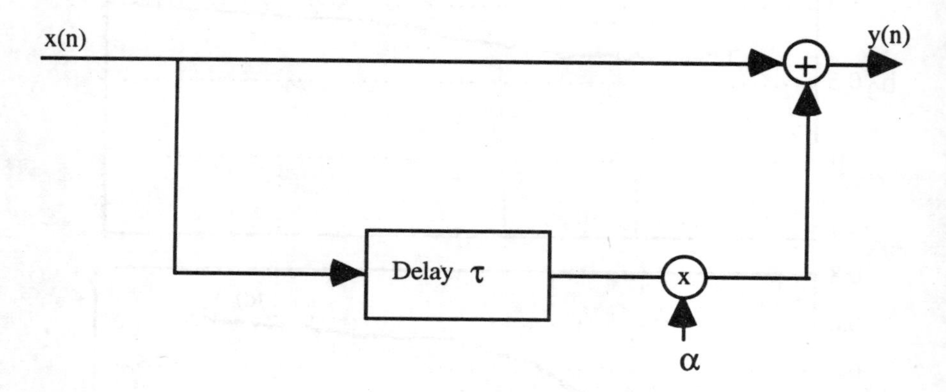

Figure 6.14: A schematic of echo signal model. α is reflection coefficient and τ is the propagation delay.

It is possible that there are more than one reflected paths bringing in additional terms on the right hand side of (6.59) but we shall not consider such a multipath propagation environment. Let us compute the z-transform of $y(n)$,

$$\begin{aligned} Y(z) &= X(z) + \alpha z^{-\tau} X(z) \\ &= (1 + \alpha z^{-\tau}) X(z) \end{aligned} \tag{6.60}$$

The purpose of echo cancellation is to remove the presence of the reflected component so that what is left behind is the incident signal, $x(n)$. This may be achieved by passing $y(n)$ through a filter whose response is given by

$$\boxed{H(z) = \frac{1}{(1 + \alpha z^{-\tau})}} \tag{6.61a}$$

which may be expressed as an infinite response (IIR) filter

$$H(z) = 1 - \alpha z^{-\tau} + \alpha^2 z^{-2\tau} - \alpha^3 z^{-3\tau} + ... \tag{6.61b}$$

For FIR implementation the infinite polynomial in (6.61b) will have to be truncated to M terms. Use of such a truncated filter on $y(n)$ will result in a residual echo in the output of the order α^M or in dB scale, 20Mlog(α). The time span of echo cancellation filter will be (M-1)τ seconds or in terms of tap count the filter length will be (M-1)τf_s where f_s is sampling rate. Let us place a bound on the residual echo power as -40dB. Then, it is easily shown that $M = \frac{-2}{\log \alpha}$. For α=0.8, M turns out to be >20. A FIR filter for echo cancellation often consists of several tens to hundreds of filter coefficients. To design an echo cancellation filter we need to estimate both α and τ from the observed signal. The simplest method is, given the transmitted signal $x(n)$, we correlate the observed signal (containing an echo) with the transmitted one. The peak at zero lag corresponds to the direct signal and the peak appearing after some delay corresponds to the reflected signal. The relative peak height and the position of the peak are the estimates of α and τ, respectively. Alternatively it is also possible to devise an LMS based method for estimation FIR filter model (Example 6.1). We like to demonstrate that the LMS approach can be used to estimate the reflection coefficient and the delay.

Example 6.5: Let the transmitted signal be a pseudo-random sequence of 512 points. The antenna receives the direct signal and a reflected signal. The coefficient of reflection is assumed to be 0.8 and the delay with respect to the direct signal is 4 sample units.

$$x(n) = \varepsilon(n) + 0.8\varepsilon(n-4) + 0.1\eta(n)$$

where $\eta(n)$ is unit variance measurement noise. We shall treat the received signal with echo as an output of a FIR filter with coefficients [1, 0, 0, 0, 0.8] driven by a psuedo- random sequence. We employ LMS algorithm to estimate the FIR coefficients as outlined in Example 6.1. The filter coefficients after 512 iterations are as listed below:

h=[0.9932 -0.0104 -0.0093 -0.0046 0.7949]

The evolution of some of the filter coefficients as adaptation proceeded is shown in fig. 6.15. The step size, μ was set to 0.01. From these figure we can conclude that at least 200 observations are required to obtain reasonably good estimates of the filter coefficients.

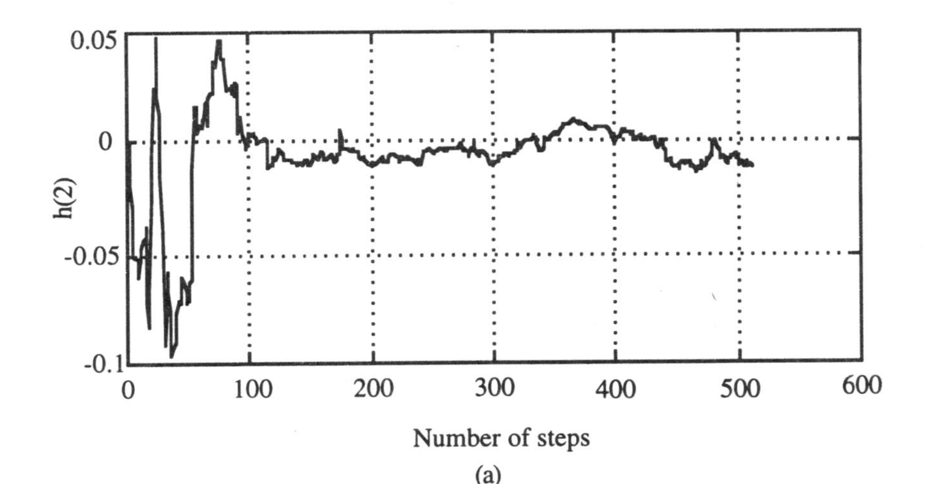

(a)

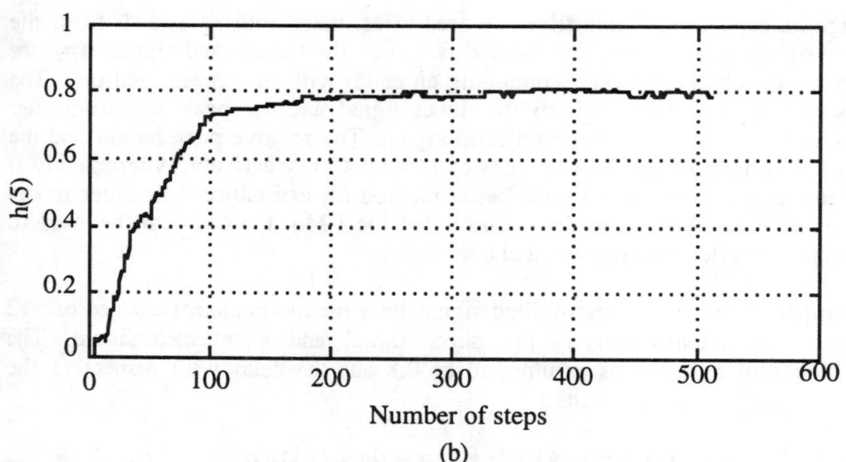

Figure 6.15: Evolution of FIR filter coefficients as a number of steps. A signal with an echo having a reflection coefficient α=0.8 and a delay τ=4.0 sample units. The evolution of other coefficients follows the above pattern.

After estimation of the coefficients of the FIR filter representing the echo model we like to construct an inverse filter given by (6.61a) but such a filter would be at least 100 points long which would be quite impractical. It is possible to overcome this difficulty by using a simple trick. In equation (6.59) we interchange the role of *x(n)* and *y(n)*. Let *y(n)* be known signal and *x(n)* is unknown. In other words, the signal with echo is a known sequence while the transmitted signal itself is unknown. Eq. (6.59) may now be looked upon as a system of linear equations. The solution of these equations is straightforward without requiring any matrix inversion. It is given by

$$\begin{aligned}
x(1) &= y(1) \\
x(2) &= y(2) \\
&\ldots \\
x(\tau) &= y(\tau) \\
x(\tau+1) &= y(\tau+1) - \alpha x(1) \\
x(\tau+2) &= y(\tau+2) - \alpha x(2) \\
&\ldots
\end{aligned} \qquad (6.62)$$

Note that the estimates *x(1), x(2),...,x(* τ*)* will be generally erroneous as we had to assume that *x(n)*=0 for $0 \geq n \geq -\tau$. The error in the initial values will affect the solution for some distance but soon the error will die down.

Example 6.6: We shall consider the echo model dealt with earlier in Example 6.5. The FIR filter coefficients estimated in that example were now used in the solution of linear equations (6.62). The small magnitude coefficients (second, third, and fourth) were rounded off to zero. The direct signal is a unit rectangular pulse of width 64 time samples and the reflected signal, also a rectangular pulse, arrives four sample units later with a reduced magnitude of 0.8. The sum of the direct and reflected signals is shown in fig. 6.16a. This is *y(n)* sequence in

(6.62). By solving the system of linear equations, (6.62), the transmitted signal has been recovered as shown in fig. 6.16b. Perfect recovery has been possible as we had a noise free output signal and the initial values were also error free.

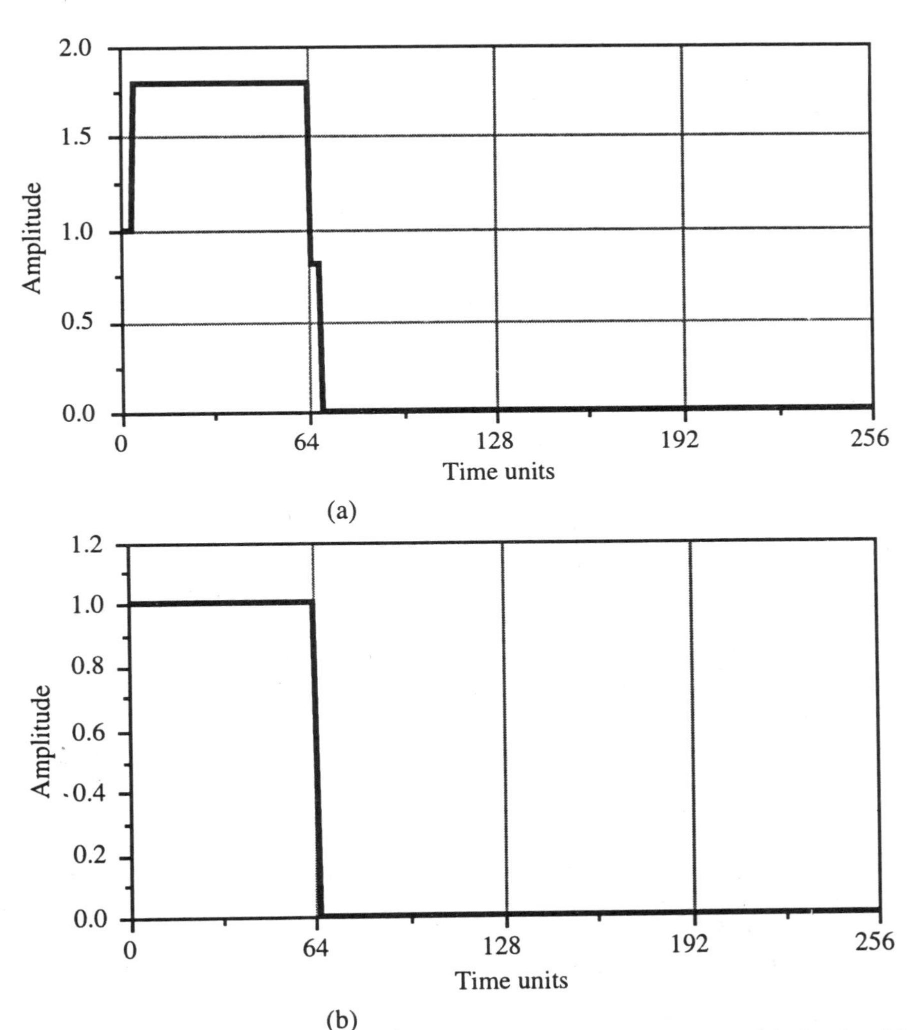

Figure 6.16: The sum of the direct and reflected signals is shown. (a) signal with echo (b) after echo cancellation .

§6.7 Practical Echo Problems

There are many practical problems where the presence of an echo can seriously affect the final result. Here we shall briefly describe a few practical situations involving the echoes. The first practical example is that of radar and sonar. A transmitted signal is returned not only by the target of interest but also by other objects, which are of no interest. The signal model is similar to that described by equation (6.59), perhaps with more than one echo. A classical example of multiple echoes is from sonar detection in shallow waters. There is sometimes a coupling between a source and a receiver, for example, between a microphone and a loudspeaker as in hands free telephone in a car. This produces an echo via a direct path from the loudspeaker to the microphone. In telephone networks an echo is generated due to an impedance mismatch between two wire and four wire circuits. In this section we like to elaborate on the basic process of echo generation in above-mentioned physical environments.

Shallow Water: There are at least two echoes resulting from reflections at water/air interface (reflection coefficient =-1) and at the sea bottom. This is schematically shown in fig. 6.17. When the separation between the source and receiver is large, the bottom, on account of total internal reflection, acts as a highly reflecting mirror like surface. As a result, multiple reflections of each ray is possible. The receiver would then be receiving many more than just three rays as depicted in fig. 6.17. In the simplest model it is possible to assume that the effect of propagation is simply to delay and scale down the transmitted signal. The receiver output may be expressed as a FIR filter output

$$y(n) = x(n) - 1.0x(n - \tau_1) + r\,x(n - \tau_2)$$

where τ_1 is the delay of the surface reflected signal with respect to the direct signal, τ_2 is the delay of the bottom reflected signal and r is the bottom reflection coefficient. The zeros of the filter response of the FIR filter may lie outside the unit circle. Then, the recovery of the transmitted signal from the received signal using the method of solving a system of linear equations described earlier (page 300) becomes unstable. Fortunately, other effects such as attenuation through water medium and scattering at the boundaries will further reduce the echo strength. When this happens, the zeros of the FIR filter may move inside the unit circle. We shall next illustrate such a situation.

Example 6.7: Consider a 50-meter deep horizontal shallow water channel. Assume that a source at a depth of 15 meters below the water surface transmits binary bits (no carrier). A receiver, also at a depth of 15 meters, is placed at a distance of 1000 meters from the source. The surface reflection coefficient is -0.7 and that at the bottom is 0.3. We like to demonstrate that we can recover the transmitted bits after canceling the echoes. We assume that there are three paths as shown in fig.6.17. The sound speed in water is 1500 m/s. The propagation delays are:

666.96 ms: Path via surface
666.66 ms: Direct path
668.30 ms: Path via bottom.

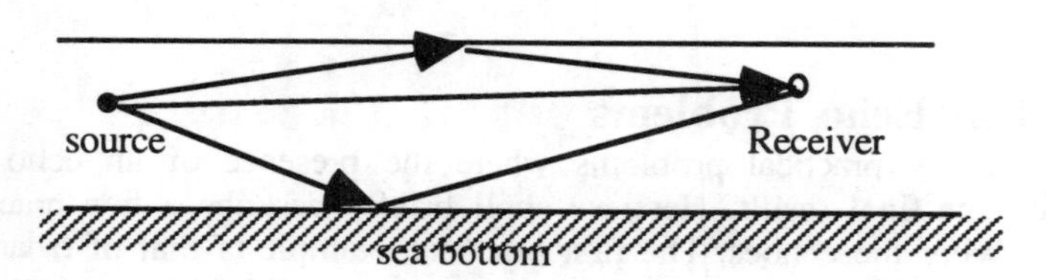

Figure 6.17: Sonar detection in shallow water. Apart from the direct path there are at least two echoes signal reflected from the water/air interface and sea bottom.

The source is assumed to transmit random binary bits at the rate of 666 bit/sec, each binary bit is 1.5 ms wide. A sample of the transmitted signal is shown in fig.6.18a. The signal was sampled at a rate of 10 samples per bit or 6660 samples/sec. The received signal at the receiver is shown in fig. 6.18b. The spikes appearing in the received signal are due to the interference between the direct and the reflected signals. By simple examination of the received signal we cannot extract the transmitted bits. We will now show how by canceling the echoes we can recover the original transmitted signal. This is known as channel equalization. The first task is to estimate the channel parameters. For this both input (pilot signal) and output are assumed to be known. We use LMS algorithm to estimate the parameters. We needed 128 bits with step size μ=0.02 to arrive at a convergent solution. The estimated channel parameters are as below:

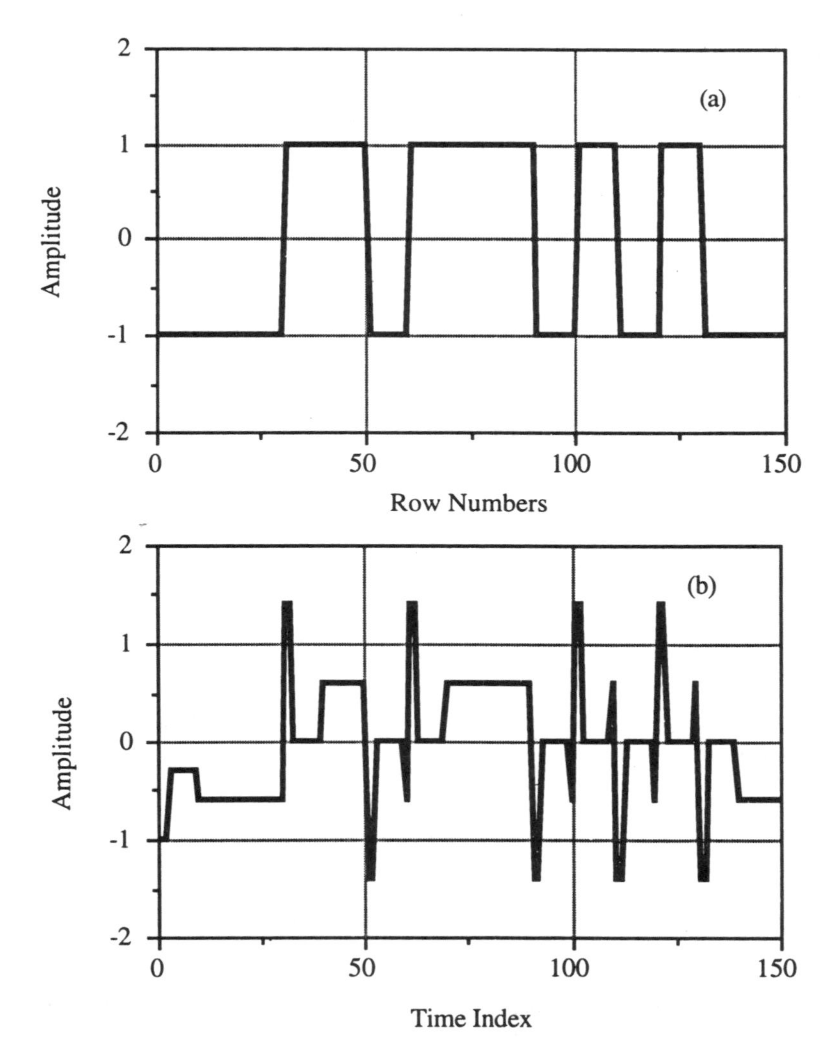

Figure 6.18: A sample of the transmitted signal is shown in (a) and the corresponding received signal is shown in (b). The spikes appearing in the received signal are due to interference between the direct and reflected signals.

Estimated Channel: {1.0006, -0.0045, -0.6925, -0.0060, 0.0025, 0.0003, 0.0006, -0.0040, 0.0069, 0.2961}
Assumed Channel: [1, 0, -0.7, 0, 0, 0, 0, 0, 0, 0.3] (For this selection the FIR filter has all its zeros are inside the unit circle). The recovered signal is shown in fig 6.19.

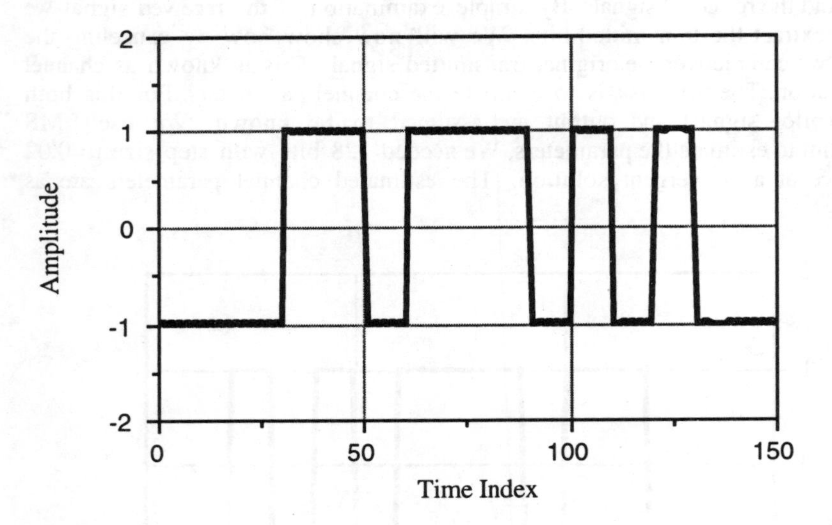

Figure 6.19: The transmitted signal has been correctly recovered.

Source-Receiver Coupling: A coupling between a microphone and loudspeaker may occur as for example in a public address system, or in a hands free telephone in a car, or a personal computer used in teleconferencing, etc. In all these examples a microphone (receiver) is close to the loudspeaker (transmitter) and therefore some of the acoustic radiation from the loudspeaker reaches the microphone directly (see fig. 6.20).

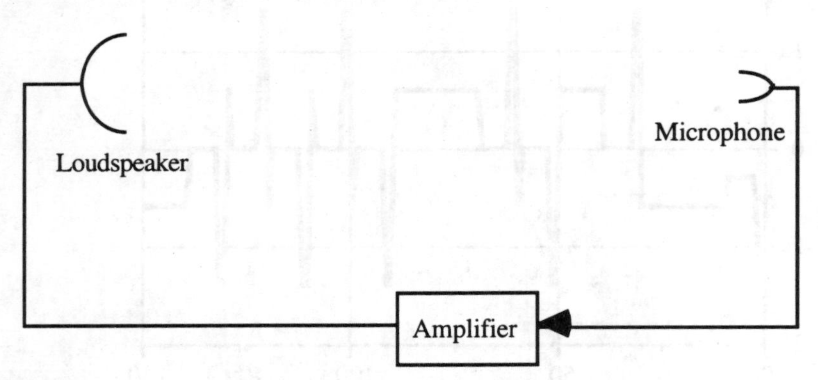

Figure 6.20: A simple example of coupling between a microphone and loudspeaker. The signal from the microphone is fed to the loudspeaker after amplification. A part of the acoustic energy radiated by the loudspeaker is picked up by the microphone and is fed back to the loudspeaker. The process continues until a stable situation is reached.

This is the echo that gets added to the speech signal at the microphone. The delay is equal to the travel time from the loudspeaker to the microphone. The echo is fed back to the loudspeaker, whose output will now be a sum of the original signal plus echo. The feedback process goes on indefinitely. The resulting output of the loudspeaker may be modeled as

$$y(t) = x(t) + r_1 x(t-\tau_1) + r_1^2 x(t-2\tau_1) + r_1^3 x(t-3\tau_1) + \ldots \quad (6.63)$$

where $x(t)$ is the speech signal, τ_1 is travel time delay and r_1 is the net gain in the loop, from the loudspeaker to the microphone and back to the loudspeaker. Taking z transform on both sides of (6.63) we obtain,

$$\begin{aligned} Y(z) &= (1 + r_1 z^{-\tau_1} + r_1^2 z^{-2\tau_1} + r_1^3 z^{-3\tau_1} + \ldots) X(z) \\ &= \frac{1}{1 - r_1 z^{-\tau_1}} X(z) \end{aligned} \quad (6.64)$$

In order that pole of the IIR filter given in (6.64) be always inside the unit circle for any delay, we must have $|r| < 1$, that is, the net gain must be less than one. In time domain (6.64) may be written as

$$y(t) = x(t) + r_1 y(t-\tau_1) \quad (6.65)$$

Given $x(t)$ (a pilot signal) and $y(t)$, the LMS algorithm may be used to estimate the channel parameters, τ_1 and r_1 (see Example 6.5).

Echo in Telephone Circuits: Two telephone users, A and B (see fig. 6.21) are connected through local exchanges. The connection from the user to the local exchange is through a two-wire circuit in local loop while the exchanges are connected through a four-wire circuit in Public Switched Telephone Network (PSTN). The device that connects a two-wire circuit to a four-wire circuit is known as Hybrid. Because of impedance mismatch in the hybrid between two pairs of circuits, a part of the received signal will leak through the Hybrid to the outpost and form an echo. The mechanism of echo generation is illustrated in fig. 6.21. The speech signal from speaker A travels over the upper arm of the four-wire circuit and at the far end of the four-wire circuit the signal goes through the hybrid to speaker B. At this point, because of impedance mismatch, an echo is generated, which travels towards the speaker A. The echo reaches the speaker A after a delay equal to the total propagation time. Long propagation delays in a telephone circuit can become perceptible to the speaker causing a great annoyance. The speaker A hears his own voice after some delay. A similar phenomenon takes place when the speaker B is active. A highly simplified model of an echo is that of a scaled and delayed replica of the transmitted signal. But in real life situation, there may be a large number of minor echoes cluttered around one major echo. Indeed, we can think of an impulse response function to describe the process of complex echo formation. For the purpose of present study, however, we shall use the simplest possible model. In terms of impulse response function the model is described by a single delta function, $r_A\delta(t-\tau_A)$ where r_A is the over all echo strength and τ_A is the total propagation delay.

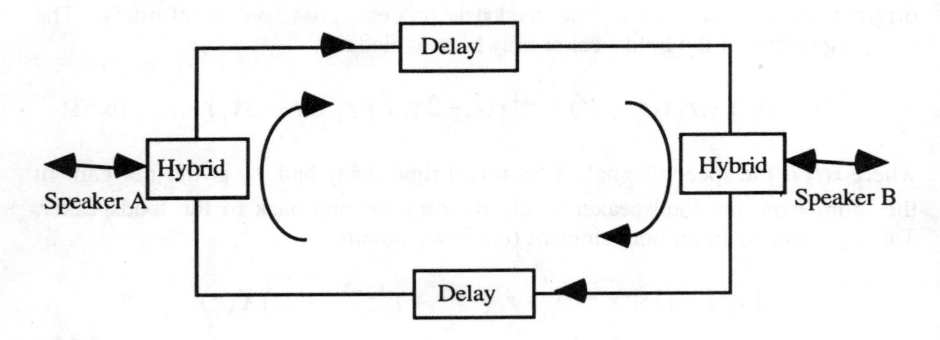

Figure 6.21: Echo generation in a long distance telephone circuits is illustrated. An echo returns after a delay which is equal to the total return propagation time.

At the far end, the returned signal may be expressed as

$$y(n) = r_A x_A(n-\tau_A) + \eta(n)$$

s(6.66)

where $x_A(n)$ is the speech signals produced by speaker A and $\eta(n)$ is noise in the circuit. Note that at the near end signal $x_A(n)$ is the known and $y(n)$ is the observed signal. The signal from the speaker A in the upper arm of the circuit is used to predict the echo in the lower arm. This is then subtracted from the returned signal removing most of the echo signal. An adaptive algorithm like LMS may be used to achieve the task of channel estimation and echo cancellation (see fig. 6.22).

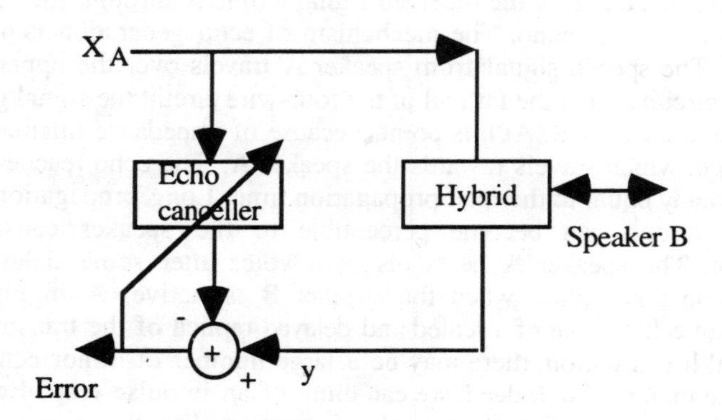

Figure 6.22: Echo canceller. The signal from the speaker A in the upper arm of the circuit is used to predict the echo in the lower arm. This is then subtracted from the returned signal removing most of the echo signal.

§6.8 Exercises

Problem:

1. Use the result that $tr\{\mathbf{c}_x\} = \sum_{m=1}^{M} \lambda_m$ where $\lambda_m, m = 1,\ldots,M$ are eigenvalues of a covariance matrix in eq(6.13) and show that the following holds

$$1/tr\{\mathbf{c}_x\} > \mu > 0$$

Since in practice the trace of a matrix is more easily computed than the maximum eigenvalue, the above result is useful.

2. Let σ_x^2 be the power of the input signal to an adaptive linear combiner. What is the range of the step size, μ? (Hint: Note that $tr\{\mathbf{c}_x\} = M\sigma_x^2$. Use the result from exercise #1).

3. In equation (6.2) let the weight vector be equal to the Wiener filter (optimum weight vector),

$$\varepsilon_n = d_n - \mathbf{h}_W^T \mathbf{x}$$

Show that ε_n is uncorrelated with the input signal, that is, $\{\varepsilon_n x_n\} = 0$. This is often known as principle of orthogonality.

4. The decay constant is numerically equal to the number of steps required to let the error decay to $\frac{1}{e}$(=)th fraction of the initial value. It is related to the step size μ and eigenvalue as in eq.(6.14). Let the eigenvalues be [10, 8, 6.5, 2.8] and μ=0.02. Compute the maximum, minimum and average value of the decay constant.

5. What is the role of forgetting factor?. We like to compute the effective length of an exponential window, $w(n) = \lambda^n, \quad 0 < \lambda < 1, \quad n = 0,1,\ldots.$ The effective length is defined as the length of a uniform window whose area is equal to the area under the exponential window. Compute the area under an exponential window defined by a forgetting factor equal to 0.99 and compute the effective length of the exponential window.

6. In eq.(6.33) let us assume that $\left(\mathbf{X}_{N-1}^T \mathbf{X}_{N-1}\right)^{-1} = \mathbf{I}$ for all n. The weight update equation (6.34), after introducing the above assumption, reduces to the normalized LMS (nLMS) algorithm.

7. Here is a model of two path echo system,

$$y(n) = x(n) + 0.85x(n-3) - 0.4x(n-8)$$

Obtain a FIR filter model for above echo system.

8. The quadratic equation (6.53a) has two possible solutions. One of these solutions is given by (6.53b). Show that the second solution is simply $\frac{1}{c}$. What is the significance of this solution?

9. In normalized LMS (nLMS) what is the range of the step size, μ. Derive the result (Ans: $1 > \mu > 0$).

10. Show that when the reference signal is scaled up or down, the weight vector in the LMS algorithm is also scaled by the same factor. (Hint: Start with the Wiener equation (6.5) for the weight vector).

11. In equations (6.40b) and (6.53b) let c be a very large constant. Show that the Kalman gain is approximately $1/c$ and the mean square prediction error is proportional to $1/c^2$.

12. In adaptive weight adjustment to minimize the mean square error, the rate at which the weight vector approaches the true weight vector depends upon the eigenvalues of the signal covariance matrix. Let the eigenvalues be [64, 24, 16, 8]. What is the maximum step size? Compute the decay constants for different terms or modes. What is the lower bound on the decay constant?

13.What is the minimum sum of the squares of the difference between the filtered and reference signals as defined in eq. (6.29)?

14.Compute the maximum misadjustment error after a large number of iterations for the signal covariance function considered in problem #12.

15. Echo strength is 0.75 times the strength of the incident signal and the delay is five sample intervals (in seconds). The FIR implementation of the echo cancellation filter has 20 terms. What is the residual echo power and the time span of the filter.

16. Let $y(n) = x(n) + 0.8x(n-3)$ where *x(n)* is the transmitted signal and 0.8x(n-3) is the reflected signal. The observed signal, *y*, is given by [0,0,1,1,1,1.8,0.8,0.8,0.8,0,0]. Compute, using eq. (6.62), the transmitted signal.

17.In Example 6.7 let the relative delays of the signal coming via the water/air interface be two sampling interval and the signal coming from the bottom be five sampling intervals. Show that some of the zeros of the FIR filter of the channel lie outside the unit circle.

18. In Exercise #17 show that the FIR filter becomes a minimum phase filter when the surface reflection coefficient is made greater than -1.0, keeping the bottom reflection coefficient sufficiently low, of the order of 0.3.

Computer Projects (CP):

1. We consider a problem of predicting the current value from the past observations. Assume the signal is a second order AR process (Chapter One)

$$x(n) + a_1 x(n-1) + a_2 x(n-2) = \varepsilon(n)$$

where $\varepsilon(n)$ is a zero mean unit variance white noise. The aim is to estimate a_1 and a_2 given $x(n)$, n=0,1,..., N-1 using LMS algorithm. Let $\mathbf{h} = (1, -h_1, -h_2)$ be a weight vector. A tapped delay model of 2nd order AR process is shown in fig. 6.23.

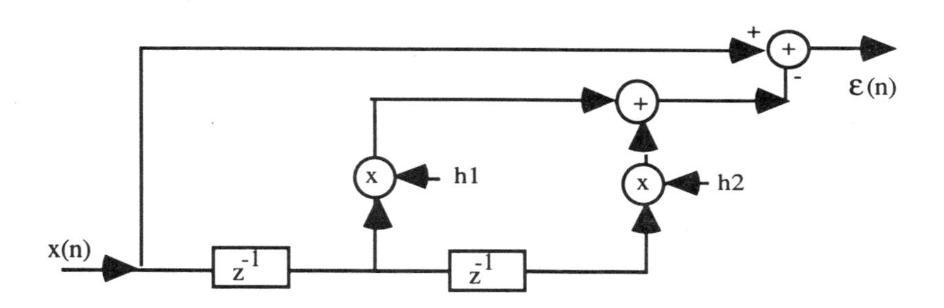

Figure 6.23: A tapped delay model of AR process.

x(n) may be generated given the AR model coefficients (e.g. a_1=1.2 and a_2=0.9). Note that the reference signal is *x(n)* and the input signals are x(n-1) and x(n-2). Write and verify a Matlab program for above prediction problem. Find an optimum value of μ and the required data length to reach the convergence.(Ans: μ=0.0005 and N>1000). It turns out that this is an example of very slow convergence as the eigenvalue spread of the input signal covariance matrix is large ($\cong 50$).

2.Develop RLS algorithm for the prediction problem described in CP#1.In particular compare the prediction error and coefficient estimates for the same data length with those obtained in CP#1.

3. The LMS algorithm for IIR filter differs from that for FIR filter in so far as the computation of gradient term is concerned. Indeed in eq. (6.56b) if we drop the summation term on the right hand side, the algorithm reduces to a standard LMS algorithm. To see the effect of this approximation we go back to Example 6.4. Write a Matlab program to implement the LMS algorithm for IIR filters and next introduce the approximation stated above. Note the effect of the approximation on the evolution of the filter coefficient estimates.

4. Let the transmitted signal be pure pseudo-random sequence of length 512 points. An echo with a relative amplitude of 0.8 and a delay of 5 sample points is present in the observed signal (see Example 6.5). Write a Matlab program to

compute a cross correlation between the transmitted signal and the observed signal (a sum of direct and reflected signal). From the cross correlation function estimate α and τ required for the design of echo cancellation filter.

5. Repeat the Example 6.5 with two reflected paths,

$$x(n) = \varepsilon(n) + 0.8\varepsilon(n-5) + 0.7\varepsilon(n-8) + 0.1\eta(n)$$

Use the LMS algorithm (M-file) developed in CP#4 after suitable modifications to estimate the coefficients of a new FIR filter for above two-echo problem. What is the minimum data length required for good estimate.

6. We often encounter a situation where a sinusoidal signal is present in the background of noise. The aim is to estimate the frequency of the sinusoid. We exploit the property that the sinusoidal signal is fully predictable while the background noise is uncorrelated. A linear filter is sought to predict the signal (sinusoid plus noise) one step ahead. Write a Matlab program to implement the LMS algorithm to compute the desired prediction filter and the frequency from the roots of the filter coefficient polynomial. Also plot a spectrum, that is, the magnitude square of the z-transform of the filter.

7. In a telephone network an echo is formed at the hybrid on account of impedance mismatch. The simplest model of an echo is a scale and delayed replica of the transmitted signal (see eq. (6.66)). Let the transmitted signal $x_A(n)$ be a known sinusoid and the observed signal *y(n)* be a sum of unknown echo and circuit noise. Let w(p), p=1,2,..., P be the desired weights, where the maximum propagation delay is P times the sampling interval. Write a Matlab code to implement the LMS algorithm for cancellation of echo in the observed signal (See Example 6.5 for related illustration).

8. Repeat the calculations in Example 6.6 with a pulse width equal to 8 and 16 time units. The other parameters remain unchanged.

7 Beamformation

Beamforming is a basic signal processing tool used in many applications such as radar, sonar, seismic exploration, earthquake monitoring and detection, radio astronomy, wireless communication and so on. In all above applications an array of sensors is used to receive wave energy from a given direction or transmit wave energy to a given direction. This is essentially the purpose of beamforming. The process of beamformation involves delaying the signal to be transmitted or the signal to be received by a specified amount and then summing up all receiver outputs. The sharpness of .the beam depends upon the sensor distribution and the number of sensors. In two dimensional (2D) space a uniformly distributed linear array is commonly used while in three dimensional (3D) space planar array (sensors on a square grid) is used. However, a circular array is a cheaper alternative to a full-scale planar array. The process of beamforming can also be carried out in near field region where the distances involved are comparable to the array size. We call such a beamformation as focused beamformation. This is of great interest in seismic exploration, synthetic aperture radar mapping, room conference, etc. Among several applications of beamforming we shall discuss the use of a linear array for cancellation of interference. The final topic is about adaptive beamformation. We show how the LMS algorithm (described in Chapter Six) can be used to derive an optimum weight vector, which enables us to cancel the unknown interference but allows the known signal.

§7.1 Wavefields

An array of sensors is often used to measure wavefields and to extract information about the sources and the medium through which the wavefield has propagated. The wavefields of interest are electromagnetic (EM), mechanical (acoustic, and low frequency seismic). The most important information we look for in a wavefield is the direction of propagation. The main interest in beamformation, therefore, is to receive power from or transmit power in a given direction.

Types of wavefields: We shall quickly review the different types of wavefields and the basic equations governing them, propagation of waves, and wavefronts. Different types of physical waves and the basic equations governing them are listed in table 7.1. It is hoped that this background shall be useful in the understanding of how an array of sensors responds to a propagating wave.

Types of sensors: A scalar sensor is used to sense a scalar wavefield such as pressure or any one of the components of the EM field or mechanical waves. The most common example of a scalar sensor is the microphone or hydrophone. A vector sensor will measure all components of the vector wavefield, three components of mechanical waves in solid or six components of EM field. Three-

Table 7.1: The basic equations governing different types of physical fields which are often used in beamformation.

Type	Governing Equation	Remarks
Scalar field: Acoustic field	$\nabla^2\phi = \frac{\rho}{\gamma\phi_0}\frac{d^2\phi}{dt^2}$ where ϕ_0 is ambient pressure, γ is ratio of specific heats at constant pressure and volume and ρ is density. $c = \sqrt{\frac{\gamma\phi_0}{\rho}} = \sqrt{\frac{\kappa}{\rho}}$	The scalar potential gives rise to longitudinal waves or pressure waves (p-waves)
Vector field: Waves in solids	$\rho\frac{\partial^2 \mathbf{d}}{\partial t^2} = (2\mu+\lambda)\mathit{grad\ div}\mathbf{d}$ $-\mu \mathit{curlcurl}\mathbf{d}$ **d** stands for the displacement vector. ϕ is scaler potential and ψ is vector potential. $\mathbf{d} = \nabla\phi + \nabla\times\psi$ $\nabla^2\phi = \frac{1}{\alpha^2}\frac{\partial^2\phi}{\partial t^2},\ \nabla.\psi = 0$ $\nabla\times\nabla\times\psi = -\frac{1}{\beta^2}\frac{\partial^2\psi}{\partial t^2}$	The vector potential gives rise to transverse waves or shear waves (s-waves). The p-waves travel with speed α and s-waves travel with speed β.
Vector field: Electromagnetic waves	$\left.\begin{matrix}\nabla\times\mathbf{E} = -\frac{\partial\mathbf{B}}{\partial t}\\ \nabla\cdot\mathbf{D} = \rho\end{matrix}\right\}$ Faraday's law $\left.\begin{matrix}\nabla\times\mathbf{H} = \mathbf{J} + \frac{\partial\mathbf{D}}{\partial t}\\ \nabla\cdot\mathbf{B} = 0\end{matrix}\right\}$ Ampere's law. $\nabla^2\phi - \varepsilon\mu\frac{\partial^2\phi}{\partial t^2} = -\frac{1}{\varepsilon}\rho$ $\nabla\times\nabla\times\psi - \varepsilon\mu\frac{\partial^2\psi}{\partial t^2} = -\mu\mathbf{J}$	Both electric and magnetic fields travel with the same speed unlike p and s waves in solids. The electric and magnetic vectors lie in a plane ⊥ to the direction of propagation. The tip of a field vector will execute a smooth curve known as a polarization ellipse.

component seismometers are sometimes used in seismic exploration. Six component EM sensors are likely to be available very soon off the shelf. Modern sensor arrays consist of several tens or hundreds of sensors. One major problem is the lack of uniformity in sensor response. In addition to above wavefield sensors, we have chemical sensors capable of detecting a very small quantity of chemicals in the vapour state. Extremely sensitive detectors of magnetic fields based on the principle of super conducting quantum interference have also appeared and have been used in magneto encephalography.

Fourier Representation of WaveField: The wave equation in a homogeneous medium is given by

$$\nabla^2 f = \frac{1}{c^2}\frac{\partial^2 f}{\partial t^2} \tag{7.1}$$

where $f(\mathbf{r},t)$ stands for any one of the wave types, for example, pressure or any one of the components of the vector field and $\mathbf{r}$ is position vector. We introduce the Fourier integral representation of the wavefield,

$$f(\mathbf{r},t) = \frac{1}{8\pi^3}\int\int_{-\infty}^{\infty}\int F(u,v,\omega)H(u,v,z)e^{j(\omega t - ux - vy)}dudvd\omega \tag{7.2}$$

in the wave equation (7.1). We observe that $H(u,v,z)$ must satisfy an ordinary differential equation given by

$$\frac{d^2H(u,v,z)}{dz^2} = (u^2 + v^2 - \frac{\omega^2}{c^2})H(u,v,z) \tag{7.3}$$

whose solution is

$$H(u,v,z) = \exp(\pm\sqrt{(u^2 + v^2 - k^2}z) \tag{7.4}$$

where $k = \frac{\omega}{c}$ is known as a wavenumber. When $\sqrt{(u^2 + v^2} > k$, we choose (-) sign for $z>0$ and (+) sign for $z<0$ so that the wavefield does not diverge. In both cases the field will rapidly decay as $|z| \to \infty$. These are known as evanescent waves. When $\sqrt{(u^2 + v^2} < k$ we get propagating waves whose integral representation reduces to

$$f(x,y,t) = \frac{1}{8\pi^3}\int\int_{-\infty}^{\infty}\int F(u,v,\omega)e^{\pm j(\sqrt{k^2 - u^2 - v^2}z)}e^{j(\omega t - ux - vy)}dudvd\omega \tag{7.5}$$

where the sign in $e^{\pm j(\sqrt{k^2 - u^2 - v^2}z)}$ is selected depending on whether the waves are diverging or converging. The convention is (-) sign for diverging waves and (+) sign for converging waves. Note that in a bounded space both diverging and converging waves can coexist and hence it would be necessary to use both signs in describing the wavefields in a bounded space.

Wavefront: From (7.5) it may be observed that the wavefield consists of a large number of plane waves propagating in different directions. An individual wave may be expressed by

$$\phi_1(x,y,z,t) = Ae^{j(\omega t - ux - vy - \sqrt{k^2-u^2-v^2}z)}$$

where the spatial frequencies u and v are related to the direction of propagation ,

$$u = k\sin\theta\cos\varphi$$
$$v = k\sin\theta\sin\varphi$$

where φ and θ are respectively azimuth and elevation angles of a plane wave and $k = \omega/c$ is wavenumber. On a plane, $ux + vy + \sqrt{k^2 - u^2 - v^2}\,z =$ constant, the phase is constant at a given time instant. Therefore, the surface of constant phase, i.e. wavefront, is a plane surface (see fig. 7.1). The wavefront moves forward with time in the direction perpendicular to the wavefront (this direction is known as a wave vector shown by an arrow in fig 7.1).

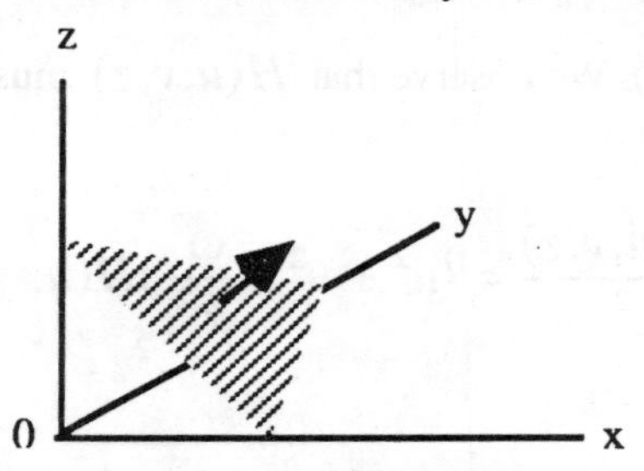

Figure† **7.1**: A wavefront is a plane surface on which the phase remains constant. The wave vector is a direction perpendicular to wavefront.

The line perpendicular to the wavefront is called a ray. The direction cosines of a ray are

$$\alpha = \sin\theta\cos\varphi$$
$$\beta = \sin\theta\sin\varphi$$
$$\gamma = \cos\theta$$

The spatial frequencies are thus related to the direction cosines, $u = (\omega/c)\alpha$ and $v = (\omega/c)\beta$. For a fixed w, the spatial frequencies u and v must lie within a circle of radius k (= ω/c), that is, a disc defined by

$$(u^2 + v^2) \le (\frac{\omega}{c})^2 \tag{7.6}$$

For spatial frequencies $(u^2 + v^2) > (\frac{\omega}{c})^2$ the wavefield gets rapidly attenuated. Using the relation between (u, v) and (α, β) we introduce the apparent speeds with which the wavefront travels along the three coordinate axes.

†Figures 7.1, 7.3-7.5, 7.7-7.10, 7.13, 7.14, 7.17-7.22 were taken from [30]

$$c_x = \frac{\omega}{k \sin\theta \cos\varphi} = \frac{\omega}{u}$$

$$c_y = \frac{\omega}{k \sin\theta \sin\varphi} = \frac{\omega}{v}$$

$$c_z = \frac{\omega}{k \cos\theta} = \frac{\omega}{\sqrt{k^2 - u^2 - v^2}} \tag{7.7}$$

Example 7.1: Interestingly the apparent speed can be much higher than the wave speed. To emphasize this point, consider three different wavefronts shown in fig. 7.2. The wavefront A travels along x-axis, hence $\varphi = 0.0$ and $\theta = \pi/2$. The wavefront B travels along z-axis, $\varphi = undefined$ and $\theta = 0$. And the wavefront C travels along the diagonal, $\varphi = 45^0$, $\theta = 45^0$. From (7.7) we compute the apparent speeds,

For wavefront A: $c_x = c, \quad c_y = \infty, \quad c_z = \infty$

For wavefront B: $c_x = \infty, \quad c_y = \infty, \quad c_z = c$

For wavefront C: $c_x = 2c, \quad c_y = 2c, \quad c_z = \sqrt{2}c$

The infinite apparent speed is physically not possible but only a mathematical concept.

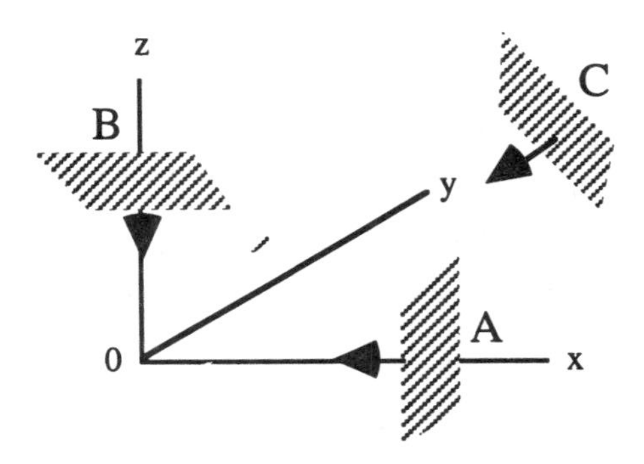

Figure 7.2: Wavefront A travels along x-axis, wavefront B along z-axis and wavefront C along the diagonal $\theta = 45^0$ and $\varphi = 45^0$.

Finally, in the (u, v, w) space, a narrowband plane wave is represented by a point (see fig.7.3) but a broadband plane wave is represented by a straight line (e.g. line ab in fig. 7.3) whose direction cosines are equal to those of the plane wavefront ■.

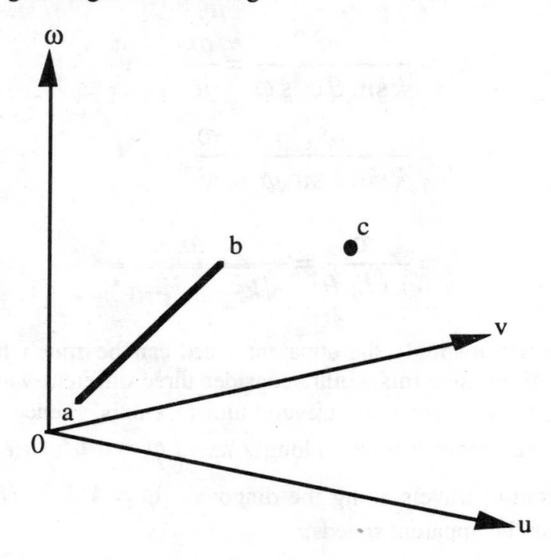

Figure 7.3: Representation of a plane wave in (u, v, ω) space. Point c represents a single frequency (narrowband) plane wave and line ab represents a wideband plane wave.

§7.2 **Uniform Linear Array** (ULA)

An array of sensors, placed on a line (linear array), or on a circle (circular array), or on a square grid (planar array), is used to receive a propagating wave with the following objectives:

1) To localize a source.
2) To receive a message from a distant source.
3) To image a medium through which the wavefield has propagated.

First, we shall consider a linear array, the simplest of the three arrays mentioned above. It is commonly used in such applications as radar, sonar, seismic exploration and tomography.

Array Response: Consider a plane wavefront, having a temporal waveform *f(t)* incident on a uniform linear array (ULA) of sensors (see fig. 7.4) at an angle θ. In signal processing literature the angle of incidence is also known as direction of arrival (DOA). Note that the DOA is always measured with respect to the normal to the array aperture, that is, the spatial extent of a sensor array; while another related quantity azimuth, which was introduced in §7.1 is measured with respect to the x-axis. In this work θ stands for DOA or angle of elevation and φ stands for azimuth. We shall assume that a source emits a stationary stochastic signal *f(t)*. Let $f_m(t)$, m=0, 1, 2,..., M-1 be the outputs of the sensors. In a homogeneous medium the signal arrives at successive sensors with an incremental delay.

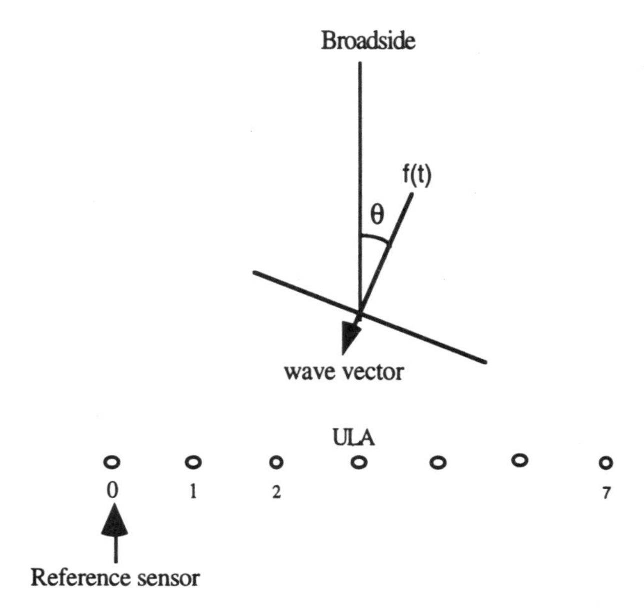

Figure 7.4: Uniform linear array of sensors. Note the convention of sensor indexing. The left most sensor is the reference sensor with respect to which all time delays are measured.

Some times it is convenient to represent the sensor output in the frequency domain

$$f_m(t) = \frac{1}{2\pi} \int_{-\infty}^{\infty} F(\omega) e^{j\omega(t - \frac{md}{c}\sin\theta)} d\omega \tag{7.8}$$

The simplest form of array signal processing is to sum all sensor outputs without any delay,

$$\begin{aligned} g(t) &= \frac{1}{M} \sum_{m=0}^{M-1} f_m(t) \\ &= \frac{1}{2\pi} \int_{-\infty}^{\infty} F(\omega) e^{j\omega t} d\omega \frac{1}{M} \sum_{m=0}^{M-1} e^{-j\omega \frac{md}{c}\sin\theta} \\ &= \frac{1}{2\pi} \int_{-\infty}^{\infty} F(\omega) H(\omega\tau) e^{j\omega t} d\omega \end{aligned} \tag{7.9}$$

where $H(\omega\tau)$ is the array response function, $\tau = (d/c)\sin\theta$, and d is sensor spacing. The array response function for a ULA is given by

$$H(\omega\tau)=\frac{1}{M}\sum_{m=0}^{M-1}e^{j\omega\frac{md}{c}\sin\theta}=\frac{\sin(\frac{M}{2}\omega\tau)}{M\sin\frac{\omega\tau}{2}}e^{j\frac{M-1}{2}\omega\tau} \tag{7.10}$$

A few examples of the array response function (magnitude only) are shown in fig. 7.5 for different values of M, that is, the array size. The response function is periodic with a period 2π. The maximum occurs at $\omega\tau=2n\pi$. The peak at n=0 is known as the main lobe and other peaks at $n=\pm1,\pm2,...$ are known as grating lobes. Since the magnitude of the array response is plotted, the period appears as π, as seen in fig. 7.5. The grating lobes can be avoided if we restrict the range of $\omega\tau$ to $\pm\pi$, that is, at a fixed frequency the direction of arrival must satisfy the relation $(d/\lambda)\sin\theta\leq0.5$ or $d\sin\theta\leq\lambda/2$. Since $|\theta|\leq\pi/2$, $\sin|\theta|\leq1$ it is sufficient to ensure that $d\leq\lambda/2$ for the grating lobes to gremain outside the range $\pm\pi/2$.

Array Steering: We have noted that the array response is maximum when the direction of arrival (DOA) is on broadside (θ=0). The maximum, however, can be changed to any other direction through a simple act of introducing a time delay to each sensor output before summation. This is known as array steering. Let an incremental delay of τ per channel be introduced. The sum output of the array is now given by

$$\begin{aligned} g(t)&=\frac{1}{M}\sum_{m=0}^{M-1}f_m(t+m\tau)\\ &=\frac{1}{2\pi}\int_{-\infty}^{\infty}F(\omega)e^{j\omega t}d\omega\frac{1}{M}\sum_{m=0}^{M-1}e^{j(\tau-\frac{d}{c}\sin\theta_0)\omega m} \\ &=\frac{1}{2\pi}\int_{-\infty}^{\infty}F(\omega)H((\tau-\frac{d}{c}\sin\theta_0)\omega)e^{j\omega t}d\omega \end{aligned} \tag{7.11}$$

where we have assumed that the DOA is θ_0. Let $\tau=(d/c)\sin\theta$. Then the array response is maximum whenever $\theta=\theta_0$. We say that the array is steered in the direction θ_0, that is, in the direction of arrival of the incident wavefront. Let $\tau=(d/c)k/M$ or the steering angle takes a set of discrete values,

$$\theta_k=\sin^{-1}(\frac{k}{M}),\quad k=0,1,...,M-1$$

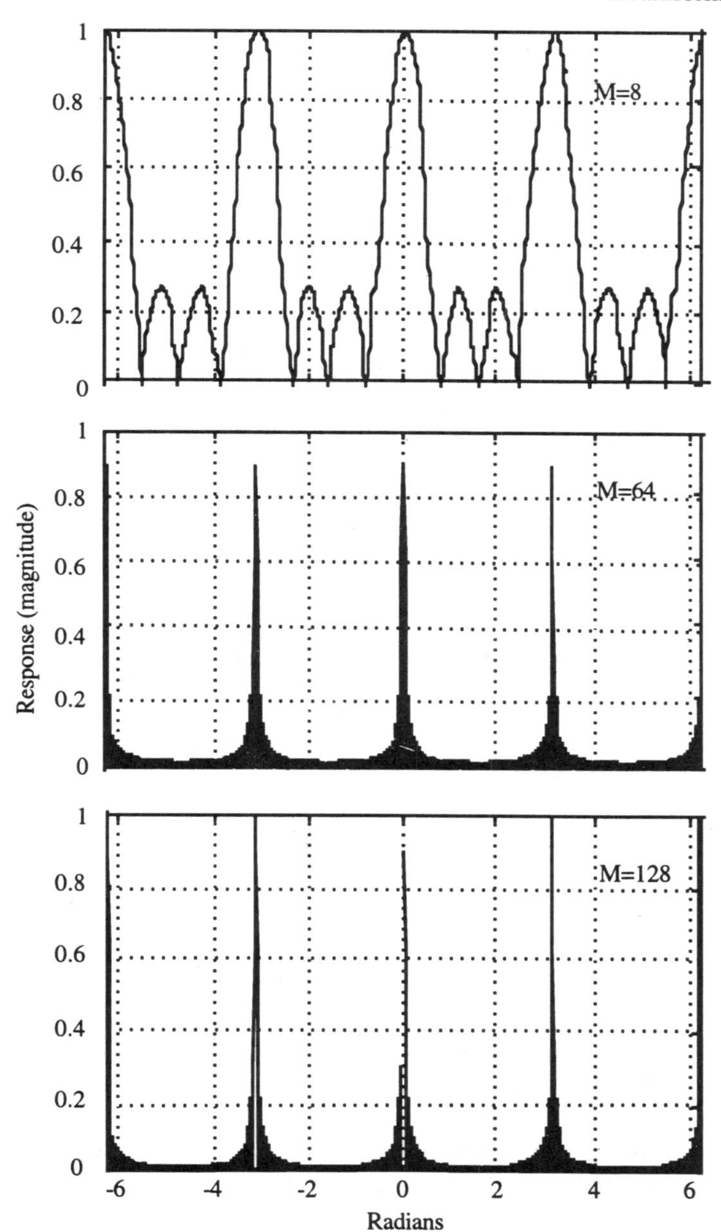

Figure 7.5: Array response function (magnitude) for different values of M. Notice that the main lobe becomes sharper as the array size is increased.

The array response in (7.11) may be expressed as a FDFT of $e^{-j\frac{2\pi d}{\lambda}\sin\theta_0 m}$, m=0,1,...,M-1

$$\begin{aligned} H(k) &= \frac{1}{M}\sum_{m=0}^{M-1} e^{j(\frac{d}{c}\frac{k}{M}-\frac{d}{c}\sin\theta_0)\omega m} \\ &= \frac{1}{M}\sum_{m=0}^{M-1} e^{-j\frac{2\pi d}{\lambda}\sin\theta_0 m} e^{j2\pi\frac{km}{M}\frac{d}{\lambda}} \\ &= FDFT\left\{ e^{-j\frac{2\pi d}{\lambda}\sin\theta_0 m}, m = 0,1,\ldots M-1 \right\} \end{aligned}$$

We can exploit this property for beamformation using FFT algorithm.

Radar/Sonar Signal: In many active detection systems a target is illuminated by a narrowband signal, for example in phased array radar. We have described in Chapter One (page 29) a simple narrowband signal whose envelop is a complex function. Here we consider a narrowband signal $f_{nb}(t)$ with a real envelop,

$$f_{nb}(t) = s_0(t)\cos(\omega_c t + \varphi_0(t)) \tag{7.12}$$

where $s_0(t)$ is a slowly varying waveform, often called envelope and $\varphi_0(t)$ is also a slowly varying phase. $\cos(\omega_c t)$ is a rapidly varying sinusoid, often known as a carrier and ω_c is the carrier frequency. Equation (7.12) may be expressed by expanding $\cos(\omega_c t + \varphi_0(t))$ as

$$f_{nb}(t) = f_i(t)\cos(\omega_c t) - f_q(t)\sin(\omega_c t) \tag{7.13}$$

where

$$\begin{aligned} f_i(t) &= s_0(t)\cos(\varphi_0(t)) \\ f_q(t) &= s_0(t)\sin(\varphi_0(t)) \end{aligned}$$

$f_i(t)$ is known as an inphase component and $f_q(t)$ is a quadrature component. The inphase and quadrature components are uncorrelated. They have, however, the same spectral density function. The inphase and quadrature components can be uniquely recovered from a narrowband signal by a process known as mixing which involves multiplication with $2\cos(\omega_c t)$ and $-2\sin(\omega_c t)$ and lowpass filtering. A complex analytical signal is defined as $f_c(t) = f_i(t) + jf_q(t)$. Consider a narrowband signal delayed by one quarter period, that is, delay $= \tau_0/4$ where $\tau_0 = 2\pi/\omega_c$. Assuming that both inphase and quadrature components are slowly varying signals we get the following approximate result:

$$\begin{aligned} f_{nb}((t-\frac{\tau_0}{4})) &= f_i(t)\cos(\omega_c(t-\frac{\tau_0}{4})) - f_q(t)\sin(\omega_c(t-\frac{\tau_0}{4})) \\ &= f_i(t)\sin(\omega_c t) + f_q(t)\cos(\omega_c t) \end{aligned} \tag{7.14}$$

Let us evaluate the Hilbert transform (see Chapter One, page 29) for the definition of Hilbert transform) of (7.13).

$$\begin{aligned} f_{nb}^{Hilb}(t) &= \int_{-\infty}^{\infty} \frac{f_i(t)\cos(\omega_c t) - f_q(t)\sin(\omega_c t)}{t-t'} dt' \\ &\approx f_i(t) \int_{-\infty}^{\infty} \frac{\cos(\omega_c t)}{t-t'} dt' - f_q(t) \int_{-\infty}^{\infty} \frac{\sin(\omega_c t)}{t-t'} dt' \\ &= f_i(t)\sin(\omega_c t) + f_q(t)\cos(\omega_c t) \end{aligned} \tag{7.15}$$

From (7.14) and (7.15) we obtain

$$f_{nb}((t - \frac{\tau_0}{4})) \approx f_{nb}^{Hilb}(t) \tag{7.16}$$

We define a complex analytical signal as

$$\begin{aligned} &f_{nb}(t) + jf_{nb}^{Hilb}(t) \\ &= f_{nb}(t) + jf_{nb}(t - \frac{\tau_0}{4}) \\ &= f_i(t)e^{j\omega_c t} + jf_q(t)e^{j\omega_c t} = f_c(t)e^{j\omega_c t} \end{aligned} \tag{7.17}$$

where

$$f_c(t) = f_i(t) + jf_q(t).$$

The process described in (7.17) is often referred to as quadrature filtering, which is illustrated in fig. 7.6. Note that the input to the quadrature filter is real but the output is complex.

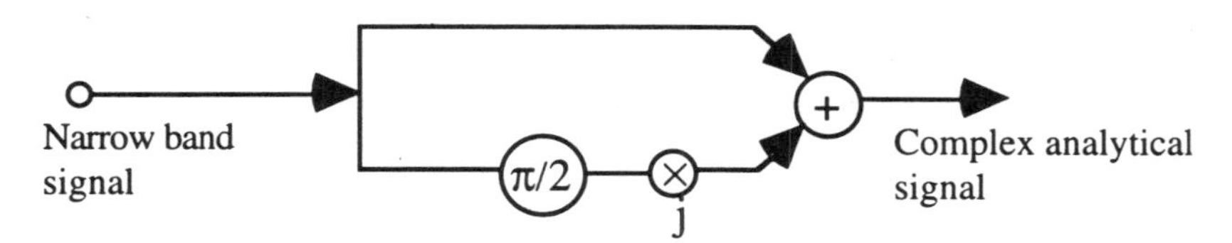

Figure 7.6: Quadrature filter structure. Since the phase change due to propagation appears in the complex sinusoid at the output it is easy to introduce additional phase delays for beamformation.

The representation given by (7.17) is useful in narrowband beamformation. Consider the mth sensor of a ULA. The complex output of the quadrature filter is

$$\begin{aligned} f_m(t) &= f_{nb}(t - m\frac{d}{c_x}) + jf_{nb}(t - m\frac{d}{c_x} - \frac{\tau_0}{4}) \\ &= f_c(t)e^{j\omega_c t - jm\omega_c \frac{d}{c_x}} \end{aligned} \tag{7.18}$$

where, we have used the apparent speed $c_x = c/\sin\theta_0$. Now consider a weighted sum of the narrowband signal and its Hilbert transform. We assume that $s_0(t)$ and $\varphi_0(t)$ are slowly varying functions,

$$\begin{aligned} &\hat{f}_m(t) \\ &= w_1 f_{nb}(t - m\frac{d}{c_x}) + w_2 f_{nb}(t - m\frac{d}{c_x} - \frac{\tau_0}{4}) \\ &\approx \begin{bmatrix} w_1 s_0(t)\cos(\omega_c(t - jm\frac{d}{c_x}) + \varphi_0(t)) + \\ w_2 s_0(t)\sin(\omega_c(t - jm\frac{d}{c_x}) + \varphi_0(t)) \end{bmatrix} \\ &= \begin{bmatrix} s_0(t)[w_1\cos(\omega_c(t - jm\frac{d}{c_x}) + \varphi_0(t)) + \\ w_2\sin(\omega_c(t - jm\frac{d}{c_x}) + \varphi_0(t)) \end{bmatrix} \\ &= s_0(t) r\cos(\omega_c(t - jm\frac{d}{c_x}) + \varphi_0(t) - \theta) \end{aligned}$$

where w_1 and w_2 are arbitrary weights. Let $w_1 = r\cos(\theta)$ and $w_2 = r\sin(\theta)$. By choosing proper weights (real) it is possible to alter both phase, $\tan^{-1}(w_2/w_1)$, and amplitude, $(= \sqrt{w_1^2 + w_2^2}$), of the quadrature filter. For example, we can cancel the phase introduced by propagation delay, namely, $m(d/c_x)\omega_c$.

Matrix Representation: A snapshot is a vector representing the outputs of all sensors taken at the same time instant, t. Let $\mathbf{f}(t) = col\{f_0(t), f_1(t), \ldots, f_{M-1}(t)\}$ be a snapshot, where $f_0(t), f_1(t), \ldots,$ $f_{M-1}(t)$ stand for the sensor outputs at time instant t. When the incident signal is narrowband, the signal varies slowly with time (we assume that the carrier has been removed). In the noise-free case, a single time shot is adequate as it contains all available information. A snapshot vector for narrowband signal may be expressed using (7.18) as

$$\mathbf{f}(t) = f_c(t)\boldsymbol{\phi}(\theta_0) \tag{7.19}$$

where

$$\phi(\theta_0) = col\left\{1, e^{-j\omega_c \frac{d}{c_x}}, \ldots, e^{-j(M-1)\omega_c \frac{d}{c_x}}\right\}$$

Further let the sensor response matrix be $\boldsymbol{\alpha}(\theta_0) = diag\{\alpha_0(\theta_0), \alpha_1(\theta_0), \ldots \alpha_{M-1}(\theta_0)\}$, in which each element represents the response of a sensor as a function of the angle of incidence of the wavefront. For an ideal sensor array we must have $\alpha_0(\theta_0) = \alpha_1(\theta_0) = \ldots = \alpha_{M-1}(\theta_0) = 1$. $\phi(\theta_0)$ represents the propagation effect of the medium on a wavefront propagating across the array. $\phi(\theta_0)$ and $\alpha(\theta_0)$ together form a direction vector $\mathbf{a}(\theta_0) = \boldsymbol{\alpha}(\theta_0)\boldsymbol{\phi}(\theta_0)$ representing the response of an array to a wavefront incident at angle θ_0 (DOA). Finally, the array output, a snapshot vector, may be expressed as follows:

$$\begin{aligned}\mathbf{f}(t) &= f_c(t)\boldsymbol{\alpha}(\theta_0)\boldsymbol{\phi}(\theta_0) \\ &= f_c(t)\mathbf{a}(\theta_0)\end{aligned} \tag{7.20}$$

The array steering can also be represented in terms of a matrix operation. To steer an array to a desired direction, θ, we form an inner product of the steering vector and the array snapshot

$$\begin{aligned}\mathbf{a}^H(\theta)\mathbf{f}(t) &= \mathbf{a}^H(\theta) f_c(t)\mathbf{a}(\theta_0) \\ &= f_c(t)\mathbf{a}^H(\theta)\mathbf{a}(\theta_0)\end{aligned} \tag{7.21a}$$

The output power is computed by squaring and statistical averaging

$$\mathbf{a}^H(\theta)E\{\mathbf{f}(t)\mathbf{f}^H(t)\}\mathbf{a}(\theta) = E\{|f_c(t)|^2\}|\mathbf{a}^H(\theta)\mathbf{a}(\theta_0)|^2$$

$$\mathbf{a}^H(\theta)\mathbf{C}_f\mathbf{a}(\theta) = \sigma_{s_0}^2 M^2 \left|H(\omega\frac{d}{c}(\sin\theta - \sin\theta_0))\right|^2 \tag{7.21b}$$

where $\mathbf{C}_f = E\{\mathbf{f}(t)\mathbf{f}^H(t)\}$ is the spatial covariance matrix (SCM). Whenever $\theta = \theta_0$, that is, when the steering angle is equal to the DOA, the left-hand side of (7.21b) equals $\sigma_{s_0}^2 M^2$ giving the power of the source.

The steering vector satisfies the following properties:

(a) $\mathbf{a}(\theta) = \mathbf{a}(\pi - \theta)$

(b) $\mathbf{a}^*(\theta) = \mathbf{a}(-\theta)$

(c) $\mathbf{a}(\theta)$ is periodic with a period $\pm\pi/2$ only if $d = \lambda/2$.

Property (a) implies a wavefront coming from the north and another symmetrically opposite from the south (a and b in fig 7.7) cannot be distinguished This is known as north-south ambiguity. Property (b) implies a wavefront coming from the east and another symmetrically opposite from the west (a and c in fig. 7.7) can be distinguished only if the signal is complex This is kown as east-west ambiguity. To show this, recall (7.20) and compare the outputs of a ULA for a real input signal (for example, $f_c(t)$ is real when $\varphi_0(t)=0$) incident at angle θ and $-\theta$. Let $\mathbf{f}_\theta(t)$ be output of a ULA for an incident angle, θ, and $\mathbf{f}_{-\theta}(t)$ be the output for an incident angle, $-\theta$. For a real signal $\mathbf{f}_\theta(t)=\mathbf{f}^*_{-\theta}(t)$ but for a complex signal $\mathbf{f}_\theta(t)\neq\mathbf{f}^*_{-\theta}(t)$. Property (c) implies that there is no grating lobe in the range $\pm\pi/2$ when the sensor spacing is $d\leq\lambda/2$.

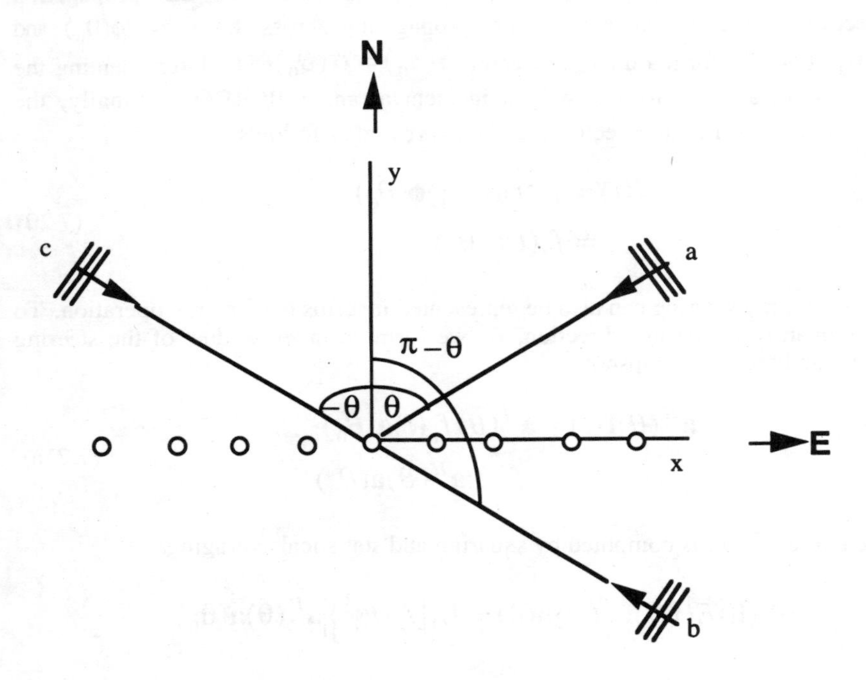

Figure 7.7: A ULA cannot distinguish wavefronts a and b (north-south ambiguity). However, it can distinguish wavefronts a and c if the signal is complex (east-west ambiguity).

§7.3 **Uniform Circular Array** (UCA)

The sensors may be placed on a plane in a polar grid. For a fixed radial distance we have a circle on which the sensors are uniformly placed, forming a circular array. A circular array can measure both azimuth and elevation angles. Additionally, it has no grating lobes as with a linear array. The sensors may also be placed over a square grid giving what is known as planar array consisting of several thousands of sensors. Planar array finds application in military where the operating environment is highly demanding. A circular array is a poor cousin of a expensive planar array.

Consider a circular array of radius 'a' with M sensors, symmetrically placed on the circumference (see fig. 7.8). Let a plane wavefront be incident on the array at angles φ and θ. The output of the mth sensor is given by

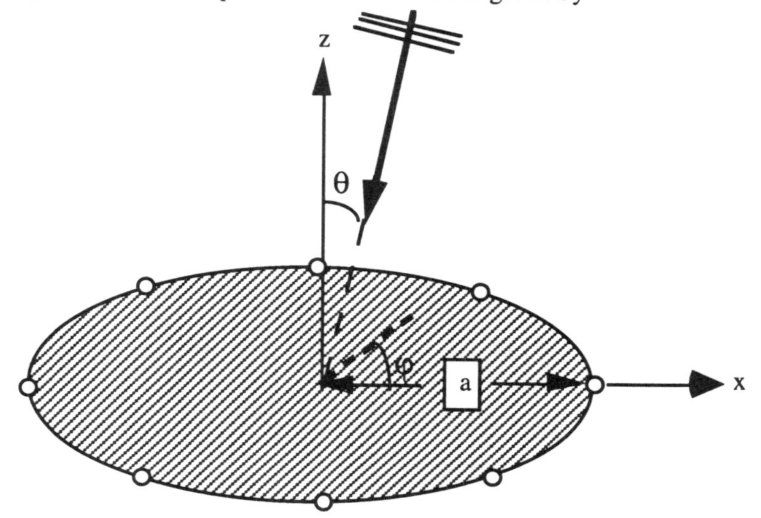

Figure 7.8: Sensors are uniformly spaced on the circumference of a circle of radius a. A plane wave is incident at an azimuth angle φ and an elevation angle θ.

$$
\begin{aligned}
f_m(t) &= \frac{1}{2\pi}\int_{-\infty}^{\infty} F(\omega) e^{j[\omega t - \frac{\omega a}{c}(\cos\varphi \sin\theta \cos\frac{2\pi m}{M} + \sin\varphi \sin\theta \sin\frac{2\pi m}{M})]} d\omega \\
&= \frac{1}{2\pi}\int_{-\infty}^{\infty} F(\omega) e^{j[\omega t - \frac{\omega a}{c}(\sin\theta \cos(\frac{2\pi m}{M} - \varphi))]} d\omega
\end{aligned}
\tag{7.22}
$$

Note that time is measured with respect to the time of arrival of the wavefront at the center of the array.

Array Response: First, we evaluate the frequency response function. The sum of all outputs in the frequency domain is given by

$$
\begin{aligned}
g(t) &= \frac{1}{M}\sum_{m=0}^{M-1} f_m(t) \\
&= \frac{1}{2\pi}\int_{-\infty}^{\infty} F(\omega) e^{j\omega t} \frac{1}{M}\sum_{m=0}^{M-1} e^{-j[\frac{\omega a}{c}(\sin\theta \cos(\frac{2\pi m}{M} - \varphi))]} d\omega \qquad (7.23a) \\
&= \frac{1}{2\pi}\int_{-\infty}^{\infty} F(\omega) H(\omega, \varphi, \theta) e^{j\omega t} d\omega
\end{aligned}
$$

where the frequency response function $H(\omega, \varphi, \theta)$ is given by

$$H(\omega,\varphi,\theta)=\frac{1}{M}\sum_{m=0}^{M-1} e^{-j[\frac{\omega a}{c}(\sin\theta\cos(\frac{2\pi m}{M}-\varphi))]} \tag{7.23b}$$

A circular array may be steered to any desired direction just like a ULA or a UPA. A delay τ_m is introduced at each sensor output before summation, where

$$\tau_m=[\frac{a}{c}(\cos\varphi\sin\theta\cos\frac{2\pi m}{M}+\sin\varphi\sin\theta\sin\frac{2\pi m}{M})]$$

and φ and θ respectively are the desired azimuth and elevation angles. The delayed outputs of all sensors are then summed.

$$g(t)=\frac{1}{M}\sum_{m=0}^{M-1} f_m(t+\tau_m)$$

$$=\frac{1}{2\pi}\int_{-\infty}^{\infty} F(\omega)e^{j\omega t}\frac{1}{M}\sum_{m=0}^{M-1} e^{-j[\frac{\omega a}{c}(\sin\theta_0\cos(\frac{2\pi m}{M}-\varphi_0)-\sin\theta\cos(\frac{2\pi m}{M}-\varphi))]}d\omega \tag{7.24a}$$

where φ_0 and θ_0 are respectively the unknown azimuth and elevation angles of the incident wavefront. Let

$$H(\frac{\omega a}{c},\theta_0,\varphi_0,\theta,\varphi)=\frac{1}{M}\sum_{m=0}^{M-1} e^{-j[\frac{\omega a}{c}(\sin\theta_0\cos(\frac{2\pi m}{M}-\varphi_0)-\sin\theta\cos(\frac{2\pi m}{M}-\varphi))]} \tag{7.24b}$$

The output power of the array, steered to any chosen direction φ and θ, is given by

$$\text{output power}=|F(\omega)|^2\left|H(\frac{\omega a}{c},\theta_0,\varphi_0,\theta,\varphi)\right|^2$$

Example 7.2: Consider a UCA of radius 4λ with M=8 or 32 sensors uniformly spaced. The array response function given by (7.24b) was computed for azimuth ranging over 0 to 2π and elevation over 0 to $\pi/2$. A plane wavefront is assumed to be incident on the array with azimuth, $\varphi_0=45^0$ and elevation, $\theta_0=45^0$. The computed response function (magnitude square) is shown in fig. 7.9 as contour maps. The left figure is for M=8 and the right figure for M=32. Inter element spacing in the first case is 3.06λ and in the second case it is 0.784λ. The following observations are pertinent:

i) The main lobe is correctly positioned, however its shape does not depend upon the number of sensors.
ii)The main lobe is much wider along the elevation axis than along the azimuth axis.

iii) There are no grating lobes even though the inter sensor spacing is greater than $\lambda/2$.

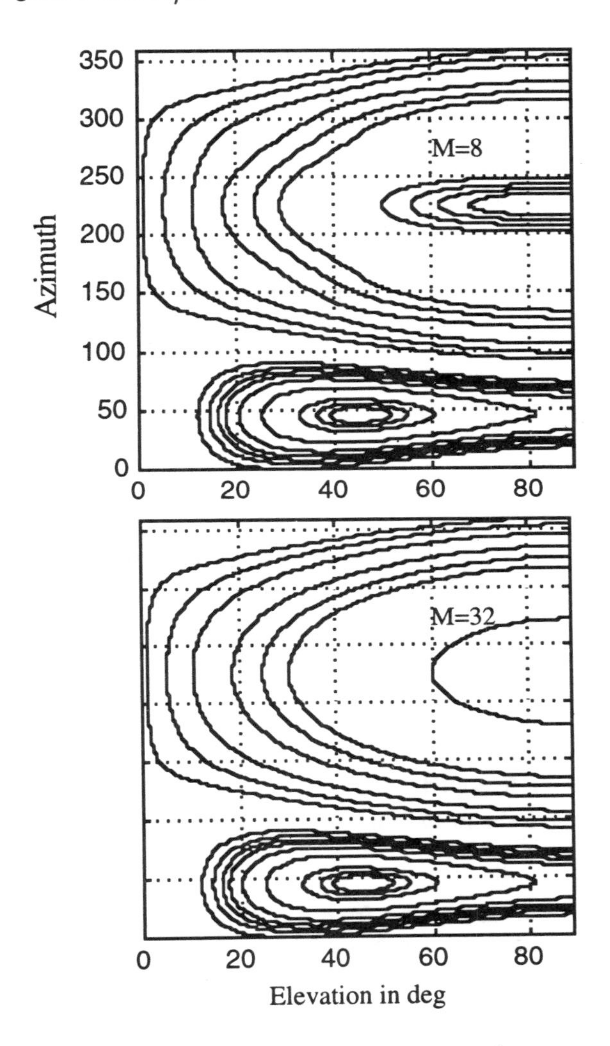

Figure 7.9: Circular array response function $\left(\left|H(\frac{\omega a}{c},\theta_0,\varphi_0,\theta,\varphi)\right|^2\right)$. Note that there are no grating lobes even though the sensor spacing is greater than $\lambda/2$ ■.

For circular array we define a steering vector as

$$\mathbf{a}(\varphi,\theta) = col\left[e^{-j[\frac{\omega a}{c}(\sin\theta\cos(\varphi))]}, e^{-j[\frac{\omega a}{c}(\sin\theta\cos(\frac{2\pi}{M}-\varphi))]}, \ldots e^{-j[\frac{\omega a}{c}(\sin\theta\cos(\frac{2\pi(M-1)}{M}-\varphi))]}\right]$$

The array output, that is, a snapshot vector may be expressed in terms of the steering vector as in (7.20) for the linear array,

$$\mathbf{f}(t) = f_c(t)\mathbf{a}(\varphi_0, \theta_0).$$

To steer the array to the desired azimuth and elevation we form an inner product of the steering vector in the direction of the desired azimuth and elevation and the array snapshot,

$$\mathbf{a}^H(\varphi, \theta)\mathbf{f}(t) = f_c(t)\mathbf{a}^H(\varphi, \theta)\mathbf{a}(\varphi_0, \theta_0)$$

The output power is computed by squaring and statistical averaging. We obtain as in the case of a linear array (see equation (7.21)),

$$\begin{aligned}\text{output power} &= \mathbf{a}^H(\varphi, \theta)\mathbf{c}_f(t)\mathbf{a}(\varphi, \theta) \\ &= \sigma_{s_0}^2 \left|\mathbf{a}^H(\varphi, \theta)\mathbf{a}(\varphi_0, \theta_0)\right|^2 \\ &= \sigma_{s_0}^2 M^2 \left|H(\frac{\omega a}{c}, \varphi_0, \theta_0, \varphi, \theta)\right|^2\end{aligned}$$

The steering vector for a circular array possesses some interesting properties, namely,

(i) $\mathbf{a}(\varphi, \theta) \neq \mathbf{a}(-\varphi, \theta)$,
(ii) $\mathbf{a}^*(-\varphi, \theta) = \mathbf{a}(\pi - \varphi, \theta)$
(iii) $\mathbf{a}(\varphi, \theta)$ is periodic in φ with period 2p, and independent of sensor spacing.

Property (i) implies a wavefront coming from the north can be distinguished from the one coming from the south (north-south ambiguity is absent). Property (ii) implies that a complex signal coming from the east can be distinguished from the one coming from the west. The properties (i) and (ii) are illustrated in fig. 7.10. Waves a and b can be distinguished but waves b and c can be distinguished only when the incident signal is complex. The property (iii) implies that, for any sensor spacing, there is no grating lobe in the range of $\pm\pi$. A circular array differs from a linear array in respect of properties (i & iii). For large M (for example, M> 48 when a=6 λ and M>32 when a=4 λ) the summation in (7.23a) may be replaced by an integral and the result is

$$H(\omega, \varphi, \theta) \approx J_0(\frac{\omega a}{c} \sin\theta)$$

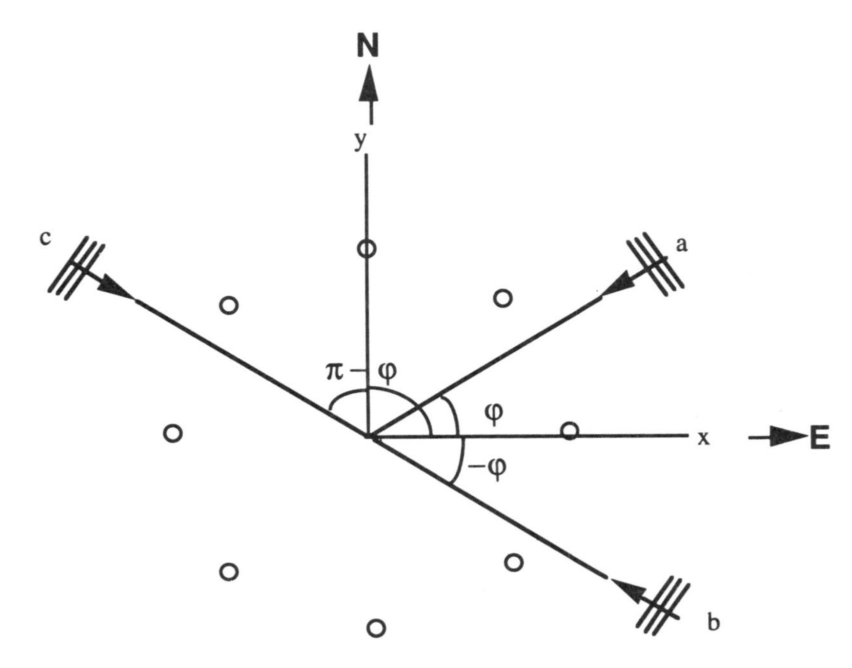

Figure 7.10: Circular array does not suffer from north-south ambiguity, that is, wavefronts a and b can be distinguished. There is no east-west ambiguity for complex signals, that is, wavefronts a and c can be distinguished. This is same as for linear array.

We shall call such a UCA a fully populated array. The most interesting property of a fully populated circular array is that the frequency response function is independent of φ. The property arises from (7.23b). Taking the distance to the first zero as the effective half width of the main lobe, the angular width will be equal to $\Delta\theta = \sin^{-1}(2.45\frac{c}{\omega a})$. The height of the first (largest) sidelobe is 0.4025 at $\theta = \sin^{-1}(0.8c/\omega a)$.

2D FDFT: Earlier we had noted that steering of an array is equivalent to the spatial Fourier transform of the array output. This result holds in a slightly different form for a circular array. We will demonstrate how the spatial Fourier transform can be used for estimation of azimuth [26] and elevation angles. Consider the spatial discrete Fourier transform of the circular array output.

$$\begin{aligned}
g_k(t) &= \frac{1}{M}\sum_{m=0}^{M-1} f_m(t) e^{-j\frac{2\pi m}{M}k} \\
&= \frac{1}{2\pi}\int_{-\infty}^{\infty} F(\omega) e^{j\omega t} \frac{1}{M}\sum_{m=0}^{M-1} e^{-j[\frac{\omega a}{c}(\sin\theta_0 \cos(\frac{2\pi m}{M}-\varphi_0))]} e^{-j\frac{2\pi m}{M}k} d\omega \\
&= \frac{1}{2\pi}\int_{-\infty}^{\infty} F(\omega) H_k(\omega,\varphi_0,\theta_0) e^{j\omega t} d\omega
\end{aligned} \tag{7.25a}$$

where

$$H_k(\omega,\varphi,\theta)=\frac{1}{M}\sum_{m=0}^{M-1}e^{-j[\frac{\omega a}{c}(\sin\theta\cos(\frac{2\pi m}{M}-\varphi))]}e^{-j\frac{2\pi m}{M}k} \tag{7.25b}$$

$$=J_k(\frac{\omega a}{c}\sin\theta_0)e^{jk\frac{\pi}{2}}e^{-jk\varphi_0} \quad \text{for large } M$$

Taking the temporal Fourier transform of (7.25a) and using (7.25b) we obtain an important result,

$$G_k(\omega)\approx F(\omega)J_k(\frac{\omega a}{c}\sin\theta_0)e^{jk\frac{\pi}{2}}e^{-jk\varphi_0} \tag{7.26}$$

which is valid for $k < k_{\max} \approx \omega a/c$ [27] and for sensor spacing approximately equal to $\frac{\lambda}{2}$ [28]. Consider the following quantity:

$$\frac{G_{k+1}(\omega)}{G_k(\omega)}=je^{-j\varphi_0}\frac{J_{k+1}(\frac{\omega a}{c}\sin\theta_0)}{J_k(\frac{\omega a}{c}\sin\theta_0)} \tag{7.27}$$

Referring to the recurrence relation of Bessel functions [27],

$$J_{k+1}(x)=\frac{2k}{x}J_k(x)-J_{k-1}(x),$$

we can write

$$\frac{J_{k+1}(x)}{J_k(x)}=\frac{2k}{x}-\frac{J_{k-1}(x)}{J_k(x)}$$

which we use in (7.27) and derive a basic result for the estimation of φ_0 and θ_0

$$-je^{j\varphi_0}\frac{G_{k+1}(\omega)}{G_k(\omega)}=\frac{2k}{\frac{\omega a}{c}\sin\theta_0}-je^{-j\varphi_0}\frac{G_{k-1}(\omega)}{G_k(\omega)} \tag{7.28}$$

Equation (7.28) may be solved for φ_0 and θ_0. As an example, we consider a 16 sensor circular array of 3λ radius and a source in a far field emitting a bandlimited random signal. The center frequency is 100Hz and the bandwidth is 10Hz. The azimuth and elevation angles of the source are respectively 10° (0.1745 rad) and 45° (0.7854 rad). The sampling rate was 500 samples/sec. The estimates were averaged over all frequency bins lying within the bandwidth. The results are shown in fig. 7.11. Notice that the standard deviation of the estimates is small when a reference sensor is used at the center. The decrease is more pronounced at a very low SNR, e. g., at 0 dB; by a factor of three or more. An analysis of errors has shown that the standard deviation is dominated by a few outliers which are caused by random noise in the array output. Unless these outliers are eliminated the mean and the standard deviation of the estimate gets severely affected. To overcome this problem median in place of mean may be considered. It was observed through computer simulation that the median is a better estimate of the azimuth than the mean.

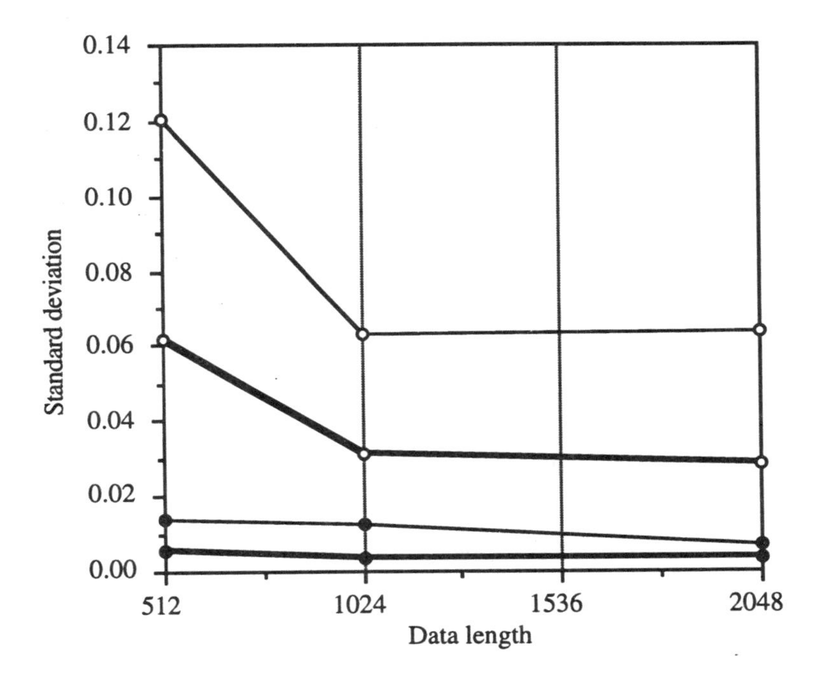

Figure 7.11: Standard deviation of azimuth and elevation estimates as a function of data length. Thick line: with a sensor at the center, thin line: without a sensor at the center, filled circle: azimuth and empty circle: elevation. SNR=10 dB.

§7.4 Beamformation

The purpose of an array of sensors is to receive or to transmit a signal from or to a desired direction In this process it is possible to suppress any interference coming from another direction. In transmit mode, the beamforming is useful in conserving power by not transmitting in unwanted direction. A beam in a desired direction is formed by introducing delays before summation. The required delay per sensor in a ULA is equal to $\tau = d/c \sin\theta$ and in a UCA the delay for the mth sensor is

$$\tau_m = \frac{a}{c}\sin\theta\cos(\frac{2\pi m}{M} - \varphi)$$

Digital Beamformation: In analog beamformation, introduction of continuously varying delays is achieved through analog delay lines, but in digital beamformation the delays can be achieved only as integral steps of sampling time units. Consider a ULA with sensor spacing d equal to $\lambda/2$ and time sampling interval, Δt, equal to d/c. There will be no aliasing, spatial or temporal, for this choice of parameters. However, we can form just one beam, namely, for $\theta = 0$ (excluding endfire beam). Clearly, to form more beams we need more samples between two Nyquist samples. Assume that we have sampled at q times the Nyquist rate, that is, we have q equispaced samples between two Nyquist samples which will enable us to form beams at angles, θ_i, i=0,1,..., q-1, where $\theta_i = \sin^{-1}(i/q)$. For example, let q=8, the beam angles are 0°, 7.18°, 14.48°, 22.04°, 30.0°, 38.68°, 48.59°, 61.04°. Evidently, only a fixed number of beams can be formed for a given oversampling rate. It is not possible to form a beam in any arbitrary direction. The situation with a UCA is far more difficult as for no direction of arrival can a uniformly sampled sensor output be used for beamformation. For example, consider a UCA of 16 sensors and $a/c = 8$ time samples. The delays to be introduced in the sensor outputs, in units of the temporal sampling interval, for $\theta = 90^0$ and $\varphi = 0$ are: 8.00, 7.39, 5.66, 3.06, 0.0, -3.06, -5.65, -7.39, -8.00, -7.39, -5.66, -3.06, 0.0, 3.06, 5.66, 7.39 (rounded to second decimal place). All these delays are with respect to a hypothetical sensor at the center of the circle. Notice that the delays are not in integral steps of the sampling interval. This leaves us with the only alternative of non-uniform sampling through interpolation of uniformly sampled sensor output. To minimize the computational load a simple linear interpolation has been suggested [29].

Narrowband: For narrowband signals the delays applied to the sensor outputs before summation may be expressed in terms of phase rotation. The output of the mth sensor is passed through a filter that removes the -ve frequencies. The result of filtering is given by

$$
\begin{aligned}
f_m(t) &\approx \frac{1}{2\pi} \int_{\omega_0 - \frac{\Delta\omega}{2}}^{\omega_0 + \frac{\Delta\omega}{2}} F_{nb}(\omega) e^{j\omega(t - m\frac{d}{c}\sin\theta_0)} d\omega \\
&= \frac{1}{2\pi} \int_{-\frac{\Delta\omega}{2}}^{\frac{\Delta\omega}{2}} F(\omega_0 + \omega') e^{j\left[\begin{array}{l}\omega_0 (t - m\frac{d}{c}\sin\theta_0) \\ +\omega'(t - m\frac{d}{c}\sin\theta_0)\end{array}\right]} d\omega' \\
&= e^{-j\omega_0 m \frac{d}{c}\sin\theta_0)} f_{nb}^{+}(t)
\end{aligned}
\tag{7.29a}
$$

where the subscript nb stands for narrowband and the superscript '+' stands for +ve frequencies. ω_c is the center frequency and $\Delta\omega$ is the bandwidth of the narrowband signal. The approximation is valid whenever the bandwidth satisfies a condition, $\Delta\omega(md/c\sin\theta_0) << 2\pi$ for all m, which implies that the time taken for a wave to sweep across the array must be much less than the inverse of the bandwidth, expressed in Hertz. In vector notation (7.29a) may be expressed as

$$\mathbf{f}(t) = \mathbf{a}_0 f^{+}_{nb}(t) \tag{7.29b}$$

where

$$\mathbf{a}_0 = \left[1, e^{-j\omega_0 \frac{d}{c}\sin\theta_0}, \ldots, e^{-j\omega_0 (M-1)\frac{d}{c}\sin\theta_0}\right]$$

is the direction vector of the incident wavefront. The delays applied to sensor outputs may be expressed in terms of a vector dot product. Define a vector,

$$\mathbf{a} = \left[1, e^{-j\omega_0 \frac{d}{c}\sin\theta}, \ldots, e^{-j\omega_0 (M-1)\frac{d}{c}\sin\theta}\right]$$

known as the steering vector, which will rotate the phase of each sensor output by an amount equal to $\omega_0 m d/c \sin\theta$ for the mth sensor. Thus, a narrowband beam is formed in the direction θ as

$$\mathbf{a}^H \mathbf{f}(t) = \mathbf{a}^H \mathbf{a}_0 f^{+}_{nb}(t)$$

or in terms of the beam power, that is, the (ω_0, θ) spectrum is given by

$$\begin{aligned} S(\omega_0, \theta) &= E\left\{\left|\mathbf{a}^H \mathbf{a}_0 f^{+}_{nb}(t)\right|^2\right\} \\ &= \left|\mathbf{a}^H \mathbf{a}_0\right|^2 \sigma^2_{f^+} \end{aligned} \tag{7.30}$$

Window: The sensor outputs are often weighted before summation, the purpose being to reduce the sidelobes of the response function just as in spectrum estimation where a window was used to reduce the sidelobes and thereby reduce the power leakage. As this topic is extensively covered under spectrum estimation (see Chapter Three), we shall not pursue this any further here. Instead, we like to explain the use of a weight vector to reduce the background noise variance or to increase the SNR. Let us select a weight vector, $\mathbf{w}$, such that the signal amplitude is preserved but the noise power is minimized.

$$\mathbf{w}^H \mathbf{a}_0 = 1$$

and

$$\mathbf{w}^H \mathbf{c}_\eta \mathbf{w} = \min \tag{7.31a}$$

where $\mathbf{c}_\eta$ is the noise covariance function. The solution to the constrained minimization problem in (7.31a) results in

$$\mathbf{w} = \frac{\mathbf{c}_\eta^{-1}\mathbf{a}_0}{\mathbf{a}_0^H \mathbf{c}_\eta^{-1}\mathbf{a}_0} \tag{7.31b}$$

It may be observed that for spatially white noise

$$\mathbf{c}_\eta = \sigma_\eta^2 \mathbf{I},$$

$\mathbf{w} = \mathbf{a}_0/M$. In other words, the weights are simply phase shifts or delays as in beamformation. The variance of the noise in the output is equal to $\sigma_{\hat{\eta}}^2 = \sigma_\eta^2/\text{M}$.

Rayleigh Resolution: When two wavefronts are simultaneously incident on an array, we would naturally like to measure their respective directions of arrival. For this to be possible the spectrum given by (7.30) must show two distinct peaks. Let $f_{nb_1}^+(t)$ and $f_{nb_2}^+(t)$ be two narrowband uncorrelated signals incident at angles θ_1 and θ_2, with the center frequencies being the same for both signals. The beam power is given by

$$s(\omega_0,\theta) = \left|\mathbf{a}^H\mathbf{a}_1\right|^2 \sigma_{f_1^+}^2 + \left|\mathbf{a}^H\mathbf{a}_2\right|^2 \sigma_{f_2^+}^2$$

In order that each signal gives rise to a distinct peak, $\left|\mathbf{a}^H\mathbf{a}_1\right|^2 \sigma_{f_1^+}^2$, when plotted as a function of θ, should not overlap with $\left|\mathbf{a}^H\mathbf{a}_2\right|^2 \sigma_{f_2^+}^2$. A condition for non-overlap is necessarily arbitrary as the array response to an incident wavefront is strictly not limited to a fixed angular range. The Rayleigh resolution criterion states that two wavefronts are resolved when the peak of the array response due to the first wavefront falls on the first zero of the response due to the second wavefront (see fig. 7.12). The first zero is located at an angle, $\sin^{-1}\lambda/(Md)$, away from the direction of arrival (broadside). Thus, two wavefronts are resolved, according to the Rayleigh resolution criterion when their directions of arrival differ by $\sin^{-1}\lambda/(Md)$.

Table 7.2: The Rayleigh resolution angle as a function of the number of sensors (ULA with $\lambda/2$ sensor spacing).

No of Sensors	Rayleigh Resolution Angle in deg.(ULA)	Rayleigh Resolution Angle in deg.(UCA)
4	30	30.61
8	14.48	12.67
16	7.18	5.84
32	3.58	2.83
64	1.79	1.39

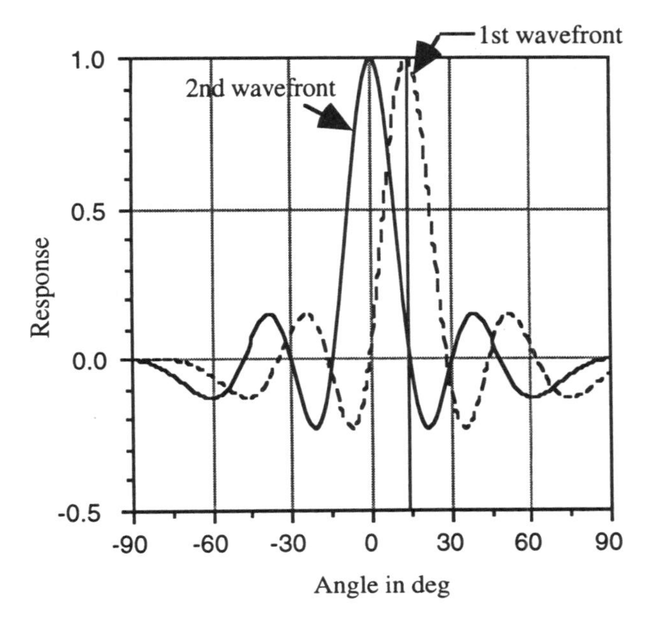

Figure 7.12: The Rayleigh resolution criterion states that two wavefronts are resolved when the peak of the array response due to the first wavefront falls on the first zero of the array response due to the second wavefront.

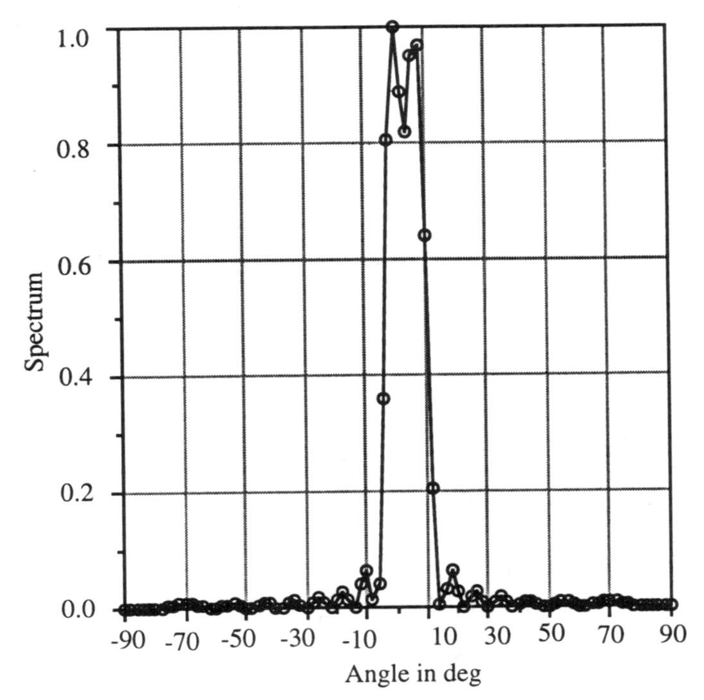

Figure 7.13: Two uncorrelated wavefronts with DOA's, 0° and 7.18^{o}, are incident on a 16-sensor ULA. The waves are clearly resolved. The DOAs were chosen to satisfy the Rayleigh resolution criterion.

Example 7.3: An example of resolution is shown in fig. 7.13. For a UCA we can derive a simple expression when it is fully populated. In this case, its response function is a Bessel function of 0th order. The first zero of the Bessel function of 0th order is at 2.45. Two wavefronts are said to be resolved according to the Rayleigh resolution criterion when the angular separation is greater than $\sin^{-1}(\frac{1.225\lambda}{\pi a})$. Let us compare the resolution properties of a ULA and a UCA. having *equal* aperture, for example, a 16 sensor ULA with 7.5l aperture and the corresponding UCA with a radius equal to 3.75l but fully populated with more than 32 sensors. The relative performance is shown in table 7.2. The performance of the UCA is marginally better than that of the ULA ●.

Beamformation in the frequency domain requires the 2D Fourier transform. For a fixed temporal frequency, the magnitude of the spatial Fourier transform coefficients is related to the power of a wave coming from a direction, which may be computed from the spatial frequency number,

$$\theta = \sin^{-1}(\frac{k}{M})$$

where k is the spatial frequency number[30]. Here too, only a finite number of fixed beams can be formed. This number is equal to the number of sensors. However, the discrete Fourier transform allows interpolation between fixed beams through a simple means of padding zeros or placing dummy sensors giving no output.

Example 7.4: Consider an example of a wavefront incident at an angle of 21.06^{o} on a 16 sensor array. The output is first subjected to the temporal Fourier transform. Then the spatial Fourier transform is performed before and after padding zeros. In fig. 7.14a the spatial Fourier transform before padding is shown. The peak appears at frequency number 6 corresponding to an angle of 22.02^{o} ($\sin^{-1}(6/16)$). Next, the sequence is padded with 48 zeros before Fourier transformation. The result is shown in fig. 7.14b where a peak that appears at frequency number 23 corresponding to an angle 21.06^{o} ($\sin^{-1}(23/64)$), is the correct peak. Note that the correct peak position lies between frequency numbers 5 and 6 (closer to 6). By padding zeros we are able to interpolate between the frequency numbers 5 and 6 and are thus able to capture the peak at its correct position. Further, the peak is better defined, however the peak width remains unchanged. It may be emphasized that by introducing dummy sensors (zeros) we cannot achieve higher resolution ●.

Sources of Error: In practical beamformation we encounter several sources of phase errors such as those caused by sensor position errors, variable propagation conditions, sensor and associated electronics phase errors, quantization error in the phase shifter, etc. The array response is highly prone to these phase errors. Nominally, the array response may be expressed as

$$H(\omega,\theta_0,\theta) = \mathbf{a}^H(\omega\frac{d}{c}\sin\theta)\mathbf{a}(\omega\frac{d}{c}\sin\theta_0)$$

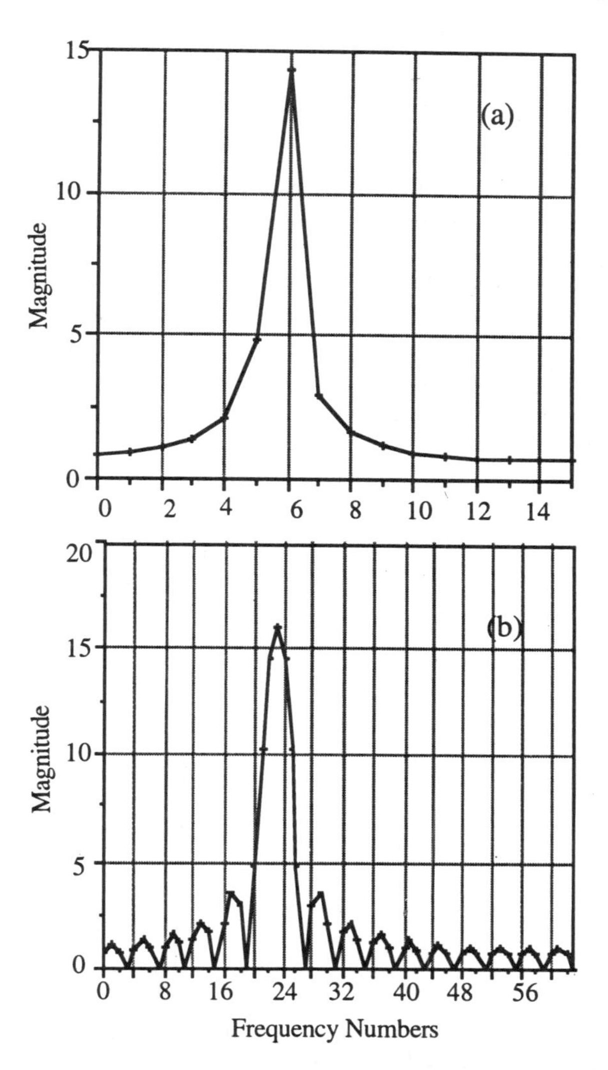

Figure 7.14: The role of padding zeros or introducing dummy sensors is to correctly position the peak (a) before padding zeros and (b) after padding zeros. The beam width remains unchanged.

where θ_0 is DOA of the incident wave and θ is the steering angle. We shall model two types of phase errors, namely, those caused by position errors and phase errors, which are caused by all other sources, lumped into one. The corrupted direction vector has the following form:

$$\tilde{\mathbf{a}} = col\left\{e^{-j\phi_0}, e^{-j(\phi_1+\omega\frac{d+\Delta d_1}{c}\sin\theta_0)}, \ldots, e^{-j(\phi_{M-1}+\omega(M-1)\frac{d+\Delta d_{M-1}}{c}\sin\theta_0)}\right\} \tag{7.32}$$

where Δd_1 is the position error of the ith sensor and ϕ_1 is the phase error. We have assumed that the first sensor is a reference sensor and hence there is no position error. We shall illustrate the effect of position and phase errors on the array response function. We assume that the ULA has 16 sensors which are equispaced but with some position error. Let $d = \lambda/2$ and Δd be a uniformly distributed random variable in the range $\pm\lambda/16$. The resulting response is shown in fig. 7.15. The array response due to phase errors, caused by other factors, is shown in fig. 7.16. The sensor position and phase errors largely affect the sidelobe structure of the response function while the main lobe position and the width remain unchanged.

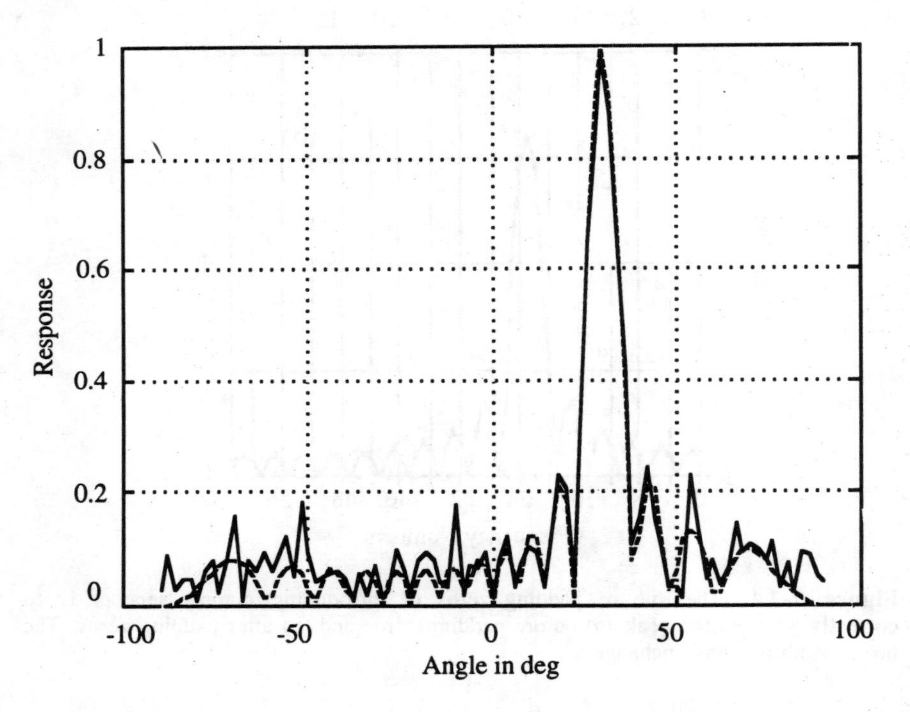

Figure 7.15: Response of a ULA with position errors which are uniformly distributed in the range $\pm\lambda/4$ (solid curve). Compare this with the response of the ULA without any position errors (dashed curve).

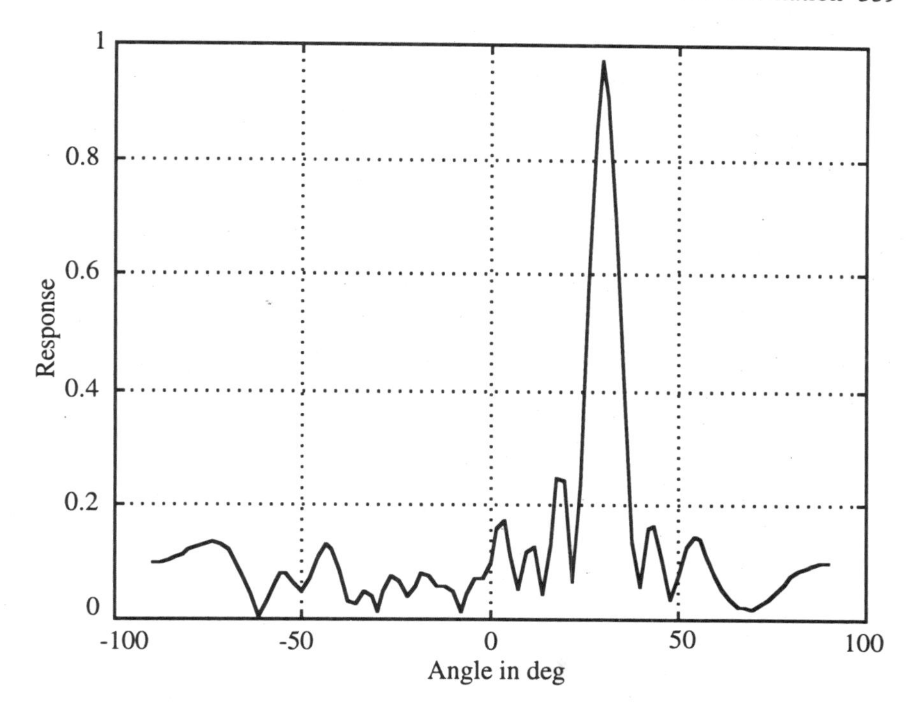

Figure 7.16: Response of a ULA with other phase errors which are uniformly distributed in the range $\pm\pi/4$.

§7.5 **Focused Beam**

When a source is in the far field, the directions of arrival (DOA), azimuth and elevation are of interest. We have already seen in the previous section how a beam is formed in a given direction. On the other hand, when a source is in the near field region, a beam may be formed to receive energy not only from a given direction but also from a given point. This is akin to focusing in an optical system. In seismic exploration the array size is of the same order as the depth of reflectors; therefore, it is often inappropriate to assume a far field or plane wavefront condition. Also, in a conference room the source may be in the near field region of a microphone array.

Focusing: To form a focused beam we must have many common depth point (CDP) gathers which are obtained by means of a specially designed source-receiver array. For example, consider a linear array where every location is occupied either by a source or a sensor. The array is fired as many times as the number of locations. From every reflecting element we obtain a number of gathers equal to the number of receivers. This is illustrated in fig. 7.17. A ULA of four sensors is headed by a source. In position (A) the source is fired and the reflected signal is received by sensor #1. The entire receiver-source array is moved laterally by half sensor spacing as in position (B) and the source is fired once again. The reflected signal is received by sensor #2. This procedure is continued

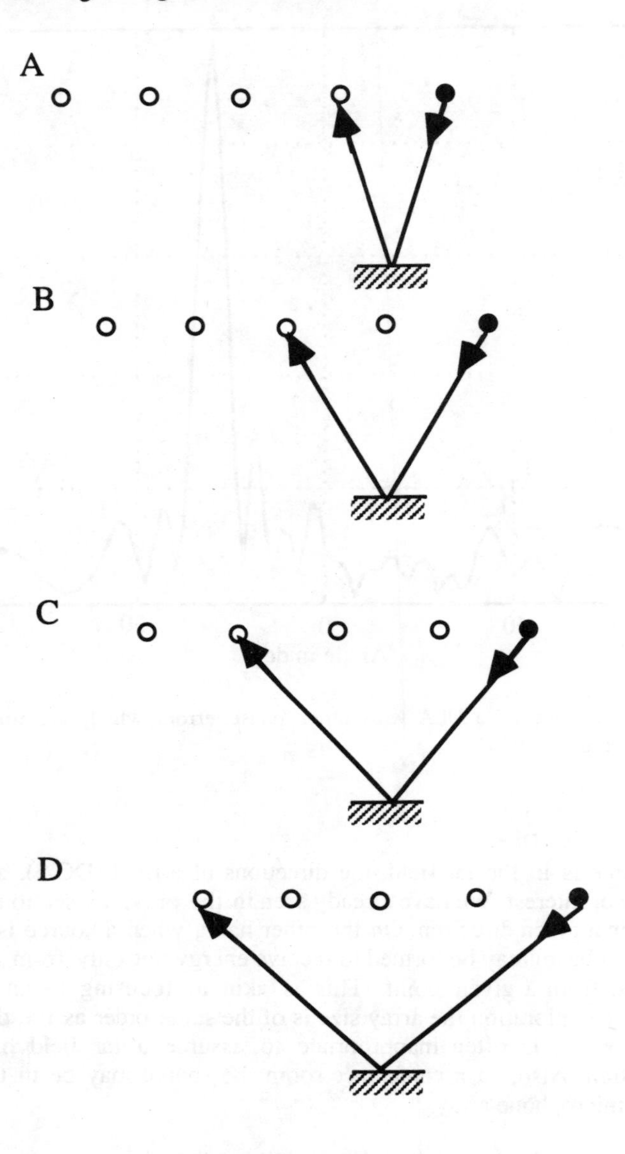

Figure 7.17: A reflecting element is illuminated by a source at different angles. The reflected signal is received by one of the sensors as shown in the figure. The entire receiver-source array is moved laterally by half sensor spacing.

as in (C) and (D). Thus, we get four CDP gathers from the same reflecting element.

Let $T_1, T_2, \ldots T_M$ be the round-trip travel time from the source to sensor #1, #2,...#M, respectively. Let $f_0(t)$ be the signal transmitted by the source and $f_i(t), i = 1, 2, \ldots M$ be the reflected signals received by four receivers. Since these are the delayed versions of $f_0(t)$ we can express them as

$$f_i(t) = f_0(t - T_i), i = 1,2,\ldots M \tag{7.33}$$

Focusing involves firstly correcting for delayed reception and secondly summing coherently after correction. Let the delay correction be given by $\hat{T}_1, \hat{T}_2, \ldots, \hat{T}_M$. These are, for an assumed depth to the reflector, given by [30, p433]

$$\hat{T}_m = \hat{T}_0 \sqrt{(1 + (\frac{\Delta tm}{\hat{T}_0})^2)} \tag{7.34}$$

where $\hat{T}_0 = 2\hat{l}/c$, $\hat{l}$ is assumed depth to the reflector and $\Delta t = d/c$. Recall that d stands for sensor spacing. We assume that the wave speed c is known. The CDP gathers are coherently summed after correction for the delays computed from (7.34). As before we shall assume that the source emits a broadband signal. The coherently summed output may be expressed as follows:

$$\begin{aligned} g(t) &= \frac{1}{2\pi} \int_{-\infty}^{\infty} F(\omega) \frac{1}{M} \sum_{m=0}^{M-1} e^{-j\omega\left[T_0\sqrt{(1+(\frac{\Delta tm}{T_0})^2)} - \hat{T}_0\sqrt{(1+(\frac{\Delta tm}{\hat{T}_0})^2)}\right]} e^{j\omega t} d\omega \\ &= \frac{1}{2\pi} \int_{-\infty}^{\infty} F(\omega) H_M(\omega) e^{j\omega t} d\omega \end{aligned} \tag{7.35a}$$

where

$$H_M(\omega) = \frac{1}{M} \sum_{m=0}^{M-1} e^{-j\omega\left[T_0\sqrt{(1+(\frac{\Delta tm}{T_0})^2)} - \hat{T}_0\sqrt{(1+(\frac{\Delta tm}{\hat{T}_0})^2)}\right]} \tag{7.35b}$$

is the filter transfer function. A numerical example of the transfer function is shown in fig. 7.18. A horizontal reflector is assumed at a depth corresponding to round-trip time equal to 5 seconds. The sensor spacing, measured in units of propagation time, $\Delta t = d/c = 0.1$ seconds.

Depth of Focus: The response function has a finite width. A sharp reflector will now appear as a diffused zone whose width is known as the depth of focus, analogous to that in optics. Ideally, one would like the depth of focus to be as narrow as possible. For the purpose of quantitative measure we shall define the depth of focus as a distance between two 3 dB points on the response function. The depth of focus, measured in the units of $2\Delta l/c$, where Δl is depth of focus, and the array aperture, also measured in terms of the propagation time $(= x/c)$, are shown in fig. 7.19. Notice that the minimum occurs when the aperture size is about four times the round-trip propagation time, in this case five seconds. With further increase in the aperture size the depth of focus rapidly deteriorates. The minimum depth of focus appears to be independent of depth to the reflector; however, the required array aperture increases rapidly as the reflector depth increases. The dependence of the depth of focus on the frequency is significant, as shown in fig. 7.20. Notice that the depth of focus becomes very narrow beyond about 50Hz.

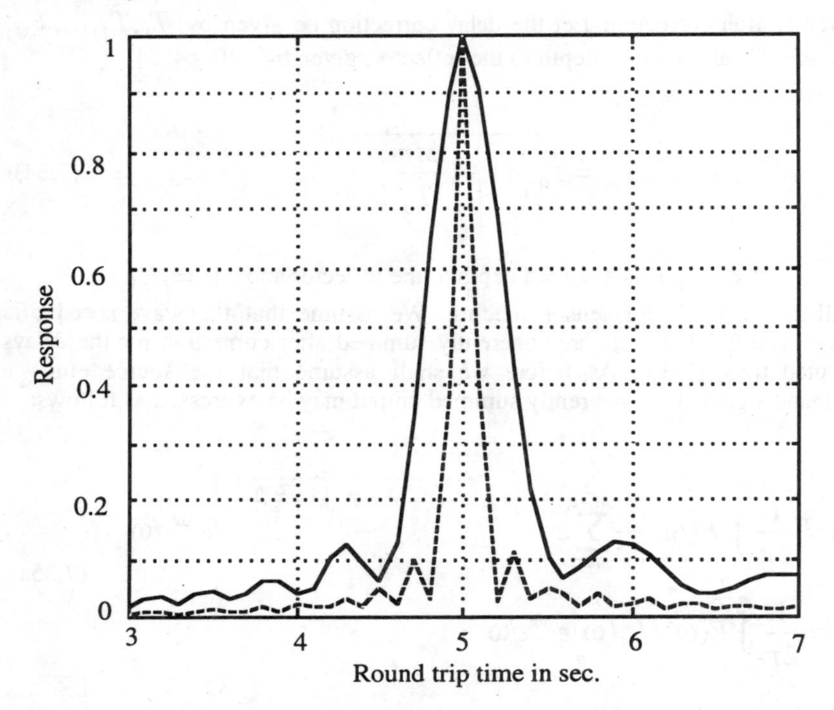

Figure 7.18: Response function of focused beamformation. The reflector is at a depth corresponding to 5 sec round trip time. The array aperture is measured in units of propagation time x/c. The solid line is for array aperture of 2.5 seconds and the dashed line is for 5.0 seconds. Further, the angular frequency is assumed to be 100 radians/sec.

§7.6 Application to Interference Cancellation

Interference cancellation is an important requirement in many applications. An array of sensors may be effectively used to cancel the interference when the signal source and interference source are spatially separated and their directions are known. In this case the array may be electronically steered simultaneously in the direction of signal and in the direction of interference. The estimate of interference obtained from the array is used to predict and then cancel the interference present in the signal. We shall demonstrate the effectiveness of this approach by considering a signal and interference as a pure tone of the same frequency but of different amplitude.

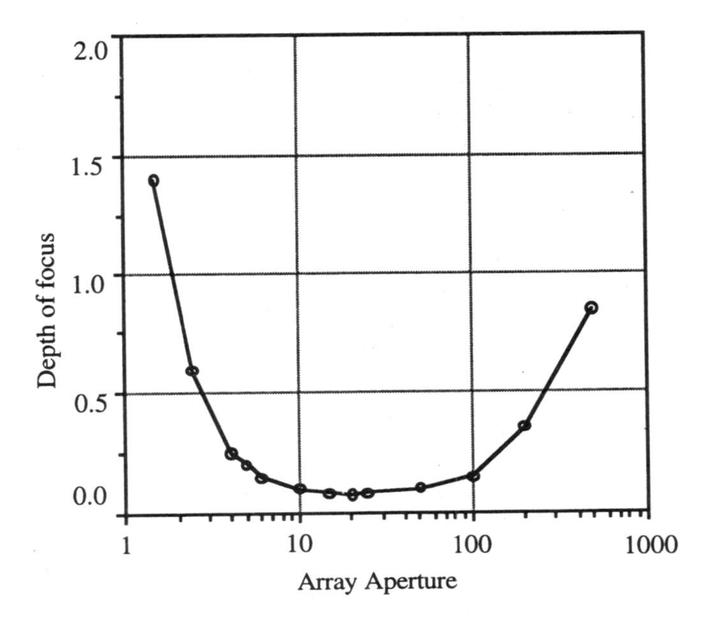

Figure 7.19: Depth of focus in units of $2\Delta l/c$ (sec), where Δl is depth of focus, as a function of array aperture, also measured in units of propagation time (sec).

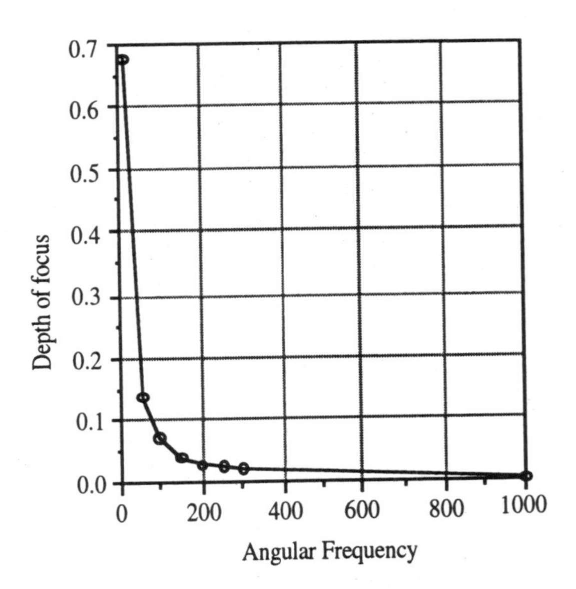

Figure 7.20: Depth of focus vs angular frequency. The reflector is at a depth of 5 seconds (round-trip travel time). The array aperture is held fixed at 20 seconds (propagation time).

Steered array: Both signal and noise sources are in the far field region but the DOAs of their wavefronts are different. Let τ_0 and τ_1 be the incremental delays produced by the signal wavefront and noise wavefront, respectively. The array can be steered to receive the signal or noise at the same time (see fig. 7.21)[31]. Let $f_1(t)$ be the output of an array when it is steered to the signal wavefront and $f_2(t)$ be the output of an array when steered to the noise wavefront. Since the array response function has finite sidelobes, some amount of wave energy will leak through the sidelobes. Hence, we model the array output as

$$f_1(t) = \xi_0(t) + \frac{1}{2\pi}\int_{-\infty}^{\infty} \mathrm{N}_1(\omega) H(\omega(\tau_0 - \tau_1)) e^{j\omega t} d\omega \tag{7.36a}$$

and

$$f_2(t) = \eta_1(t) + \frac{1}{2\pi}\int_{-\infty}^{\infty} \Xi_0(\omega) H(\omega(\tau_1 - \tau_0)) e^{j\omega t} d\omega \tag{7.36b}$$

where $\Xi_0(\omega)$ is the Fourier transform of the signal and $\mathrm{N}_1(\omega)$ is that of the noise. Note that in equation (7.36) $H(\omega(\tau_0 - \tau_1))$ is the response of the array in the direction of the noise when the array is tuned to the signal. Similarly, $H(\omega(\tau_1 - \tau_0))$ is the response of the array in the direction of signal when the array is tuned to the noise. We note that $f_1(t)$ is the signal with a small component of noise and $f_2(t)$ is the noise with a small component of signal. Therefore, $f_2(t)$ may be used to predict the noise component in $f_1(t)$. Let $h_{pred}(t)$ be the prediction filter which may be found by minimizing

$$E\left\{\left| f_1(t) - \int_0^{\infty} f_2(t' - t) h_{pred}(t') dt' \right|^2\right\} = \min$$

The result of above minimization is, in frequency domain [30],

$$H_{pred}(\omega) = \frac{S_{12}(\omega)}{S_2(\omega)} \tag{7.37a}$$

where $S_{12}(\omega)$ is the cross spectrum between $f_1(t)$ and $f_2(t)$ and $S_2(\omega)$ is the spectrum of $f_2(t)$. We compute $S_{12}(\omega)$ and $S_2(\omega)$ from (7.36) and obtain

$$H_{pred}(\omega) = \frac{S_{\xi_0}(\omega) H^*(\omega(\tau_1 - \tau_0)) + S_{\eta_1}(\omega) H(\omega(\tau_0 - \tau_1))}{S_{\xi_0}(\omega) |H(\omega(\tau_1 - \tau_0))|^2 + S_{\eta_1}(\omega)} \tag{7.37b}$$

where $S_{\xi_0}(\omega)$ is signal spectrum and $S_{\eta_1}(\omega)$ is noise spectrum. Note that, since

$$H(\omega(\tau_0 - \tau_1)) = H^*(\omega(\tau_1 - \tau_0)) \tag{7.37c}$$

reduces to

$$H_{pred}(\omega) = \frac{1 + SNR_{input}}{1 + SNR_{input}\left|H(\omega(\tau_1 - \tau_0))\right|^2} H(\omega(\tau_0 - \tau_1)) \tag{7.37d}$$

where

$$SNR_{input} = \frac{S_{\xi_0}(\omega)}{S_{\eta_1}(\omega)}$$

Let us now consider a few special cases:

(a) When $SNR_{input} >> 1$ and $SNR_{input}\left|H(\omega(\tau_1 - \tau_0))\right|^2 >> 1$

$$H_{pred}(\omega) \approx \frac{1}{H(\omega(\tau_1 - \tau_0))} \tag{7.37e}$$

When this filter is used on $f_2(t)$ (see eq.(7.36b)) for predicting the noise in $f_1(t)$, the signal component will be restored causing the cancellation of the signal.

(b) $SNR_{input}\left|H(\omega(\tau_1 - \tau_0))\right|^2 << 1$

$$H_{pred}(\omega) \approx (1 + SNR_{input})H(\omega(\tau_0 - \tau_1))$$

When this filter is used on $f_2(t)$ (see eq.(7.36b)) for predicting the noise component in $f_1(t)$ the noise component will be largely canceled without canceling the signal.

Example 7.5: As an illustration, we consider two pure sinusoidal signals (of same frequency) arriving with different DOAs (0° and 5.7^o) at a ULA of 16 sensors spaced at $\lambda/2$ spacing. The second sinusoid arrives after 50 time units with an amplitude of 0.8. Fig. 7.22(a) shows a sum of the two tones as received by the first sensor. The array is steered in the direction of the first sinusoid and at the same time in the direction of the second sinusoid. The array outputs are given by (7.36), which is now considerably simplified for pure sinusoidal inputs.

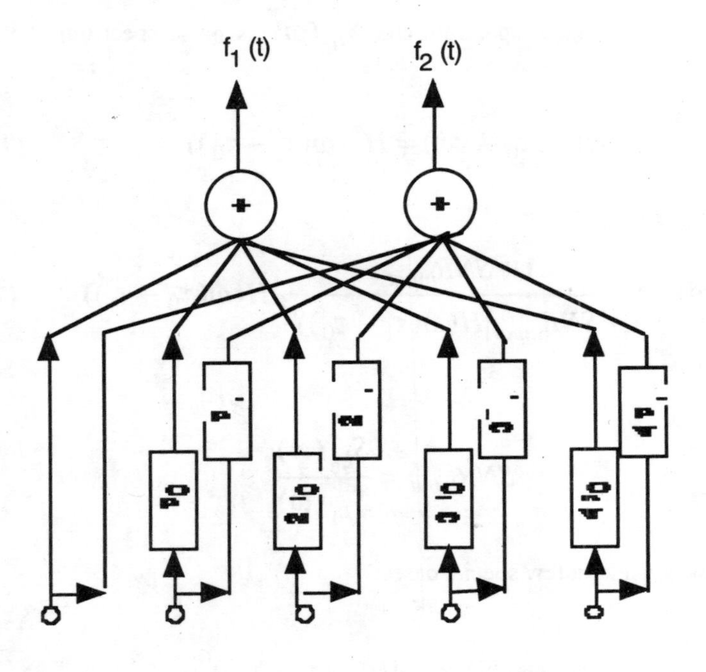

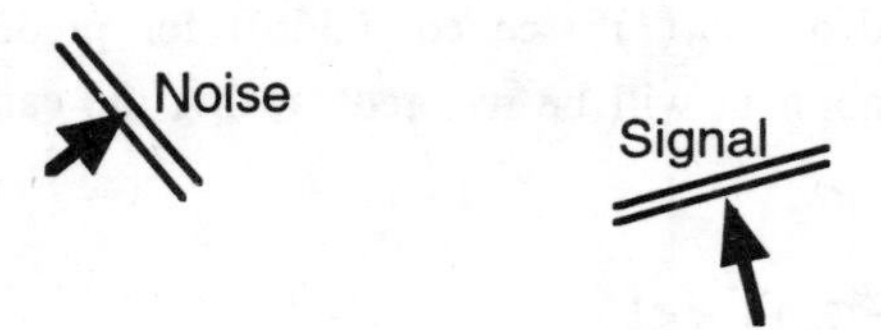

Figure 7.21: An array of sensors can be steered simultaneously in the direction of the signal and in the direction of noise. When the array is steered in the direction of signal the output $f_1(t)$ is mostly signal and when it is steered in the direction of noise the output $f_2(t)$ is mostly noise.

$$\begin{aligned} f_1(t) &= s_1(t) + s_2(t)H(\omega_0(\tau_0 - \tau_1)) \\ f_2(t) &= s_2(t) + s_1(t)H(\omega_0(\tau_0 - \tau_1)) \end{aligned} \tag{7.38}$$

where $s_1(t)$ and $s_2(t)$ are the first and the second sinusoid, respectively and ω_0 is the frequency of the sinusoids. Solving (7.38) we obtain, for $|H(\omega_0(\tau_0 - \tau_1))|^2 < 1$,

$$s_1(t) = \frac{f_1(t) - f_2(t)H(\omega_0(\tau_0 - \tau_1))}{1 - H^2(\omega_0(\tau_0 - \tau_1))}$$

$$s_2(t) = \frac{f_2(t) - f_1(t)H(\omega_0(\tau_0 - \tau_1))}{1 - H^2(\omega_0(\tau_0 - \tau_1))} \tag{7.39}$$

The results are shown in figs. 7.22 (b & c).

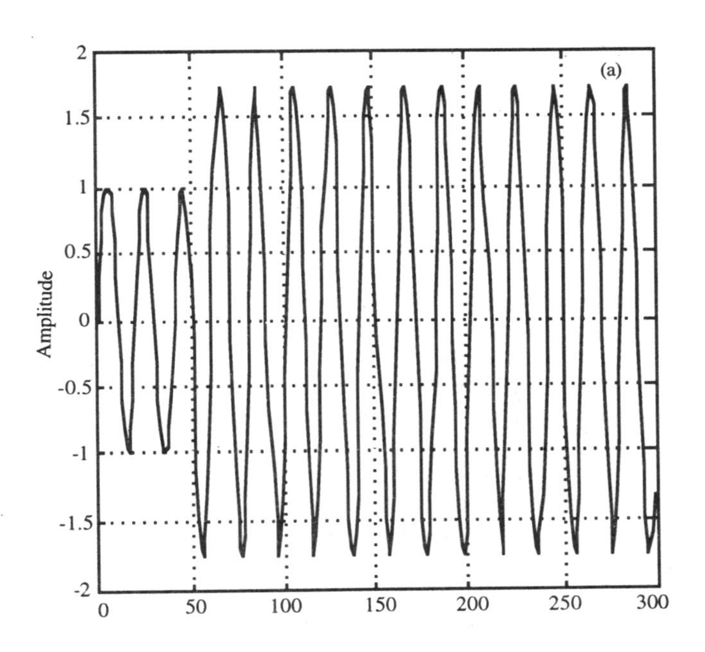

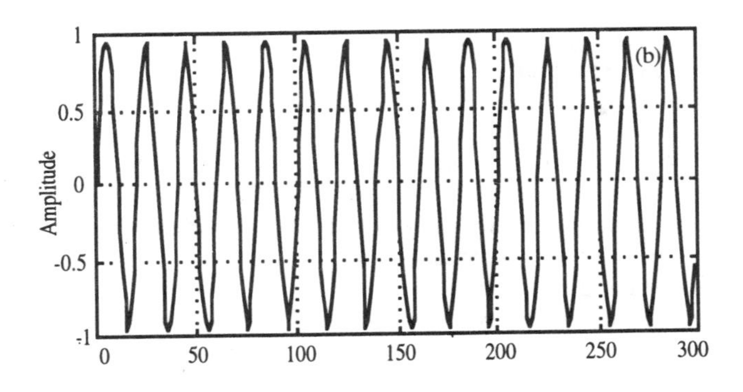

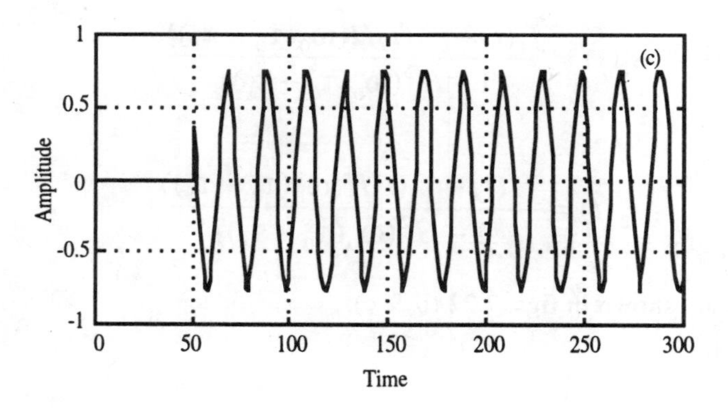

Figure 7.22: (a) sum of two sinusoids, (b) first sinusoid after subtraction and (c) second sinusoid after subtraction .

Adaptive Sidelobe Canceller: The simplest form of adaptive antenna is the sidelobe canceller originally developed by Howells[40] in late 1950 and later by Applebaum [32]. It consists of a primary sensor pointing to the signal source of interest and another secondary sensor to receive the interference. The two sensors are separated physically and hence the relative strengths of the signal and noise are different at the two locations. While the primary sensor has more signal power the secondary sensor has more interference power. The output of the primary sensor is similar to the output of an array steered to the signal source and the output of the secondary sensor is similar to the output of an array steered to the interference source. The interference in the primary output is due to the leakage of the interference power through the sidelobes of the array response function. The output of the secondary sensor is now used for the predictive cancellation of the interference in the primary output. A schematic of the Howells-Applebaum sidelobe canceller is shown in fig. 7.23.

The adaptive filter, upon convergence, is able to predict the interference in the primary channel. Upon subtraction the system output is close to the signal which is assumed to be uncorrelated with the interference. The presence of small signal power in the secondary channel does not seem to affect the performance. The adaptation is largely controlled by the dominant interference component. We have for the sake of simplicity ignored the independent receiver noise present in each channel.

Example 7.6 The effectiveness of the adaptive sidelobe canceller is demonstrated in the example given here. We consider a pure sinusoid as a signal arriving simultaneously at both the sensors. The interference is a narrowband (NB) stochastic process with a center frequency equal to the signal frequency. The interference arrives at the primary sensor with a delay of 0.6 sec. The signal-to-interference power ratio in the primary channel is equal to 0.65dB and that in the secondary channel is equal to -40dB. Naturally, the interference is uncorrelated with the signal. Recursive least-squares (RLS) algorithm (see Chapter Six where

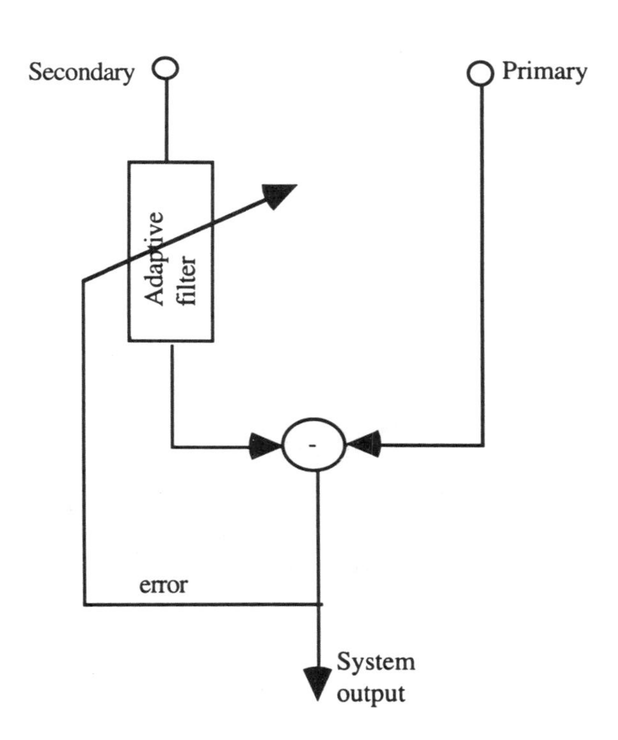

Figure 7.23: Howells-Applebaum adaptive sidelobe canceller.

RLS algorithm is described) was used. A finite impulse response (FIR) filter of length six coefficients was adaptively generated using a data length of 1000 points. The filter coefficients are: 0.1075, -0.1793, 0.0945,-0.1691, 0.0567, -0.0716. The primary sensor outputs before and after noise cancellation are shown in fig. 7.24. Much of the interference has been removed. This is more dramatically shown in the frequency domain. (Carry out the computer project # 1). There is a very low frequency modulation riding over the cleaned up signal. It is observed that the low frequency modulation is data length dependent and disappears at very large data lengths

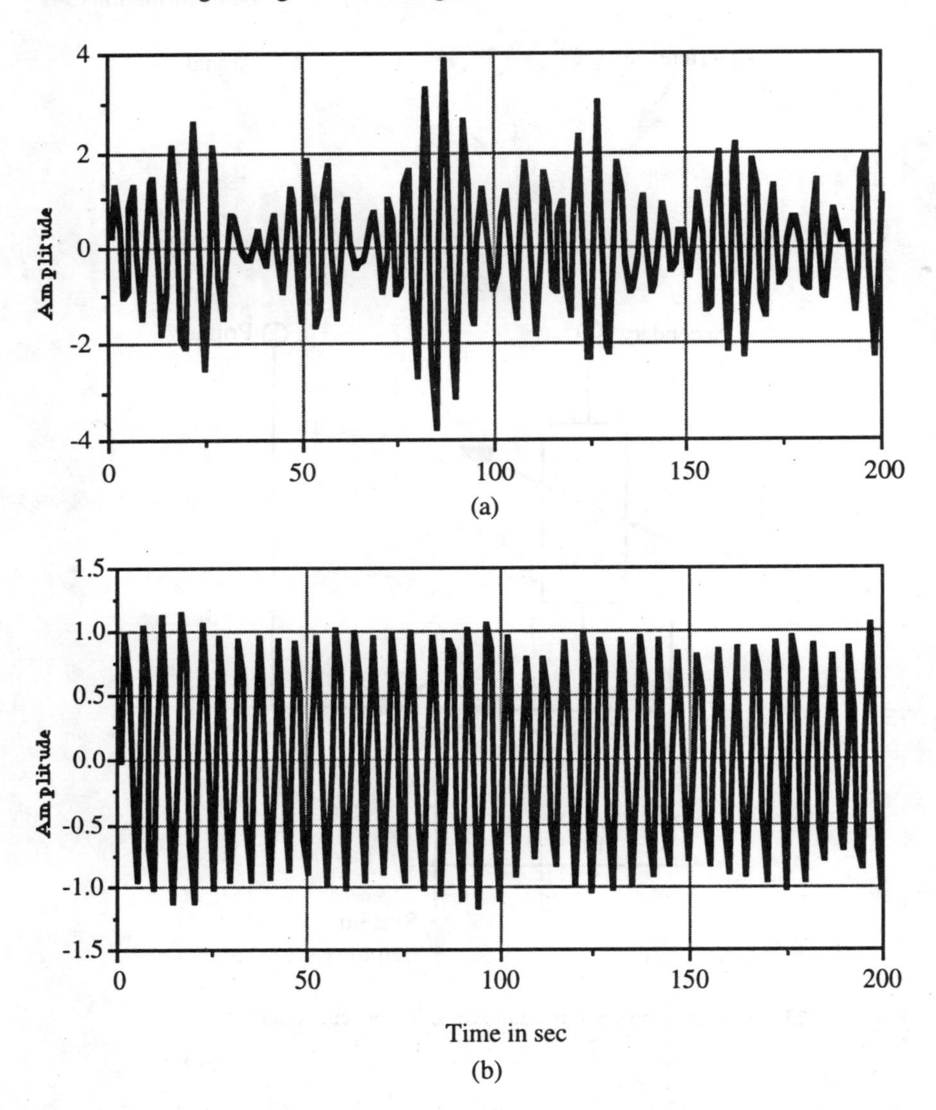

Figure 7.24: Signal waveform is a pure sinusoid (of frequency 0.2 Hz) and the interference is a narrowband (NB) signal with a center frequency 0.2 Hz and a bandwidth equal to 0.1Hz. (a) Signal observed by the main antenna (b) signal after noise cancellation .

§7.7 Adaptive Beamformation

In beamformation the array response is pointed in the desired direction, that is, the main lobe of the response function is pointed in the direction of the desired source. The sidelobes and nulls are largely controlled by the array geometry and individual sensor response. Consequently, if there is an interference, that is, a source of no interest it will come in the way of the desired source unless one of the nulls of the response function happens to be in the direction of the interference. It is, however, possible to position one of the nulls of the array response in the direction of a known interference by changing the geometry of the array or the weights (window function). It is indeed impractical to alter the array geometry each time a source of interference is encountered. On the other hand, it is possible to design a new set of weights that will null one or more known interferences. Whenever the position of interferer is unknown or changing rapidly, the process of designing appropriate weights becomes impractical. We have to explore other alternatives where the nulls are automatically positioned in the direction of all unknown sources.

Adaptive Nulling: We would now like to show that a weight vector of length M can be found such that its response function has nulls in the direction of unknown interferences. Consider an ULA of M sensors. For simplicity assume that a signal wavefront is incident on broadside (DOA=0^0) and two interference wavefronts are incident at an angle θ_1 and θ_2 (DOA),

$$\mathbf{f}(t) = \mathbf{a}(0)f_s(t) + \mathbf{a}(\theta_1)f_{i_1}(t) + \mathbf{a}(\theta_2)f_{i_2}(t) \tag{7.40}$$

where $\mathbf{a}(0) = [1,1,...,1]$ is the direction vector of a wavefront incident at broadside; similarly $\mathbf{a}(\theta_1)$ and $\mathbf{a}(\theta_2)$ are direction vectors of wavefronts incident at angles θ_1 and θ_2, respectively. Let $f_r(t)$ be the known reference signal. We seek a weight vector $\mathbf{w}$ such that

$$E\{(\mathbf{w}^H\mathbf{f}(t) - f_r(t))^2\} = \min. \tag{7.41a}$$

The left-hand side of (7.41a) may be written as

$$\text{lhs}=\sum_{i=1}^{M} w_i^2\sigma_{f_s}^2 + \left|\mathbf{w}^H\mathbf{a}(\theta_1)\right|^2\sigma_{f_{i_1}}^2 + \left|\mathbf{w}^H\mathbf{a}(\theta_2)\right|^2\sigma_{f_{i_2}}^2 + \sigma_{f_r}^2 -$$
$$2\sum_{i=1}^{M} w_i E\{f_s(t)f_r(t)\} \tag{7.41b}$$

We have assumed that the signal is uncorrelated with the interferences but it is strongly correlated with the reference signal. Further, assume for simplicity that

$$\sum_{i=1}^{M} w_i^2 = 1 \qquad \text{and} \qquad \sum_{i=1}^{M} w_i = 1$$

Eq.(7.41a) may be reduced to

$$\left|\mathbf{w}^H\mathbf{a}(\theta_1)\right|^2\sigma_{f_{i_1}}^2 + \left|\mathbf{w}^H\mathbf{a}(\theta_2)\right|^2\sigma_{f_{i_2}}^2 = \min \tag{7.42}$$

Since both terms on the left-hand side of (7.42) are positive, to minimize the sum, each term must be minimized. For an ULA with omnidirectional sensors the direction vector from (7.19) can be written as follows:

$$\mathbf{a}(\theta) = col\left\{1, e^{-j\omega_c \frac{d}{c_x}}, \ldots, e^{-j(M-1)\omega_c \frac{d}{c_x}}\right\}$$

$$= col\{1, z^{-1}, z^{-2}, \ldots, z^{-(M-1)}\}$$

where

$$z = e^{j\omega_c \frac{d}{c}\sin(\theta)}$$

Consider a polynomial in z whose coefficients are the elements of the weight vector. We select the weight vector such that the roots of the above defined polynomial are equal to $z_1 = e^{j\omega_c \frac{d}{c}\sin(\theta_1)}$ and $z_2 = e^{j\omega_c \frac{d}{c}\sin(\theta_2)}$ Then, the terms on the left-hand side of (7.42) would vanish. Conversely, given the direction vectors or the roots we can determine the coefficients of the polynomial. Since a polynomial of order M-1 has M-1 roots (assumed to be all different) M-1 interferences can, in principle, be nulled. But, in practice, to reach this theoretical limit we may require long iteration times.

LMS Algorithm: In chapter six we have shown that the LMS algorithm offers a simple and easy solution to minimization problem. Not only that, applications, where interference suppression is vital, also demand real-time solution. The LMS algorithm is most suited for such applications. Here we describe a computer experiment on the design of a weight vector. A seven-sensor array is assumed. A known signal source is incident on broadside and two interferers are at 30^0 and 45^0 angles (DOA) with respect to the broadside. All three signals are narrowband uncorrelated stochastic signals. The center frequency is 100Hz and the bandwidth is 10Hz. The sensors are spaced at $\lambda/2$. The signal, which is incident on broadside, is assumed to be known or it is treated as a pilot signal. The interference to signal power ratio is 10 dB per source. The output (real) of each sensor is passed through a quadrature filter (see fig. 7.6) whose outputs are inphase and quadrature phase real signals. Thus, seven sensors will give rise to in all fourteen real outputs and hence a weight vector of length fourteen (see fig. 7.25). The signals are sampled at a rate of 250 samples per second. The LMS algorithm is set up with the pilot signal as the reference signal and μ=0.00001. The first experiment involved only one interfeıence incident at 45^0.

The weights at the end of 1024 iterations and at the end of 2048 iterations are tabulated in table 7.3. The difference is of the order of 2%. The directivity response of the weights (after 1024 iterations) is shown in fig. 7.26. The pilot signal is incident at broadside. The direction response of uniform weights is shown by thin curve. For comparison the directivity response of uniform weights is also shown in the same figure (thin curve). Note that the x-axis in fig. 7.26 (and also in fig. 7.28) is expressed in terms of sin(θ), that is, sine of the DOA. As designed, the directivity response of the weight vector has a deep null at $\theta = 45^0$ where as for uniform weight vector directivity response has a side lobe at this location. The square of the error between the reference and the estimated as iterations proceed is shown in fig. 7.27. There is a steep fall in the error after about 300 iterations. The weight coefficients begin to stabilize at this stage.

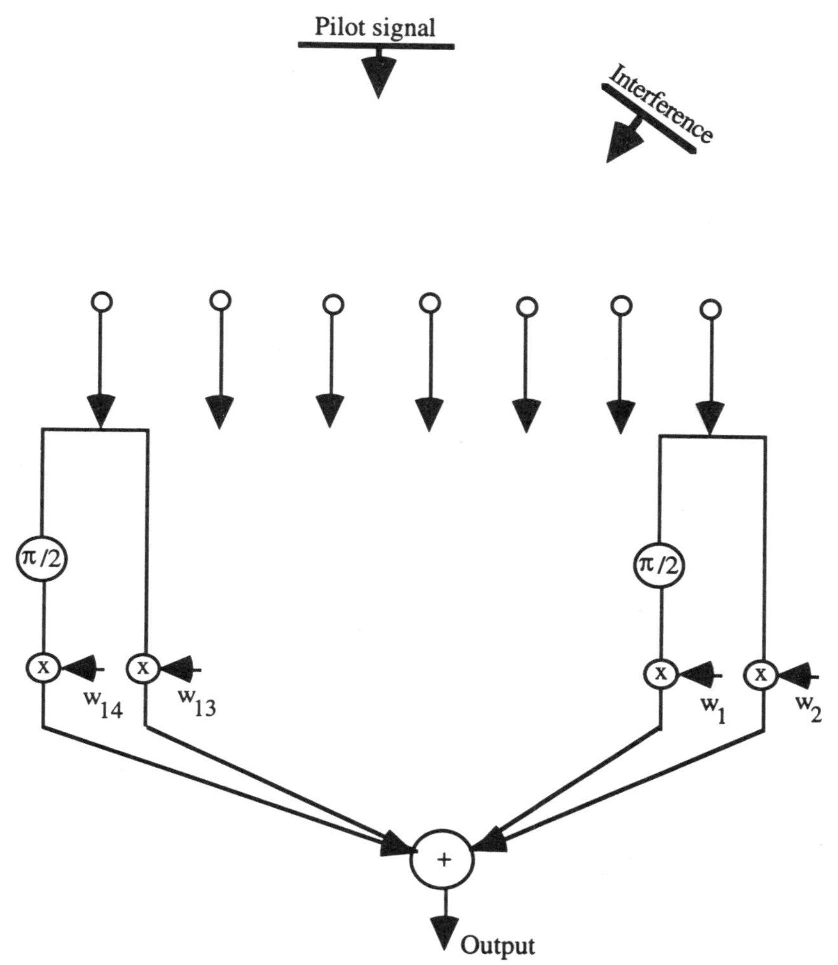

Figure 7.25: A seven sensor uniform linear array (ULA). The output of each sensor is a real narrowband signal which is divided into inphase and quadrature phase components. A pilot signal is incident at broadside and two interference sources are incident at 30^0 and 45^0 (but only one source is shown). A weight vector consisting of fourteen elements is obtained to suppress the interferences but allow the pilot signal.

In the second experiment we have assumed two independent interferences at 30^0 and 45^0. Both interferences are of equal magnitude, ten times the signal power. All other parameters remain the same as in the previous experiment. The directivity response of the weights after 2048 iterations is shown in fig. 7.27. The two nulls at the positions of the interferences are marked by vertical lines. Interestingly, the result at the end of 1024 iterations was not very conclusive. It was also noted that the weights had not stabilized even after 1024 iterations. It would thus appear that as the number of interferences increases the convergence becomes slow.

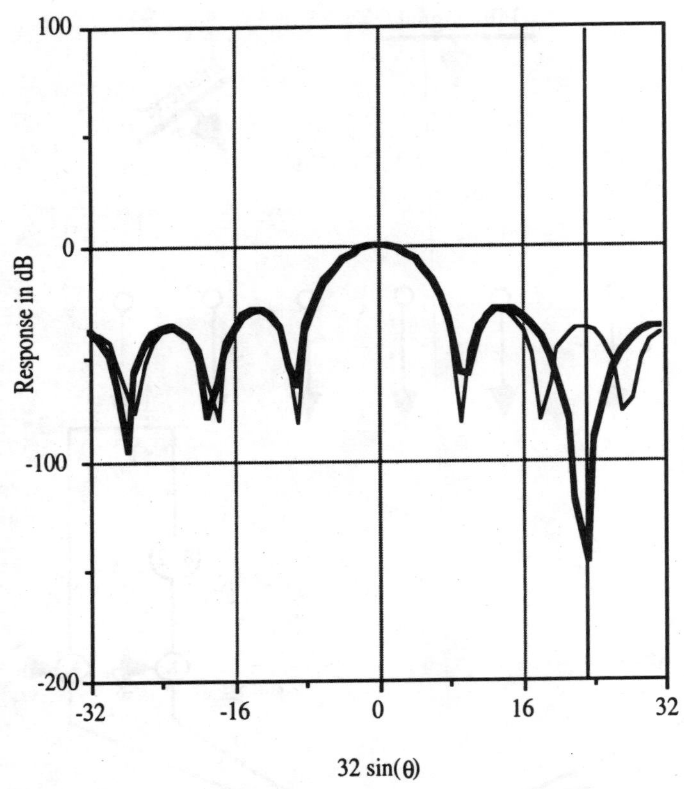

Figure 7.26: Directivity response of the weight coefficients (thick curve) listed in table 7.3: A deep null at $\theta=45^0$ is for suppressing a strong coherent interference.

Table 7.3: A list of weight coefficients after 1024 and 2048 iterations.

Weights	After 1024 steps	After 2048 steps
w_1	0.1235	0.1249
w_2	-0.0075	-0.0087
w_3	0.1497	0.1528
w_4	0.0234	0.0225
w_5	0.1584	0.1607
w_6	-0.0162	-0.0185
w_7	0.1216	0.1233
w_8	0.0009	0.0000
w_9	0.1575	0.1607
w_{10}	0.0198	0.0185
w_{11}	0.1509	0.1528
w_{12}	-0.0202	-0.0225
w_{13}	0.1230	0.1249
w_{14}	0.0093	0.0087

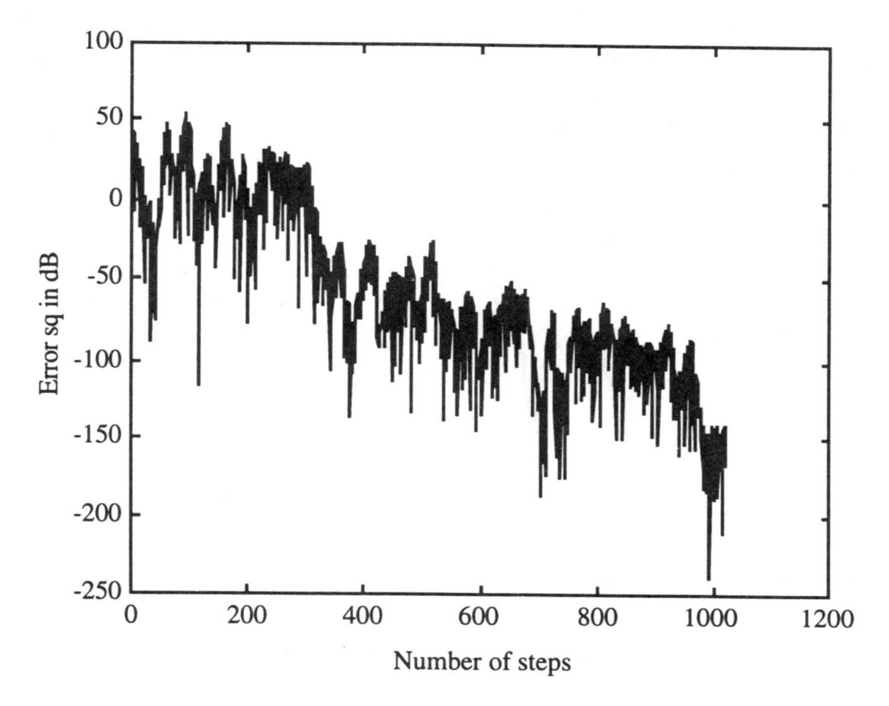

Figure 7.27: Error square (in dB) as a function of the number of iterations. Only one interference is assumed.

§7.8 Exercises

Problems

1. We like to design a ULA. It is known that the direction of arrival (DOA) of the incident signal can never exceed $\pm 60^0$. How would you use this information in optimizing the array? Remember that the array must be as large as possible for a fixed number of sensors.

2. Consider an equilateral triangular array. The sensors are at the corners of the triangle, 4λ apart (see fig. 7.29). A plane wave signal is incident on the array, $\theta = \frac{\pi}{2}$ and $\varphi = \frac{\pi}{3}$. Compute the relative delays (with reference to the fictitious sensor at the center).

3. In sonar signal processing we commonly encounter following situation: Since the water/air interface is highly reflective in addition to the direct path we have a ray coming via the reflecting interface (see fig. 7.30). Compute the directions of signal arrival and the relative delay of the two paths shown in fig. 7.30. Assume suitable numbers for the horizontal distance, depth to source and depth to sensor

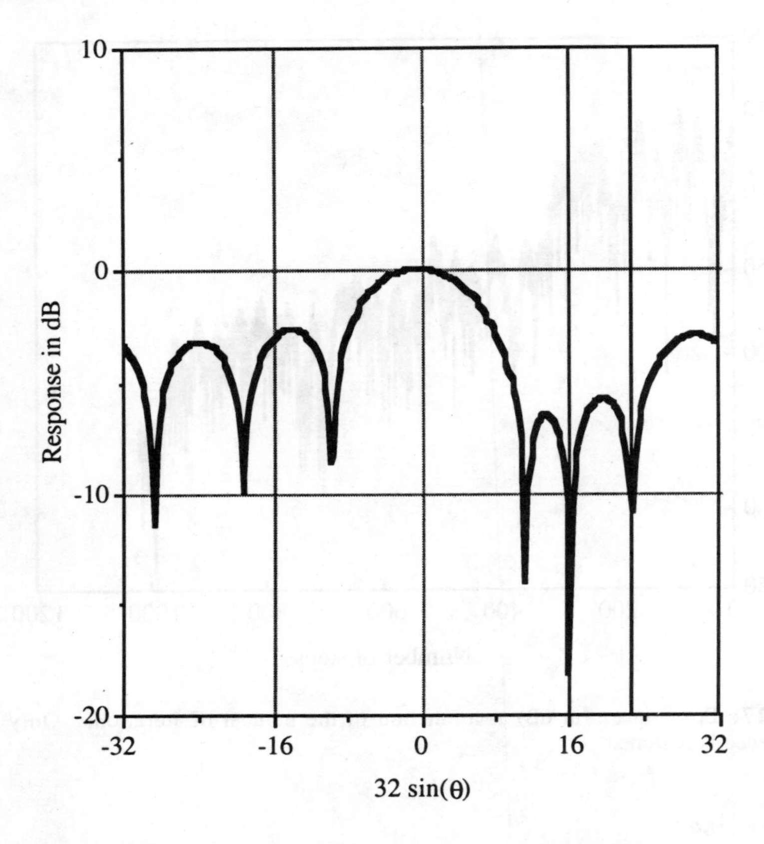

Figure 7.28: Direction response of the weight coefficients (thick curve). Two strong interferences are incident at 30^0 and 45^0. The pilot signal is incident at broadside. The result shown above was obtained after 2048 iterations.

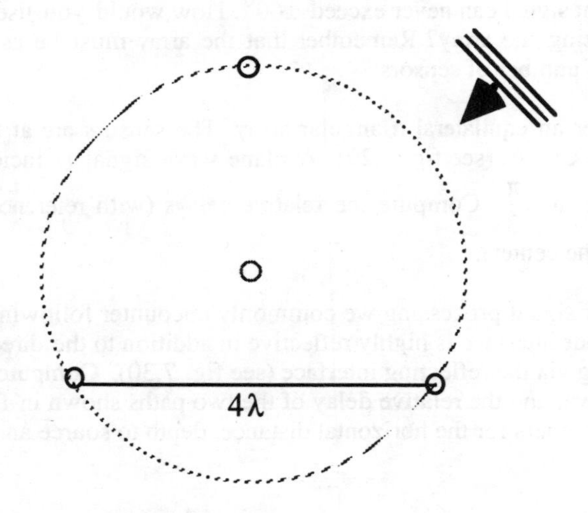

Figure 7.29: A triangular array. A fictitious sensor is placed at the center for the purpose of time reference.

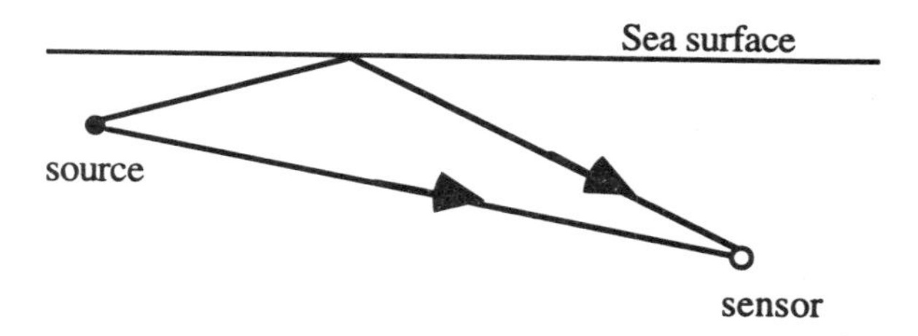

Figure 7.30: Signal propagation in the ocean medium.

4. In normal beamformation the output of each sensor is first delayed and then summed together. But in adaptive beamformation we preferred to form a weighted linear combination. Is there any particular reason for this preference?

5. How would you modify the adaptive beamformation scheme when the pilot signal is incident at some angle (not broadside) but known angle?

6. What is apparent speed? A wavefront is incident with azimuth angle=40^0 and elevation angle 30^0. Compute the apparent speeds along the three axes of coordinates. For what value of the elevation angle does the apparent speeds become infinite? What is the physical significance of this phenomenon?

7. From eq. (7.29a) show that the narrowband approximation for an ULA with M sensors is $\Delta\omega << 2\pi \big/ (M\frac{d}{c}\sin\theta_0)$ where d is sensor spacing. Let the center frequency be 200p, M=16, c=1500 m/s and $\theta_0 = 30^\circ$. Find the maximum permissible bandwidth of the array.

8. Two plane waves having same temporal frequency of 100 Hz. are propagating in a homogeneous medium with a speed of 1500 m/s. The first wave is traveling in the direction with azimuth=60^0 and elevation=30^0. The second wave is traveling in the direction with azimuth=120^0 and elevation=45^0. Represent the above waves in the frequency-wavenumber space.

9. A narrowband signal, incident on a ULA, is over sampled at 12 times the Nyquist rate. How many beams can be formed in digital domain? What are the directions of arrival of the beams?

10. What is the major limitation of a circular array in digital beamformation? How is this limitation overcome in practice?.

11. The steering vector of a circular array possesses some interesting properties as listed on page 323, and illustrated in fig. 7.10. Show that, in addition to what is shown in fig. 7.10, wavefronts traveling in opposite directions can be distinguished.

12. Consider a two-sensor array (e. g. human ears) spaced at $\frac{\lambda}{2}$ apart. Simplify eq. (7.10) for M=2 and show that the nulls of the response function are located at $\theta = \pm\frac{\pi}{2}$ (i.e. endfire diretion).

13. When does the quadratic component of a narrowband signal vanish? Show that for such a signal we encounter the so-called east-west ambiguity.

Computer Projects

1. Re-do the Example 7.6. Use the program you have developed for implementation of RLS algorithm in Chapter Six. Let all parameters remain the same as in the example. Now compute the spectrum (simply magnitude square of the FDFT coefficients) of the output of the primary sensor before and after the noise cancellation. A narrowband stochastic process can be generated by passing white noise through a bandpass filter. Alternative method, as a sum of large number of sinusoids, is described in [30].

2. Map the real weights listed in table 7.3 into complex weights. All even numbered weights form the real part and all odd numbered weights form the imaginary part of the complex weights. Compute the roots of a polynomial whose coefficients are the complex weights. Use the m-file "roots" from Matlab. Plot the roots in the *z*-plane. Which root(s) represents the deep null shown in fig. 7.28 ?

3. This problem looks into the effect of failed sensors. Consider a 64 sensor ULA. Ten sensors may be assumed to have failed at any time instant but the failed sensors are randomly distributed. Carry out a Monte Carlo experiment to bring out the effect of the failed sensors.

4.The effect of position errors on the response of ULA was shown in fig. 7.15. The position errors were assumed to be uniformly distributed in the range $\pm\lambda/4$. We like to carry out a similar study for the circular array. Assume that the sensors are misplaced in the radial direction only. Further, assume that $\theta_0 = \pi/2$ and $\varphi_0 = 0$. Let the radial position errors be uniformly distributed in the range $\pm\lambda/4$. Compute the array response function as a function of φ. How does this compare with that shown in fig. 7.15 for ULA.

8 Digital Signal Processors

Typical signal processing tasks may be grouped into three types: (a) Digital filtering (b) Signal analysis (FDFT, autocorrelation, power spectrum etc.) and (c) Singular value decomposition (SVD) of the correlation or spectral matrix. In this work we shall confine to (a) and (b) types of signal processing tasks. The SVD is generally a part of advanced signal processing. All these tasks are routinely implemented on a desk top computer or a large computer installation. This arrangement works well where the signal is stored and the processing is done off-line as in geophysical signal processing, biomedical signal processing, etc. But, there are many applications where real-time processing is required, for example, in communication both in wireline or wireless mode, in radar and sonar (excluding mapping tasks) or even in intelligent toys. In many of these on-line applications there is a further need for lightweight (for easy mobility) and low power consumption as the processor has to depend upon a battery for power supply. Hence, the most important attributes of a signal processor are (i) high processing speed, (ii) light weight, (iii) low power consumption and of course (iv) low cost.

The microprocessors used in personal computers are optimized for tasks involving data movement and inequality testing. The typical applications requiring such capabilities are word processing, database management, spread sheets, etc. When it comes to mathematical computations the traditional microprocessor are deficient particularly where real-time performance is required. Digital signal processors are microprocessors optimized for basic mathematical calculations such as additions and multiplications. Furthermore, a typical digital signal processor will have multiple arithmetic and logical units (ALU), memory units and independent buses to facilitate exploitation of parallel and pipeline structure in a DSP algorithm. There are two different types of digital signal processors:

(a) Application specific integrated circuit (ASIC): A piece of hardware performing a single specified task, for example, an FIR filter with tunable coefficients.

(b) Fully programmable microprocessor capable of performing a wide variety of signal processing tasks including FIR, IIR filters, FFT, spectrum analysis, auto and cross correlation, etc.

§8.1 Basic Computations in Signal Processing

The most important computational tasks in many common DSP applications are sum-of-products in digital filtering and butterfly evaluation in fast Fourier transform (FFT). A digital signal processor will have to be optimized to evaluate these mathematical computations.

Sum-of-Products: Let us first consider the FIR filter equation (4.6a), which we shall rewrite to emphasize the sum-of-product character of the computational task

$$y(n) = \sum_{m=0}^{M-1} h(m)x(n-m) = \begin{bmatrix} h(0)x(n) + h(1)x(n-1) + \\ h(2)x(n-2) + ... + h(M-1)x(n-M+1) \end{bmatrix} \tag{8.1a}$$

There are M real multiplications and M real additions. A linear phase FIR filter coefficients are symmetric. For M odd

$$h(0) = h(M-1), h(1) = h(M-2), ..., h(\frac{M-1}{2} - 1) = h(\frac{M-1}{2} + 1)$$

Hence equation (8.1a), using above coefficient symmetry may be written as

$$\begin{aligned} y(n) &= h(0)[x(n) + x(n-M+1)] \\ &+ h(1)[x(n-1) + x(n-M+2)] + ... \\ &+ h(\frac{M-1}{2} - 1)[x(n - \frac{M-1}{2} + 1) + x(n - \frac{M-1}{2} - 1)] \\ &+ h(\frac{M-1}{2})x(n - \frac{M-1}{2}) \end{aligned} \tag{8.1b}$$

You may now note that (8.1b) requires only $(M+1)/2$ real multiplications and but M real additions. The FIR filter uses the present sample and M-1 past consecutive samples. We can also express the sum of products as an inner or dot product of coefficient vector,

$$\mathbf{h} = [h(0), h(1), h(2), ... h(M-1)]^T$$

and data vector,

$$\mathbf{x}_n = [x(n), x(n-1), x(n-2), ..., x(n-M+1)]^T.$$

$$y_n = \mathbf{h}^T \mathbf{x}_n \tag{8.1c}$$

We have shown in the previous chapter that beamformation is also basically a dot product between the array snapshot and direction vector.

Next, let us consider IIR filter given by (4.3) which we rewrite as a sum of products involving the present and the past input signal samples as well as the past output samples,

$$\begin{aligned} y(n) &= \sum_{k=1}^{p-1} a_k y(n-k) + \sum_{i=0}^{q-1} b_i x(n-i) \\ &= a_1 y(n-1) + a_2 y(n-2) + \ldots + a_{P-1} y(n-p+1) \\ &+ b_0 x(n) + b_1 x(n-1) + \ldots + b_{q-1} x(n-q+1) \end{aligned} \quad (8.2a)$$

Define a coefficient vector

$$\theta = \left[a_1, a_2, \ldots, a_{p-1}, b_0, b_1, \ldots, b_{q-1}\right]^T$$

and a data vetor

$$\mathbf{y}_n = \left[y_{n-1}, \ldots, y_{n-p+1}, x_n, \ldots, x_{n-q+1}\right]^T$$

The IIR filter output is given by a dot product,

$$y_n = \theta^T \mathbf{y}_n \quad (8.2b)$$

There are p+q-1 real multiplications and p+q real additions. In so far as the computational task is concerned equations (8.2) and (8.1) are similar. Both involve a sum-of-products type of computational task.

Another instance of signal processing task requiring a sum of products is autocorrelation and cross-correlation functions defined in (1.4a) and (1.4b), which we rewrite as a sum of products

$$\begin{aligned} r_{xx}(\tau) &= \sum_{n=0}^{N-\tau} x(n)x(n+\tau) \\ &= x(0)x(\tau) + x(1)x(\tau+1) + \ldots + x(N-\tau)x(N) \\ r_{xy}(\tau) &= \sum_{n=0}^{N-1} x(n)y(n+\tau) \\ &= x(0)y(\tau) + x(1)y(\tau+1) + \ldots + x(N-\tau)y(N) \end{aligned} \quad (8.3)$$

Note that there are $N-\tau+1$ real multiplications and the same number of real additions, assuming that the signals are real. It is sufficient if we compute the autocorrelation function for $0 \le \tau \le \tau_{\max}$ but the cross correlation function needs to be computed for $-\tau_{\min} \le \tau \le \tau_{\max}$ where $\tau_{\max}$ and $\tau_{\min}$ are integers.

Circular Buffer: Digital signal processors must work in real-time, that is, it must produce an output before the next sample arrives. To calculate an output sample we must have access to M most recent samples, which includes the newly arrived sample and M-1 immediate past samples. These M samples must be stored in memory and continuously updated in a shortest possible time as new samples reach the processor. The circular buffer affords a means to achieve this efficiently. A circular buffer may be imagined as M samples uniformly distributed on a circle. A pointer tracks the newest sample location. M-1 immediate past samples are placed on the circle in an anti clockwise direction.

Note that the newest and the oldest samples occupy side by side locations. This arrangement is shown in fig. 8.1a. When a new sample arrives it replaces the oldest sample and the pointer is moved one step down to the newest sample (see fig. 8.1b). No other sample is moved from its earlier position. The circular buffer is efficient, as only one value needs to be changed when a new sample arrives. In practice a circular buffer consists of M consecutive memory locations with a pointer showing the start of the buffer and another pointer showing the newest arrival.

We shall now look at the steps required to implementing an FIR filter using the circular buffers for both input signal and filter coefficients. The efficient handling of these steps is what separates a digital signal processor from a traditional microprocessor. For each new sample all following steps need to be evaluated:

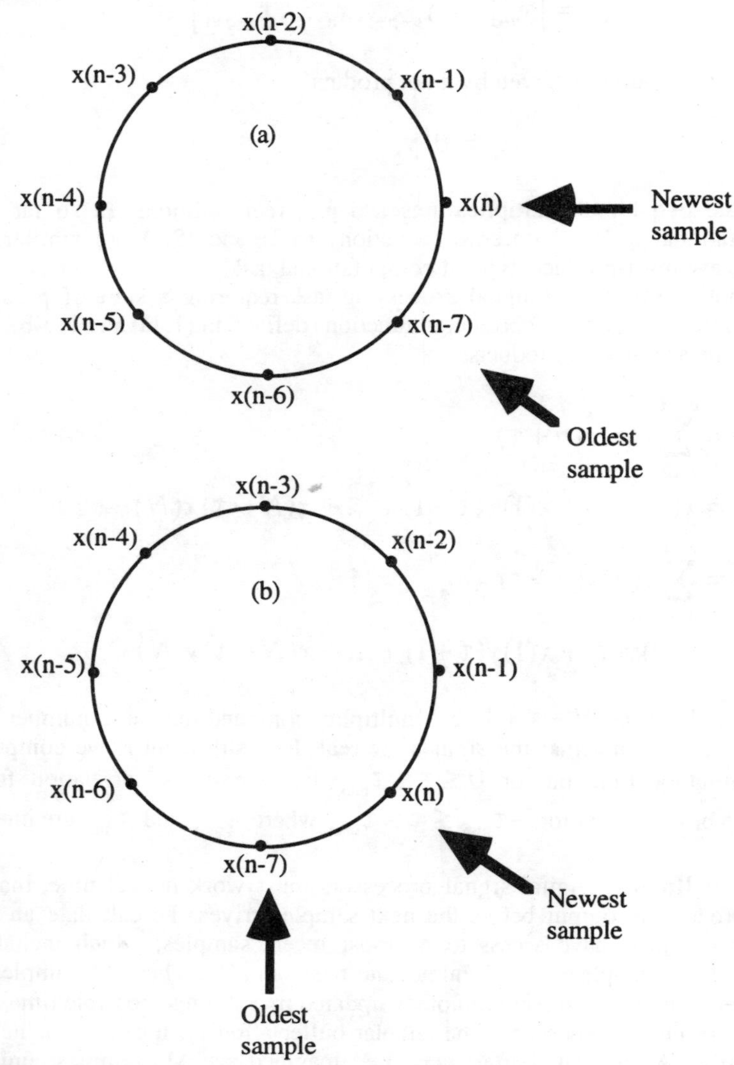

Figure 8.1: Circular buffer is used to store the most recent sample of a continually updated signal. (a) Buffer as it appears at one time instant and (b) one sample later.

1. Obtain a sample from ADC. Generate an interrupt
2. Detect and manage the interrupt.
3. Move the sample into input signal's circular buffer
4. Update the pointer for the input signal's circular buffer
5. Zero accumulator
6. Control the loop through each of the coefficients
7. Fetch the coefficients from the coeffcient's circular buffer
8. Update the pointer for the coeffcient's circular buffer
9. Fetch the sample from the input signal's circular buffer
10 Update the pointer for the input signal's circular buffer
11.Multiply the coefficient by the sample
12 Add the product to the accumulator
13. Move the output sample (accumulator) to a holding buffer
14. Move the output sample from holding buffer to data memory or to DAC.

Since the steps 6-12 are repeated many times (once for each coefficient and hence M times), special attention must be given to these operations. The traditional microprocessor will generally carry out these 14 steps in serial form while the digital signal processors are designed to carry out in parallel, often in a single clock cycle. To achieve this it must be possible to independently access the data memory, coefficient memory and the program memory. The traditional von Neumann architecture with a single memory and a single bus will not permit simultaneous access. We need other architectures when very fast processing is required. In the next section we shall look into the Harvard architecture where separate memories with separate buses are used to store the data and program instructions.

Butterfly Computation: Finite discrete Fourier transform (FDFT) plays a central role in DSP applications. It is evaluated using a fast algorithm known as fast Fourier transform (FFT). In Chapter Two we have covered the FFT algorithm. It turns out that a basic computing element in FFT algorithm is the evaluation of butterfly type of computational graph. We shall clarify this point by referring to fig. 8.2 where an eight point FFT algorithm is illustrated. There are three stages of computations where each stage consists of four butterflies, an example of this is shown in bold and is also shown separately in fig. 8.3. There are in all twelve butterflies In each butterfly there are two complex multiplications and two additions (see eq. 8.4).

$$\begin{aligned} u(m+1) &= u(m) + v(m)W^p \\ v(m+1) &= v(m)W^q + u(m) \end{aligned} \tag{8.4}$$

where $u(m)$ and $v(m)$, m=0,1,2 are complex inputs to mth stage butterfly, $u(m+1)$ and $v(m+1)$ are corresponding complex outputs and

$$W = e^{-j\frac{2\pi}{8}}$$

is a twiddle factor. p and q are integers. The twiddle factor can be pre computed and stored in the memory. In N-point FFT there are $\log_2 N$ stages and $N/2$ butterflies, thus there are $N/2 \log_2 N$ butterflies in all. The DFT computation simply reduces to evaluation of $N/2 \log_2 N$ interconnected butterflies. Another interesting observation is that the output of a butterfly can be stored in the same memory locations as the input, which are no longer required. This is called in-place computation. Thus, the eight point FFT will require only eight complex storage locations. The input to the FFT algorithm is shuffled; it is indeed in a bit reversed address format.

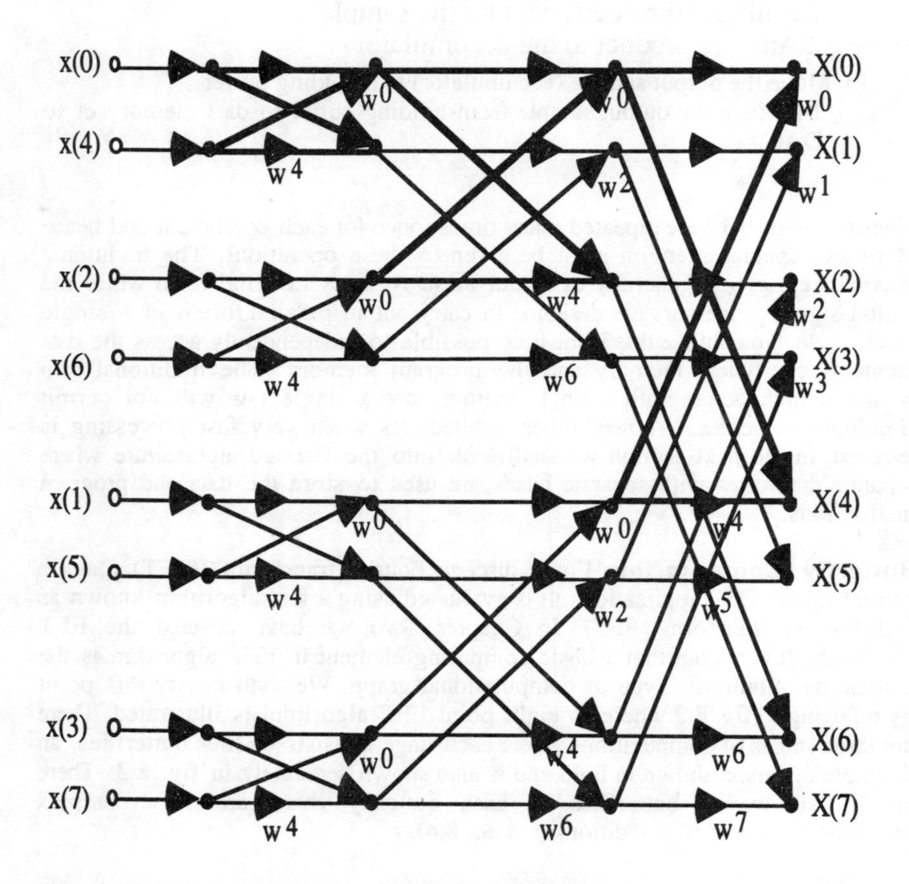

Figure 8.2: Signal flow graph of 8-point FFT algorithm (radix 2 and decimation in time). Each stage consists of four overlapping Butterflies. One Butterfly from each stage is highlighted in bold.

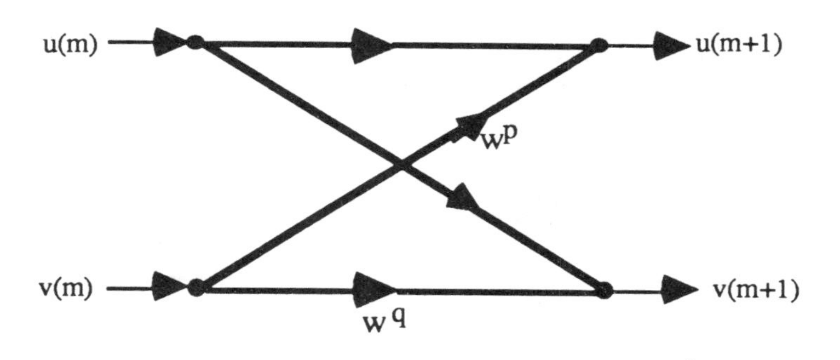

Figure 8.3: Basic computation step is Butterfly. $W = e^{-j\frac{2\pi}{8}}$ is a twiddle factor and p and q are integers.

§8.2 Signal Processor Architecture

The signal processor *architecture* basically deals with the data flows, arithmetic capabilities, memory capabilities, I/O capabilities, programmability and the instruction set. The processor *resources* are: memory capacity in words and word length, data transfer rate, processing capacity in cycles per second and instructions per second. The signal processor works under the overall control of a *operating system*. An *instruction* is an operation to be performed by the signal processor at each machine cycle. One or more number of instructions may be executed per machine cycle. An *instruction set architecture* is the total number of instructions that the processor is capable of performing. It includes all instructions for i/o, memory addressing, arithmetic processing, data manipulation, and logical operations.

Data Flow: One of the biggest bottlenecks in executing DSP algorithms is in transferring data to and from memory. This includes input signal and filter coefficients, as well as instructions from the program sequencer. The traditional microprocessor has a *single instruction single data stream (SISD) architecture*, which is the familiar von Neumann architecture (see fig. 8.4a). It has a single memory and a single bus to transfer the data and instructions. Suppose we need to multiply two numbers, which reside somewhere in the memory. The traditional microprocessor will require three clock cycles to accomplish the task. Thus the data flow becomes a limiting factor.

In *multiple instructions multiple data streams (MIMD) architecture* the multiple instructions across multiple data channels are executed in a single clock cycle. For this to be possible we need more than one independent memory and more than one bus. In Harvard architecture, which is shown in fig. 8.4b, we have separate memories for data and program instructions, with separate buses for each. Since the buses operate independently, the program instructions and the data can be fetched at the same time. This improves the speed of MIMD over SISD architecture. Most present day digital signal processors use this dual bus architecture.

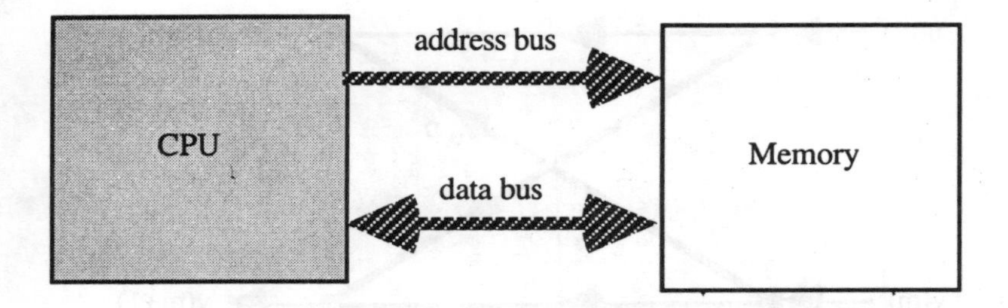

(a) von Neumann architecture

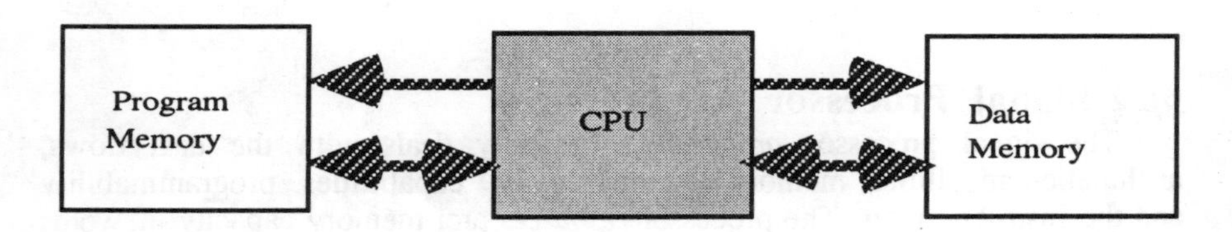

(b) Harvard architecture

Figure 8.4: The von Neumann architecture uses a single memory to store both data and instructions. On the other hand, the Harvard architecture uses separate memories for data and instructions.

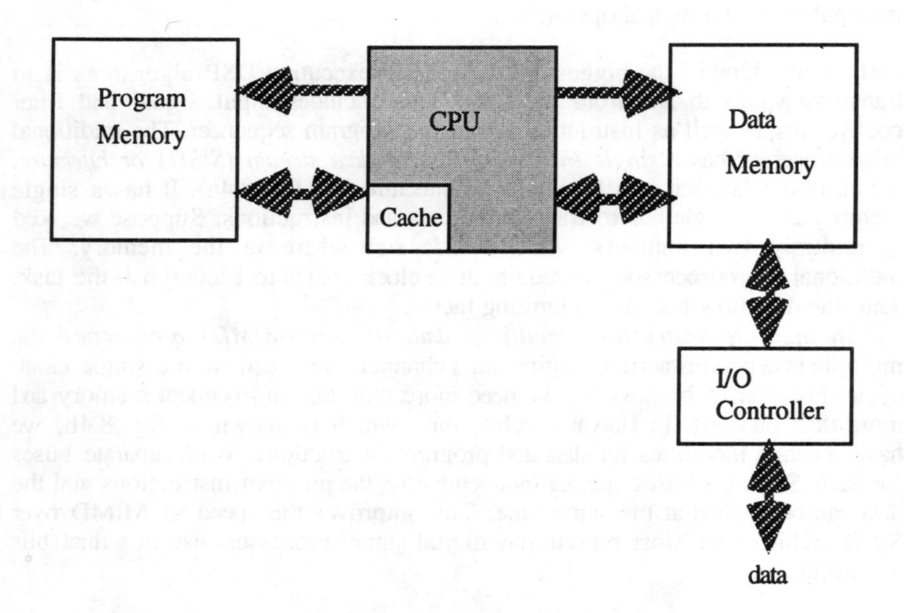

Figure 8.5: Super Harvard architecture. There is on board cache memory for storing the repeatedly used instructions. The data memory directly accessed through I/O controller without any intervention of the cpu.

The Harvard architecture has been further improved upon by introducing additional features, for example, in super Harvard architecture of Analog devices an instruction cache in CPU and an I/O controller to data memory have been added to improve its throughput. An illustration of the super Harvard architecture [33] is shown in fig. 8.5

Algorithm Structure: The existence of highly structured algorithms in signal processing computations is a major factor which drives the design of specialized signal processing machines. We have shown in §8.1 that in FIR and IIR filters the sum of the product or dot product is the fundamental computational step. Likewise, in FFT algorithm the computation of Butterflies forms the fundamental computational step. A fast multiplier-accumulator, preferably in hardware, can speed up the throughput. There are also other possibilities, for example, a filter may be factored into two or more second order sections which may be implemented in a cascade form or in a parallel form. The cascade form is ideally suited for pipelining. In pipelining the execution proceeds in assembly line style, so that many stages are in progress at any given time (see fig. 8.6). At each successive stage in the pipeline an operation (multiply, add, shift, delay etc.) is performed. The entire computation proceeds at a basic clock rate of the hardware.

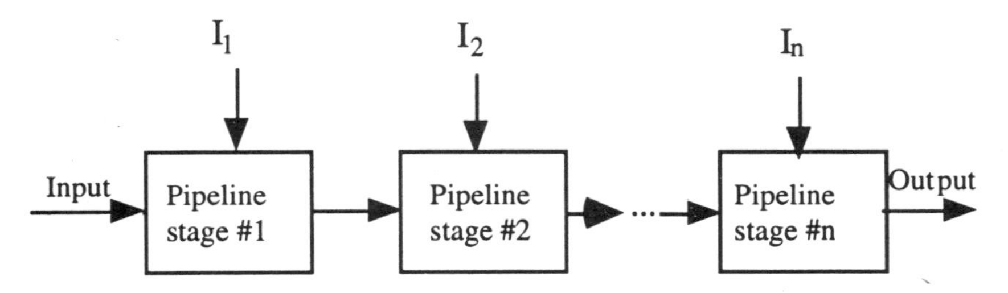

Figure 8.6: Signal processing algorithm can often be factored into smaller sections which may then be implemented in a cascade form. Each section is a stage in the pipeline. The output of the first stage is an input to the second stage and so on.

In *parallel signal processor* presence of innerent parallelism in a DSP algorithm is exploited. For example, an IIR filter may be expressed as a sum of second order section (see Chapter Five). In the most general case all processors in a parallel configuration can receive different data streams and execute different instructions at each clock cycle (see fig. 8.7). As an example, consider the FFT evaluation the computational graph for 8-point FFT algorithm is shown in fig. 8.2. There are three stages ($\log_2 N$) and in each stage there are four (N/2) independent butterflies. If we had (N/2) butterfly processors, all (N/2) butterflies could be computed in parallel. The entire FFT evaluation will then require $\log_2 N$ butterfly times plus overhead. If we had $\log_2 N$ butterfly processor all $\log_2 N$ stages can be horizontally pipelined. The entire FFT evaluation will require N/2 butterfly times. Finally, if we had $N/2 \log_2 N$ butterfly processors the entire FFT evaluation could be completed in one butterfly time which may be as fast as one clock cycle when a fast multiplier is used. It is clear from the above that there is a great deal of opportunity for concurrent computation in FFT computation.

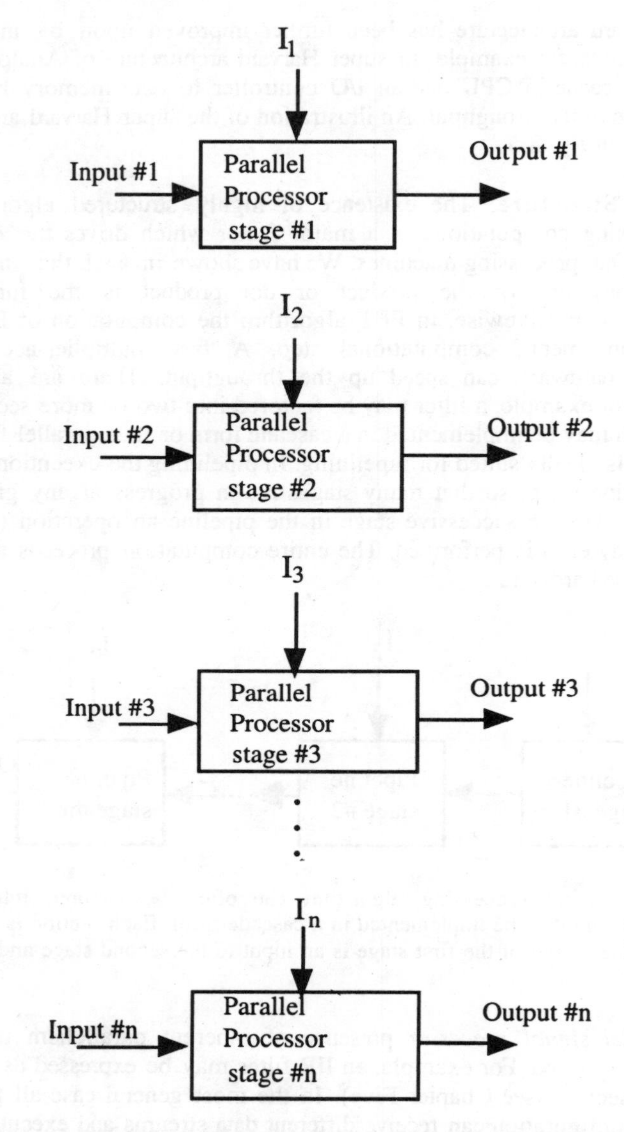

Figure 8.7: Parallel processing. Each box may represent a butterfly processor or a second order IIR filter.

Digital Signal Processor vs Microprocessor: A traditional microprocessor is fundamentally different from a digital signal processor. The microprocessor is based on sequential Neumann architecture while the digital signal processor is based on parallel Harvard architecture. A digital signal processor is a powerful number crunching machine. On the other hand a microprocessor is good at data manipulation and control. The essential architectural differences are summarized in table 8.1.

Table 8.1: What makes a digital signal processor different from a traditional microprocessor is enumerated above.

Digital Signal Processor	**Traditional Microprocessor**
Multiple memories with independent buses. (Harvard Architecture)	Single memory with single bus. (von Neumann Architecture)
Dedicated address generators	No separate address generator
Specialized addressing modes: auto increment, circular, and bit reversed.	General purpose addressing mode
2-4 memory accesses per cycle	Typically one access per cycle
Very long instruction word. Multiple operations per instruction	One operation per instruction
Specialized hardware performs all key arithmetic operations in one cycle	Multiplication takes > 1 cycle Shift takes>1 cycle
Pipeline and parallel operations	Sequential operations
Low power consumption, as the processor has to depend upon a battery for power supply.	This is not a requirement
Light weight for mobile applications	This is not a requirement

§8.3 Generic Signal Processor

A fast communication link is an essential part of any signal processor. It must transfer the data from the data and program memories to the CPU. This must take place in a shortest possible time without any bottlenecks or collisions. For this purpose independent memories and buses are standard features of digital signal processors. The input to a processor is through serial and parallel ports from external or on board digital-to-analog converter (DAC). It should be possible to access the data memory without having to route through the CPU. A generic signal processor configuration is shown in fig. 8.8. The principal components are:

1. Central Processing Unit (CPU)
2. Program memory
3. Data memory
4. I/O controller
5 DAC and ADC

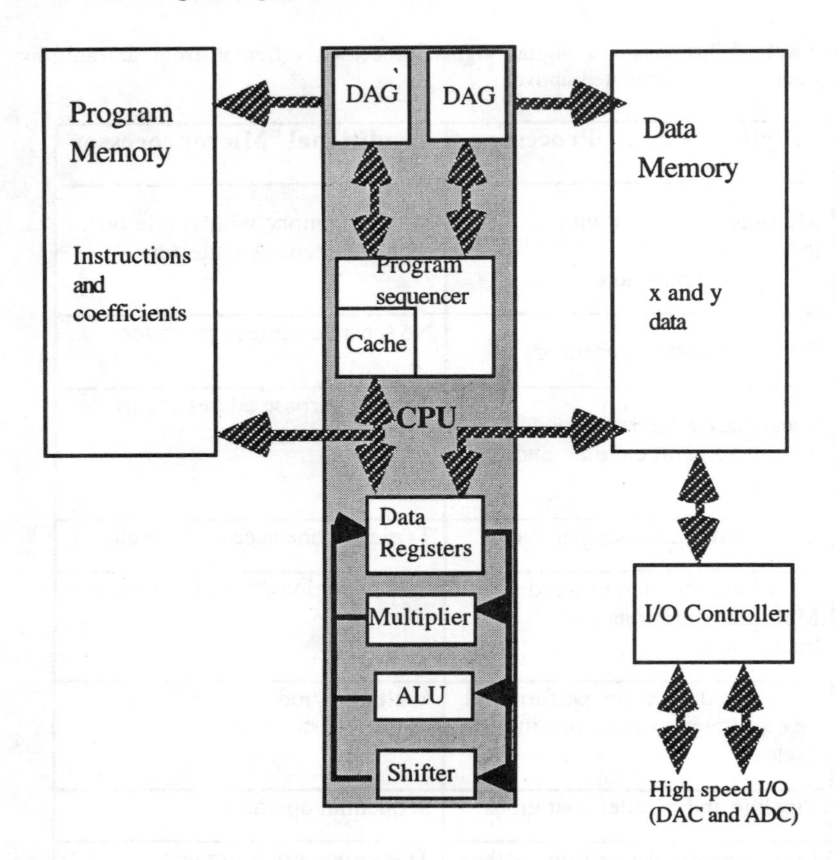

Figure 8.8: Generic signal processor. There are separate program and data memory each with its own bus and data address generator (DAG). Cache on CPU is meant for storing repeatedly used instructions such as in a loop.

Central Processing Unit (CPU): The central processing unit consists of math unit, data registers, program sequencer and two data address generators (DAG), one for the program memory and the other for data memory. The data address generators control the addresses sent to the data and program memories specifying where the information is to be read from or written to. In traditional microprocessors the program sequencer handles this task. The circular buffers, which are essential part of a digital signal processor architecture, are controlled by the DAGs. A large number of buffers are normally required whenever an algorithm is implemented as a cascade of second order sections. The DAGs are also required to generate bit reversed addresses into the circular buffers, a necessary part of the FFT algorithm.

The math (mathematics) unit consists of multiplier often with an accumulator, arithmetic logic unit (ALU) and a barrel shifter. The multiplier draws the values from two registers and multiplies them. It places the result in another register or places in the accumulator accompanying the multiplier. This feature is useful in FIR implementation. A large number of registers are usually available. There are four types of registers, namely accumulators, data registers, program registers and control registers (like program counter, stack pointer, loop register). The width of an accumulator (in bits) is usually two times that of the data to accommodate the result of product computations. The ALU performs addition, subtraction, absolute value, logical operations, conversion between fixed-point and floating-point formats, etc. While performing repeated multiply and accumulate operations in DSP algorithm the data tends to grow beyond the width of the registers or data bus. To limit data size, shift and round operations become an integral part of the architecture. A dedicated barrel shifter is required for above binary operations. All three units may be accessed in parallel.

Program and Data Memories: The instructions as well as some data such as the filter coefficients are stored in the program memory. The data address generator for the program memory generate an address from where an instruction is to be read. This instruction is passed on the program data bus into instruction cache. The input digital signal is stored in the data memory. The I/O controller can directly access the input memory. It takes the data through the serial and parallel ports and places it in the data memory. On board analog-to-digital converter (ADC) or external ADC through serial parallel ports are used to read in the external signal for processing. Similarly the output can be sent to an external device through either on board digital-to-analog converter (DAC) or external DAC through serial or parallel ports. When the CPU is busy performing arithmetic tasks the I/O controller can keep fetching the data from I/O ports. This amounts to asynchronous parallelism, where processing units are working independent of each other. The data from the data memory is passed on the data memory bus under control of data memory DAG.

Fixed-point vs Floating-point: There are two types of digital signal processors, namely, fixed-point and floating-point. In fixed-point processor a number is represented in fixed-point format, usually 16 bits and in floating-point processor a number is represented in 32 bit floating-point format. In fixed-point format a fraction, in the range -1 to $1\text{-}2^{-15}$, is represented in two's complement which consists of a sign bit and remaining fifteen bits for data. A positive fraction is represented as

$$X = 0.b_1, b_2, \ldots, b_{15}$$

where b_1, b_2,...b_{15} are data bits. Given a binary number its value is $X = b_1/2 + b_2/4 + b_3/8 + \ldots + b_{15}/2^{15}$. A negative fraction is represented by forming two's complement of the corresponding positive number and adding one least significant bit (LSB).

$$X_{2c} = 1.\bar{b}_1, \bar{b}_2, \ldots, \bar{b}_{15} + 00\ldots01$$

where $\bar{b}_1$ is complement of b_1.

$$X_{2c} = -1 + \frac{\bar{b}_1}{2} + \frac{\bar{b}_2}{4} + \frac{\bar{b}_3}{8} + \ldots + \frac{\bar{b}_{15}}{2^{15}}$$

For example, $13/16$ is represented as 0.1101 and $-13/16$ as 1.0011.

In floating-point format a number is represented (IEEE standard 754-1985) as

$$X = (-1)^s M.2^{E-127}$$

where s stands for sign bit (0 or 1), M stands for mantissa, a 23 bit fraction and E for exponent, an eight bit integer. The mantissa represents the fractional part of the number. It has the following form:

$$M = 1.b_1 b_2 b_3 \ldots b_{23}$$

$$= 1 + \frac{b_1}{2} + \frac{b_2}{4} + \frac{b_3}{8} + \ldots + \frac{b_{23}}{2^{23}}$$

The leading digit is always one and hence it is not represented. The largest and smallest numbers are $\pm 3.4 \times 10^{38}$ and $\pm 1.2 \times 10^{-38}$, respectively. The values are unequally spaced between these two numbers, such that the difference between the adjacent numbers is approximately 10^7 times the value of the number. Consequently, small values are closely spaced while large values are widely spaced. The trade-offs between fixed-point and floating-point digital signal processors are summarized in table 8.2.

Table 8.2: The trade-offs between fixed-point and floating-point digital signal processors.

Fixed-point	Floating-point
All registers and buses are 16 bit wide	All registers and buses are 32 bit wide
Smaller instruction set	Larger instruction set as it handles both floating and fixed-point numbers
Small dynamic range	Large dynamic range
More difficult to program as the programmer has to worry about overflow, under flow and round-off errors	These problems do not exist The development costs are low.
Simpler internal structure	Internal structure is more complicated
Much cheaper	More expensive
Large quantization noise	Low quantization noise
More popular in competitive consumer products.	Better suited where high performance is needed and cost is not important

Given these trade-offs, the question one faces is how do you select which to use? The important deciding factor is the resolution of the ADC and DAC used in the application. In many applications 12-14 bits per sample are considered sufficient. Evidently, a fixed-point digital signal processor will suffice. For instance, the television and other video signals typically use 8 bit ADC and DAC. In contrast, in professional audio applications we sample signals with as high as 20 or 24 bits. A floating-point digital signal processor is then more appropriate.

§8.4 Programmable Digital Signal Processors

Programmable digital signal processors evolved out of two DSP trends in the late 1970. The first trend was the development of dedicated single chip processor and the second trend was the development of multichip processor for high performance military applications, which used bit slice technology to create special purpose solutions. These two trends plus the development of the microprocessors lead to the invention of programmable digital signal processors.

Digital Signal Processors: It has been pointed out that the computational speed of a standard microprocessor is inadequate for real-time digital signal processing. The processor architecture has to be changed in order that the microprocessor can perform the signal processing tasks in real-time. The traditional von Neumann sequential architecture is replaced by the Harvard architecture. Present day digital signal processors have the following special features:

(a) Fast multiplication: Single cycle multiply-accumulate, Hardware multiplier-accumulator (MAC).

(b) Multiple data processing units: Separate MAC, ALU, Comparator and Barrel shifter.

(c) Separate data and program buses (Harvard architecture): supports concurrent memory access to data memory and program memory.

(d) Specialized program control, Addressing mode, and Instruction sets.

(e) On-chip memory.

(f) Multiple instructions per clock cycle.

(g) On board A/D and D/A converters.

(h) Low power consumption.

(i) Low cost.

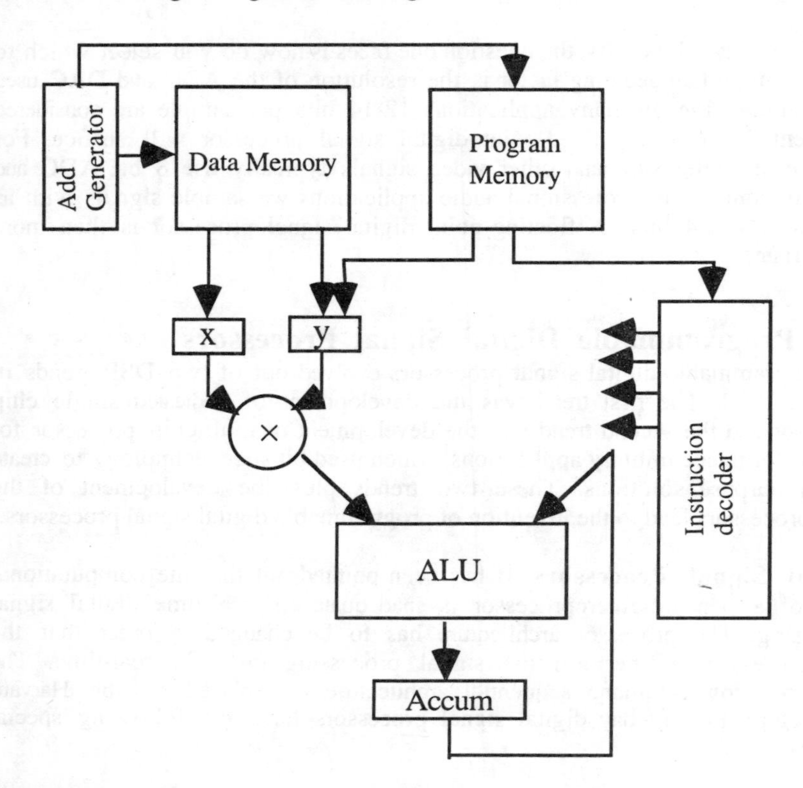

Figure 8.9: Classical digital signal processor architecture.

The early digital signal processors had a single hardware multiplier, ALU and accumulator but separate program and data memories with separate buses but single address generator for both program and data memories. The clock speed was in the range of a few tens of MHz. The multiplier-accumulator took about 390 nanoseconds. A typical schematic of the classical processor is shown in fig. 8.9. The examples of classical processors are: NEC 7722, TMS32010.

The modern day digital signal processors have more than one multiplier, ALU and barrel shifter, and separate memories for data and program with separate buses and address generators. The data memory is further partitioned into two sections, x-data and y-data. A large number of registers are provided for ease of programming. The clock speed is in the range a few hundreds of MHz. The multiplier-accumulator requires less than three nanoseconds for one multiplication and addition. A schematic of a modern highly parallel digital signal processor is shown in fig. 8.10 and typical examples of modern digital signal processors are TMS320C62, DSP56307, etc.

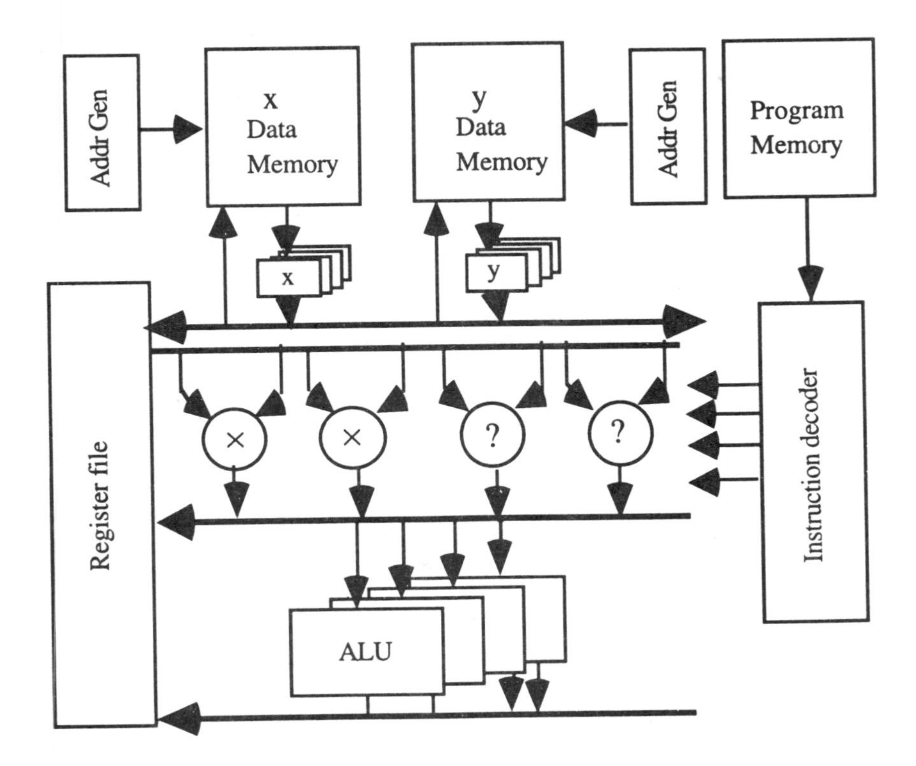

Figure 8.10: Modern highly parallel digital signal processor. It has multiple arithmetic units and two separate data memories, one for x data and the other for y data with independent buses and one for program.

Speed of Digital Signal processors: The speed of digital signal processor is measured in terms of millions of integer operations per second (MIPS) or millions of floating-point operations (MFLOPS). The measure of speed without any reference to particular application does not appear to be of much significance. Consider an FIR filter with 25 to 400 coefficients. Let d be the speed of the processor in MHz. The processor takes one clock cycle to complete one multiplication and accumulation. The time taken to output one sample is $25/d10^{-6}$ to $400/d10^{-6}$ seconds. Therefore, the output rate or the throughput varies from $d/400$ to $d/25$ mega samples per second. For IIR filter the throughput is much higher as the number of filter coefficients is small, in the range 5 to 17. The throughput varies from $d/20$ to $d/8$ mega samples per second including the overhead. For Fourier transform calculation a highly optimized code in assembly language is usually provided by the manufacturer of a processor, for example, ADSP-21062 SHARC processor, which requires for a 1024 point complex FFT 18221 clock cycles or 17.8 clock cycles per sample. This figure varies slightly with longer or shorter segment length. For instance, for a segment length of 256 points it is 14.2 clock cycles per sample and for a segment length of 4096 points it is 21.4 clock cycles per sample [33].
Let us now estimate the throughput for convolution using the frequency domain approach (Chapter Two). In the overlap-and-add method let the segment length be

512 and the filter also of same length. We pad the data segment and the filter coefficients with 512 zeros and form a complex input to FFT algorithm. The product of FDFT of the signal and that of filter coefficient may be obtained in 1024 complex multiplications and 2048 complex additions (Exercise #7, Chapter Two). The inverse FDFT of the product is next evaluated using the same number of clock cycles as the forward FDFT. Thus, we need approximately 75 clock cycles per convolution output. The throughput for the convolution in the frequency domain is $d/75$ mega samples per second. In fig. 8.11 we have shown the throughput for different signal processing tasks for a representative modern digital signal processor with 100 MHz clock rate, that is, d=100.

The next question is how fast do we need to process in real-time for different applications? Consider the telephone quality speech signal. The frequency range is 100 Hz to 3.2 KHz. The required sampling rate is 8 KHz. For high fidelity music the frequency range is 20Hz to 20KHz which requires a sampling rate of 44.1 KHz or 88.2 KHz for stereo music. A low quality video signal for video phone (352x288 pixels with 3 colours per pixel and 30 frames per second) requires 9.1 MHz sampling rate. At the other extreme, the high definition TV (1920x1080 pixels) requires a sampling rate of 186 MHz. The modern digital signal processors, as may be seen from fig 8.11, can meet the requirements of both telephone quality speech and Hi Fi music requirements. But they are still inadequate for video applications. Several processors may have to be combined into a single system known as multiprocessor in order to enhance the throughput.

History of Digital Signal Processors: The first commercially available digital signal processor, NEC 7720, was introduced in 1980. It had a hardware multiplier and Harvard architecture. Once the software is validated, the program is hand coded into on-chip ROM for mass production. It was more like one time programmable special purpose digital signal processor. In 1982 Texas Instruments (TI) introduced the TMS32010. It had a hardware multiplier and Harvard architecture with separate on-chip buses for data memory and program memory. This was the first programmable digital signal processor to support executing instructions from off-chip RAM. This feature brought programmable digital signal processor closer to the microprocessor programming model. In addition, TIs emphasis on development tools and libraries led to wide spread use. Next milestone was adoption of floating-point, which was first introduced in the HD61810 from Hitachi. Thereafter, different developers brought out a series of digital signal processors.

A brief account of developments in digital signal processors is given in table 8.3. From this table, we note that we have already reached 600 MHz clock speed and 4800 MIPS. The multiplication-accumulation (MAC) time is as low as 0.5 nanoseconds, that is, 2000 megaMACs, which is quite consistent with the predicted trend shown in fig. 8.12. The technology is growing at such a fast rate that the above information may become obsolete by the time the book comes out of press.

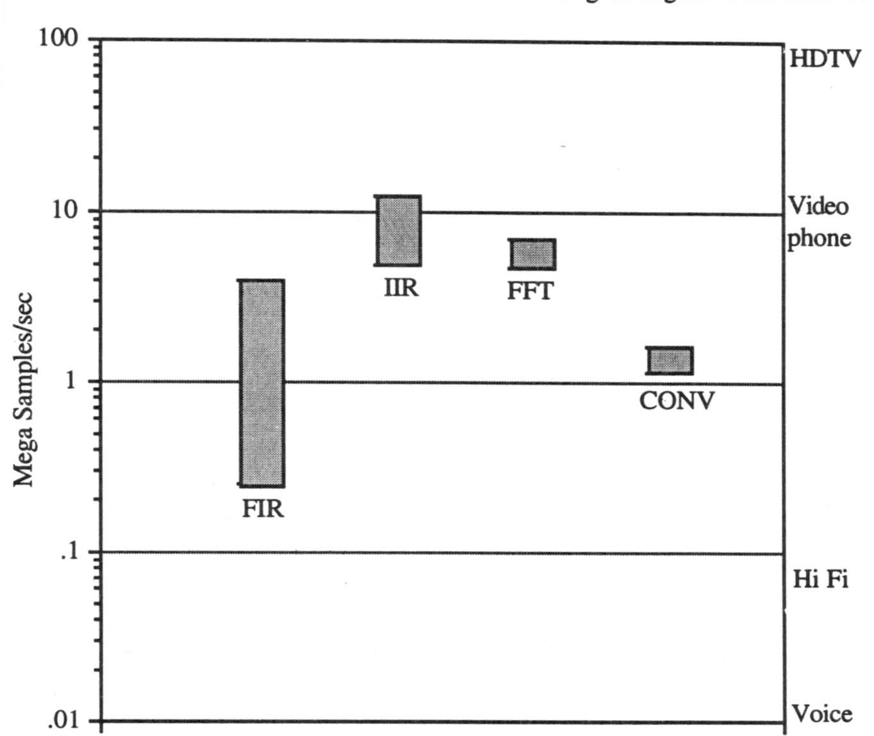

Figure 8.11: The throughput for different signal processing tasks for a representative modern digital signal processor with 100 MHz clock rate, that is, d=100. Also shown on the right the throughput required in four different applications [33].

Table 8.3: The history of digital signal processor since its inception [34, 35].

Name of the Processor	Year	Brief description	Remarks
AMI S2811	1978 (Not available for several years)	Hardware multiplier and Harvard architecture. It could execute instruction from on-chip memory or from I/O pins one at a time.	Designed to be a peripheral part to a micro-processor
NEC 7720	1980	Hardware multiplier Harvard architecture, 16 bit word size	More like a dedicated signal chip DSP
Intel 2920	1980	On-chip A/D and D/A converters but no hardware multiplier Continued on next page	Lacked digital, interface

TI TMS32010	1982	Hardware multiplier and Harvard architecture. Capable of executing instructions from off-chip program RAM. 16 bit fixed-point	Extensive development tools and libraries 390 ns MAC time, 5Mips
Hitachi HD61810	1982	Floating-point arithmetic with 12 bit mantissa and 4 bit exponent format.	First floating-point DSP
AT&T	1984	32 bit floating-point arithmetic. 24 bit mantissa and 8 bit exponent.	First 32 bit floating-point DSP
AT&T DSP16A	1988	Same as above	Fastest digital signal processor running at 30MHz.
Mot 56001	1987	3 memory spaces, P, X,Y Modulo Addressing, 75 ns MAC Ns: nanosecond, i.e. 10^{-9} sec.	Second generation
ADSP2100	1987		Second generation
TI TMS320C40		32 bit float DSP capable of 60 MFLOPS	
Motorola DSP56307	1995 (?)	On board 64K 24bit memory. Enhanced filter co-processor, Pipelined 24x24 bit parallel MAC, 20 ns MAC, Low power,	Clock rate 100MHz, Third generation
Philips TriMedia	1996	27 functional units. Five instructions per word	
TI TMS320C62xx	1997-98	Very-long-instruction-word architecture (VLIW). It supports multiple execution units. Eight 32 bit instructions per cycle 3ns MAC, 1200 Mips	Clock rate 200MHz Fourth Generation
TMS320c64xx	2000-02	32 bit, Fixed-point, Dual MAC, 4800 MIPS	Clock rate 600 MHz. Fifth Generation
ADSP TigerSHARC Ts101s	2000-02	Dual MAC, 0.5 ns MAC, Three 64 K byte on board memory, 1800 MFLOPS	Clock rate 300 MHz. Fifth Generation

Evolution of Digital Signal Processors: The digital signal processors have evolved over past two decades from a humble beginning as a traditional microprocessor equipped with a hardware multiplier-accumulator (MAC) to modern digital signal processor with multiple fast MACs and multiple independent memories. This growth in digital signal processors is captured in table 8.4 where we have grouped according to incremental growth in the features. Representative examples of processors belonging to each group or generation are shown in the last column.

Table 8.4: Evolution of digital signal processors over past two decades.

Generation	Features	Examples
0 (1980)	Von Neumann architecture	DSP-1 (AT&T)
1 (1982)	Basic Harvard architecture 6 MIPS	TMS320C10(TI) NEC7720
2 (1987)	1 data/program bus, 1 data bus, Modulo addressing 13 MIPS	TMS320C50 (TI) DSP16A (AT&T) DSP56001 (Mot) ADSP2100
3 (1990-94)	Extra addressing modes, extra functions 2 data buses, 1 program bus, co-processor 50 MIPS	TMS320C5X (TI) DSP16XX (AT&T) TMS320C54X (TI) DSP56307
4 (1997-98)	Two data buses and one program bus, 200MHz 32 bit, fixed-point, 1200MIPS	TMS3206201
5 (1999-Now)	Three memories 2 data buses, 1 program bus, 2 MACs, 0.5 ns 300 MHz, 1800 MFLOPS	TigerSHARC Ts101s TMS320c67xx

There are two other attributes of a digital signal processor where there has been a significant development over the past two decades. These are: power consumption measured in units of milliwatts per MIPS (million instructions per second) and performance measured in units of megaMACS The growth of the digital signal processors with respect to above two attributes is illustrated in fig. 8.12. It may be noted that the performance, on the average, has doubled every two years and the power consumption in mW/MIPS, on the average, reduced by a factor of 1000 every two years. The power reduction in DSPs is achieved by using three design techniques: low voltage, gated clock and sleep modes. By lowering supply voltage the power consumption can be reduced but only at the cost of increased delay. The power consumption depends on the square of the operating voltage (energy due to capacitor charging and discharging). So lowering operating voltage reduces power consumption. However, the transistor source-drain saturation current is proportional to the square of the operating voltage. So lowering operating voltage lowers the current. But a smaller current implies that we need a longer period of time to charge and discharge capacitors. The charge on the capacitor is directly related to the logic value it holds. For a logic value of 1, the charge required is some value, Q, independent of the operating voltage. Therefore the circuit has higher delay at a lower operating voltage. The loss of performance, however, can be regained by adding more

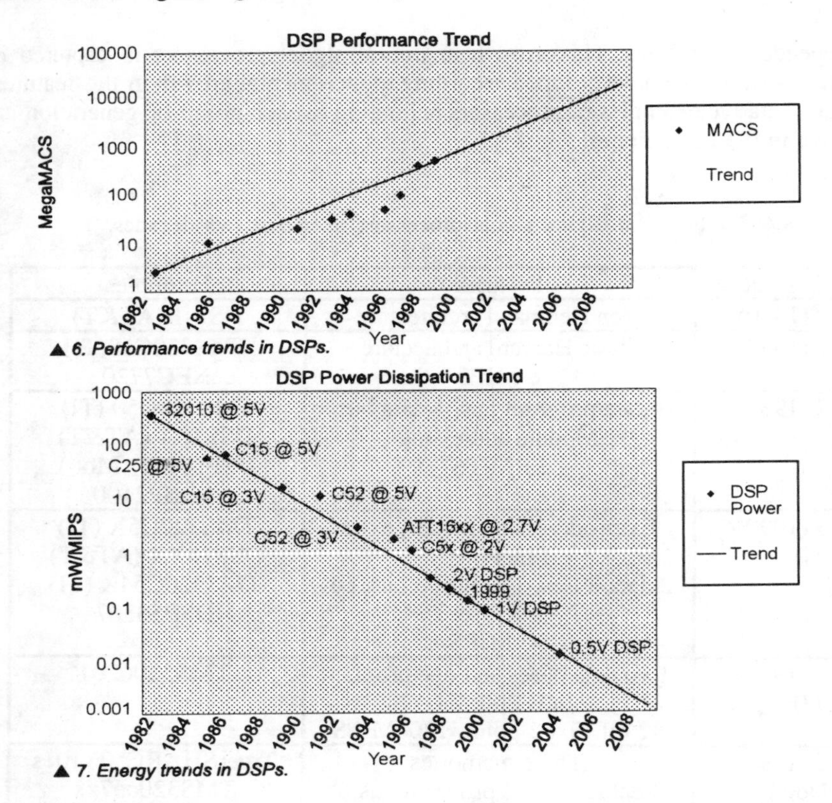

Figure 8.12: Growth of digital signal processors with respect to (a) performance and (b) power consumption (Adapted from [34] with permission).

parallelism. Gated clocks can be used to only enable the segments of data paths and the section of memory required to execute each instruction. This is possible because of the highly parallel nature of digital signal processors. The third approach to reduce the power consumption is to provide multiple sleep modes. As many as three different sleep modes are available on state-of-the-art digital signal processors. The sleep modes vary just from disabling the cpu to disabling the entire processor.

§8.5 Finite Word Length Effect

An important consequence of the use of digital signal processors is loss of precision due to finite word length of the processor. The analog signal is sampled and then quantized to a fixed number of bits, b plus one bit for sign, making the word length equal to b+1. This is the familiar analog-to-digital (A/D) conversion. The difference between the analog signal, $x(n)$ and quantized signal, $\hat{x}(n)$ is the quantization noise,

$$\hat{x}(n) - x(n) = \varepsilon(n) \tag{8.5}$$

The characteristics of the quantization error depends upon the particular form of number representation (fixed-point or floating-point) and on whether the quantization is due to rounding or truncation. In rounding a quantity is rounded to the nearest integer. Here the quantization noise is modeled as an uncorrelated uniformly distributed random variable. The probability density function of the quantization noise due to round off is given by

$$p(\varepsilon) = \frac{1}{\delta} \qquad -\frac{\delta}{2} \le \varepsilon \le \frac{\delta}{2}$$
$$= 0$$

where $\delta = 2^{-b}$ is called quntization step. The quantization noise has zero mean and variance equal to $\sigma_\varepsilon^2 = \delta^2/12$.

The truncation of arithmetic operation, in particular multiplication of two numbers, leads to another type of quantization error. The output of a product of a signal rounded to b bits and a filter coefficient (also rounded to b bits) is of length 2b bits. This has to be truncated to b bits so as to fit it into a register or send it over a data bus. The quantization error due to truncation depends upon the type of representation: fixed-point or floating-point, sign-magnitude or two's complement. The fixed-point representation is commonly used in digital signal processors. But we have a choice between sign-magnitude and two's complement. The statistical model shown in (8.5) remains valid for both types of representations but the probability density function of the error depends upon the type. The probability density function of the truncation error for two's complement is given by

$$p(\varepsilon) = \begin{cases} \dfrac{1}{\delta} & -\delta \le \varepsilon \le 0 \\ 0 \end{cases}$$

and that for the sign magnitude is given by

$$p(\varepsilon) = \begin{cases} \dfrac{1}{2\delta} & -\delta \le \varepsilon \le \delta \\ 0 \end{cases}$$

The input-output relations of three above quantizers are illustrated in fig. 8.13.

Quantization is a non-linear operation whose analysis becomes extremely involved, particularly in a large system containing many multipliers and summation nodes. Non-linearity leads to oscillations known as limit cycles in a recursive digital filter. We shall not take this approach. Instead, we shall concentrate on the stochastic model which, in effect, results in a linear model for the filter. The linear model leads to an interesting phenomenon of shift in the position of poles and zeros, which points to a possibility of large quantization error, should a pole drift away from the unit circle.

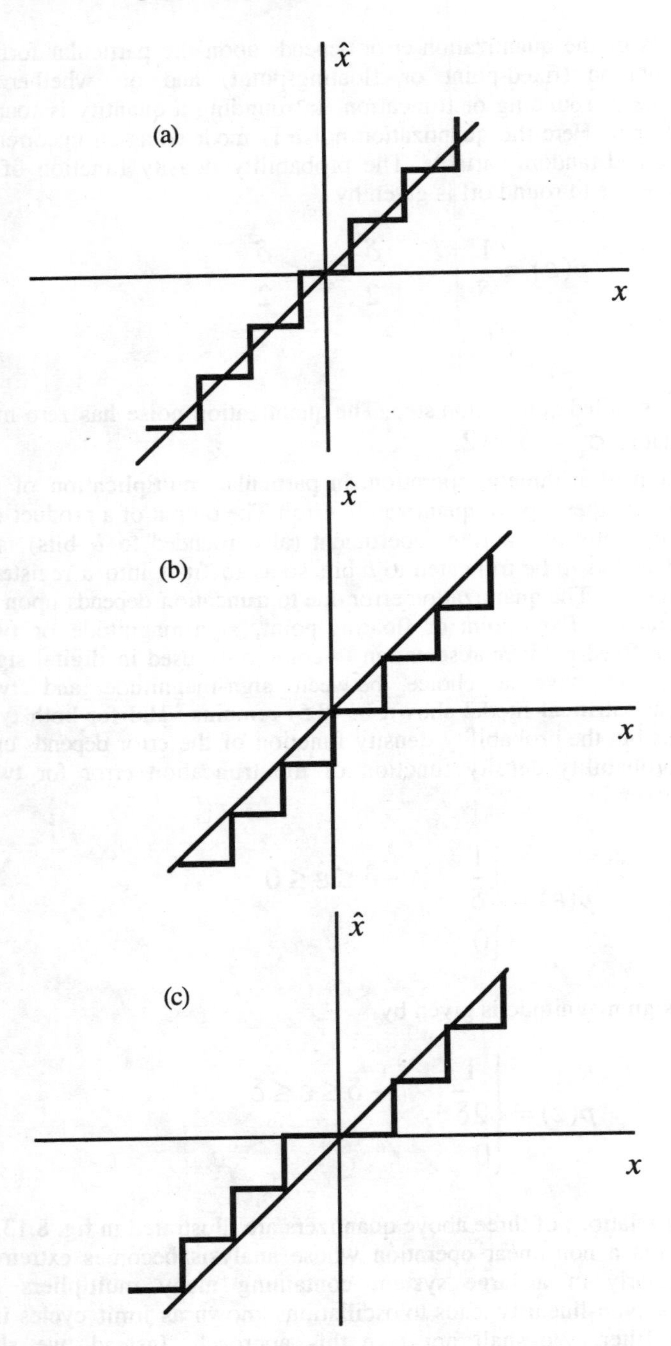

Figure 8.13: The input-output relations of three types of quantizers, (a) rounding, (b) truncation, two's complement, (c) truncation, sign-magnitude. Step size: δ.

Coefficient Quantization in FIR Filters: Consider a Mth order FIR filter, *h(m)*, *m*=0,1,...,*M*-1. The filter coefficients are quantized to b+1 bits (b data bits and one sign bit). As in (8.5) the quantized coefficient may be modeled as

$$\hat{h}(m) = h(m) + \varepsilon(m) \tag{8.6a}$$

where the quantization noise may be considered as uncorrelated uniformly distributed random number. Using the quantized filter coefficients we obtain a new transfer function

$$\hat{H}(z) = H(z) + \mathrm{E}(z) \tag{8.6b}$$

where $\mathrm{E}(z) = \sum_{m=0}^{M-1} \varepsilon(m) e^{-j\omega m}$. Thus, a quantized filter may be considered as a parallel connection of two filters as shown in fig. 8.14. $\mathrm{E}(z)$ is also an FIR filter with random coefficients (quantization noise). We like to study the frequency characteristics of the random filter. The mean can be shown to be zero,

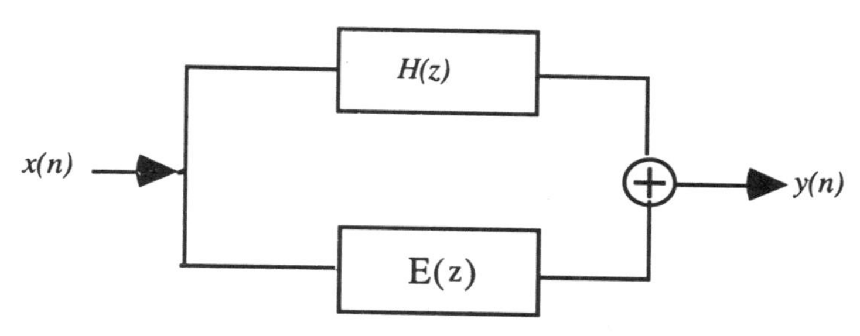

Figure 8.14: An FIR filter with quantized coefficients is modeled as a parallel connection of two FIR filters, one with unquantized coefficients and the other with random (quantization noise) coefficients.

$$E\{\mathrm{E}(\omega)\} = \sum_{m=0}^{M-1} E\{\varepsilon(m)\} e^{-j\omega m} = 0$$

as the quantization noise is a zero mean random variable. Next let us compute the variance of the random filter. Note that since $\mathrm{E}(z)$ is also a FIR filter, the random coefficients will satisfy the symmetry property of an FIR filter. Hence, we can write its response following equation (4.4a) for M odd as

$$\mathrm{E}(\omega) = \left[2 \sum_{m=0}^{\frac{M-1}{2}} \varepsilon(m) \cos(\omega(\frac{M-1}{2} - m)) + \varepsilon(\frac{M-1}{2}) \right] e^{-j\omega \frac{M-1}{2}} \tag{8.7}$$

To compute the variance of the frequency response of the random filter we compute expected value of the magnitude square on both sides of (8.7). We have

$$\sigma_{\mathrm{E}}^2 = E\left\{\mathrm{E}(\omega)\mathrm{E}^H(\omega)\right\}$$

$$= \sigma_\varepsilon^2\left[1+4\sum_{m=0}^{\frac{M-1}{2}}\cos^2\left(\omega\left(\frac{M-1}{2}-m\right)\right)\right] \tag{8.8a}$$

$$= \sigma_\varepsilon^2\left[M-1+\frac{\sin(M\omega)}{\sin(\omega)}\right]$$

Or the relative influence of the quantization noise is given by

$$\frac{\sigma_{\mathrm{E}}^2}{\sigma_\varepsilon^2} = \left[M-1+\frac{\sin(M\omega)}{\sin(\omega)}\right] \tag{8.8b}$$

A plot of the quantity inside the square bracket is shown in fig. 8.15

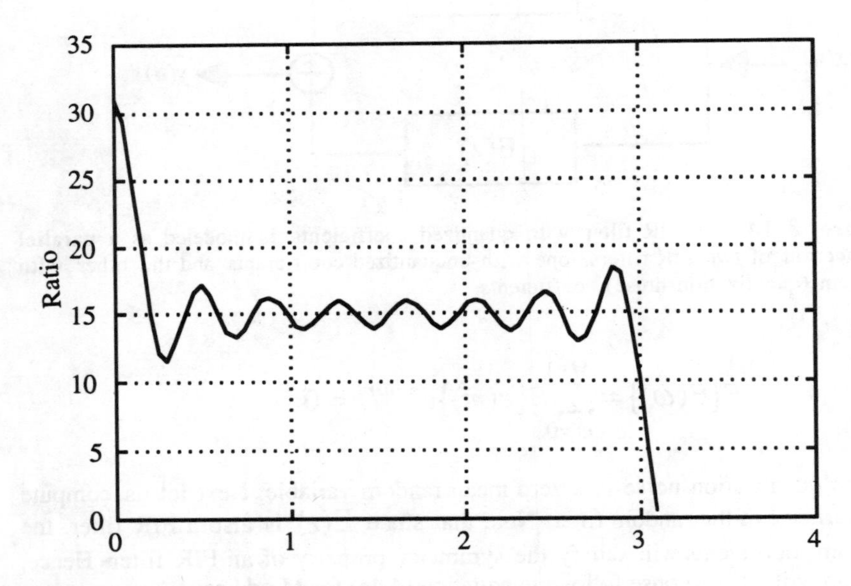

Figure 8.15: A plot of the ratio defined in eq.(8.8b).

From fig.8.15 it may be concluded that the ratio is bounded from above as

$$\boxed{\frac{\sigma_E^2}{\sigma_\varepsilon^2} \leq [2M-1]} \tag{8.9a}$$

For uniform quantization error, $\sigma_\varepsilon^2 = \frac{\delta^2}{12}$ (for round-off error) and (8.9) reduces to

$$\sigma_E^2 \leq [2M-1]\frac{\delta^2}{12} \tag{8.9b}$$

It has an interesting implication. For an acceptable error in the frequency response function, the quantization step must be reduced as filter length is increased; indeed for every doubling of the filter length, the word length must be increased by one bit. Conversely, for short filters the quantization step can be large. This opens up a possibility of using many short filters in cascade form in place of one large high order filter.

Example 8.1:We shall now examine the effect of the rounding-off of the filter coefficients on the frequency response of a lowpass filter whose coefficients are listed in Table 4.1. The original filter coefficients had seven-digit precision but they were rounded off to two-digit precision. The frequency response before and after round-off is plotted in fig. 8.16. While the passband response is practically unchanged there is a noticeable deterioration of the response in the stop band region. There is more than 10 dB increase in the stop band ripple magnitude. This is the worst case of round-off deterioration. A two-digit precision will correspond to four to five binary digits. Evidently a 16 bit processor would do much better ♠.

Shift in Position of Zeros: Alternate way of looking at the effect of quantization of the filter coefficients is to study the shift in the position of the zeros. The FIR filter response can be expressed either in terms of its filter coefficients or the position of zeros,

$$H(z) = 1 + \sum_{m=1}^{M-1} h(m) z^{-m}$$

$$= \prod_{m=1}^{M-1} (1 - \beta_m z^{-1})$$

where M stands for the filter length and β_m stands for position of mth zero. We have assumed for convenience that $h(0) = 1$. A small error in $h(m)$ will naturally introduce a small perturbation in β_m. The perturbation may be expressed as

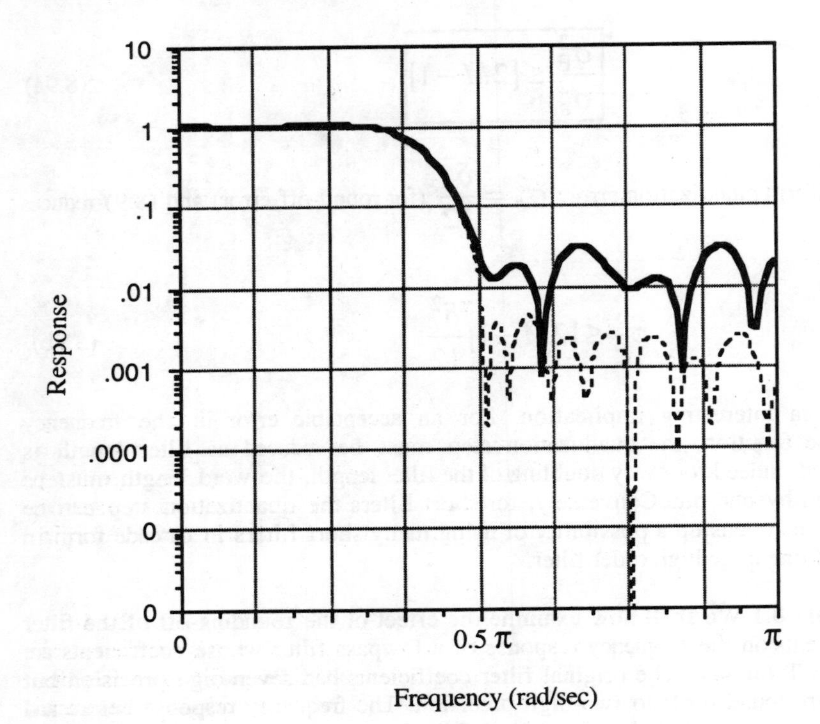

Figure 8.16: The frequency response function was computed for two types of precision. Solid curve is for 2 digit precision and dashed curve for 7 digit precision. The filter coefficients were taken from table 4.1.

$$\Delta\beta_m = \sum_{k=1}^{M-1} \frac{\partial\beta_m}{\partial h(k)} \Delta h(k) \tag{8.10}$$

where the partial derivative $\partial\beta_m / \partial h(k)$ can be obtained from the following

$$\left(\frac{\partial H(z)}{\partial h(k)}\right)_{z=\beta_m} = \left(\frac{\partial H(z)}{\partial z}\right)_{z=\beta_m} \frac{\partial\beta_m}{\partial h(k)}$$

Or

$$\frac{\partial\beta_m}{\partial h(k)} = \frac{\left(\frac{\partial H(z)}{\partial h(k)}\right)_{z=\beta_m}}{\left(\frac{\partial H(z)}{\partial z}\right)_{z=\beta_m}} \tag{8.11a}$$

To simplify (8.11a) further we note the following

$$\frac{\partial H(z)}{\partial h(k)} = z^{-k}$$

and

$$\frac{\partial H(z)}{\partial z} = \frac{\partial \prod_{m=1}^{M-1}(1-\beta_m z^{-1})}{\partial z} = \sum_{k=1}^{M-1} -\frac{\beta_k}{z^2} \prod_{\substack{l=1 \\ l \neq k}}^{M-1} (1-\beta_m z^{-1})$$

We obtain

$$\frac{\partial \beta_m}{\partial h(k)} = \frac{z^{-k}\Big|_{z=\beta_m}}{\sum_{k=1}^{M-1} -\frac{\beta_k}{z^2} \prod_{\substack{l=1 \\ l \neq k}}^{M-1} (1-\beta_l z^{-1})\Bigg|_{z=\beta_m}} \tag{8.11b}$$

It may be noted that the only non-zero term in the denominator of (8.11b) is when $\beta_k = \beta_m$. Hence the result is

$$\frac{\partial \beta_m}{\partial h(k)} = \frac{-\beta_m^{M-1-k}}{\prod_{\substack{l=1 \\ l \neq m}}^{M-1} (\beta_m - \beta_l)} \tag{8.11c}$$

Substituting (8.11c) into (8.10) we obtain an expression for perturbation in β_m,

$$\Delta \beta_m = -\sum_{k=1}^{M-1} \frac{\beta_m^{M-1-k}}{\prod_{\substack{l=1 \\ l \neq m}}^{M-1} (\beta_m - \beta_l)} \Delta h(k) \tag{8.12a}$$

which may be written in a matrix form as

$$\begin{bmatrix} \Delta\beta_1 \\ \Delta\beta_2 \\ \vdots \\ \Delta\beta_q \end{bmatrix} = \mathbf{\Gamma} \begin{bmatrix} \Delta h(1) \\ \Delta h(2) \\ \vdots \\ \Delta h(q) \end{bmatrix} \tag{8.12b}$$

where

$$[\Gamma]_{m,k} = -\frac{\beta_m^{M-1-k}}{\prod_{\substack{l=1 \\ l \neq m}}^{M-1} (\beta_m - \beta_l)} \tag{8.12c}$$

The perturbation magnitude will be large whenever the zeros come very close to each other. This will happen when a zero is close to the unit circle as its reciprocal will also be close to the unit circle. Let

$$\beta_m = 0.95e^{j\theta}$$

then its reciprocal is at $1.0526e^{j\theta}$. Further, when the filter size is large (q>>1) the numerator of (8.12) will become large for the zeros lying outside the unit circle. One way to minimize the quantization effect is to express a normal FIR filter as a cascade of shorter filters. In selecting such filters we must bear in mind that the zeros are located as far apart as possible.

Example 8.2: To illustrate these points consider an FIR filter consisting of four zeros, $\left\{0.95e^{j30^0}, 0.95e^{-j30^0}, 1.0526e^{j30^0}, 1.0526e^{-j30^0}\right\}$. The transfer function in z-domain is given by

$$H(z) = H_1(z)H_2(z)$$

where

$$H_1(z) = (1 - 1.0526e^{j30^0} z^{-1})(1 - 1.0526e^{-j30^0} z^{-1})$$
$$H_2(z) = (1 - 1.0526e^{j30^0} z^{-1})(1 - 1.0526e^{-j30^0} z^{-1})$$

where $H_1(z)$ and $H_2(z)$ are shorter filters, each containing two zeros. $H_1(z)$ contains zeros which lie inside the unit circle while $H_2(z)$ contains zeros which lie outside the unit circle at reciprocal position (See fig. 8.17). We are thus able to keep apart zeros and their reciprocal, which in this example are very close to each other.

We shall now compute the Γ matrix, which maps the coefficient quantization errors into errors in the position of zeros. From table 8.5 we observe that the elements of the Γ matrix for the composite filter are about a factor of ten greater than those for the short filters. Therefore, the shift in the position of zeros of the short filters would be decreased by the same factor.

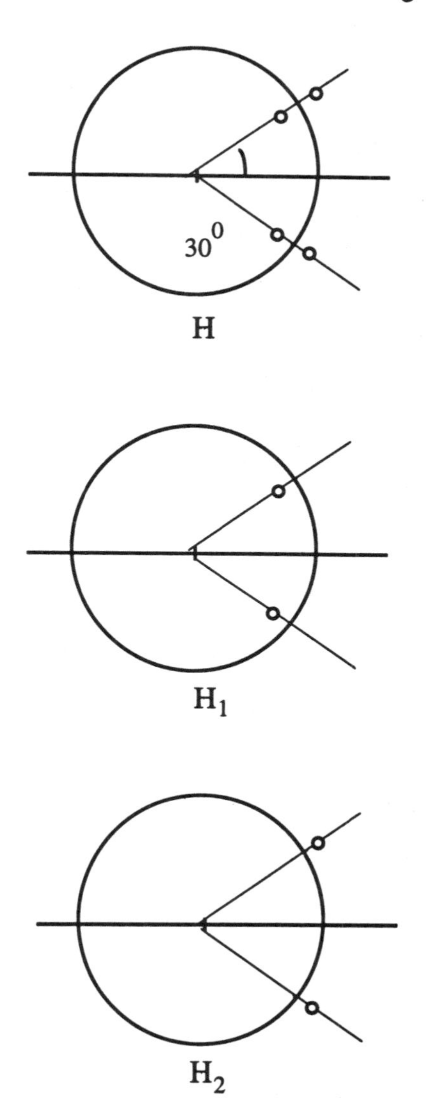

Figure 8.17: A linear phase FIR filter H is factored into two filters. H_1 has zeros inside the unit circle and H_2 has zeros outside the unit circle. The distance between the zeros in H_1 and H_2 is much larger than that in H. Hence the quantization error in H_1 and H_2 is much lower than that in H.

The effect of coefficient quantization in the IIR filter is similar to that in the FIR filter. The shift in the position of zeros and poles on account of coefficient quantization may be analyzed following the procedure used above for FIR filters. In IIR filter there is a further possibility for a pole, close to the unit circle, to move out of the unit circle and to cause instability.

Table 8.5: Matrix Γ in Eq. (8.7). H(z) is composite matrix with four zeros as shown in fig. 8.17 and H_1 and H_2 are smaller factored filters each with two zeros. Observe that the elements of the Γ matrix for the composite filter are about a factor of ten greater than those for the short filters.

G: H(z)	$-\begin{bmatrix} -8.35-3.88j & -9.66+0.86j & -8.35+5.86j & -4.53+9.74j \\ -8.35+3.88j & -9.66-0.86j & -8.35-5.86j & -4.53-9.74j \\ 8.35+5.86j & 9.66+0.86j & 8.35-3.88j & 5.03-7.16j \\ 8.35-5.86j & 9.66-0.86j & 8.35+3.88j & 5.03+7.16j \end{bmatrix}$
G: H_1(z)	$-\begin{bmatrix} 0.5-0.866j & -1.052j \\ 0.5+0.866j & 1.052j \end{bmatrix}$
G: H_2(z)	$-\begin{bmatrix} 0.5-0.866j & -0.95j \\ 0.5+0.866j & 0.95j \end{bmatrix}$

Product Quantization in IIR Filters: We now consider the product quantization effects on one-sided IIR filter with the impulse response function, $h(n), n = 0,1,\ldots,\infty$ (h(0)=1.0). An IIR filter can be expressed as a recursive filter (see Eq.4.3) which is reproduced here for convenience

$$\begin{aligned} &y(n) - a_1 y(n-1) - a_2 y(n-2) - \ldots - a_p y(n-p) \\ &= b_0 x(n) + b_1 x(n-1) + \ldots + b_q x(n-q) \end{aligned}$$

The product quantization is modeled as

$$Q\{a_1 y(n-1)\} = a_1 y(n-1) + \varepsilon_1(n)$$

where $\varepsilon_1(n)$ is quantization noise. For simplicity we shall consider a simple two-pole filter (zeros at origin),

$$y(n) - a_1 y(n-1) - a_2 y(n-2) = x(n) \tag{8.13}$$

The product quantization in (8.13) results into the following,

$$\hat{y}(n) - a_1 \hat{y}(n-1) - a_2 \hat{y}(n-2) = x(n) + \varepsilon_1(n) + \varepsilon_2(n) \tag{8.14}$$

where $\hat{y}(n) = y(n) + \theta(n)$ and $\theta(n)$ is the error induced into the output of the filter due to the product quantization. We assume that $\varepsilon_1(n)$ and $\varepsilon_2(n)$ are uncorrelated with $x(n)$. The product quantization and its role in two-pole IIR filter is illustrated in fig. 8.18.

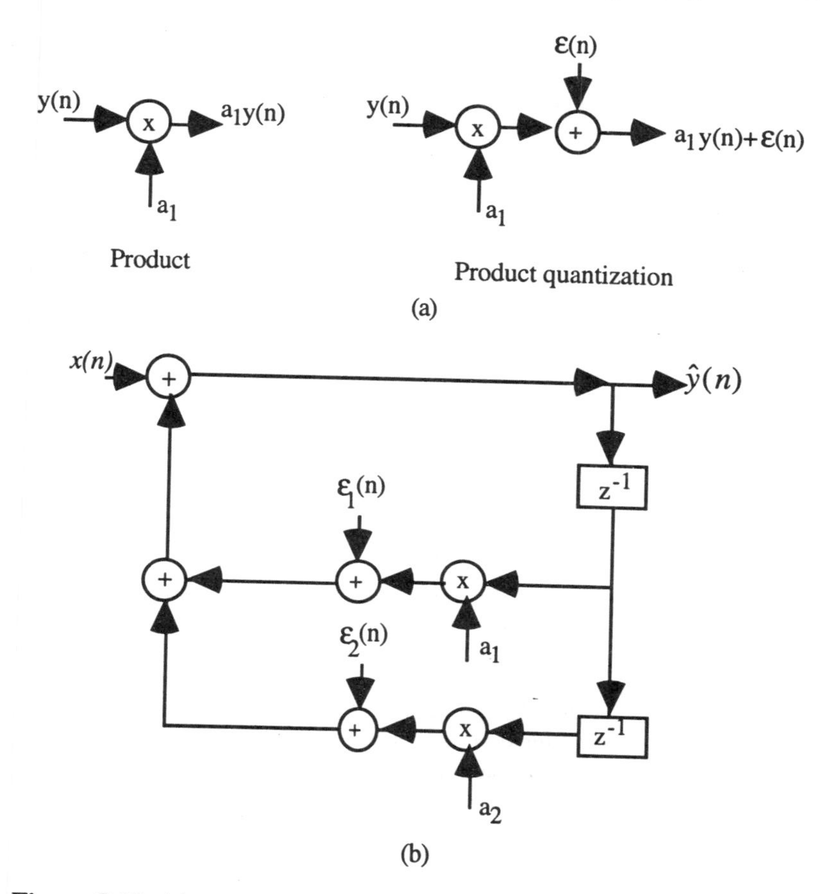

Figure 8.18: (a) Additive noise model of product quantization and (b) two-pole IIR filter with product quantization.

Equation (8.14) reduces to two independent equations,

$$y(n) - a_1 y(n-1) - a_2 y(n-2) = x(n) \tag{8.15a}$$

$$\theta(n) - a_1\theta(n-1) - a_2\theta(n-2) = \varepsilon_1(n) + \varepsilon_2(n) \tag{8.15b}$$

The quantization noise $\varepsilon_1(n)+\varepsilon_2(n)$ acts as an input to the same filter and the resulting output is the error in the output of the filter with product quantization. Note that although $\varepsilon_1(n)+\varepsilon_2(n)$ is uncorrelated noise, $\theta(n)$ is a correlated stochastic process whose variance may be shown to be, after some tedious algebra,

$$\sigma_\theta^2 = \frac{(1+r_0^2)}{(1-r_0^2)}\frac{\sigma_\varepsilon^2}{(1-2r_0^2\cos(2\omega_0)+r_0^4)} \tag{8.16}$$

where $\sigma_\varepsilon^2 = \sigma_{\varepsilon_1}^2 + \sigma_{\varepsilon_2}^2$ and the roots of (8.13) are at $r_0 e^{\pm j\omega_0}$. It may be observed that for $\omega_0 = 0$ (lowpass) (8.16) reduces to

$$\sigma_q^2 = \frac{(1+r_0^2)\sigma_\varepsilon^2}{(1-r_0^2)^3}$$

and for $\omega_0 = \pi/2$ (highpass) it reduces to

$$\sigma_q^2 = \frac{\sigma_\varepsilon^2}{(1-r_0^4)};$$

both tending to infinity as $r_0^2 \to 1$, but much faster in the first case. Hence, the product quantization noise poses a serious threat for lowpass filter than for highpass filter.

Shift in Pole Position: The coefficient quantization in an IIR filter will lead to shift in both zeros and poles. We have already examined the effect of coefficient quantization on the shift of position of zeros of a FIR filter. The same will hold good for the zeros of IIR filter. In place of the impulse response coefficients in (8.12a) we need to substitute the numerator coefficients of the IIR filter. The effect of the denominator coefficient quantization may be evaluated exactly in the same fashion as we have done for FIR filter. Here we give the final expression

$$\Delta\alpha_i = -\sum_{k=1}^{p} \frac{\alpha_i^{p-k}}{\prod_{\substack{l=1 \\ l \neq i}}^{p} (\alpha_i - \alpha_l)} \Delta a_k \tag{8.17}$$

where $\alpha_i, i = 1,2,\ldots,p$ are complex position of poles and $\Delta\alpha_i$ is the error in i^{th} pole position due to quantization errors in the denominator coefficients.

$$\begin{bmatrix} \Delta\alpha_1 \\ \Delta\alpha_2 \\ \vdots \\ \Delta\alpha_p \end{bmatrix} = \Gamma \begin{bmatrix} \Delta a_1 \\ \Delta a_2 \\ \vdots \\ \Delta a_p \end{bmatrix} \tag{8.18}$$

where Γ is defined as in (8.12c). To avoid a large shift in the pole position the poles must not be clustered nor be close to the real axis. It is recommended that a higher order denominator polynomial may be factored into elementary second order polynomials. Then, the higher order filter can be realized as a cascade of many second order IIR filters.

§8.6 Exercises

Problems

1. A quantized signal is passed through an FIR filter. The noise in the output will be the result of passing the quantization noise through the filter. The quantization noise is usually modeled as an uncorrelated uniformly distributed stochastic process. What can you say about the noise in the filter output. In particular, compute the covariance and power spectrum of the output noise.

2. Two FIR filters, each with M coefficients, are connected in parallel. The filter coefficients are quantized to b+1 bits. Draw the filter configuration with quantized coefficients. Show that the variance of the error in the frequency response is given by

$$\sigma_E^2 = 2\sigma_\varepsilon^2\left[M - 1 + \frac{\sin(M\omega)}{\sin(\omega)}\right]$$

Now connect above two filters in series form. Draw the filter configuration with quantized coefficients. Estimate the variance of the error in the resulting composite frequency response.

3. Compute the variance of the truncation error in sign-magnitude and in two's complement type of number representation.

4. Show that the variance of the product quantization noise given by eq. (8.16) is maximum at $\omega_0 = 0$ and minimum at $\omega_0 = \pi/2$.

5. Suppose we have four math units, which can work in parallel. The FIR filter equation may be parallelized as follows:

$$y(n) = x(n)c(0) + x(n-1)c(1) + \mathbf{x(n-2)c(2)} + \ldots + x(n-N+1)c(N-1)$$
$$y(n+1) = x(n+1)c(0) + x(n)c(1) + \mathbf{x(n-1)c(2)} + \ldots + x(n-N+2)c(N-1)$$
$$y(n+2) = x(n+2)c(0) + x(n+1)c(1) + \mathbf{x(n)c(2)} + \ldots + x(n-N+3)c(N-1)$$
$$y(n+3) = x(n+3)c(0) + x(n+2)c(1) + \mathbf{x(n+1)c(2)} + \ldots + x(n-N+4)c(N-1)$$

The products shown in bold can be performed together as they use the same coefficient. Further, assume that loading of input data, coefficient and math operation can be performed in one clock cycle. How many clock cycles are required for one output sample?

6. How are digital signal processors different from the traditional microprocessors?

7. Write down the steps required to compute in real-time the cross-correlation between two signals on a digital signal processor. What part of the program may be stored in the cache on CPU?

8. The contents of a circular buffer at the current time (t=0) is as follows:

$$\begin{Bmatrix} 2.3456 & 1.3429 & -1.9823 & 3.4510 & 3.8702 & 1.5619 & 7.6421 & 2.9871 \\ t=0 & t=1 & t=2 & & \ldots & & & t=7 \end{Bmatrix}$$

At next sample instant a new value, 4.1105, arrives and it needs to be store in the circular buffer. Show the contents of the circular buffer after storage.

References

1. J. G. Proakis and D. G. Manolakis, Digital signal processing: Principles, algorithms and applications, Prentice Hall of India pvt. ltd, New Dehli, 1995.
2. S. K. Mitra, Digital signal processing: A computer based approach, Tata McGraw-Hill publishing Co. ltd, New Delhi, 1998
3. S. Salivahanan, A. Vallavaraj and G Gnanapriya, Digital signal processing, Tata McGraw-Hill publishing Co. ltd, New Delhi, 2000.
4. D. J. Defatta, J. G. Lucas and W.S. Hodgkis, Digital signal processing: A system design approach, John Wiley & Sons, New York, 1988.
5. A. Papoulis Probability, random variables and stochastic processes, McGraw-Hill International Edition, Third Edition, 1991.
6. P. S. Naidu, Modern spectrum analysis of time series, CRC Press, Boca Raton, Fl. USA, 1996.
7. T. S. Parker and L.O. Chau, Chaos: A tutorial for engineers, Proc. IEEE, vol. 75, pp. 982–1008, 1987.
8. T. Lo, H. Leung, J. Litva, and S. Haykin, Fractal characterization of sea scattered signals and detection of sea surface targets, Proc. IEE, part F, vol. 149, pp. 243–250, 1993.
9. J. D. Farmer and J. J. Sidorowich, Predicting chaotic time series, Phy Rev. Lett, vol. 59, pp. 845–848, 1987.
10. C. A. Pickover and A. L. Khonrasani, Fractal characterization of speech waveform graphs, Comput & Graphics, vol. 10, pp. 51–61, 1986.
11. Runge, C. and Konig, H. Die Grundlehren der mathematischen wissenschaften vorlesungen uber numerische rechen, vol. 11, Springer Verlag, Berlin, 1924.
12. Cooley, J. W., Lewis, P.A. W. and Welch, P. D., Historical note on the fast Fourier transform, Proc. IEEE, vol. 55, pp. 1675, 1967.
13. Cooley, J. W. and Tukey, J. W.: An algorithm for machine calculation of complex Fourier series, Mathematics of Computation, vol. 19, pp. 297, 1965.
14. Cochran, W. T. et al, What is fast Fourier transform?, IEEE Trans., ASSP 15, pp. 45–, 1967.
15. Solodovnikov, V. V. Introduction to the Statistical Dynamics of Automatic Control Systems, Dover Publication, Inc, New York, 1960.
16. A. V. Oppenheim and A. S. Wilsky, Signals and Systems, Prentice-Hall of India Pvt Limited, New Delhi, 1997.
17. P. Stoica and R. Moses: Introduction to spectral analysis, Prentice-Hall, Upper Saddle River, NJ, USA.

18. D. Slepian, H. O. Pollack, and H. J. Landau, Prolate spheroidal wave functions, Fourier analysis and uncertainty, Bell Sys. Tech. J., vol. 40, pp. 43–84, 1961.
19. Y. C. Kim and E. J. Powers, Digital bispectral analysis and its applications to nonlinearwave interactions, IEEE Trans, Plasma Science, PS-7, pp. 120–131, 1979.
20. J. S. Bendat and A. G. Piersol, Random Data: Analysis and Measurement Procedure, Wiley Interscience, New York, 1971
21. L. R. Rabiner, B. Gold and C. A. McGonegal, An approach to the approximation problem for nonrecusrive digital filters, IEEE Trans AU-18, pp. 83–105, 1970.
22. K. E. Atkinson, An introduction to numerical analysis, John Wiley and Sons, New York, (p.192)23. A. Antoniou, Digital filters: Analysis and Design, Tata McGraw-Hill Publishing Co. Ltd, New Delhi, 1980.
23. A. Antoniou, Digital filters: Analysis and Design, Tata McGraw-Hill Publishing Co. Ltd, New Delhi, 1980.
24. C. F. N. Cowan and P. M. Grant (Editors), Adaptive filters, Prentice-Hall, Inc., Englewood Cliffs, NJ, USA, 1985.
25. S. M. Kay, Modern Spectrum estimation, Prentice Hall, Englewood Cliffs, NJ, USA, 1988.
26. M. P. Moody, Resolution of coherent sources incident on circular antenna array, Proc. IEEE, vol. 68, pp. 276–277, 1980.
27. M. Abramowitz and I. A. Stegum (Editors), Handbook of Mathematical Functions, National Bureau of Standards, Applied mathematics Series 55, 1964.
28. C. P. Mathews and M. D. Zoltowaski, Eigen structure technique for 2D angle estimation with uniform circular arrays, IEEE Trans SP-42, pp. 2395–2407, 1994.
29. I. D. Rathjen, G. Boedecker and M. Siegel, Omnidirectional beam forming for linear antennas by means of interpolated signals, IEEE Jour. of Ocean Eng., vol. OE-10, pp. 360–368, 1985.
30. P. S. Naidu, Sensor Array Signal Processing, CRC Press, Boca Raton, Fl., 2000.
31. A. Cantoni and L. C. Godara, A Performance of a post beamformer intereference canceller in presence of broadband directional signals, J. Acoust Soc. of Am., vol. 76, pp. 128–138, 1984.
32. S. P. Applebaum Adaptive arrays, IEEE Trans. Antennas and Propag, AP-24, pp.585–596, 1976
33. S. W. Smith, The Scientist and Engineer's guide book on digital signal processing, www.dspguide.com/pdfbook.htm, 1997
34. Report of the Technical Committee on Design and Implementation of Signal Processing Systems (DISPS), IEEE Signal Processing Mag, pp. 22–37, 1998.
35. www.analog.com/DSP and www.ti.com.
36. W. A. Gardner, The spectral correlation theory of cyclostationary time series, Signal Processing, vol. 11, pp.13–86, 1986.

37. C. G. Fox and D. E. Hays, Quantitative methods for analyzing the roughness of the seafloor, Reviews of Geophysics and Spece Physics, vol. 23, pp.1–48,1985.
38. R. N. Bracewell, The Fourier transform and Its Applications, 2nd Edition, McGraw-Hill International Editions, 1986.
39. F. J. Harris, On the use of windows for harmonic analysis with discrete Fourier transform, Proc IEEE, vol.66, pp. 51–83, 1978.
40. P. F. Howells, IF sidelobe canceler, US Patent 3,202,990, Aug 24, 1965.

Index